Functional Nucleic Acids Detection in Food Safety

Wentao Xu

Functional Nucleic Acids Detection in Food Safety

Theories and Applications

Springer

Wentao Xu
Beijing Advanced Innovation Center
for Food Nutrition and Human Health,
College of Food Science & Nutritional
Engineering
China Agricultural University
Beijing, China

Beijing Laboratory for Food Quality
and Safety, College of Food
Science & Nutritional Engineering
China Agricultural University
Beijing, China

ISBN 978-981-10-9402-6 ISBN 978-981-10-1618-9 (eBook)
DOI 10.1007/978-981-10-1618-9

Softcover reprint of the hardcover 1st edition 2016

Printed on acid-free paper

This Springer imprint is published by Springer Nature
The registered company is Springer Science+Business Media Singapore Pte Ltd.

Preface

Food safety is usually a nonnegotiable attribute for consumers in their decision-making process as well as for manufacturers in the global trade. It is of great significance to detect the potential danger that may harm the quality of food to control and regulate food safety in an official and legal way. Contributed by the progress in analytical chemistry and molecular biology, types of methods appeared to fulfill the specific and sensitive detection. The applied detection methods have all been focused on target characteristic properties, varying in their physical, chemical, and biological activities. But there exists a fact that all these hazard exposures are related to a magic molecule – nucleic acid. Nucleic acids are the most important biological macromolecules, essential for all known forms of life; each is found in abundance in all living things, where they play a role in encoding, transmitting, and expressing genetic information. What genetic central dogma tells us is that information is conveyed through the nucleic acid sequence, or the order of nucleotides within a DNA or RNA molecule. On the other hand, single-stranded DNA can fold into complex tertiary structures and play an active role in molecular recognition and catalysis such as mycotoxins, the whole bacteria and heavy metals, etc. Once recognized, the complex becomes much more activated, which will be helpful in their physical, chemiluminescence, and electrochemiluminescence detection. Thus, it is indicated obviously that nucleic acids will be enough for the detection of all foodborne potential hazard factors, and research on nucleic acids is becoming profound in food safety assurance. This book aims to offer a summarized and comprehensive representation of nucleic acid detection methods in food safety for professionals interested in this area and the public.

This book is divided into two parts. Part I, consisting 13 chapters, discusses the basic principles of nucleic acid detection owing to various detecting demands. Chapters 1 and 2 serve as an introduction to target nucleic acids and their corresponding extraction methods. Chapter 3 introduces the basic mechanism of nucleic acid detection. Chapters 4 and 5 discuss the necessary nucleic acid detection compositions – standard materials and endogenous gene. Chapters 6, 7, 8, 9, and 10 summarize unknown gene exploration, digital PCR, quantification, and high-throughput and isothermal detection methods. Chapters 11, 12, and 13 try to connect aptamer, lateral strips and biosensors to detection methods. Part II is about the *novel* application of detection methods in genetically modified organisms, dead–alive microorganism identification, microbial diversity, heavy metal detection, gene toxicity, and noncoding RNA identification.

The book was fulfilled through the joint efforts of all contributors: Chenguang Wang, Ying Shang, Pengyu Zhu, Yuancong Xu, Nan Cheng, Longjiao Zhu, Mingzhang Guo, Boyang Zhang, Liye Zhu, Wenying Tian, Wenjin Xiang, Xingtian Cui, Yuting Chen, Jingjing Tian, and Xiangli Shao. I am very grateful for their suggestions and help in preparing the manuscripts. Special thanks go out to the National Science and Technology Major Project, the Beijing New-Star Plan of Science and Technology, and the Beijing Advanced Innovation Center for Food Nutrition and Human Health because without their support, some parts of this work would not have been accomplished.

Beijing, China
April 2016

Wentao Xu

Contents

Chapter 1
Introduction: A Brief Guide to Targets and Strategies of Functional Nucleic Acids Detection in Food Safety

Abstract To safeguard the food supply from potential dangers requires comprehensive knowledge and effective control measures. Central to this solution is the need to accurately detect and identify the origin of such hazards. As the fundamental base in the development of living organisms, nucleic acids function to deliver genetic information or to act as signal molecules for catalytic activity. Therefore, detection techniques based on nucleic acids can be universally applied in determining contaminant hazards in food caused by the harvesting and manufacturing processes. This review will emphasize the importance of nucleic acids in food safety and give an introduction to nucleic acid detection techniques.

Keywords Food safety • Consumer • Danger • Nucleic acids • Target strategies

1.1 Introduction

Life can be defined as a biological, naturally occurring, and self-reproducing device that affects controlled manipulations of matter, energy, and information. Human beings are a part of the prosperous life within the world. For millions of years, our ancestors have adapted to their surroundings and transmitted important genetic information to subsequent generations.

Human beings cannot live without food. Efforts to locate food are an obligatory human requirement to survive in this unfamiliar world for energy intake [1, 2]. With their hunger satiated, humans began to concentrate on making food both flavorsome and healthy. Attributed to their thirst for longer lives and creativity, many measures have been implemented to minimize illness by way of food safety. Cultivation processes, which emerged in the agricultural sector, increase the total amount of edible materials [3]; heating processes, which are the most obvious action to prevent illness, attempt to eliminate pathogens [4]; and watering processes are attempt to clean our food, making it free from soil among other contaminants.

From the time when our caveman ancestors gnawed bones in front of smoking fires to the present days when we enjoy meals on tables spread with snow-white linen and sparkling silver, we have a history of changing our eating habits from simple satisfaction of hunger to a delightful meal and sometimes rather complex, healthy, and social experience. Since the Industrial Revolution in the Western

W. Xu, *Functional Nucleic Acids Detection in Food Safety*,
DOI 10.1007/978-981-10-1618-9_1

world, when raw materials, such as grains, fruits, vegetables, and meats, are started to be processed into food products in a systematic fashion, food started to be under-manipulated to maintain healthy level in a scientific and standard way [5].

Since then, the nature of food has developed into multifunctional substances in daily life. Indeed, modern societies have changed our lifestyle, as well as diet. Consumers require food not only as necessities for starvation but also as a functional product by way of easy accessibility, long storage, quick preparation, and even attractiveness [6, 7]. Food is also an economic means for civilian subsistence. Food that is bought from the market comes from retailer transportation, and the retailer stores food products with good quality from the manufacturer, where factory processing occurs and raw ingredients are exchanged from farmers. Collectively, this progression results in a new industrial trade: food industry. The whole process provides various job opportunities to a considerable proportion of the world's population [8, 9]. In addition, food industry serves as an engine for international import and export businesses, therefore fostering economic development and prosperity [10]. Thus, overcoming certain food safety issues, particularly where legislation is not feasible or where management is not effective, can have negative consequences in terms of compromising the livelihoods of many people and corrupting economic development.

The birth of food industry is intended to serve consumers and the social development [10]. Food industry aims to satisfy the following three main expectations of its customers: (1) to trust the food dealer's reputation from whom they buy their food products, (2) to trust the manufacturer's ability to ensure the safety of their products, and (3) to get knowledge of corresponding solutions in the event of a crisis [11]. Evidence obviously shows that consumers only expect intrinsically safe food to be sold on the market, with some consumers reporting that they would never consume unsafe food or food beyond its shelf life [12, 13]. There is always the expectation that food in the market is safe for the consumers and nutrition has improved a lot [14, 15]. However, consumers do not care about food safety under normal conditions, but an occurrence of the food safety incident may cause consumer concern and even raise consumers' anger.

Governments and official authorities play an important role in food safety management as they are responsible for drafting applicable policy and control measures to ensure people's right to pursue subsidy and health. Food safety is a discipline with a formal definition as follows: food safety is the assurance that food will not cause harm to the consumer when it is prepared and/or eaten according to its intended use, as defined in 1997 by Codex Alimentarius Commission (CAC) [16].

Although many levels of authority exist to ensure food safety against harmful contaminants, however, several natural hazards exist that can adhere to agricultural products on which more artificial efforts have been penetrated into food. Delighted to see that modern manufacturing processes have succeeded in decreasing natural dangers to humans, as evidenced by significant scientific efforts and technological progress in food safety [17–22], still, industrial food products are exposed to more hazardous chemicals than those from previous generations, despite improvements in quality and safety control. Indeed, the increased use of pesticides and herbicides can poison people, and excessive chemical additives can potentially bioaccumulate. Additionally, adulteration heavily damages the rights of consumers [23].

Multiple concerns threat the food safety in the ways of causing a variety of foodborne diseases and even death, such as the pathogens, toxins that may have a chronic latent time, chemical pollution, and the adulteration and scientific progress in the form of genetically modified organisms. Disappointedly, massive increases in our exposure to food pollutants began in the last century during rapid economic growth. Additionally, some chemicals that have been banned for decades continue to persist in the environment and contaminate our food or are illegally added to the food chain for economic benefit [24].

Time flies to the twenty-first century, along with the medical achievements to elongate human life, foodborne diseases directly compromise advances in medicine as they negatively affect human life. The environment has been polluted for hundreds of years and varies dramatically in climate. Moreover, it is continually experiencing microbial and ecological changes, which indirectly affect the food supply [25–27]. This is not the end. Human involvement in urbanization, complex migration patterns, international tourism and trade, poor management, illegitimate business behavior, and degradation of sanitation places extreme burdens on food supply and may cause pathogens to spread transnationally [15].

More specifically, food safety issues can be summarized as follows (Fig. 1.1):

1. Adulteration. Adulteration aims to add another substance in the raw material to increase the quantity of food item and to lower the retail price for economical profit, thereby resulting in the loss of the actual quality of food. These substances may be either relative food additives or even nonedible substances. In 2013, the horse meat scandal in Europe revealed the weaknesses in the food safety system

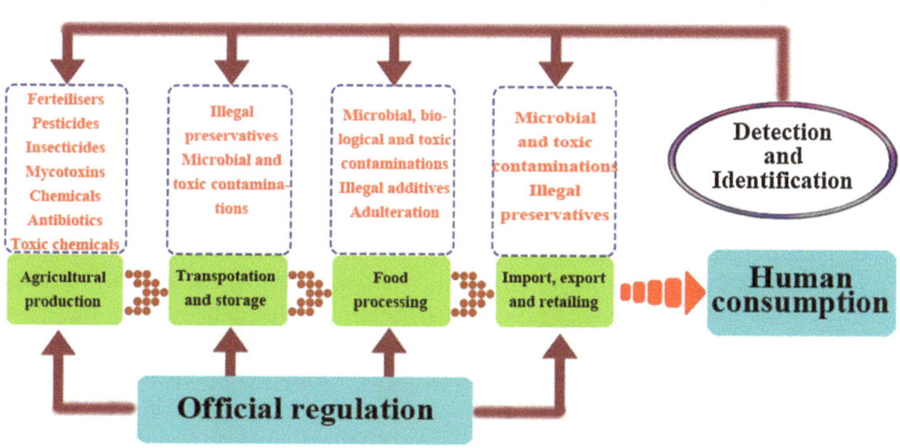

Fig. 1.1 Flowchart of the relationship between regulation, consumption, and scientific detection. Potential dangers exposed on the stages of food processing are shown in red. Consumption and regulation both influence the food safety, which is decided by detection technologies

and contributed to a decrease of confidence in the food industry [28]. Fraudulent practices, such as the presence of undeclared horse DNA in food products labeled as containing beef, were confirmed in 4.66 and 0.61 % of controlled foods in 2013 and 2014, respectively [29]. A recent survey in the USA indicated that a rate of 33 % of investigated seafood samples were mislabeled according to the US Food and Drug Administration (FDA) guidelines [30].

2. Internal biological pollution. Microorganisms play important roles in the production process, either by transforming the food into the desired products (e.g., fermentation to make beer and wine, fermented meat, and various fermented dairy products) or by spoiling or contaminating the food. However, when these microorganisms negatively affect the food to cause serious illness issues or even undesired death, these internal attributors become pathogens. Pathogen pollution has been a frequent reminder of the food safety issues that urge the overall manipulation all over the world. Many pathogens are persistent in animal reservoirs and stored in raw materials, such as flesh, milk, and eggs, contaminating our food supply. Some pathogens contaminate food supply in the excreta of infected animals. Further, some pathogens can contaminate foods we eat via complex worldwide pathways that reflect the variety of ecosystems that make up our food supply [31–36]. Other pathogens are mixed into food processing step because of human careless activity, represented by poor sanitation, illicit ingredient trade, and unhealthy storage conditions for the products.
3. Toxins. Toxins are poisonous substances produced within living cells or organisms. According to the International Committee of the Red Cross review of the Biological Weapons Convention, toxins are inanimate and not capable of reproducing themselves, which makes them different from internal pathogens [37]. Sources of some toxins are pathogens that can secrete poisonous small molecules, whereas others are from foodstuffs, like tetrodotoxin (TTX) and solanine. Additionally, mycotoxins are secondary toxic metabolites produced by different fungi that contaminate food and feed products, such as nuts and cereals. Some mycotoxins have been shown to be possibly mutagenic, teratogenic, and carcinogenic [38, 39].
4. Chemical pollutants. The rapid development of industry and society has created wealth and a sense of happiness, as well as some undesirable chemical pollution. Industrial wastes found in rural areas, such as heavy metals and chemical additives, continue to contaminate the industrial products and the upstream agricultural land that provides food ingredients and manufacturing elements. Cadmium is a metal that is implicated in a series of severe health risks ranging from renal failure to even some cancers, some of which resisted in the food owing to the polluted environment and industrial waste. Another major threat of chemical substances for public health and food safety is the misuse and excessive application of agrochemicals, which is represented by pesticides to prevent devastation from pests, herbicides to inhibit the fast-growing weeds, and veterinary drugs to treat serious animal diseases. Pesticides can lead to unsafe residues in the human body that may cause intestinal disorder and gastritis, and veterinary drugs, represented by antibiotics and hormones, pose a threat to

human neurotoxicity and even death to some extent. Finally, illegal chemicals used as food additives for economical profit have made headlines in terms of food safety in Europe and now in fast-developing countries such as China and Brazil. All these chemical pollutions will lead to human health hazards, a lack of social confidence in industry, and even government distrust in the regulatory system [40–48].

5. Genetically modified organisms. There has been a worldwide trend of the rapid development of genetically modified organisms (GMOs) during the past 30 years throughout the world, as reported by the International Service for the Acquisition of Agri-biotech Applications (ISAAA). In 2015, 18 million farmers in 28 countries planted more than 1.79 billion hectares of GMOs, which have been growing for the nineteenth consecutive year of commercialization [49]. However, the adventitious contamination of GMOs in the foods and the environment is inevitable and nearly impossible to avoid. Consumers are more concerned about the health safety issues of consuming GMOs. There is currently debate on whether GMOs represent a potential risk factor to foods and whether they are equivalent to traditional agricultural products [50].

Reasons for the dangerous exposure of hazardous contaminants to foods are multiple, several of which can be clarified as follows:

- Complexity of food itself that is related to inorganic, organic, and biological activities
- Complexity and lack of effective manipulations of manufacturing operations
- Complexity of the food supply chain from the factory to the retail market
- Human influence

Identifying such causes is the prerequisite for taking measures to mitigate the potential risks, but not the strategies that are utilized by researchers and manufacturers. Detection and identification are the primary objectives in the beginning of food safety manipulation on the society. In practice, however, transforming existing systems may not be straightforward. Compared to increasing numbers of global food safety issues, the underreporting and misreporting of incidents are common, especially since the reported number of incidents is not credible even in industrialized countries like the USA, based on the US Centers for Disease Control and Prevention [51]. The misreporting of incidents in the USA and other countries should decrease, allowing for urgent emergence of more specific and sensitive detection methods.

As a rapidly developing country in the world, China is predicted to reach 14 billion in population in 2025. Concomitant with this growth is the transition from mere economic growth to awareness of diet consumption and from population migration to products trade. The increasing volume of China's food production and consumption, which ensures the daily life of citizens and improve the national economy indirectly, makes food supply a major focus in international manufacturers and businessmen [52, 53].

The world has witnessed a remarkable growth in food production and conquered the hunger problems in the past half century, allowing for a dramatic decrease in the proportion of the world hunger, despite a doubling of the total population [54]. Now one of the main problems faced by human beings is the struggle against illness and to live better. Food safety must be addressed with conviction to uncover the pressing issues that will continue to emerge so that effective mitigation strategies might be proposed for the betterment of humanity's future.

1.2 Significance of Nucleic Acids in Food Safety Detection

It is meaningful in the food supply to detect the hazardous factor prior to the final products reaching the consumer. In this regard, an increasing number of scientific achievements have been developed. Indeed, it is because of scientific progress, especially in the fields of chemistry and biology, that food safety is easy to be determined under chemical instruments, enlarged cultures, and biological reactions. Of the plethora of methods, without question, nucleic acids are the universal targets or mediums in the food safety detection.

The rise of nucleic acids can be dated back to times of Charles Darwin's book *On the Origin of Species by Means of Natural Selection*. In short, species within and between populations compete with each other. Those that are well adapted to their environment survived in the competition and can reproduce subsequent generations [55]. Other scientists at the same periods proposed a similar theory, such as Alfred Russel Wallace and Georges Cuvier [56, 57]. The core of natural selection can be stated as the adaptation of individuals or species to a changing environment and the survival of evolved and competition-dominant individuals for heredity supremacy.

Gregor Mendel, who has gained posthumous fame as the founder of the modern science of genetics, concluded a paradigm that characteristics are passed from one generation to the next in pairs, some of which are dominant over others [58]. The significance of Mendel's work is to build a genetic relation between offspring and their parents, no matter if the offspring resembles one parent or both or even none of them. Later research identified these units of inheritance as genes. Sir Archibald Garrod, a British doctor, developed this theory on his research of alkaptonuria that was published in *The Lancet* in 1902 [59]. He found the link between the inheritance of particular behavior and the activity of protein. The work of Mendel and Garrod was not paid attention until this link was seen again by two American scientists (George Beadle and Edward Tatum) [60].

Once the basic principle of inheritance was established, scientists were focused on searching for substances that pass from one generation to the next. In 1944, Oswald Avery, Colin MacLeod, and Maclyn McCarty at the Rockefeller University Hospital conducted an experiment in which deoxyribonucleic acid, known as DNA later, is transferred from one generation to the other [61]. Since that time, DNA has become the central record of genetic information for all living organisms in the world.

It has been known that functional information of an organism is genetically stored as DNA and then transformed into other types of small molecules in a linear chain [62]. DNA is comprised of four deoxyribonucleotides (i.e., adenine, thymine, cytosine, and guanine), RNA is comprised of four ribonucleotides (i.e., adenine, uracil, cytosine, and guanine), and proteins are created from 20 amino acids [63, 64]. These sequences can be compared and analyzed to determine similarities within species to evacuate valuable genes or between species to find some functional information. In addition, these sequences can also be the targets considered in detection methodologies.

1.2.1 Targets Focused on the Microscopic View

The central dogma of molecular biology states that DNA acts as a template to replicate itself and transcribe into RNA. RNA is then translated into protein (Fig. 1.2). This dogma summarizes the function of the genome and the transfer drift of genetic information [63, 64]. Genetic information is conserved in the form of double-helix structure and replicates a new identical DNA for the next generation. Genetic information is conserved as the form of a strain of DNA to produce many metabolic constituents, some of which possess significant roles in life regulation. The process from DNA to metabolite involves multiple processes, including transcription and translation. These transitions functioned at the structural, biochemical, and cellular levels. However, all the information is organized in the form of DNA, a double-stranded helix that is composed of thousands of genes. In another

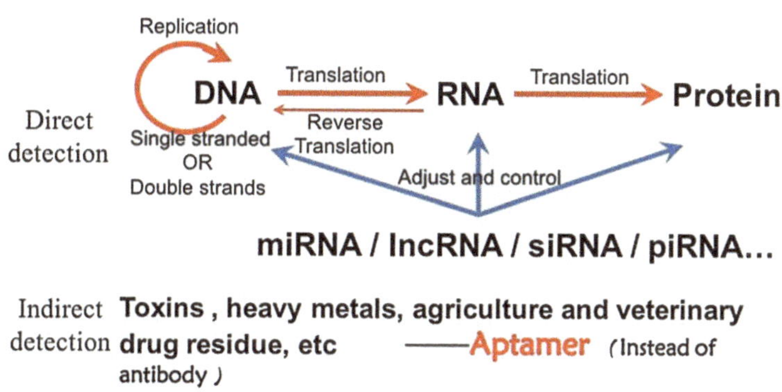

Fig. 1.2 Detected targets on the whole process of central dogma of molecule biology. Direct detection is based on the central dogma, which means that DNA, RNA, and other types of nucleic acids (miRNA, siRNA, etc.) from pathogens can be detected directly. Indirect detection is the way to detect targets based on nucleic acid-specific recognition. The most obvious example is aptamer, which binds to the special small molecule

way, DNA is life's blueprint. Without DNA, organisms would not be able to replicate themselves [65]. The unusual molecular structure of DNA affords its special properties. These properties allow the genetic information to be stored in the structure and passed to the next generation, from parents to their offspring [66]. Therefore, DNA can be used as a detection target for all organisms, regardless of whether it be alive or dead. Additionally, certain types of RNA (e.g., mRNA, tRNA, microRNA, small RNA, noncoding RNA, etc.) can be used as detection target, as being the genetic information transfer in the life form.

Nucleic acids can also be used as functional detection medium in the food safety detection [67]. Some small molecules, such as mycotoxins and small RNA, will bind to particularly selected oligonucleotides. The most significant part of oligonucleotides is called aptamers. Aptamer is first described by Thomas Cech from the University of Colorado and Sidney Altman from Yale University who both received the 1989 Nobel Prize in Chemistry for discovering catalytic activity between intron I in the *Tetrahymena* and RNA sequence of RNase P [68, 69]. Aptamers are useful in analytical and bioanalytical applications because they offer unique recognition properties for small molecules such as proteins and mycotoxins [70, 71]. Additionally, aptamers possess significant advantages over antibodies because they can be prepared easily just like primers, possess desirable storage properties, and elicit little or no immunogenicity in therapeutic applications. Aptamers have then been widely applied for detection fields for the purpose of biosensing and catalysis [72, 73]. Other functional nucleic acids, such as DNA probes and DNAzymes, are also viewed as applicable food safety detection media and signal indicators [74].

Nucleic acids cannot be seen by the naked eye, but ubiquitous in the food. They can be recognized as danger positive through sensitive biological techniques, regardless of whether DNA sequence or catalytic aptamer sequence information is available, which is the main aim of this book.

1.2.2 Targets Focused on the Macroscopic View

The entire food manufacturing process can be summarized as follows: raw materials are being sent to the factory, and processed and prepared the final food products are done step by step [75]. Take cookies as example. As a favorite food for children, cookies are a baked product that are made of grain, usually containing eggs, sugars, flour, and some types of fat (oil or butter) and often include extra ingredients such as oats, raisins, nuts, or chocolate chips. All the ingredients are processed cooking materials, except eggs that can be obtained from a chicken farm. Wheat harvested from the field is processed into flour; sugarcane is extracted for sugar making; soybean and olive can be leached to make edible oil. Such products can be termed semi-products (Fig. 1.3). Food supply is a commercial problem, which is not within the scope of this book. What this book wishes to illustrate is that the entire process, from the raw wheat to the cookies, from potatoes to the French fries, and from tomatoes to the tomato sauce for use at Kentucky Fried Chicken

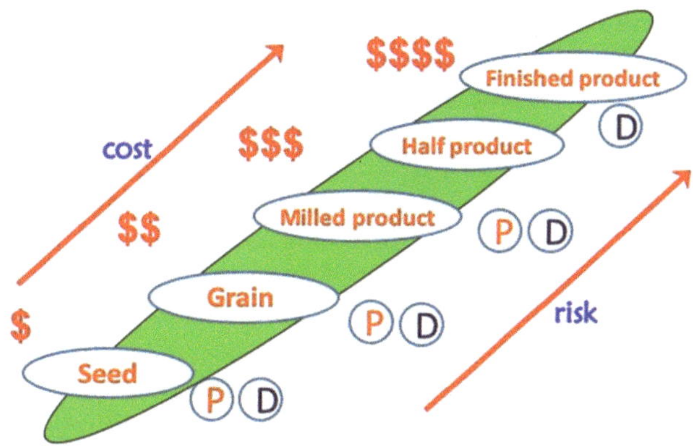

Fig. 1.3 Detected target focused on the macroscopic view. Food production process is from raw seed material to finished product. The deeper the status of food is, the cost is higher. Correspondingly, the detection risk is higher. *D* means difficulty in detecting potential hazard. P means price in detectin potential hazard

(KFC) restaurants, can be manipulated through nucleic acid methods. Ingredients and pathogens that may be included in the food can be detected with nucleic acid scope using molecular biological methods. Functional nucleic acids can help to identify additives processed into food.

Tough work that should be done to identify deep-processed foods appeals to the scientists to develop universal, specific, and ultrasensitive nucleic acid-based detection methods, when nucleic acid assays are specific (Chaps. 3 and 13), sensitive (Chaps. 7 and 8), high throughput (Chap. 9), and simple (Chaps. 10 and 12), that have been applied in medical diagnoses, gene expression profiles, and other life science fields.

Additionally, detection through the manufacturing process can also help to achieve food safety tracing system. Indeed, formulating a supply chain source identification system via biological methods will offer a comprehensive knowledge of hazardous critical control points in the supply chain [76–78].

1.3 Proper and Applicable Strategies on Nucleic Acid Detection in Food Safety

The development of various regulations and the corresponding auxiliary equipment are essential to protect the food safety. In a national administration, the task usually consists of four to five components, ranging from agriculture, health, environment, and trade to food. A leader in this regard is the American that built the US Food and Drug Administration (FDA), founded in 1906 (http://www.fda.gov). In 2002, the European Union created the European Food Safety Authority (EFSA) in 2002 (http://www.efsa.europa.eu) after serious food safety issues occurred in Belgium and the UK in the late 1990s [17, 79]. As a developing country, China rebranded State Food and Drug Administration (SFDA) to China Food and Drug Administration (CFDA) in 2013 after melamine-contaminated milk caused outrage among

Chinese people, especially parents (http://www.sda.gov.cn/WS01/CL0001/) [47]. Various governmental administrations emphasized the importance of identifying and mitigating risks in the food supply.

From a scientific viewpoint, it is vital to establish a surveillance system for foodborne diseases and to monitor food for contaminants. Proper technologies have to be developed and applied in different situations for possibly different aims, including identifying the presence of danger, preventing the entry of pathogens into food, inactivating them if necessary, and preventing their growth [80].

In this book, we have reached up to a conclusion that all foodborne dangerous factors can be detected via nucleic acids. All currently available detection methods have all been focused on the target characteristic properties, which vary in their physical, chemical, and biological activities. But there exists a fact that all these hazard exposures are all related to the magic molecule—nucleic acids. Nucleic acids are the most important biological macromolecules, being essential for all known forms of life. What genetic central dogma states is that genetic information is stored and conveyed via the nucleic acid sequence, in the form of DNA or RNA. On the other hand, single-stranded DNA can fold into complex tertiary structures and play an active role in molecular recognition and catalysis, involving mycotoxins, the whole bacteria, heavy metals, etc. Once recognized, the complex becomes much more activated, which will be motivated in their physical, chemiluminescence, and electrochemiluminescence detection. Thus, it is indicated obviously that nucleic acids will be enough for detection of all foodborne potential hazard factors, and research on nucleic acids is becoming profound in food safety assurance. Essentially, mainstream of food safety detection can be summarized as follows:

1. Detection based on the entire genome or other genetic information carriers (e.g., tRNA, small RNA, or noncoding RNA). Food adulteration is the most obvious case. Food consists mostly of agricultural products, which are regulated by DNA or RNA. Adulterated components in the food have no particular smell or color, although they do possess the basic ingredients for life. Only the nucleic acid-based method can solve the problem simply and accurately. Furthermore, internal biological pollution, such as pathogens, viruses, fungi, and parasites, can be detected directly using nucleic acid detection, rather than laborious enlarged culture [32, 81]. Exogenous insertion elements of GMOs in the samples can only be identified through polymerase chain reaction (PCR), which will be detailed in Chap. 14.
2. Detection based on recognition between small nucleic acids and target. Toxic pollutants, which are metabolites of fetal pathogens or intrinsically released from living organisms, can be identified through functional nucleic acids. Aptamer is the most obvious example, which will be detailed in Chap. 13. Small nucleic acids can also spontaneously recognize external risk factors, represented by pesticide, veterinary drug, and heavy metals, and be able to catalyze reactions necessary for fluorescence and biosensing.

3. Nucleic acid analytical methods are applied to gain knowledge of the relation between the environment and food safety. Microbial diversity provides clues of the environmental and climatic influence on food quality. These issues will be detailed in Chaps. 16, 17, 18, and 19.

Of all the above strategies, the following two issues should not be forgotten: reference gene and standard materials. Reference gene is a conserved gene that provides an internal reaction control. Reference gene is usually expressed in all cells with stable copy number. Standard materials are technically official precise molecular mixtures that are required for the calibration and validation of the detected target.

1.4 Scope of the Book

The book in my vision consists of two parts. Part I comprises the basic principle of nucleic acid detection methods. This chapter and Chap. 2 introduce the detection target and extraction method introduction. Chapter 3 introduces the basic mechanism. Chapters 4 and 5 discuss the necessary nucleic acid detection compositions—standard materials and endogenous gene. Chapters 6, 7, 8, and 9 summarize unknown gene exploration, quantification, high throughput, and isothermal detection. Chapter 10 provides a deeper analysis on target amplification improvement. Chapters 11, 12, and 13 attempt to connect aptamer, strips, and biosensors to food safety detection. Part II is novel applications of detection methods in genetically modified identification, noncoding RNA identification, heavy metal detection, microbial diversity, dead–alive microorganism identification, etc.

In this book, we want to offer a summarized and comprehensive representation of nucleic acid detection methods in food safety. The purpose of the publication is to serve a wider public audience. Our upcoming publication will be the first monograph about nucleic acid-based detection methods. Nearly all hazardous factors that concern food safety will be identified in terms of detecting nucleic acids. The book will also be applied as a systematic, authoritative, and professional reference manual by readers.

Acknowledgments This work is supported by the Ministry of Science and Technology of Beijing (XX2014B069) and the National Science and Technology Major Project (2016ZX08012-001). Many thanks to Chenguang Wang for his kind help in the manuscript conception and preparation.

References

1. Food Standard Agency, Institute of Food Research, Public Health England. McCance and Widdowson's the composition of foods. Cambridge: Royal Society of Chemistry; 2014.

2. Lappe FM, Collins J, Fowler C. Food first. Beyond the myth of scarcity. New York: Houghton Mifflin Co; 1977.
3. Georgievskii VI, Annenkov BN, Samokhin VT. Mineral nutrition of animals: studies in the agricultural and food sciences. New York: Elsevier; 2013.
4. Earle RL. Unit operations in food processing. New York: Elsevier; 2013.
5. Nestle M. Food politics: how the food industry influences nutrition and health, vol. 3. Berkeley: University of California Press; 2013.
6. Motarjemi Y, Lelieveld H. Food safety management: a practical guide for the food industry. Waltham: Academic; 2013.
7. Potter NN, Hotchkiss JH. Food science. New York: Springer Science & Business Media; 2012.
8. Maloni MJ, Brown ME. Corporate social responsibility in the supply chain: an application in the food industry. J Bus Ethics. 2006;68(1):35–52.
9. Card D, Krueger AB. Minimum wages and employment: a case study of the fast-food industry in New Jersey and Pennsylvania: reply. Am Econ Rev. 2000;90:1397–420.
10. Brownell KD, Horgen KB. Food fight: the inside story of the food industry, America's obesity crisis, and what we can do about it. Chicago: Contemporary Books; 2004.
11. Stank TP, Goldsby TJ, Vickery SK. Effect of service supplier performance on satisfaction and loyalty of store managers in the fast food industry. J Oper Manag. 1999;17(4):429–47.
12. Rouvière E, Caswell JA. From punishment to prevention: a French case study of the introduction of co-regulation in enforcing food safety. Food Policy. 2012;37(3):246–54.
13. Yeung RMW, Morris J. Food safety risk: consumer perception and purchase behaviour. Br Food J. 2001;103(3):170–87.
14. Verbeke W, Frewer LJ, Scholderer J, De Brabander HF. Why consumers behave as they do with respect to food safety and risk information. Anal Chim Acta. 2007;586(1):2–7.
15. Fedoroff NV, Battisti DS, Beachy RN, Cooper PJM, Fischhoff DA, Hodges CN, Knauf VC, Lobell D, Mazur BJ, Molden D. Radically rethinking agriculture for the 21st century. Science (New York, NY). 2010;327(5967):833.
16. Alimentarius C. Food hygiene basic texts. Rome: Food and Agriculture Organization of the United Nations World Health Organization; 1997.
17. Ryser ET, Marth EH. Listeria, listeriosis, and food safety. Boca Raton: CRC Press; 2007.
18. Techer C, Baron F, Delbrassinne L, Belaïd R, Brunet N, Gillard A, Gonnet F, Cochet MF, Grosset N, Gautier M. Global overview of the risk linked to the Bacillus cereus group in the egg product industry: identification of food safety and food spoilage markers. J Appl Microbiol. 2014;116(5):1344–58.
19. Duncan TV. Applications of nanotechnology in food packaging and food safety: barrier materials, antimicrobials and sensors. J Colloid Interface Sci. 2011;363(1):1–24.
20. Tuladhar E, Bouwknegt M, Zwietering MH, Koopmans M, Duizer E. Thermal stability of structurally different viruses with proven or potential relevance to food safety. J Appl Microbiol. 2012;112(5):1050–7.
21. Franz CMAP, Huch M, Abriouel H, Holzapfel W, Gálvez A. Enterococci as probiotics and their implications in food safety. Int J Food Microbiol. 2011;151(2):125–40.
22. Rendueles E, Omer MK, Alvseike O, Alonso-Calleja C, Capita R, Prieto M. Microbiological food safety assessment of high hydrostatic pressure processing: a review. LWT-Food Sci Technol. 2011;44(5):1251–60.
23. Pierson MD. HACCP: principles and applications. New York: Springer Science & Business Media; 2012.
24. Müller L, Stacey NH. Precaution principle versus danger prevention in toxicology. In: Regulatory toxicology. Heidelberg: Springer; 2014. p. 463–7.
25. Miraglia M, Marvin HJP, Kleter GA, Battilani P, Brera C, Coni E, Cubadda F, Croci L, De Santis B, Dekkers S. Climate change and food safety: an emerging issue with special focus on Europe. Food Chem Toxicol. 2009;47(5):1009–21.
26. Tirado MC, Clarke R, Jaykus LA, McQuatters-Gollop A, Frank JM. Climate change and food safety: a review. Food Res Int. 2010;43(7):1745–65.

27. Vermeulen SJ, Campbell BM, Ingram JSI. Climate change and food systems. Annu Rev Environ Resour. 2012;37(1):195.
28. Griffiths AM, Sotelo CG, Mendes R, Pérez-Martín RI, Schröder U, Shorten M, Silva HA, Verrez-Bagnis V, Mariani S. Current methods for seafood authenticity testing in Europe: is there a need for harmonisation? Food Control. 2014;45:95–100.
29. Montowska M, Alexander MR, Tucker GA, Barrett DA. Authentication of processed meat products by peptidomic analysis using rapid ambient mass spectrometry. Food Chem. 2015;187:297–304.
30. Tennyson JM, Winters KS, Powell K. A fish by any other name: a report on species substitution. In: 22nd annual meeting of Seafood Science the Technology Society of the Americas, Biloxi, Mississippi. 1997.
31. Behravesh CB, Williams IT, Tauxe RV. Emerging foodborne pathogens and problems: expanding prevention efforts before slaughter or harvest. Washington, DC: National Academies Press; 2012.
32. Meng J, Doyle MP. Introduction. Microbiological food safety. Microbes Infect. 2002;4 (4):395–7.
33. D'Mello JPF. Food safety: contaminants and toxins. Cambridge, MA: CABI; 2003.
34. Söderström A, Österberg P, Lindqvist A, Jönsson B, Lindberg A, Blide Ulander S, Welinder-Olsson C, Löfdahl S, Kaijser B, De Jong B. A large Escherichia coli O157 outbreak in Sweden associated with locally produced lettuce. Foodborne Pathog Dis. 2008;5(3):339–49.
35. Brugha R, Vipond IB, Evans MR, Sandifer QD, Roberts RJ, Salmon RL, Caul EO, Mukerjee AK. A community outbreak of food-borne small round-structured virus gastroenteritis caused by a contaminated water supply. Epidemiol Infect. 1999;122(01):145–54.
36. Frank C, Werber D, Cramer JP, Askar M, Faber M, an der Heiden M, Bernard H, Fruth A, Prager R, Spode A. Epidemic profile of Shiga-toxin–producing Escherichia coli O104: H4 outbreak in Germany. N Engl J Med. 2011;365(19):1771–80.
37. Nixdorff K. Biological weapons convention. In: Verifying treaty compliance. Berlin: Springer; 2006. p. 107–34.
38. Dall'Asta C, Berthiller F. Masked mycotoxins in food: formation, occurrence and toxicological relevance. Cambridge: Royal Society of Chemistry; 2015.
39. Sherif SO, Salama EE, Abdel-Wahhab MA. Mycotoxins and child health: the need for health risk assessment. Int J Hyg Environ Health. 2009;212(4):347–68.
40. Lam H-M, Remais J, Fung M-C, Xu L, Sun SS-M. Food supply and food safety issues in China. Lancet. 2013;381(9882):2044–53.
41. Cheng S. Heavy metal pollution in China: origin, pattern and control. Environ Sci Pollut Res. 2003;10(3):192–8.
42. Järup L. Hazards of heavy metal contamination. Br Med Bull. 2003;68(1):167–82.
43. Zhang X, Zhong T, Liu L, Ouyang X. Impact of soil heavy metal pollution on food safety in China. PLoS ONE. 2015;10(8):e0135182.
44. Zhang W, Jiang F, Ou J. Global pesticide consumption and pollution: with China as a focus. Proc Int Acad Ecol Environ Sci. 2011;1(2):125–44.
45. Van Boxstael S, Habib I, Jacxsens L, De Vocht M, Baert L, Van de Perre E, Rajkovic A, Lopez-Galvez F, Sampers I, Spanoghe P. Food safety issues in fresh produce: bacterial pathogens, viruses and pesticide residues indicated as major concerns by stakeholders in the fresh produce chain. Food Control. 2013;32(1):190–7.
46. Zhang M, Zeiss MR, Geng S. Agricultural pesticide use and food safety: California's model. J Integr Agric. 2015;14(11):2340–57.
47. Pei X, Tandon A, Alldrick A, Giorgi L, Huang W, Yang R. The China melamine milk scandal and its implications for food safety regulation. Food Policy. 2011;36(3):412–20.
48. Oplatowska-Stachowiak M, Elliott CT. Food colours: existing and emerging food safety concerns. Crit Rev Food Sci Nutr. 2015; (just-accepted):00–00.
49. James C. Global status of commercialized biotech/GM crops: 2015. Ithaca: International Service for the Acquisition of Agri-biotech Applications (ISAAA); 2015.

50. Aldemita RR, Reaño IME, Solis RO, Hautea RA. Trends in global approvals of biotech crops (1992–2014). GM Crops Food. 2015;6(3):150–66.
51. Cdc CDC. Estimates of foodborne illness in the United States. Atlanta: Center for Disease Control and Prevention (CDC); 2011.
52. Wu Y, Chen Y. Food safety in China. J Epidemiol Community Health. 2013;67(6):478–9.
53. Broughton EI, Walker DG. Policies and practices for aquaculture food safety in China. Food Policy. 2010;35(5):471–8.
54. Desa UN. World population prospects: the 2010 revision, highlights and advance tables. Working Paper No ES/P/WP. 220. New York: United Nations, Department of Economic and Social Affairs, Population Division. 2011.
55. Darwin C. On the origin of species by means of natural selection. 1859. London: Murray; 1968. p. 502.
56. Wallace AR. On the physical geography of the Malay Archipelago. J R Geograph Soc Lond. 1863;33:217–34.
57. Outram D. Georges Cuvier: vocation, science, and authority in post-revolutionary France. Manchester: Manchester University Press; 1984.
58. Mendel G. Experiments in plant hybridisation. New York: Cosimo, Inc; 2008.
59. Garrod A. The incidence of alkaptonuria: a study in chemical individuality. Lancet. 1902;160 (4137):1616–20.
60. Beadle GW, Tatum EL. Genetic control of biochemical reactions in neurospora. Proc Natl Acad Sci. 1941;27(11):499–506.
61. Avery OT, MacLeod CM, McCarty M. Studies on the chemical nature of the substance inducing transformation of pneumococcal types induction of transformation by a desoxyribonucleic acid fraction isolated from pneumococcus type III. J Exp Med. 1944;79 (2):137–58.
62. Lewin B, Dover G. Genes v, vol. 299. Oxford: Oxford University Press; 1994.
63. Crick FH. On protein synthesis. In: Symposia of the society for experimental biology. Cambridge: Company of Biologists Ltd; 1958. p. 138.
64. Crick F. Central dogma of molecular biology. Nature. 1970;227(5258):561–3.
65. Bell SD, Méchali M, DePamphilis ML. DNA replication. Cold Spring Harbor: Cold Spring Harbor Laboratory Press; 2013.
66. Calladine CR, Drew H. Understanding DNA: the molecule and how it works. London: Academic; 1997.
67. Breaker RR. Natural and engineered nucleic acids as tools to explore biology. Nature. 2004;432(7019):838–45.
68. Guerrier-Takada C, Gardiner K, Marsh T, Pace N, Altman S. The RNA moiety of ribonuclease P is the catalytic subunit of the enzyme. Cell. 1983;35(3):849–57.
69. Bass BL, Cech TR. Specific interaction between the self-splicing RNA of Tetrahymena and its guanosine substrate: implications for biological catalysis by RNA. Nature. 1984;308 (5962):820–6.
70. Tuerk C, Gold L. Systematic evolution of ligands by exponential enrichment: RNA ligands to bacteriophage T4 DNA polymerase. Science. 1990;249(4968):505–10.
71. Ellington AD, Szostak JW. In vitro selection of RNA molecules that bind specific ligands. Nature. 1990;346(6287):818–22.
72. Cho EJ, Lee J-W, Ellington AD. Applications of aptamers as sensors. Annu Rev Anal Chem. 2009;2:241–64.
73. Tan W, Donovan MJ, Jiang J. Aptamers from cell-based selection for bioanalytical applications. Chem Rev. 2013;113(4):2842–62.
74. Liu J, Cao Z, Lu Y. Functional nucleic acid sensors. Chem Rev. 2009;109(5):1948–98.
75. Fellows PJ. Food processing technology: principles and practice. New York: Elsevier; 2009.
76. Kelly S, Heaton K, Hoogewerff J. Tracing the geographical origin of food: the application of multi-element and multi-isotope analysis. Trends Food Sci Technol. 2005;16(12):555–67.

77. McEntire JC, Arens S, Bernstein M, Bugusu B, Busta FF, Cole M, Davis A, Fisher W, Geisert S, Jensen H. Traceability (product tracing) in food systems: an IFT report submitted to the FDA, volume 1: technical aspects and recommendations. Compr Rev Food Sci Food Saf. 2010;9(1):92–158.
78. Beulens AJM, Broens D-F, Folstar P, Hofstede GJ. Food safety and transparency in food chains and networks relationships and challenges. Food Control. 2005;16(6):481–6.
79. Lang T. Food industrialisation and food power: implications for food governance. Dev Policy Rev. 2003;21(5-6):555–68.
80. Frankenfeld CL, Leslie TF, Makara MA. Diabetes, obesity, and recommended fruit and vegetable consumption in relation to food environment sub-types: a cross-sectional analysis of Behavioral Risk Factor Surveillance System, United States Census, and food establishment data. BMC Public Health. 2015;15(1):491.
81. Patel PD. (Bio) sensors for measurement of analytes implicated in food safety: a review. TrAC Trends Anal Chem. 2002;21(2):96–115.

Chapter 2
Adaptable Methods to Extract Nucleic Acid Targets and Evaluate Quality

Abstract Extraction of nucleic acid targets (DNA and RNA) is a basic method that is used in molecular biology. In recent years, nucleic acid analysis has become an important analytical tool in many fields, demanding high-quality DNA or RNA. Nucleic acid targets can be extracted from various materials for subsequent processes, for example, PCR test. To date, many different ways have been developed to extract nucleic acid, such as column-based and solution-based methods. For the success of subsequent tests, extracting protocols should result in high purity, yield, and reproducibility of the nucleic acid targets, and the accuracy, speed, and reliability of the protocols should be maximal, while the contamination should be minimized.

Keywords Nucleic acid extraction • Sample preparation • Cell lysis • Purification • DNA quantification

2.1 The Importance of Nucleic Acid Extraction

2.1.1 DNA Extraction

The molecular biology of DNA is relatively a new field and has undergone significant changes in a short time. During the last few years, an important development of molecular biology techniques has been observed, and the availability of DNA is, consequently, becoming essential for studies. Several DNA extraction procedures have been proposed. However, because the quality of the extracted DNA is highly variable, standardization of the storage of samples and of extraction procedures becomes essential.

Because of the importance of the DNA molecule in biological studies, it has become important to be able to define it explicitly and precisely, notwithstanding the fact that more than one type exists in any given DNA sample. To date, many studies have been focused on the properties of a DNA sample, but it is obviously that if someone needs to do a research for molecular interpretation of biological function, it is necessary that the heterogeneity of the assembly of DNA studied be understood in all chemical and physical respects.

The sampling and preparation methods, sample sizes, unexpected inhibitors, and other factors may affect the quantity and quality of DNA extracted from various

W. Xu, *Functional Nucleic Acids Detection in Food Safety*,
DOI 10.1007/978-981-10-1618-9_2

samples. The application of a quantification and detection of DNA-based method also relay on the quantity and quality of the extracted DNA. There are plenty of factors that affect the reliability and applicability of the quantitative and qualitative detection processing. Sample preparation involving high temperature, mechanical stress, pH variations, fermentations, and enzymatic activity affects the basic DNA structure and results in oxidation, hydrolysis, or deamination of the sample [1, 2]. Food processing will result in significant degradation of DNA or totally destroy the DNA in the sample. So the processing will reduce the sensitivity of the analysis and influence limits of quantification and detection [3–5], which may alter the result of the analysis [6, 7].

Moreover, the test result could be affected by the extraction method we choose, making it necessary to remove the PCR inhibitors, including plant polysaccharides and polyphenolics, feed additives, or reagents used in extraction procedures.

2.1.2 RNA Extraction

RNA is an unstable molecule and has a very short duration once extracted from cells or tissues [8]. There are several types of RNA including ribosomal RNA (about 80–90 %), transfer RNA (tRNA), and messenger RNA (mRNA, about 2.5–5 %). RNA isolation needs more patience and care because RNA degrades easily. RNA is very unstable when RNases are present, which are always present in many samples, such as most fungi and bacteria in the air and many tissues [8]. Inhibiting endogenous RNases is necessarily used in intact RNA isolation by using strong denaturants. Successful extraction of high-quality RNA is contingent on RNase-free environment and skilled experimental operation. RNAse could resist heat and it is difficult to be deactivated as it does not require cofactors. The commonly used isolation methods can be summarized into two classes: the utilization of SDS (sodium dodecyl sulfate)/phenol and guanidinium thiocyanate.

2.2 The Principle of Nucleic Acid Extraction

In order to keep the function of nucleic acids, the structural integrity of the DNA samples is the basic requirement because all of the genetic information is stored in the primary structure, which is also important for its advanced structure and other biological macromolecules and in how they bond to each other. So the nucleic acid should be based on the following three requirements: (1) the organic solvents and high concentration of metal ions do not exist in the nucleic acid sample, which have inhibitory effect on enzyme; (2) other biological macromolecules, such as proteins, polysaccharides, and lipid molecules, should be down to a minimum; and (3) eliminate the pollution of other nucleic acid molecules, and RNA must be removed from DNA molecules and vice versa.

DNA quality is determined by its average molecular weight, fragment length, and the degree of damage, so it varies according to the type of samples under examination, the DNA extraction method used, and the degree of processing [1, 9, 10]. PCR test is normally done to assess DNA quality for an endogenous DNA sequence that is known to be present in this genome.

The determination of the A260/A230 and A260/A280 ultraviolet absorption ratios is always used to assess the purity of extracted DNA by using a spectrophotometer. When the 260/280 nm absorption ratio is between 1.8 to 2.2 and the 260/230 nm absorption ratio is more than 1.7, the DNA extracted is considered to be available for subsequent PCR analysis [11]. The purity of DNA will be badly influenced by many contaminants in the sample processing. The choice and optimization of the DNA extraction methods that remove interfering substances and possible inhibitory components should be very important for the success of every kind of detection method [10, 12]. For one thing, contaminants may come from the sample examined, for example, polyphenols, lipids, proteins, and polysaccharides [13, 14]. Additionally, CTAB, SDS, phenol, EDTA, chloroform, ethanol, and isopropanol which are used during the DNA extraction procedures can result in inhibition of the Taq polymerase to inhibit the PCR [15–17]. A large number of carbohydrates, salts, and other compounds that are frequently used in buffer solutions also reduce the performance of PCR [18].

2.3 General Steps of DNA Extraction Methods

2.3.1 Preparation of the Sample

To extract DNA, the first step is the preparation of the samples. As for a uniform sample, such as bacteria, meat, or one ingredient food product, these samples could be considered to have only one extracted material. However, in case of a heterogeneous sample, sufficient homogenization is particularly critical to ensure a representative sample, especially if a subsequent quantitative analysis is needed [19]. For blood samples, it is important to remove red blood cells and protein; for fecal samples, fecal matter is usually washed by PBS buffer to remove impurities; and for muscle tissue and organ samples, they are often ground in liquid nitrogen to preliminarily break cell walls and increase the contact area of lysis buffer and samples. In brief, different types of samples require different preparation methods before DNA extraction. The main purposes of preparation can be summarized as: (1) removing impurities of samples to decrease the influence of the impurities on following extracting process and (2) increasing the effect of the cell lysis. Once a sample has been obtained and prepared, the next step is to choose proper methods for DNA extraction.

2.3.2 Cell Lysis

For extraction of genomic DNA, cell lysis is a necessary step and can yield high-quality DNA. Cell lysis is used in laboratories to break open cells and purify or further study their contents. Currently, there are three cell lysis types (or called membrane disruption process) used: (1) chemical, (2) physical, and (3) enzymatic disruption. Chemical lysis often has been used either alone or in combination with physical methods. Cetyltrimethylammonium bromide (CTAB) and SDS might be the most common reagents. Many methods are using CTAB extraction buffer on the precipitation of the sample. These methods are considered efficient for many different samples, in particular for the samples that contain polysaccharides, which can be separated by CTAB [20–22]. CTAB is cheaper than commercial kits because of less use of costed chemicals [23]. Physical treatments will destroy cell wall structure. The most commonly used physical preparation methods are freezing–boiling (or called freezing–thawing) [24] and bead-mill homogenization [25]. Bead-mill processing can homogenize the sample with high speed and reduced extraction time, but resulting in DNA shearing. Lysozyme, for example, is always used as enzymatic disruption buffer for cell lysis of many bacteria.

2.3.3 Purification

After cell lysis, the next step is purification. The sample purification is done by using organic solvent (phenol and chloroform), followed by isopropanol-, polyethylene glycol-, or ethanol-assisted precipitation. Process-introduced phenol, lithium chloride, and EDTA are all PCR inhibitors. PCR test will be inhibited when phenol is more than 0.2 %. Phenolic compounds from the extracted materials or that come from DNA extraction processing will inhibit PCR test by denaturing the polymerase. Generally, phenolic compounds from plant (e.g., caffeic acid, chlorogenic acid) can alter various properties of the biomolecule by reacting with proteins or enzymes [26]. For potato, chlorogenic acid with concentration about 0.3 μg/μL will inhibit RT-PCR process [27]. 0.5 mM or higher concentration of EDTA has been proved to influence PCR test. In the last step of extraction, DNA is usually dissolved in TE buffer (containing 1 mm EDTA and10 mm Tris-HCl). The EDTA should be reduced or removed completely. We can also use only 0.01 M Tris-HCl to dissolve DNA that is used for subsequent real-time PCR. Distilled sterile water will also work. However, DNA cannot be stored for a long period of time in the absence of stabilizing compounds. High concentration of protein has been revealed to influence PCR. Ionic detergents such as CTAB and SDS are reported to have stronger effects than nonionic detergents, such as Triton X-100. High concentration (over 0.005 %) of SDS or CTAB will inhibit PCR test. However, SDS or CTAB is used early in DNA extraction protocols, and their residue is often washed by the subsequent washing regent. 2-Propanol and ethanol are often used for DNA precipitation, while they could also inhibit PCR test when the concentrations of them are over 1 % in the DNA product. Some commercial kits are now available for both extraction and purification of DNA target, and some kits are only used for purification of DNA. We can reduce the inhibitor concentration by dilution of the DNA and thus, enhance PCR efficiency; however, PCR sensitivity will be decreased when DNA concentration is excessive low. When the extracted DNA is used for real-time PCR amplification or other subsequent tests, inhibition tests should be performed to determine the effectiveness by using internal controls or evaluation of the linearity of the calibration curves [28].

2.4 The Classification of DNA Extraction Method by Cell Lysis

2.4.1 *Chemical Methods*

2.4.1.1 Cetyltrimethylammonium Bromide (CTAB) Method

Among all the cell lysis methods, chemical methods are used extensively. The strong associative behavior displayed by DNA and cationic surfactant systems is well known, and it has been widely used. In 1967, a procedure that used CTAB to precipitate DNA for the extraction and purification was first described. The cetyltrimethylammonium bromide (CTAB) method, with some kind of important modifications over past few years, has been still largely used for extraction of DNA from many samples, such as food and feed samples, micrioorganisms, leaves, seeds and grains. But the protocol needs more time and uses toxic chemicals, such as chloroform and phenol. The commonly used CTAB extraction buffer consists of 2 % (w/v) CTAB, 20 mM EDTA, 1.4 M NaCl, and 100 mM Tris-HCl, pH 8.0 [28], and some modifications will also be applied . Cell lysis is a convenient tool to isolate certain macromolecules that exist primarily inside of cells. Cell membranes consist of hydrophilic and lipophilic end groups. Therefore, detergents are often used to dissolve these membranes because they interact with both polar and nonpolar end groups. CTAB has emerged as the preferred choice for biological use because it maintains the integrity of precipitated DNA during isolation. EDTA is widely used for scavenging metal ions; ion depletion is commonly used to deactivate metal-dependent deoxyribonuclease, suppress its reactivity, and decrease its damage to DNA. A high concentration of NaCl solution can contribute to better DNA solubility. The original protocol [29] showed the use of CTAB extraction buffer for lysis of cell walls and organic buffer (contains chloroform and isoamyl alcohol) to remove impurities such as polysaccharides and proteins. Some modifications for the CTAB method have been developed, including the addition of mercaptoethanol and proteinase K to the CTAB extraction buffer and the use of organic buffer for further purification after the first use of organic buffer (phenol, chloroform, isoamyl alcohol (25:24:1)) [30]. Some concentration of

N-phenacylthiazolium bromide was reported to be added into the CTAB buffer to release DNA, which could be combined with sugar-derived condensation products [31, 32]· For RT-PCR, RNase T1 or RNase A is added to remove RNA in the solution. Since many modifications have been applied, different CTAB extraction buffer components with different concentrations are used for different samples. It is also important to test the pros and cons before determining which CTAB extraction method should be used.

CTAB extraction buffer is the most widely used reagent for DNA extraction. Cells typically have high concentrations of macromolecules, such as glycoproteins and polysaccharides that coprecipitate with DNA during the extraction process, causing the extracted DNA to lose purity. The positive charge of the CTAB molecule allows it to denature these molecules, which would interfere with the isolation. Thus, the CTAB extraction buffer has better effects on plant samples, which are very rich in secondary metabolites (such as polysaccharide and polyphenol). However, DNA extracted by CTAB buffer needs further more purification steps to be used for subsequent real-time PCR or other tests according to published reports. The A260/A230 ratio was an average of 1.29 of DNA product extracted by CTAB, which may be resulted by contamination, such as polysaccharides, and these compounds will inhibit qPCR process [30]. CTAB-extracted DNA purified with commercial kits can result in a reduction of impurity concentration and produce the expected values. Dilution of the DNA product can also result in a reasonable quantification, for the concentration of impurities is also reduced. Additionally, phenol–chloroform is harmful to human health and the environment. Special manipulation and personal protective equipment must be used in the disposal of phenol–chloroform. Moreover, phenol in DNA extraction inhibits enzymatic reactions in downstream applications. Thus, a protocol that does not require phenol–chloroform is generally safer.

2.4.1.2 Sodium Dodecyl Sulfate (SDS)

The basic SDS extraction buffer always consists of 1 % SDS; 150 mM Tris-HCl; 50 mM EDTA, pH 8.0; 10 mM 2-mercaptoethanol; and 0.5 M NaCl. The concentrations of the SDS extraction buffer contents may alert according to different samples. After preparation, samples are always heated to 65 °C with extraction buffer for several hours. SDS is an organic compound and an anionic surfactant. It is an organosulfate consisting of a 12-carbon tail attached to a sulfate group, and SDS is often used to lyse cells during DNA extraction processing. The 2-mercaptoethanol is used to reduce disulfide bonds, and it can act as a biological antioxidant to reduce DNA degradation. Proteins and polysaccharides are removed using organic solutions, just like CTAB method. SDS extraction buffer has always been used in combination with organic buffer (phenol and chloroform) to successfully remove impurities. However, DNA purification is also needed.

2.4.1.3 Polyvinylpyrrolidone (PVP) Methods

The PVP extraction method is commonly used for samples containing a large amount of phenolic compounds [33]. It is often used to absorb polyphenols during DNA extraction. Polyphenols are common in many plant tissue samples and, if not removed, will deactivate proteins and therefore inhibit many subsequent reactions such as PCR. The PVP method includes a cell lysis in high temperature with the buffer containing CTAB (or SDS) and Tris-HCl followed by the addition of 1 % PVP to remove the contaminants such as phenolic compounds or polysaccharides. However, PVP can result in a loss of DNA.

Addition of CTAB and PVP has been widely reported. Humic compounds can be removed by both CTAB and PVP content. CTAB can combine with the polysaccharides, denatured proteins, and cell residual to form insoluble complexes. PVP was shown to be not effective in cell lysis, but it is efficient to be used during the purification step of DNA.

2.4.2 Physical Methods

Physical treatments will destroy cell structure and serve as an assistant helper on cell lysis. The most widely used physical methods of cell disruption are freezing–boiling or freezing–thawing [34] and bead-mill mix [25]. Bürgmann et al. [35] established a method that increased DNA yield with higher speed, longer beating times, and reduced extraction buffer volumes, but this method will result in DNA shearing. Frostegard et al. [36] reported that the lysis efficiency will greatly improve by using a drying treatment before grinding. Microwave thermal shock, ultra-sonication grinding, and mortar mill grinding in liquid nitrogen have also been reported and tested. In summary, cell disruption of samples of small cells, spores, and vegetative forms has shown more efficiency with physical methods, but at a cost of significant DNA shearing.

2.4.3 Enzymatic Lysis Methods

Enzymatic lysis has been developed in many extraction processing. Lysozyme treatment is the most commonly used method. Lysozyme hydrolytic action on glycosidic or other humic components may lead to the improvement of DNA purity. Achromopeptidase is another enzyme that was commonly used to improve the lysis of the cell wall of Gram-positive bacteria, while proteinase K was used to degenerate proteins and reduce contamination.

2.5 Classification of DNA Extraction Methods by Samples

2.5.1 Plants

Plant genome DNA extraction generally includes breaking the cell wall; cracking the cell membrane; removing polysaccharides, polyphenols, and precipitation of proteins and nucleic acids; and removing RNA to obtain the DNA. Plant DNA extraction methods are basically mature, but for different plants, because of the differences in their biological and chemical composition and structure, there are substantial differences in the specific procedures. Yan-hong et al. [37], for instance, have noted that different plants and even different parts of the same plant have their own characteristics: some have higher degrees of tissue lignification, some have thicker cell walls, and some contain phenolics, while others are easily degraded. It is difficult to use a single method to extract the DNA of different plants. Therefore, studies should examine how the plant of interest differs and explore and improve methods of extracting DNA. Peng Rui et al. [38] found that using the same method to extract the DNA of different species of *Dendrobium* stem plants results in poor stability. Hong [39] also noted that red algae and brown algae in a high-viscosity sugar solution and different kinds of algae, different parts of the same strain of algae, and the same kind of algae at different stages of cell wall composition show different characteristics, so extracting DNA from algae should be performed using different protocols according to the specific biochemical properties.

2.5.1.1 The Processing of Experimental Materials

The processing of experimental materials has two aspects: the choice of the experimental materials and their storage and pretreatment. Generally speaking, in the selection of materials, one should try to choose tender parts of the plant, but other factors include experimental conditions such as the plant growth cycle, the distance from the laboratory to the growth site, and the biochemical properties of the plant. Generally, plant's leaves are taken for experimental material, and some of the older leaves should be stored at 4 °C in the dark for one day to allow the consumption of starch and other polysaccharide substances [40].

To obtain high-quality genomic DNA samples, the material should be protected from oxidation by mechanical damage and browning, and the DNA should not be degraded by endogenous nucleic acid enzyme. There are two methods commonly used to preserve the samples, low temperature and dehydration. According to many studies, these two methods show almost no difference for DNA extracted from fresh leaves.

2.5.1.2 The Choice of Cell Lysis Method

In the extraction of genomic DNA, cell lysis is a necessary step. To obtain high-quality DNA, the details of this step are very important. Among them, the composition of the pyrolysis buffer and the warm bath time and temperature are directly or indirectly related to the results. The cell lysis buffer generally contains surfactant, often with CTAB and SDS. The buffer composition should be based on general factors regarding the aims, the follow-up study on the quality of the products and quantity of DNA with different selections and adjustments, the buffer pH, protective agents, and the type and content of surfactant for overall optimization. In general, CTAB buffer is suitable for DNA extraction from herbaceous plants with little or no phenol content. However, for woody plants and phenol-rich plants, increasing the CTAB concentration can also obtain a higher quality of DNA. Xiong Guang Ming [9] used lysis buffer containing 2 % CTAB to extract mandarin DNA and obtained high-quality DNA for use in further analysis.

Tea DNA was extracted with lysis buffer containing 2 % CTAB by Changchun Yuan [41], and high-quality plant DNA was again obtained. However, there are still some deficiencies we can improve in the CTAB method to obtain high-quality DNA. When the CTAB concentration is too high, the heat-preserved material and CTAB buffer are difficult to blend fully, causing an uneven mixture. Additionally, extending the time of heat preservation may increase the risk of DNA degradation.

2.5.1.3 Nucleic Acid Separation and Purification

In the nucleus, nucleic acid is usually combined with proteins into a ribonucleoprotein complex, and DNA itself exists in a nucleoprotein form. Cell lysis must separate protein and nucleic acid for nucleic acid extraction. Phenol and chloroform (generally phenol–chloroform–isoamyl alcohol at 25:24:1) are used after cell lysis to remove protein. Alternatively, 5 mol/L potassium acetic acid is added, and the mixture is centrifuged to remove denatured protein. Proteinase K is also used for the catalytic hydrolysis of protein. The protein removal can be performed by different approaches in practice: one is using potassium acetate or chloroform–isoamyl alcohol (volume ratio of 24:1) individually; the second is to use the two reagents in combination [42]; and the third is to increase the number of chloroform–isoamyl alcohol extractions [43]. In addition, there are reports of increasing the warm bath time to complete protein denaturation.

2.5.2 *Animal*

The use of nucleic acids to identify animal species, examine the origin of species, evaluate diversity, and explore the relationships of system evolution is currently the most effective means of molecular biology research in animals. However, animal tissue is rich in protein, fat, and other substances that can easily combine with DNA and form coprecipitates during the extraction process, which can seriously inhibit the activity of restriction enzymes and Taq polymerase, causing the failure of enzyme digestion and PCR amplification. Therefore, in the process of extracting DNA, removing impurities such as proteins is key for accurate molecular biology identification.

2.5.2.1 Invasive Sampling (Muscle Tissue and Organs)

For muscle tissue and organ sample pretreatment, some researchers grind tiny amounts of tissue in a mill. Because muscle has a certain toughness, the effect is not ideal. The grinding process can produce a large amount of heat, which may contribute the degradation and fracturing of DNA, reducing its integrity. Therefore, liquid nitrogen should be used to freeze the material for grinding, effectively preventing the fracturing of DNA and saving time [43].

Commonly, CTAB and SDS extraction buffer are available for cell lysis after sample pretreatment, and DNA purification is performed. Some protocols can be used to extract DNA and RNA at the same time. The method involves extraction of tissues in buffered phenol and guanidine isothiocyanate solutions followed by extraction with chloroform, which facilitates separation of the aqueous and organic phases demarcated by an interphase. RNA is recovered from the upper aqueous phase, and genomic DNA and proteins are extracted from the lower two phases.

2.5.2.2 Noninvasive Sampling (Hide/Blood/Feces)

Since the mid-1980s, because of the wide application of PCR technology, we have been able to detect very small amounts of genetic material and create a new class of sampling methods for noninvasive sampling. The sample source can be an animal hide, a lost hair, feces, urine, etc. In some situations, extracting DNA from these samples is necessary.

Hide

A hide is composed of animal skin that has undergone a complex series of chemical treatments called tanning. Because of these harsh manufacturing treatments, the DNA is frequently highly degraded, chemically modified, and co-extracted with inhibitors, which makes it quite resistant to further analysis. At present, there have been many scholars [44–47] in certain subject areas that have successfully extracted DNA from hides, but the protocols are often very cumbersome. Therefore, many scholars hope to find a better extraction solution. Rao Gang et al. [45] used Chelex (chelating resin) in the process of extraction, resulting in improved yield of the DNA. Chelex is a type of chelating resin with high affinity for multivalent metal ions that can chelate these metal ions to prevent the degradation of DNA, which can obtain high molecular weight DNA from trace samples. Additionally, to digest the material better, Lan Hong et al. [47] utilized a PBS digestive buffer containing collagen and trypsin; this is because the collagen can be folded to the natural 3D helical structure of collagen hydrolysate under appropriate conditions, and trypsin can selectively hydrolyze protein, catalyzing the hydrolysis of peptide bonds, breaking down proteins into smaller peptides, and dissolving the digestive denaturation of protein. Merheb et al. [48] added leather samples to 10 ml of extraction buffer, which contained 0.5 M EDTA (pH 8), 1 % of N-lauroylsarcosine, and 0.5 mg/mL of proteinase K, and incubated it for 16 h at 37 °C under rotation. The remaining leather or glue powder was collected by centrifugation, and only the supernatant was further used in the silica extraction protocol.

Blood

The key to extract DNA from blood samples is to remove red blood cells and protein. The conventional method is to obtain lymphocytes first and then digest them with proteinase K and SDS and remove impurities using phenol–chloroform and ethanol precipitation. However, this method is complex, and the reagent is harmful to the body; thus, many scholars have made further improvements.

Zhaoquan et al. [49] used Triton X-100 to break open red and white blood cells and to obtain the white blood cells and then depolymerized the nucleoprotein by $NaClO_4$ and SDS instead of proteinase K and removed impurities with PCl (phenol–chloroform–isoamyl alcohol at 25:24:1) and ethanol precipitation. This method has greatly improved the removal of red blood cells. Jin-long Huo et al. [50] put forward a method of using an isotonic solution of NH_4Cl for the removal of red blood cells. This method was effective for removing red blood cells, but the operation is very complex. Wu Qingmin et al. [51] made a modest improvement by using a buffer containing NH_4Cl and Tris-HCl for sample processing, and at the same time proteinase K was added for protein digestion, incubating the sample for 2 h in a 54 °C water bath and adding 6 mol/L NaCl; a pure DNA product was obtained after centrifugal precipitation and washing.

Fecal Samples

Recently, gut microbiota has been given more attention because the microbiota could be viewed as a metabolic "organ" that performs functions that our body cannot perform. These functions include the ability to process otherwise indigestible components of our diet, such as dietary fiber and plant polysaccharides. Researching the signaling pathways regulated by the microbiota will provide many new therapeutic targets for promoting health. Fecal DNA not only contains many gut microbes, it also contains undigested food, digestive enzymes, mucus, choline, bilirubin, etc. [52]; there are also many herbivorous animal waste plant polysaccharides in it, and these impurities will not only cause the degradation of target DNA but also can inhibit PCR reactions [53]. It is difficult to extract fecal DNA of high quality and quantity, and the risk of contamination is higher [54]. Therefore, how to effectively remove metabolic impurities (digestive enzymes, bile salts, etc.) or food components (polysaccharides, pigments, proteins, fat, etc.) from fecal samples will be key to successful experimentation [54].

CTAB is always used to handle some predatory animal waste to effectively remove harmful wastes, such as polysaccharides and proteins, which are the main impurities that affect the acquisition of high-quality DNA for PCR amplification. Constable et al. [55] and Savill [56] used proteinase K to crack cells joined with CTAB, and the DNA products they obtained could be successfully used for PCR amplification. Zhong Hua, Baowei Zhang, and Jianyuan Zhao [54, 57, 58] pretreated feces before cracking with acetone, ethanol, or water waste, respectively, and removed PCR inhibitors such as pigments, polysaccharides, some organic salts, and so forth; then, the conventional organic solvent extraction method was used to purify the DNA. The DNA product can be used for PCR amplification and further genetic analysis, and the effect is better than using untreated fecal DNA.

2.5.3 *Microorganisms*

Microorganisms are ubiquitous in all environments. However, no more than 1 % of the total microorganisms exist in the environments are cultivable in lab [59]. The diversity analysis of bacteria in different environments has depended on DNA extraction [59] and mainly on 16S rDNA amplification method. However, DNA from samples without inhibitory contaminants is needed to apply these methods [60].

DNA extracted from microorganism in soil samples was developed by Torsvik [61] and Bakken [62], which followed on these steps: (1) homogenization of soil samples; (2) separation of the bacteria from the soil samples by centrifugation depending on buoyant density, sedimentation speed, or both; (3) cell lysis of the target microorganisms; and (4) the purification process of DNA [63].

An extraction method of microorganisms has been largely used for many years. This method will lyse all the target microorganisms, which commonly provides the highest yields of DNA within relatively short processing time. The original procedure from Ogram et al. [64] summarized the method to two main steps. The first step is disrupting the cell wall, which leads to the release of all DNA from all microorganisms or the location of microorganisms. In the second step, DNA is separated.

2.5.3.1 Extracting DNA from Bacteria

The cell wall is outside of the cell membrane in bacteria. The cell wall and membrane consist of the barrier of bacteria. Peptidoglycan is the most common material of a bacterial cell wall, which is made from polysaccharide chains cross-linked by peptides containing D-amino acids [65]. Bacterial cell walls are different from the plants and fungi cell walls, which are made of chitin and cellulose, respectively [66]. The cell wall of bacteria is also different from that of Archaea, which does not contain peptidoglycan. Many bacteria need cell wall to survive, and the antibiotic penicillin is able to kill bacteria by inhibiting a step in the synthesis of peptidoglycan [66].

Although guanidine thiocyanate treatment and silica column purification methods are widely used in recent years and they are satisfactory for DNA extraction from nearly all kinds of bacteria, the preparation of bacterial samples with mechanical disruption or enzymatic lysis is necessary for the extraction of some thick-cell-wall bacteria. This is usually a big problem when many different types of bacteria are presented in the sample, such as in urine or stool, and sometimes it is important to extract different bacteria with equal efficiency so as not to influence or alert the relative populations of the target microbiota. Some studies have proven that different structures of the cell wall in bacteria are the root of these problems. Gram-positive bacteria have thicker layer of peptidoglycan in the cell walls, thus making them more difficult to lyse than Gram-negative bacteria in DNA extraction processing [67]. Detergents have often been used in combination with chelating agents, such as Chelex 100, EDTA, and sodium phosphate or Tris buffers. In several studies, increasing the concentration of EDTA can increase yields but will result in lower purity of isolated DNA. It can be concluded that it is important to choose the proper buffer to get the expected DNA quantity and the required DNA purity.

2.5.3.2 Extracting DNA from Fungi

Fungi have a cell wall. It contains the biopolymer chitin in addition to glucans and other typical components. More complicated cell walls of fungi make it more difficult to lyse than Gram-positive and Gram-negative bacteria. Thus, methods of cell lysis are always used in combination (physical, chemical, and enzymatic disruption). More and cheaper next-generation sequencing and methods and advancements in microbial diagnostic microarray technologies have increased culture-independent studies in both number and resolution. As most prokaryotes remain uncultivable in laboratory, the development of biology technologies has greatly helped us in understanding microbial diversity and its functioning [68]. However, these molecular techniques require high-quality genomic DNA; thus, a modified and effective extraction method is important.

Because of the need to lyse the cell wall and remove impurities (carbohydrates and proteins), fungal DNA extraction protocols are often time-consuming. These procedures typically require hazardous chemicals and need many time-consuming steps of sequence-based identification. Spin column or other DNA binding materials can replace organic solution in some methods. However, these protocols have shortcomings and result in the lack of a simple, universal procedure that is suitable for the efficient isolation of DNA from phylogenetically diverse fungi samples [69]. However, many improvements have been achieved. A. M. Romanelli et al. [69] developed and standardized a rapid, inexpensive DNA extraction protocol applicable to the major fungal phyla. They used PrepMan Ultra Lysis Reagent (Applied Biosystems, Inc., Foster, CA, USA) and heated the sample at 99 °C for 10 min for cell lysis. Many commercial kits are also effective for fungal DNA extraction.

2.5.4 Deep-Processed Food

Food safety and quality have become more and more important in modern society for consumers, industry manufactures, as well as the governments [30]. In particular, the DNA quality of food samples is highly influenced by different ways: the grade of damages of the extracted DNA, the contaminations that will inhibit PCR test in the samples, and the average molecular length of the DNA products extracted. These influences come from several ways: (1) the sample itself, (2) the chemical and physical characteristics of the extraction method used, and (3) the food production processing carried out. DNA damage will happen when DNA is exposed to heat, and the fragment length and molecular weight will reduce. Physical or chemical treatments will also result in the breakage of DNA strands, thus influencing the average fragment size of DNA [70]. Because of the mechanism of hydrolytic degradation of DNA, it is very sensitive to alkaline agents and acid. For example, when pH is low, purines will be removed from the DNA backbone as a result of the cleavage of N-glycosidic bonds between deoxyribose residues. Next, hydrolysis action will happen in the adjacent phosphodiester linkages, resulting in the breakage of DNA strands [71, 72]. Some industrial food processing with high temperatures will allow the condensation of carbonyl groups of reducing sugars with primary amines, which will result in the changes of many molecules, producing some impurities that will prevent DNA extraction processing or amplification process [72, 73]. Some fruit and vegetable samples, which are of acidity, will accelerate the acid-catalyzed reactions when the heating procedures are done. For the other things, some food processing needs high pH, such as the initial stage of the preparation of tacos and other similar products from maize. Many processing factors affect the DNA quality. DNA will breakdown under high temperature or low pH.

Many successful protocols have been tested to extract DNA from deep-processed food. The CTAB–PVP method was used to extract DNA from chocolate, and the product was of high quality for PCR amplification [74]. Commercial kits are also commonly used for the DNA extraction from deep-processed food.

2.6 DNA Quantification Technologies

2.6.1 Agarose Gel Electrophoresis

To assess DNA quality and get the amount of DNA in extracted product solution, the method of agarose gel electrophoresis is often used to separate DNA molecules according to different molecule sizes. DNA is separated by the electric field to push the DNA molecules through the agarose gel, because the main factor that affects its movement through the gel is the DNA molecular length [75]. Shorter DNA molecules have lower molecular weight so that they move faster through gels compared with fragments with higher molecular weight. One should choose different percent concentrations of gel for different sizes of DNA fragments. Lower agarose gel concentrations could separate larger DNA fragments, while smaller DNA molecules need higher concentrations of agarose. The concentration of most agarose gels used is between 0.8 and 2 % in the electrophoresis buffer. For quantification of genomic DNA, concentration between 0.8 and 1 % is commonly used. Agarose concentration of 2 % is used for PCR products. The DNA in the gel can be viewed by dyeing with the fluorescent dye, for example, EB (ethidium bromide); when electrophoresis is finished, the EB–DNA complex can be viewed under UV light. The DNA products with good quality and quantity are intact without severe degradation, showing an intense band and high molecular weight in the gel. Damaged or degraded DNA with RNA contamination can be viewed as a long stripe toward the lower molecular weight portion in the gel [30]. A DNA ladder is often run together with DNA product to estimate the fragment length of the DNA.

2.6.2 Ultraviolet Spectrophotometry

Concentrations of the nucleic acid (DNA and RNA) solutions can be determined by UV spectrophotometry with the absorbance at 260 nm (A260), as well as their purity (A260/A280 ratio), for an analytical instrument used in molecular biology. This analysis is based on the principle that DNA or RNA absorbs ultraviolet light in a specific pattern and they can absorb the ultraviolet light at the 260 nm wavelength. The optical density (OD) value can be calculated as

$$\text{OD} = \text{Log}(\text{Intensity of Incident Light}/\text{Intensity of Transmitted Light}) \quad (2.1)$$

For double-stranded DNA (dsDNA), one OD refers to 50 μg of dsDNA per mL in a sample. For single-stranded DNA and RNA (ssDNA and ssRNA), one OD unit corresponds to 37 and 40 μg of product, respectively [76]. For protein contamination in the product solution, the A280 value that equals an OD refers to 1 μg/mL of protein in the sample [77]. A260/A280 ratio is often used to access protein contamination of DNA product, and the product is considered acceptable when the ratio value is higher than 1.7 [75].

Contaminants can contribute to the final signal intensity of the A260 measurement including phenols, RNA, and ssDNA, which cannot be distinguished from dsDNA. The DNA degradation will alert the ratio of dsDNA to ssDNA in the product, and the A260 value will increase as dsDNA denatures. The DNA product is routinely considered acceptable when A260/A280 ratio is less than 2.0. In degraded DNA samples, strand separation of ssDNA and the unstacked base pair will contribute to the increase of the A260 absorbance signal than pure dsDNA. Double-helix structure of DNA will disappear when DNA denatures into single strands, and the interactions of the base pairs will become disrupted; thus, more resonance structures will contribute to increases in A260 value [78, 79].

UV spectrophotometry DNA quantification method usually requires large amounts of DNA, about 10–50 μg/mL of DNA per sample [75, 80]. In conclusion, precision and accuracy of UV spectrophotometry to quantify DNA can be summarized by (1) the size distribution of DNA in solution; (2) other contaminations that contribute to A260 nm, such as ssDNA and RNA; and (3) the ratio of dsDNA to ssDNA.

2.6.3 Fluorometry

Fluorometry is a rapid, convenient, and economic method to quantify dsDNA concentration. To determine the DNA concentration, the sensitivity of fluorometry is similar to chemiluminescence, and these two techniques are more sensitive than spectrophotometry. UV spectrophotometry is less sensitive (the detection limit is about 0.05 μg/mL), but the measurements are not influenced by plasmid and molecule size. The linear standard curve is constructed by the values of the standards, and the concentration of the test sample will be estimated by the standard curve. The linear concentration range for DNA quantification extends over four magnitude (always from 25 pg/mL to 1 μg/mL) using one dye concentration.

Accuracy and precision of fluorometry can be influenced by the following factors: (1) the distribution of the DNA size in solution, (2) the sensitivity of the assay to photobleaching, (3) the sensitivity for different dyes and their best ranges of the working concentration, (4) the DNA integrity, (5) the exclusivity of dyes, and (6) the DNA purity [79, 80].

2.7 Prospects

Many advanced molecular biology techniques, such as sequencing techniques, PCR test, and mass spectrometry, have been optimized to enhance the detection and study of many organisms. DNA products with high content and purity are the best guarantee for further study and analysis, such as PCR reactions. For many different organisms, the quantity and quality of extracted DNA from samples will change a lot depending on the extraction method chosen, and thus different methods should be used for different types of sample and conditions. In conclusion, the DNA extraction method should be fast and accurate and produce a high-quality DNA product without using hazardous chemicals. Furthermore, different samples require different methods, and the improvement of current methods should be carried out with the development of new technologies.

Acknowledgments This work is supported by the National Science and Technology Major Project (2016ZX08012-004). Many thanks to Wenying Tian and Xingtian Cui for their kind help in manuscript conception and preparation.

References

1. Meyer R. Development and application of DNA analytical methods for the detection of GMOs in food. Food Control. 1999;10:391–9. doi:10.1016/S0956-7135(99)00081-X.
2. Kharazmi M, Bauer T, Hammes WP, Hertel C. Effect of food processing on the fate of DNA with regard to degradation and transformation capability in Bacillus subtilis. Syst Appl Microbiol. 2003;26(4):495–501. doi:10.1078/072320203770865774.
3. Weighardt F. GMO quantification in processed food and feed. Nat Biotechnol. 2007;25 (11):1213–14. doi:10.1038/nbt1107-1213c.
4. Lipp M, Shillito R, Giroux R, Spiegelhalter F, Charlton S, Pinero D, Song P. Polymerase chain reaction technology as analytical tool in agricultural biotechnology. J AOAC Int. 2005;88 (1):136–55.
5. Spiegelhalter F, Lauter FR, Russell JM. Detection of genetically modified food products in a commercial laboratory. J Food Sci. 2001;66(5):634–40. doi:10.1111/j.1365-2621.2001. tb04613.x.
6. Gryson N, Dewettinck K, Messens K. Detection of genetically modified soy in doughs and cookies. Cereal Chem. 2007;84(2):109–15. doi:10.1094/Cchem-84-2-0109.
7. Cankar K, Stebih D, Dreo T, Zel J, Gruden K. Critical points of DNA quantification by real-time PCR – effects of DNA extraction method and sample matrix on quantification of genetically modified organisms. BMC Biotechnol. 2006;6. doi:Artn 37 10.1186/1472-6750-6-37.
8. Phillips† DAJ. Gavin Brooks, Biotechnology in Healthcare, An introduction to biopharmaceuticals. J Pharm Pharmacol. 2002;(No.10):1433.
9. Guangming X, Luolu L, Yanyong, Suqiong X, Wuqing, Xiqing L, Jiangdong. DNA extraction method for AFLP analysis in citrus. J Fruit Sci. 2002;04:267–8.
10. Anklam E, Gadani F, Heinze P, Pijnenburg H, Eede GVD. Analytical methods for detection and determination of genetically modified organisms in agricultural crops and plant-derived food products. Eur Food Res Technol. 2002;214(1):3–26.
11. Matsuoka T, Kuribara H, Akiyama H, Miura H, Goda Y, Kusakabe Y, Isshiki K, Toyoda M, Hino A. A multiplex PCR method of detecting recombinant DNAs from five lines of genetically modified maize. J Food Hyg Soc Jpn. 2001;42(1):24–32. doi:10.3358/shokueishi.42.24.
12. Meyer R, Candrian U. PCR-based DNA analysis for the identification and characterization of food components. Food Sci Technol-Leb. 1996;29(1–2):1–9.
13. Van Hoef AMA, Kok EJ, Bouw E, Kuiper HA, Keijer J. Development and application of a selective detection method for genetically modified soy and soy-derived products. Food Addit Contam. 1998;15(7):767–74.
14. Kuiper HA. Summary report of the ILSI Europe workshop on detection methods for novel foods derived from genetically modified organisms. Food Control. 1999;10(6):339–49. doi:10. 1016/S0956-7135(99)00072-9.
15. Rossen L, Norskov P, Holmstrom K, Rasmussen OF. Inhibition of Pcr by components of food samples, microbial diagnostic assays and DNA-extraction solutions. Int J Food Microbiol. 1992;17(1):37–45. doi:10.1016/0168-1605(92)90017-W.
16. Tinker NA, Fortin MG, Mather DE. Random amplified polymorphic DNA and pedigree relationships in Spring Barley. Theor Appl Genet. 1993;85(8):976–84.
17. Zimmermann A, Lüthy J, Pauli U. Quantitative and qualitative evaluation of nine different extraction methods for nucleic acids on soya bean food samples. Z Lebensm Unters Forsch. 1998;207(2):81–90.
18. Hammes† PDWP, Hertel‡ DC. Mit Hilfe der Gentechnik erzeugte Lebensmittel: Novel Foods und die Problematik ihres Nachweises. Biologie in unserer Zeit 1995;(No.4):246–55.
19. Zimmermann A, Luthy J, Pauli U. Quantitative and qualitative evaluation of nine different extraction methods for nucleic acids on soya bean food samples. Z Lebensm Unters F A. 1998;207(2):81–90. doi:10.1007/s002170050299.

20. Di Bernardo G, Del Gaudio S, Galderisi U, Cascino A, Cipollaro M. Comparative evaluation of different DNA extraction procedures from food samples. Biotechnol Progr. 2007;23(2):297–301. doi:10.1021/bp060182m.
21. Olexova L, Dovicovicova L, Kuchta T. Comparison of three types of methods for the isolation of DNA from flours, biscuits and instant paps. Eur Food Res Technol. 2004;218(4):390–3. doi:10.1007/s00217-004-0872-y.
22. Gryson N, Dewettinck K, Messens K. Influence of cocoa components on the PCR detection of soy lecithin DNA. Eur Food Res Technol. 2007;226(1–2):247–54. doi:10.1007/s00217-006-0533-4.
23. Kakihara Y, Matsufuji H, Chino M, Takeda M. Extraction and detection of endogenous soybean DNA from fermented foods. Food Control. 2006;17(10):808–13. doi:10.1016/j.foodcont.2005.05.006.
24. Gryson N, Messens K, Dewettinck K. Evaluation and optimisation of five different extraction methods for soy DNA in chocolate and biscuits. Extraction of DNA as a first step in GMO analysis. J Sci Food Agr. 2004;84(11):1357–63. doi:10.1002/jsfa.1767.
25. Simonet P, Capellano A, Navarro E, Bardin R, Moiroud A. An improved method for lysis of Frankia with achromopeptidase allows detection of new plasmids. Can J Microbiol. 1984;30(10):1292–5.
26. Rohn S, Rawel HM, Kroll J. Inhibitory effects of plant phenols on the activity of selected enzymes. J Agr Food Chem. 2002;50(12):3566–71. doi:10.1021/jf011714b.
27. Singh RP, Singh M, King RR. Use of citric acid for neutralizing polymerase chain reaction inhibition by chlorogenic acid in potato extracts. J Virol Methods. 1998;74(2):231–5. doi:10.1016/S0166-0934(98)00092-5.
28. Demeke T, Jenkins GR. Influence of DNA extraction methods, PCR inhibitors and quantification methods on real-time PCR assay of biotechnology-derived traits. Anal Bioanal Chem. 2010;396(6):1977–90. doi:10.1007/s00216-009-3150-9.
29. Demeke T, Ratnayaka I, Phan A. Effects of DNA extraction and purification methods on real-time quantitative PCR analysis of roundup ready (R) soybean. J AOAC Int. 2009;92(4):1136–44.
30. Di Bernardo G, Galderisi U, Cipollaro M, Cascino A. Methods to improve the yield and quality of DNA from dried and processed figs. Biotechnol Progr. 2005;21(2):546–9. doi:10.1021/bp049710p.
31. Bernardo GD, Gaudio SD, Galderisi U, Cascino A, Cipollaro M. Comparative evaluation of different DNA extraction procedures from food samples. Biotechnol Progr. 2007;23(2):297–301.
32. Corbisier P, Broothaerts W, Gioria S, Schimmel H, Burns M, Baoutina A, Emslie KR, Furui S, Kurosawa Y, Holden MJ, Kim HH, Lee Y, Kawaharasaki M, Sin D, Wang J. Toward metrological traceability for DNA fragment ratios in GM quantification. 1. Effect of DNA extraction methods on the quantitative determination of Bt176 corn by real-time PCR. J Agr Food Chem. 2007;55(9):3249–57. doi:10.1021/jf062931l.
33. Smith DS, Maxwell PW, De Boer SH. Comparison of several methods for the extraction of DNA from potatoes and potato-derived products. J Agr Food Chem. 2005;53(26):9848–59. doi:10.1021/jf051201v.
34. Miller DN, Bryant JE, Madsen EL, Ghiorse WC. Evaluation and optimization of DNA extraction and purification procedures for soil and sediment samples. Appl Environ Microbiol. 1999;65(11):4715–24.
35. Burgmann H, Pesaro M, Widmer F, Zeyer J. A strategy for optimizing quality and quantity of DNA extracted from soil. J Microbiol Methods. 2001;45(1):7–20. doi:10.1016/S0167-7012(01)00213-5.
36. Tien CC, Chao CC, Chao WL. Methods for DNA extraction from various soils: a comparison. J Appl Microbiol. 1999;86(6):937–43. doi:10.1046/j.1365-2672.1999.00775.x.
37. Yan-hong G, Jun L. Experiments on genomic DNA isolation from the leaves of litsea coreana lévl.var.lanuginosa. Sichuan For Sci Technol. 2003;(02):47–50.

38. R P, HY S, QS L, Y. W. Extraction and characterization of total DNA from Dendrobium. Zhongguo Zhong Yao Za Zhi 2003;(No.12):1129–31.
39. Hong YK, Sohn CH, Lee KW, Kim HG. Nucleic acid extraction from seaweed tissues for polymerase chain reaction. J Mar Biotechnol. 1997;5(2–3):95–9.
40. Ling J, lihong C. A method for extracting DNA of Ginkgo biloba. Plant Physiol Commun. 2000;04:340–2.
41. Changchun Y, Suhua S, Chuangxing Y. Total DNA extracted from pure tea phenols rich in plant leaves. Sun Yatsen Univ For 2001;(03):1–4.
42. Martín MP, Winka K. Alternative methods of extracting and amplifying DNA from lichens. Lichenologist. 2000;32(2):189–96.
43. Li Z, Fei L, Gang W. Overview of animal DNA extraction. Anim Husb Vet Med. 2008;03:66–8.
44. Shiyan, Xiaobing W, Peng Y, Zhe Z. A method for DNA extraction from tanned leather and scales of Chinese alligator. Curr Zool. 2004;02:297–301.
45. Gang R, Ming L, Yidong N, Jing W, Fuwen W, Shengguo F. A new method for DNA extraction from the dried skins. Chin J Zool. 2001;04:53–7.
46. Bajorath J, Raghunathan S, Hinrichs W, Saenger W. Long-range structural-changes in proteinase-K triggered by calcium-ion removal. Nature. 1989;337(6206):481–4. doi:10.1038/337481a0.
47. Hong L, Wen W, Liming S. DNA extraction and PCR Muntiacus animal skin sample amplification. Zool Res. 1995;02:146–52.
48. Merheb M, Vaiedelich S, Maniguet T, Hanni C. Mitochondrial DNA, restoring Beethovens music. Mitochondrial DNA. 2016;27(1):355–9. doi:10.3109/19401736.2014.895988.
49. Quan Z, Gang F, Yongmei W. A simple and fast method for extracting DNA from blood and tissues. Chem Life. 1994;04:26–7.
50. Jinlong H, Guyue L, Juan Z, Mei Z, Yangzhi Z. Study on the extraction of high quality genomic DNA from porcine blood. Shanghai J Anim Husb Vet Med 2004;(03):23–22.
51. Qingmin W, Qiaohong L, Yun T, Lingxun S, Hui W, Yan C, Zhenrong Y. Comparison of methods for DNA extraction from peripheral blood. Chin J Lab Med. 2004;07:50–1.
52. Deuter R, Peitsch S, Hertel S, Muller O. A method for preparation of fecal DNA suitable for Pcr. Nucleic Acids Res. 1995;23(18):3800–1. doi:10.1093/nar/23.18.3800.
53. Gerloff U, Schlotterer C, Rassmann K, Rambold I, Hohmann G, Fruth B, Tautz D. Amplification of hypervariable simple sequence repeats (microsatellites) from excremental DNA of wild living bonobos (Pan-paniscus). Mol Ecol. 1995;4(4):515–18. doi:10.1111/j.1365-294X.1995.tb00247.x.
54. Baowei Z, Fuwen W, Ming L, Xiaoping L. A simple protocol for DNA extraction from faeces of the giant panda and lesser panda. Acta Zool Sin. 2004;03:452–8.
55. Constable JJ, Packer C, Collins DA, Pusey AE. Nuclear-DNA from primate dung. Nature. 1995;373(6513):393–3. doi:10.1038/373393a0.
56. Savill MG, Murray SR, Scholes P, Maas EW, McCormick RE, Moore EB, Gilpin BJ. Application of polymerase chain reaction (PCR) and TaqMan PCR techniques to the detection and identification of Rhodococcus coprophilus in faecal samples. J Microbiol Methods. 2001;47(3):355–68.
57. Hua Z, Xunlong L, Rongping W, Zhonglai L. An improved method for extracting DNA from the feces of giant pandas. Acta Zool Sin. 2003;05:670–4.
58. Jianyuan Z, Jinhua L, Yang L, Xuhua Y. Research on DNA extraction from old faeces of Macaca thibetana. Acta Theriol Sin. 2005;04:100–3.
59. Torsvik V, Goksoyr J, Daae FL. High diversity in DNA of soil bacteria. Appl Environ Microbiol. 1990;56(3):782–7.
60. Wilson IG. Inhibition and facilitation of nucleic acid amplification. Appl Environ Microbiol. 1997;63(10):3741–51.
61. Torsvik VL, Goksoyr J. Determination of bacterial DNA in soil. Soil Biol Biochem. 1978;10(1):7–12. doi:10.1016/0038-0717(78)90003-2.

62. Bakken LR, Lindahl V. Recovery of bacterial cells from soil. Berlin: Springer; 1995.
63. Tanner MA, Goebel BM, Dojka MA, Pace NR. Specific ribosomal DNA sequences from diverse environmental settings correlate with experimental contaminants. Appl Environ Microbiol. 1998;64(8):3110–13.
64. Ogram A, Sayler GS, Gustin D, Lewis RJ. DNA adsorption to soils and sediments. Environ Sci Technol. 1988;22(8):982–4. doi:10.1021/Es00173a020.
65. van Heijenoort J. Formation of the glycan chains in the synthesis of bacterial peptidoglycan. Glycobiology. 2001;11(3):25r–36r.
66. Koch AL. Bacterial wall as target for attack: past, present, and future research. Clin Microbiol Rev. 2003;16(4):673-+. doi:10.1128/Cmr.16.4.673-687.
67. Vingataramin L, Frost EH. A single protocol for extraction of gDNA from bacteria and yeast. Biotechniques. 2015;58(3):120–5. doi:10.2144/000114263.
68. Antony-Babu S, Murat C, Deveau A, Le Tacon F, Frey-Klett P, Uroz S. An improved method compatible with metagenomic analyses to extract genomic DNA from soils in Tuber melanosporum orchards. J Appl Microbiol. 2013;115(1):163–70. doi:10.1111/jam.12205.
69. Romanelli AM, Fu J, Herrera ML, Wickes BL. A universal DNA extraction and PCR amplification method for fungal rDNA sequence-based identification. Mycoses. 2014;57 (10):612–22. doi:10.1111/myc.12208.
70. Watson A, Ramstad PE. Corn: chemistry and technology. Food Chem. 1988;(No.1):410–18.
71. Anklam E, Neumann DA. Method development in relation to regulatory requirements for detection of GMOs in the food chain. J AOAC Int. 2002;85(3):754–6.
72. H Y, K S. Two detection methods of genetically modified organisms in agricultural crops and plant-derived food products. Eur Food Res Technol. 2003;214:3–26.
73. Poinar HN, Hofreiter M, Spaulding WG, Martin PS, Stankiewicz BA, Bland H, Evershed RP, Possnert G, Paabo S. Molecular coproscopy: dung and diet of the extinct ground sloth Nothrotheriops shastensis. Science. 1998;281(5375):402–6. doi:10.1126/science.281.5375.402.
74. Costa J, Melo VS, Santos CG, Oliveira MBPP, Mafra I. Tracing tree nut allergens in chocolate: a comparison of DNA extraction protocols. Food Chem. 2015;187:469–76. doi:10.1016/j.foodchem.2015.04.073.
75. Ageno M Fau – Dore E, Dore E Fau – Frontali C, Frontali C. The alkaline denaturation of DNA. Biophys J (0006-3495 (Print)). doi:D – NLM: PMC1367631 EDAT- 1969/11/01 MHDA- 1969/11/01 00:01 CRDT- 1969/11/01 00:00 AID - S0006-3495(69)86452-0 [pii] AID - 10.1016/S0006-3495(69)86452-0 [doi] PST – publish. 1969.
76. Vella F. Experimental biochemistry – Stenesh. J Biochem Educ. 1984;(No.3):142–3. 1984.
77. Sambrook J, Russell DW. Molecular cloning: a laboratory manual by Joseph Sambrook; David W. Russell. Q Rev Biol. 2001;(No.3):348–9.
78. Freifelder D. The DNA molecule, structure and properties. Biochem Educ. 1978;(No.2):46
79. Holden MJ, Haynes RJ, Rabb SA, Satija N, Yang K, Blasic JR. Factors affecting quantification of total DNA by UV spectroscopy and PicoGreen fluorescence. J Agr Food Chem. 2009;57 (16):7221–6. doi:10.1021/jf901165h.
80. Singer VL, Jones LJ, Yue ST, Haugland RP. Characterization of PicoGreen reagent and development of a fluorescence-based solution assay for double-stranded DNA quantitation. Anal Biochem. 1997;249(2):228–38. doi:10.1006/abio.1997.2177.

Chapter 3
Basic Rules in Nucleic Acid-Mediated Amplification and Hybridization Methods in Food Safety Detection: A Review

Abstract Detection methods used in food safety can be classified into two categories: amplification and hybridization. Their purpose is to accumulate or to recognize target sequence specifically using a proper reaction buffer in vitro. Polymerase chain reaction (PCR) is a valuable tool for monitoring gene expression, for quantifying foodborne pathogens, and for use in clinical diagnostics. Hybridization methods offer a new frontier in the identification of targets in a simple and high-throughput way. This review mainly describes and compares the basic rules of PCR and hybridization, including their formatting source, characteristics, and applications.

Keywords Principle • Base pairing • Recognition • PCR • HCR

Molecular biology is to study the processes of replication, transcription, translation, and cell function in the organism in the molecular level. Among these processes, researchers found valuable processes and the corresponding chemical substances occurring in live organisms [1]. In addition to study their function and mechanism in controlling living cycles, scientists also find it applicable to manipulate in vitro, which means the living steps can be imitated artificially for molecular biology study [2]. As for food detection methods, most of which are based on the DNA formation and duplication, basic principles catered to the target identification. On the other hand, nucleic acids, target in this book, must be analyzed through biochemical or molecular ways. A thorough look into basic rules of nucleic acid detection methods on food safety is necessary and should start with well-known rhythms in the life cycles.

3.1 Basic Theory on Molecular Diagnoses

All the molecular diagnostic techniques are optimized with living systems in vivo in regard to DNA replication, transcription, and translation. Among them, base pairing and DNA replication are two of the main theories that facilitate the innovative specific and sensitive detection in food safety, which also afford the opportunity to evaluate the biological contribution to human beings.

W. Xu, *Functional Nucleic Acids Detection in Food Safety*,
DOI 10.1007/978-981-10-1618-9_3

3.1.1 Watson–Crick Base-Pairing Principle

DNA is composed of four nucleotides (guanine, cytosine, adenine, and thymine), all of which have the same three constituents: phosphates, sugars, and bases [3, 4]. The first two of them are soluble in water. DNA bases are not soluble in water, but rather in weak acid. Under the neutral pH surroundings of living organisms and neutral organisms-imitated experimental environments, bases have to tuck themselves into the very center of a folded structure to avoid water [5, 6]. The Watson–Crick base pairing states that, within any DNA base, negative electric charge undergoes a small surplus on nitrogen and oxygen atoms if they are not attached to hydrogen. Meanwhile positive charge undergoes a small surplus of on the same atoms when they are attached to a hydrogen [7, 8]. Thus, three nitrogens on adenine and two oxygens on thymine in the DNA structure have a surplus of negative charge. In contrast, one nitrogen on adenine and one nitrogen on thymine have a surplus positive charge. Together, these positive and negative charges create the hydrogen bonds. To simplify, only the purine–pyrimidine pair can be incorporated into the double helix at the proper hydrogen-bonding distance, and only guanine–cytosine or adenine–thymine pairs have moieties in the proper locations for hydrogen bonding to occur [9].

Watson–Crick base pairing involves the interactions between each nucleotide in the single-stranded nucleic acid to form a stable double-helix structure [2]. This means that single-stranded DNA in the solution is not stable and is eager to search for a complementary sequence to form a stable structure. If the sequence of a gene of interest is known, it can be detected by designing a complementary sequence to capture it under the principle of Watson–Crick base pairing [10, 11].

Based on this theoretical principle of DNA pairing structure, some several tips can be summarized below:

- DNA stringency should be taken into consideration in the study of molecular. High stringent structure can be stable in the normal condition not to be apart. DNA stringency is largely dependent on temperature, salt concentrations, and pH. High stringency conditions involve high temperatures and low salt concentrations, and DNA hybridizations proceed slowly but in a sequence-specific manner, whereas low stringency conditions involve cooler temperatures and high salt concentrations, and DNA can form double-stranded complexes even in the event that sequences are not complementary [12–15].
- DNA duplexes can be denatured, and the base-pairing formation of single strands can be damaged [16].
- The annealing temperature (*T*m) of a DNA molecule is defined as the temperature at which half of it is constitutive, and bases are no longer paired to their complementary partner on the opposite strand. Base pairing is of no use when DNA is denatured at a high temperature. It lies in a balance between *T*m of a DNA molecule and base pairing in special conditions [17–19].

- It is applicable to design an artificial oligonucleotide to hybridize the complementary sequence with several bases deleted, which means base pairing between different oligonucleotides can be achieved.
- Under the aid of DNA polymerase, the structure mentioned above can be extended, which is similar to DNA replication in the living organisms.

These tips, that we prefer to call bonuses, are the beginning of molecular diagnostic techniques. PCR and hybridization chain reaction (HCR) are all based on these bonuses that Watson and Crick gave us.

3.1.2 DNA Replication In Vivo

DNA is replicated in vivo as an essential occurrence in all living organisms and is the basis for biological inheritance. DNA replication in vivo is performed under the optimal choice of active proteins, polymerase and helicase, and other functional oligonucleotides [5, 20, 21]. The process is initiated at particular points in the DNA, which are targeted by initiator proteins. Helicase then breaks the hydrogen bonds holding the two DNA strands together and exposes a DNA replication origin. Once exposed, bases of sequences were bind by the complementary precursor and were replicated under the help of DNA polymerase. This precursor, usually called a "priming" oligonucleotide, acts as the primer for DNA replication. Once hybridized, this short RNA molecule acts as a starting site for DNA polymerase [22]. The new strand was synthesized by the polymerase with completely complementary bases and helped form a new double-stranded DNA. Therefore, one DNA molecule is replicated into two identical molecules that are passing into the offspring.

Much similarity is obtained after browsing the whole process of DNA replication. Scientists want to copy the replication mechanism in vitro to replicate a target with a faster speed than in vivo. "It is, however, possible that upon cooling after denaturation of the DNA duplex, renaturation to form the original duplex would predominate over the primer–template complex formation. If this tendency could not be circumvented by adjusting the concentrations of the primers, clearly on would have to resort to the separation of the strands and then carry out repair replication. After every cycle of repair replication, the process of strand separation would have to be repeated." This is a classical saying from Kleppe and Gobind Khorana, two Nobel laureates that imagined the clear principle of DNA amplification in vitro [16]. Now it is well known as polymerase chain reaction (PCR).

3.1.3 Recognition Between Target and Probe

Recognition is a frequently appeared term in many studies of analytical chemistry, molecular biology, and biochemistry. Recognition usually refers to the molecular interactions with changed free energy and directional chemical equilibrium toward stable structure, through hydrogen bonding, metal coordination, hydrophobic forces, van der Waals forces, π-π interactions, halogen bonding, and electrostatic and/or electromagnetic effects [23–27]. This type of interaction is specific if it has large size of complementarity and specific structure, usually in a static or dynamic state.

Processes of recognition can be dynamic. In the case of dynamic molecular recognition, the binding of the first guest to the first binding site of a host affects the association constant of a second guest with a second binding site [28]. The recognition between two molecules is specific, which inclined to bind or disassociate with each other only in extreme state (high temperature or pH) or a more specific and competitive molecular. The recognition can also be limitless, namely, a molecule initiates a recognition chain once it binds to the complementary molecule. This dynamic molecular recognition can be easily discriminated between several competing targets via this proofreading mechanism. Therefore, dynamic molecular recognition can be applied in the molecular diagnosis to determine the possible molecule.

As for detection in the food, it is fundamental to recognize a specific target among thousands of substances mixed in the food samples. Scarce amounts of the target typically interfere with molecular inhibitors that prevent molecular interactions between the desired probe and target in some regard [29]. In that case, recognition can be divided into two categories: amplification and hybridization. Amplification usually plays the role of enriching detected target such that it can be easily identified, while hybridization aims to directly detect the target. Polymerase chain reaction (PCR) and hybridization take the task.

3.2 Polymerase Chain Reaction (PCR)

PCR is developed by Kary Mullis in 1983 and has becoming an indispensable technique for molecular diagnosis [30–33]. PCR is functional in a series of fields, including molecule cloning, DNA or RNA sequencing, gene expression analysis, disease diagnosis, DNA fingerprinting, and the detection of target. In 1993, Mullis was awarded the Nobel Prize in Chemistry along with Michael Smith for his work on PCR [34].

Typically, PCR consists of 20–40 repeated temperature cycles. Each cycle makes two to three temperature gradient steps (Fig. 3.1). PCR is often begun by a high temperature to denature template and being processed in each cycle for target accumulation, following an extension temperature. The temperatures and primer length used in PCR depend on a variety of parameters, including DNA polymerase, the melting temperature (*T*m) of the primers, and the concentration of divalent ions and dNTPs in the reaction [35, 36]:

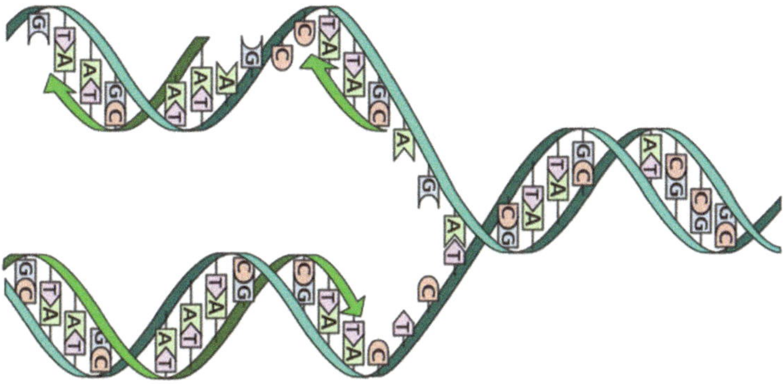

Fig. 3.1 Mechanism of DNA replication in vivo. A complementary set of bases on the strand initiates to synthesize a new strand that is represented by the *arrow*. *A*, *G*, *C*, and *T* represent adenine, guanine, cytosine, and thymine, respectively

- Initiation step: This step is the prerequisite for the PCR amplification. The template is heated to the 95 °C for 5 min. Some polymerases also need this high temperature for activation [37].
- Denaturation step: This step is the first regular cycling event and consists of heating the reaction to 94–98 °C for 30 s. It causes DNA melting of DNA template by disrupting the hydrogen bonds between complementary bases, yielding single-stranded DNA molecules.
- Annealing step: The reaction temperature is lowered to 50–65 °C for approximately 30 s allowing annealing of the primers to the single-stranded DNA template [38]. The annealing temperature is set to maintain the binding of the primer binds to the specific complementary target, which means to locate several degrees below and above the melting temperature of the primer. If the temperature is too low, the primer may bind to other unspecific sites, otherwise the primer will not be able to bind to the target. Normally, the annealing temperature is approximately 3–5 °C below the *T*m of the primers [39]. The polymerase adheres to the binding site to recognize the 3′ hydroxyl group and starts to extend the primer. It must be significant to determine annealing temperature because of its influence on PCR efficiency and specificity. Usually, the gradient of the annealing temperature is the recommended choice for annealing temperature optimization.
- Extension/elongation step: Primer is extended to duplicate complementary DNA strand under a stable temperature. DNA Taq polymerase has its optimum activity temperature at 75–80 °C, and 72 °C is the best option. DNA polymerase recognizes the 3′ end of the primer and synthesizes a new DNA strand complementary to the DNA template strand in a 5′–3′ direction, linking the 5′-phosphate group of the dNTPs and the 3′-hydroxyl group at the end of primer to formulate a phosphodiester bond. The extension time depends both on the length of template and the DNA polymerase. Usually the speed of extension is a thousand bases per

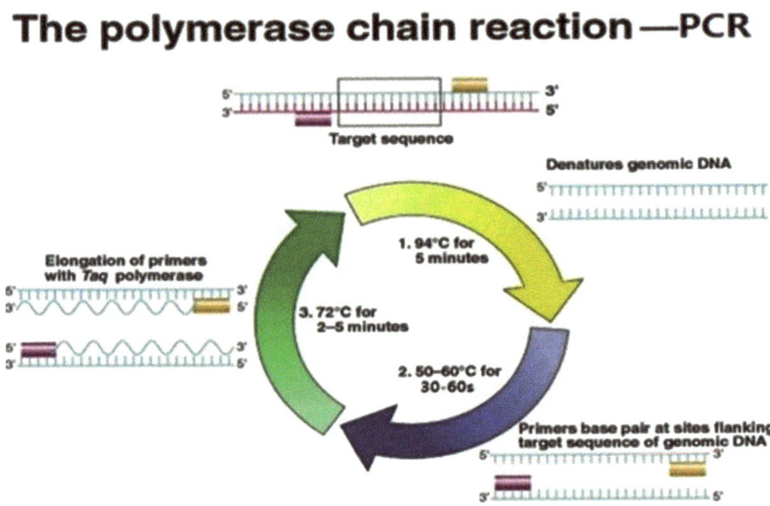

Fig. 3.2 Basic principles of PCR. The *arrows* indicate the process of one cycle of PCR, starting from the *yellow arrow* to *green*. *Primers* are indicated as *yellow* and *purple* short strands

minute. Therefore, 0.5–1 min is the appropriate choice for extension cycle [40, 41].

- Final elongation: This step is performed at a temperature of 72 °C for 5–15 min after the last PCR cycle to ensure full extension of the remaining single-stranded DNA. The elongation time is dependent on the DNA polymerase used in the reaction.
- Final hold: DNA is stable at a low temperature. After final elongation, amplicon products are typically stored at 4 °C for future analysis. Target isolated from cell of tissues was amplified in vitro under the appropriate environment. Above all, PCR permits DNA amplification with a higher efficiency than that can be achieved in vivo, providing an efficient method of target accumulation (Fig. 3.2).

3.2.1 Key Components in PCR

Basically, PCR requires several components and reagents to be performed. These components include:

- DNA template. The template often contains specific target sequence of interest for detection. Templates under detection must be purified so that amplification will not be inhibited. Template preparation usually includes tissue homogenization, cell permeabilization, cell lysis, protein and nucleases degradation, protein removal, precipitation of nucleic acids, and purification. A protease is often added to the template to degrade protein and enzymes and even chaperone nucleoproteins in the host cell. The proteinases can be removed by the organic solution or heat inactivation [42]. More detailed explanations about DNA isolation and purification are illustrated in Chap. 2.

- Two primers specifically bind to the complementary site of sense and antisense target, respectively. Primers are intended to be artificially designed to specifically bind to the target sequence and extend under catalyze of DNA polymerase. Strict requirements, usually the melting temperatures and base conformation, are needed for primer design to ensure the fitness of primer at certain temperature. Primers are now synthesized from a commercial company with highly purified content. The detailed information about primers is discussed below:
- DNA polymerase. A thermostable DNA polymerase makes successful and efficient PCR amplification. Initially the Klenow fragment of DNA-dependent DNA polymerase I, which is involved in replication and repair in the bacterium *Escherichia coli*, was employed in PCR amplification [31, 43, 44]. The Klenow fragment polymerase is a hydrolytic product of the native *E. coli* DNA-dependent DNA polymerase which lacks a 5′–3′ exonuclease activity. However, Klenow fragment DNA polymerase is performed better at 37 °C and is not advantageous to the PCR reaction. A breakthrough was achieved later, when Alice Chien found a 94 kD DNA polymerase derived from a *Thermus aquaticus* in 1976, whose optimal annealing temperature is at 70–75 °C. This polymerase, later called Taq polymerase, has strong homology at the amino acid level on the 3′-OH nucleotide addition site, the dNTP/DNA binding sites, and the exonuclease sites of the tow enzymes, which is similar with *E. coli* DNA polymerase [41]. It took 2 years before Taq polymerase was applied in the PCR amplification in the published articles [32], and now the Taq polymerase has become a gold standard of PCR amplification polymerase.
- Deoxynucleotides triphosphates (dNTPs). dNTPs are the building blocks from which the DNA polymerase synthesizes a new DNA strand and are necessary for new strand extension. The four individual deoxynucleotides (deoxyadenosine triphosphate, dATP; deoxythymidine triphosphate, dTTP; deoxycytidine triphosphate, dCTP; and deoxyguanosine triphosphate, dGTP) that make up a DNA sequence are usually added to PCR mixed in equal amounts. Sometimes if the dATP/dTTP (AT ratio) or mainly dGTP/dCTP (GC ratio) is extremely high or low, dNTPs composition may be altered to balance the ratio of four deoxynucleotides. Ten to 100 μM of each dNTP concentration can be sufficient for a 2.6 μg template to be amplified [45]. Theoretically, dNTPs can be used up in the PCR reaction, while during a 40 cycle PCR program, the amplification efficiency and polymerase activity may alter the process.
- Buffer solution. PCR buffer provides a suitable chemical environment for optimum activity and stability of the DNA polymerase while simultaneously allowing optimum PCR primer annealing to occur [46]. To optimize thermostable DNA polymerase activity, settle down the normal pH conditions, and balance the chemical structure of primer and template, PCR buffer should maintain them over a broad range of temperatures [41]. Evidence shows that

simple Tris-HCl buffers are stable for PCR, with a reduction of 0.1 pH unit occurring for every 5 °C rise in temperature [45]. In this way, the Tris-HCl buffers are formulated to exhibit a pH 8.4 at room temperature and a pH 7.4 at the 74 °C temperature, which is suitable for PCR amplification. Nuclease-free bovine serum albumin (BSA) is often used to stable the activity of DNA polymerase. Nonionic detergents may be helpful in ensurement of correct physical conformation of DNA polymerase including Tween-20 and Triton X-100 [47, 48]. In some circumstances, dimethyl sulfoxide (DMSO) is also included in the reaction to help prevent mispriming [49]. Single-stranded DNA-binding protein (SSB), betaine, pyrophosphatase, and dithiothreitol (DTT) are sometimes included in the PCR buffers for inhibitor prevention and amplification efficiency guarantee and are often used in the isothermal amplification [50–52]. Currently, it is common for manufacturers to provide a list of PCR reaction buffers (10× concentration) containing necessary components according to different PCR conditions, and researchers only need to choose as their personal need.

- Cations. Some cations, such as potassium (K^+), sodium (Na^+), ammonium (NH_4^+), and magnesium ions (Mg^{2+}) ions, can improve the performance of DNA polymerase and balance the electro-repulsive forces to ensure the binding of target and primer [53, 54]. Some hot-start PCR reactions depended on the Mg^{2+} to initiate the specific binding at a higher appropriate temperature [55]. However, elevated Mg^{2+} concentrations make PCR inhibited because the double-helix structure is stabilized and cannot be denatured enough for amplification [38]. Therefore, it is necessary to optimize Mg^{2+} concentrations in different conditions.

PCR is performed in a 10–200 μl reaction by a thermal cycler. The thermal cycler is able to alter the temperature to heat and cool the tubes for the required condition in PCR. A heated lid is necessary for PCR reactions to prevent condensation at the top of the reaction tube and facilitate high-speed temperature alternation to guarantee the amplification efficiency.

3.2.2 Kinetics

Thanks to the development of quantitative PCR (qPCR) and the involvement of fluorescent dye or molecular probes, the amplification curve of PCR can be easily obtained (Fig. 3.3). The curve suitably matches the exponential stage in PCR amplification [56].

The amplification curve is essentially a sigmoidal line indicated by the fluorescence, which is relational to the DNA concentration. The reaction begins as a linear model, with several extended targets accumulated. After 10–15 cycles, the reaction begins to amplify in an exponential stage, reaching to a plateau at the end of amplification. As a result, samples can be qualified or quantified in the exponential

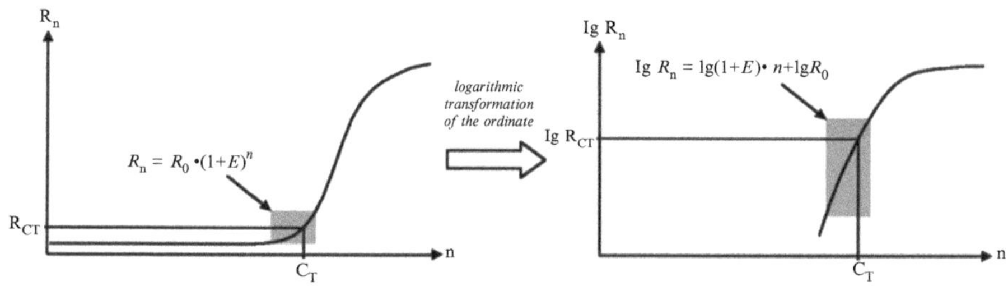

Fig. 3.3 qPCR amplification plot. Both graphs show a typical PCR kinetic curve with *R n* plotted against *n* before and after logarithmic transformation of *Rn*; the *y* axis is linear (*left*) and *lg* scaled (*right*) (Reprinted from Ref. [58], with kind permission from Springer Science + Business Media)

phase, and samples with high concentrations of target will be amplified earlier than that with low concentrations [57, 58]. A typical amplification plot is shown in Fig. 3.2, on both a linear and a logarithmic scale. Target quantification is achieved after regulating the baseline and the threshold. Baseline is the background noise of the reaction, normally identified as the mean fluorescence between 3 and 15 cycles. The threshold is an arbitrary fluorescence to demonstrate the beginning of the exponential amplification or the initial stage that can be quantified. The Ct is the PCR cycle at which each amplification plot reaches this threshold.

DNA template is replicated in every cycle. For example, quantity of a DNA sample with a five Ct value is eight times more than that with a two Ct value [59]. However, the efficiency of amplification is not always as same as theory, causing Ct variation fluctuate in a considerable state. Therefore, the efficiency of a primer–template combination should be assessed in serial dilutions of DNA template to create a standard curve of the change in (Ct) with each dilution. The slope of the linear regression is then used to determine the efficiency of amplification [60, 61]. The cycle threshold method is therefore sufficient in gene expression analysis and molecular diagnosis [58].

3.2.3 Analysis of PCR Products

The suitable way to analyze PCR products is through gel electrophoresis. Agarose gel electrophoresis is often used in PCR to separate a mixture of DNA [62]. After the power is turned on, the negative charge of phosphate backbone of DNA moves the DNA toward the positively charged anode during electrophoresis. Shorter amplicons in the reactions move faster than the larger amplicons and can be discriminated from the gel [63].

Polyacrylamide gel electrophoresis has the same principle as the agarose gel, but is more sufficient in analyzing small molecule with several bases difference, with higher pixels than agarose gel. However, polyacrylamide gel has higher voltage than agarose gel for DNA, which may undermine its application in the DNA analysis.

Besides gel electrophoresis, fluorescence dye and molecular probe can also be used to monitor the PCR amplification [64–66] and will be discussed in Chap. 6.

3.2.4 Principles of Primers and Probe Design

Several studies have been conducted on primer selection and evaluation [67–70]. Generally, primer and probe design are aimed at obtaining a balance between two goals: specificity and efficiency. Specificity is defined as the possibility of binding between primer and complementary target bind. Poor primers tend to amplify unrelated PCR products with incorrect length, which can be easily identified from the gel. Efficiency is defined as whether a primer can bind to the corresponding target and extend in every cycle to reach a theoretical twofold increase.

The primer design work attempts to achieve these two goals by setting variables for computation and evaluation. These parameters can be adjusted to cater to the PCR condition in every primer design work, some even through an experimental validation [71]. Several parameters are needed to be considered.

3.2.4.1 Location

Locations are not the restricted parameter in primer design. However, primer should not locate a GC-rich region because of its high binding energy [72]. Another general agreement on primer location is to avoid any stretch of polypurines or polypyrimidines within the expected amplicon [73]. Moreover, when using mRNA as a template, primer should not be located on the 3′ untranslated region (3′UTR) because of its complicated secondary structure [74].

3.2.4.2 Primer Composition

In general, each of four bases that compose a good primer should be in equal proportion [75]. Although some researches have shown that primers should have a GC content similar to that of the expected amplicon [73], they can be designed to contain 40–60 % GC content. Additionally, GC content at the 3′ side of primer is important for PCR reaction. 3′ site of primers with a low GC content may not bind properly to start the extension. In contrast, high GC content at the 3′ site region of primers may initiate the PCR reaction without entirely binding to the target [76]. However, if the *T*m of primer is calculated using the nearest-neighbor method, the hybridization between primer and target is more reliable than the 3′ GC content of primer.

3.2.4.3 Primer Length

Primers are related to the annealing temperature of PCR reaction, their melting temperature themselves, and the specific recognition with target. Primers should be 18–24 bases long and possess a melting temperature (*T*m) at 55–60 °C. *T*m is defined as the dissociation temperature of half of the primer/template duplex. The annealing temperature of the PCR reaction is set within a few degrees of the primer melting temperature to ensure the specificity of primer. Primers with a *T*m of 60 °C will perform better in most PCR conditions because this *T*m provides a preferred 18–24 bases oligonucleotide containing approximately 50 % GC content [76]. That length recommendation results from the fact that the random combination of 17 bases ($4^{17} = 17$ billion) will be hit less than once per total genomic sequences (3 billion bp) [71]. On the other hand, longer primers (more than 30 bases) are not specific enough to ensure the amplification [75], thereby making nearest-neighbor *T*m calculation less reliable.

3.2.4.4 Amplicon

It is obviously easier to amplify short length of targets than targets with a long length, owing to the ability of polymerase reaction and secondary structure interference. Furthermore, it is more likely for long amplicons to hybridize with other primers to initiate a branch PCR, resulting in an uncompleted amplicon. However, PCR products should be more than 100 bp to be easily distinguished from primer dimers when using gel electrophoresis as the analysis tool. In real-time PCR, detection based on the TaqMan probe allows to amplify a less than 100 bp target. Shorter amplicons in the probe system can improve the specificity and probe activity in the real-time identification.

3.2.4.5 Melting Temperature Determination

It is acknowledged that the *T*m of primers, or free hybridization energy (ΔG), is determined by the nearest-neighbor model [13, 18, 39, 77, 78]. The neighboring bases and the interactions require full evaluation to ensure the primer feasibility. The nearest-neighbor model treats a DNA helix as a string of interactions between "neighboring" base pairs. The whole hybridization energy of a primer is the sum of every neighbor bases interaction. However, identifying the exact *T*m of each primer in one PCR cycle is difficult because of decreasing concentration of primer during the amplification process. Therefore, it is advisable to calculate the overall melting temperature firstly using the appropriate nearest-neighbor coefficients. Basic theory of the two-state model is easy and has been developed to many primer designing softwares (Box 3.1).

Box 3.1: Two-State Model to Determine Melting Temperature of Primers

DNA duplex structures are the sum of their nearest-neighbor pairwise interactions thermodynamically and so is the free energy and the tendency for binding. Consequently, in the nearest-neighbor DNA thermodynamic model, an oligonucleotide's *T*m can be calculated, and thereby stability can be predicted.

Equation 3.1 shows an example of the calculation:

$$\Delta G^{\circ}_{37}(\text{total}) = \Delta G^{\circ}_{37\ \text{initiation}} + \Delta G^{\circ}_{37\ \text{symmetry}} + \sum \Delta G^{\circ}_{37\ \text{stack}} + \Delta G^{\circ}_{\text{AT terminal}}$$

Taking a sequence as an example, 5′-CGTTGA-3′:

$$\Delta G^{\circ}_{37}(\text{total}) = \Delta G^{\circ}_{37\ \text{initiation}} + \Delta G^{\circ}_{37\ \text{symmetry}} + \text{CG}$$
$$+\text{GT} + \text{TT} + \text{TG} + \text{GA} + \text{AT}_{\text{terminal}} \Delta G^{\circ}_{37}(\text{total}) = 1.96$$
$$+0 - 2.17 - 1.44 - 1.00 - 1.45 - 1.30 + 0.05 = -5.35\text{kcal/mol}$$

Data can be obtained from table below.

Nearest-neighbor thermodynamic parameters for DNA

Sequence	ΔH (kcal/mol)	ΔS (e.u.)	ΔG (kcal/mol)
AA/TT	−7.6	−21.3	−1.00
AT/TA	−7.2	−20.4	−0.88
TA/AT	−7.2	−21.3	−0.58
CA/GT	−8.5	−22.7	−1.45
GT/CA	−8.4	−22.4	−1.44
CT/GA	−7.8	−21.0	−1.28
GA/CT	−8.2	−22.2	−1.30
CG/GC	−10.6	−27.2	−2.17
GC/CG	−9.8	−24.4	−2.24
GG/CC	−8.0	−19.9	−1.84
Initiation	+0.2	−5.7	+1.96
Terminal AT penalty	+2.2	+6.9	+0.05
Symmetry correction	0.0	−1.4	+0.43

This predictive ability should prove valuable in a number of important biochemical applications, such as calculating the minimum length of a probe oligomer required to form a stable duplex with a target gene at a given hybridization temperature, estimating the melting temperature of a duplex structure formed between an oligomeric probe and its complementary gene segment, identifying potential sites of local melting within a polymer duplex by predicting the sequence-dependent melting temperatures of local DNA domains, and predicting the influence of a specific transition or transversion on the stability and melting temperature of a DNA sequence.

Single-stranded DNA sequences are actually folded molecules in the liquid system and must be unfolded to expose the binding site. Such process will alter the hybridization energy calculated by the two-state model. Additionally, mismatch hybridization can give an undesired signal that results in false-positive assays. All these considerations reach to a conclusion that a two-state model cannot be correct in primer designing, especially folded structure designing. A multistate model algorithm considers the multistate stage of primer, which is more accurate than nearest-neighbor model in the complex reaction system (Box 3.2). This model is successfully applied in the molecular beacon design and evaluation in isothermal amplification [79]. The most important concept is that *Tm* must be calculated taking the competitive equilibrium into account in a multistate model [80, 81].

Box 3.2: Multistate Model to Predict *Tm* and Free Energy

Taking a molecular beacon as an example.

In the multistate model, a molecular beacon exists in three statuses:

$$\text{MBT} \rightarrow \text{MBfold} + \text{T} \rightarrow \text{MBcoil} + \text{T},$$

where MBT is molecular beacon-template duplex, MBfold is the beacon that forms stem-loop structure, and MBcoil is the beacon in the form of random coil.

The overall fluorescence value is

$$(\text{RFU}) = \alpha \frac{[\text{MBT}]}{[\text{MB}]} + \beta \frac{[\text{MBfold}]}{[\text{MB}]} + \gamma \frac{[\text{MBcoil}]}{[\text{MB}]} \tag{3.1}$$

where α, β, γ are the RFU of molecular beacon that binds to the template, folded molecular beacon, and random coil molecular beacon, and

$$[\text{MB}] = [\text{MBT}] + [\text{MBfold}] + [\text{MBcoil}].$$

Then, the equilibrium constant (K) of each state is decided by the equation:

$$K_{12} = \frac{[\text{MBclosed}][\text{T}]}{[\text{MBT}]}, \tag{3.2}$$

$$K_{23} = \frac{[\text{MBcoil}]}{[\text{MBclosed}]}, \tag{3.3}$$

where K_{12} is the equilibrium constant of unfolding of molecular beacon-template duplex, and K_{23} is the equilibrium constant of unfolding of molecular beacon.

(continued)

Box 3.2 (continued)

Substituting these calculations to Eq. (3.1), RFU can be described by another equation:

$$\mathrm{RFU} = \frac{\alpha T + \beta K_{12} + \gamma K_{12} K_{23}}{T + K_{12} + K_{12} K_{23}}. \tag{3.4}$$

In the absence of template, namely, $T = 0$:

$$\mathrm{RFU} = \frac{\beta K_{12} + \gamma K_{12} K_{23}}{K_{12} + K_{12} K_{23}}, \tag{3.5}$$

$$K_{23} = \frac{\mathrm{RFU} - \beta}{\gamma - \mathrm{RFU}}. \tag{3.6}$$

Based on the equation of Gibbs free energy and Van't Hoff equation, unfolding of molecular beacon is decided by the equation

$$\Delta G = \Delta G0 + \mathrm{R}\theta \ln K = \Delta H - \theta \Delta \mathrm{S}, \tag{3.7}$$

where $\Delta G0$ is ΔG at absolute temperature, R is the gas constant, and θ is the temperature in Kelvin.

The equation could be transformed to a straight line having the equation

$$R \ln\left(\frac{\mathrm{RFU} - \beta}{\gamma - \mathrm{RFU}}\right) = -\frac{1}{\theta} \Delta H + \Delta S, \tag{3.8}$$

where $\ln\left(\frac{\mathrm{RFU}-\beta}{\gamma-\mathrm{RFU}}\right)$ is in linear relationship with $\frac{1}{\theta}$, and ΔH and ΔS are the slope and y-intercept of the line, respectively.

In the presence of template, Eq. (3.4) can be transformed to Eq. (3.9):

$$K_{12} = \frac{(\alpha - \mathrm{RFU})T}{(\mathrm{RFU} - \beta) + (\mathrm{RFU} - \gamma)K_{23}}, \tag{3.9}$$

where K_{23} can be calculated by the data acquired from Eq. (3.8), α, β, γ, and RFU is screened from the experimental data. Thus K_{12} at any temperature could be calculated; melting temperature (Tm) is the temperature where K_{12} equals T-0.5B.

Table 3.1 shows the commonly used methods used to calculate the melting temperature of a short sequence. In the conventional PCR, the nearest-neighbor model can accurately predict the melting temperature of a primer. However, some isothermal amplifications may take a multistate model into account.

Table 3.1 Melting temperature (*Tm*) based on different algorithms

	Basis	Formula	References
1	Number of bases	$Tm = 2°C(A+T) + 4°C(G+C)$	[82]
2	G/C content, DNA length, and salt concentration	$Tm = 81.5 + 16.6\log\left[\frac{[SALT]}{1+0.7[SALT]}\right] + 0.41(\%G+C) - \frac{500}{L}$	[82]
3	Nearest-neighbor interactions in two-state model	$Tm = \frac{T\Delta H}{\Delta H - \Delta G + RT\ln(C)} + 16.6\log\left[\frac{[SALT]}{1+0.7[SALT]}\right] - 269.3$	[83]
4	Multistate model	*Tm* is the temperature at which K_{12} equals T-0.5B	[81]

3.2.4.6 Probe Design

Principles of probe design are similar with that of primer design work, except considerations on thermodynamic parameters must be made [84]. A good probe should satisfy with two more commends that binds to the target before the primer and locates near the upstream primer. The recommended melting temperature of probe is 68–72 °C, which is 10 °C higher than the primer. There is no strict need on probe length, although low GC content may result to longer sequence.

Melting temperature determination is another aspect that needs to be reconsidered in the probe design work. Some probes, especially molecular beacons, are spontaneously folded, making it difficult to assess the accurate quantity of target that bounds by the probe and how specific a probe can bind to the desired target [65]. The designing work requires the multistate equilibrium calculations considering the probe and target concentration in the solution, salt condition, and *Tm*, fully accounting for all types of probe status in the reaction [80, 81, 85, 86].

Other primer and probe design work has the basic principle stated above, with specific demands that improve specificity and sensitivity. References can be seen from self-quenched primers [87, 88], dual priming oligonucleotides primers [89], locked nucleic acid [90], and self-probing amplicons [91].

3.3 Common Precautions on Reaction Conditioning

There are requirements on the process of PCR processing. DNA template must be purified and has no interfering effect on the activity of polymerase. Primers and probes also need to be purified. However, target may not be amplified well even in these conditions, and the reaction needs extra precautions.

3.3.1 Nested PCR

Nested PCR is suitable for trace mount of target amplification. Nested PCR needs two pairs of primers to locate different regions of target. Template is amplified using one pair of primer in the first round of PCR and acts as the new template in the next round of PCR. Primers used in the second round of PCR are located in the inner region. This design is mainly for specificity improvement and trace target accumulation, which is significant for analysis of scarce samples. Semi-nested PCR is also welcomed by the researchers if the target sequence is hard to design two pairs of primers, which use the same primer on the 5′ or 3′ site in two rounds of PCR. Moser has developed two rounds of PCR for a more specific detection of exogenous EPO [92]. Compared with other detection methods, this assay demonstrates high sensitivity. However, nested PCR increases the chance to contaminate laboratory and nearby tubes because the assay involves the manipulation of samples for two different rounds of PCR.

3.3.2 *Touchdown PCR*

PCR is often compromised by the mispriming or improper annealing temperature for the detection of different samples and therefore tends to be more laborious in parameter optimization. The universal degenerate primers may overcome the problem of mispriming, which are designed to locate common region but impair specificity to some extent. Another method, termed touchdown PCR, aims to circumvent spurious priming if the primer is not strictly specific [93, 94]. The touchdown PCR sets the higher temperature than the normal annealing temperature in the annealing step to block the unspecific hybridization in the first few cycles, followed by gradual reduction of annealing temperature for full extension.

3.3.3 *The PCR Inhibitors*

PCR can be inhibited from many sources, such as the bloods, fabrics, fats, or even glove powder in the laboratory. The inhibitors generally interfere with one or more steps in the reaction [95]. Inhibitors have the ability of escaping from the DNA extraction through binding to the DNA, or they can protect cell from being lysed. Many inhibitors can be influenced in the activity of DNA polymerase and blocking DNA from being recognized by primers. Therefore, PCR regents must be purified thoroughly to ensure that sufficient quantity of DNA still cannot be amplified.

3.4 Hybridization Chain Reaction

The basic principle of hybridization is based on the Watson–Crick base-pairing principle. A DNA microarray is the most obvious application [96–103]. In a DNA microarray, the target is detected through probe hybridization and identification and then indicated in the microfluidic chip. It is a lab-on-a-chip method that has been applied in various fields of pathogen detection. However, the normal hybridization is not so sensitive that small amounts of target will not be detected in the food. More progressive improvements have then been developed, borrowed from polymerase chain reaction and isothermal strand displacement to perform highly sensitive detection. The widespread progress is hybridization chain reaction (HCR).

Dirks and Pierce firstly developed HCR in 2004 [104]. HCR replace PCR for its free of enzyme activity and isothermal identification in some conditions [105]. HCR is more like a probe accumulation technique to produce repeated hybridization for indirect identification rather than amplify specific DNA fragments.

3.4.1 Working Principle

Single-stranded DNA can be programmed to be self-assembled into complex double-stranded two- or three-dimensional structures driven by the free energy of base pair formation [106–111]. The mechanics of DNA hybridization can be powered by Watson–Crick base pairing and the free energy change or by strand displacement interactions with the help of active polymerase [112–114]. Typically, various DNA strands begin to associate as soon as they are mixed together. DNA can be bound to its complementary oligonucleotide spontaneously, except an artificial looped oligonucleotide to block the binding site [115]. In the system described here, DNA and probes mixed in the same reaction do not hybridize at once, but need an initiator. An initiator exposes its binding site and triggers a chain reaction of hybridization through probes mixed in the reaction.

Figure 3.4 depicts the working principle of HCR. Initiator can be devised as a single-stranded DNA with known sequence. In the absence of the initiator, the artificial oligonucleotides (hairpins) protect their inherent stem-loop structure from opening up to release free energy. However, when the initiator (target) is added, one of the hairpins (H1) makes its complementary site exposed and binds to the initiator. H1 contains a complementary region to H2, and it triggers a following reaction to open up the loop of H2. H2 also contains a complementary region to H2 and will open up another H1. The continuous hybridization leads to the chain reaction and accumulation of oligonucleotides. Free energy gained from base pairing provides the driving force for this reaction [104]. The reaction eventually makes up several long nicked ds-DNA strand with various lengths until the hairpins are used up [114]. The molecular weight of the resulting ds-DNA strand is inversely proportional to the amount of the initiator provided. The number of initiators can be varied. However, fewer initiator strands lead to fewer ds-DNA strands, but each is longer with a higher molecular weight [104].

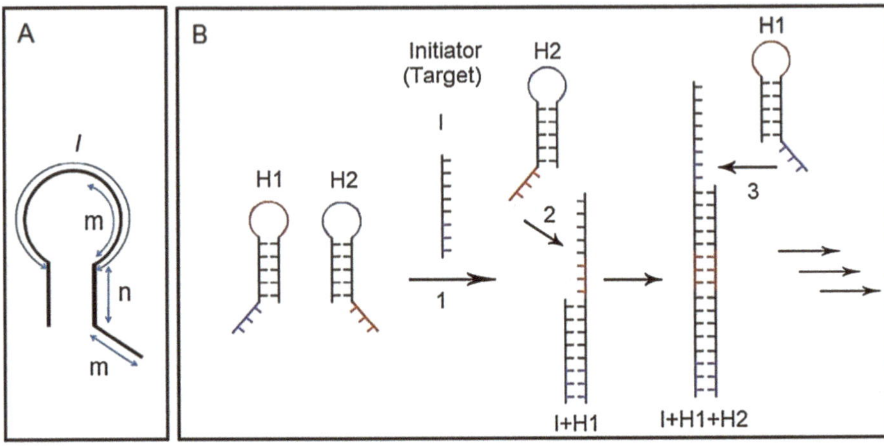

Fig. 3.4 (**a**) DNA hairpin shown with the loop, neck, and toehold region labeled *l*, *n*, and *m*, respectively. (**b**) The principal steps involved in HCR. One of the hairpins (*H1*) makes its complementary site exposed and binds to the initiator. H1 contains a complementary region to H2, and it triggers a following reaction to open up the loop of H2. H2 also contains a complementary region to H1 and will open up another H1. The continuous hybridization leads to the chain reaction and accumulation of oligonucleotides (Reprinted from Ref. [114], Copyright 2015, with permission from Elsevier)

3.4.2 Real-Time HCR

HCR can also be monitored in real time through fluorescence, which is explored by Chemeris in 2007 [105]. Real-time HCR is similar to DNA quantification using qPCR. One of the DNA hairpins (A-F/Q) is devised to contain a fluorescent dye (F) on the 5′ site and a quencher (Q) near the 3′ site of the hairpin. In normal conditions, hairpins are folded and fluorescence is quenched because of fluorescence resonance energy transfer (FRET). When HCR initiates, F and Q are apart to emit fluorescence that can be captured by the thermal cycler, therefore allowing the monitoring (Fig. 3.5).

3.4.3 Designing Principle

HCR relies on the use of a pair of complementary DNA hairpins to propagate the chain reaction of hybridization events. All parts of H1, H2, and initiator can be changed with known sequence, such as loop, neck, and toehold [116].

The hairpins are designed using a combination of criteria, such as minimal length to stable the double-stranded structure, the possibility to form a second structure at equilibrium, and kinetics when hybridizing to the target [104]. The rate of spontaneous hybridization between complementary hairpins can be

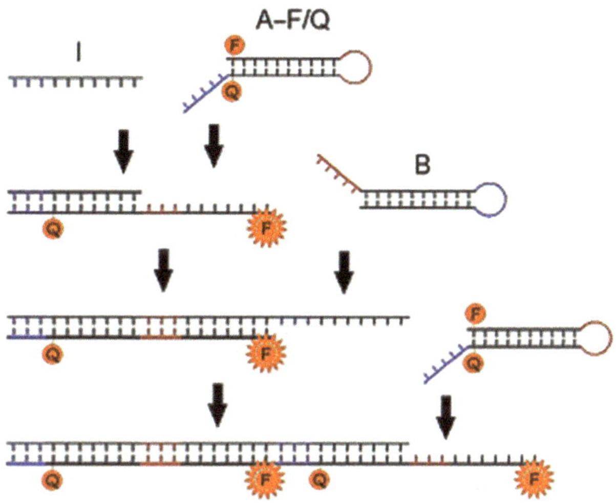

Fig. 3.5 The working principle of real-time HCR. *I* represents the initiator strand, and the two hairpins are labeled *A-F/Q* and *B*, respectively. *A-F/Q* is the hairpin containing the fluorophore and quencher (Reprinted from Ref. [105], with kind permission from Springer Science + Business Media)

monitored by increasing the neck length or decreasing the loop length. Hairpins with larger loops rapidly form long lived are easier to be opened up in the presence of initiator, while small hairpin loops may be opened by strand displacement activity that the initiator contains half of the loop sequence and half of toehold sequence. It has been found loop openings via an external toehold to be 10–100 times faster than via an internal toehold, and the rates of loop opening are at least 1000 times faster than the spontaneous interaction between hairpins [117].

The length of neck is the regulator of hairpin looping and exposure, which is much related to the reaction rate in HCR. Long hairpin neck forms duplexes slowly, but could be used for specific hybridization for long hours. It has been found that hybridization of unpaired bases contributes approximately $\Delta G_{022} = 59$ kcal/mol and -20 kcal/mol, respectively, to the overall free energy change for complete hybridization of the complementary pairs of hairpins with 40-base and 12-base loop domains [107].

3.5 Comparison Between PCR and HCR

PCR and HCR are all recognized as methods to better identify the existence of a target that elucidates the corresponding ingredients or additives in the food. Both of the methods have distinct characteristics, unique properties for molecular diagnosis, and simplicity compared to instrumental analysis methods, despite their different achieving process. Table 3.2 is the comparison between these two methods. Generally, PCR is more capable of directly identifying a target rather than signal transformation that HCR is willing to do. HCR adopted as an amplification strategy

Table 3.2 Comparison between PCR and HCR

	PCR	HCR
Target	ssDNA/RNA	ssDNA/RNA
Length	Up to 10 kb	Less than 1000 bp
Primer	At least two	Two capture hairpin (H1 and H2)
Enzyme	DNA polymerase	No
Reaction time	1.5–2 h	8–12 h
Reaction speed	Fast	Low
Isothermal	No	Yes
Temperature	Denaturation at 95 °C	Room temperature
	Annealing at 60 °C	
	Extension at 72 °C	
Sensitivity	High	Low

cannot achieve sufficient sensitivity, but has the advantage of detecting trace amounts in the food.

3.6 Prospect

PCR is inevitable in the process of molecular diagnostics, as well as pathogen detection in the food, while HCR is developing to be a signal identification method in the molecular diagnosis. The combination of HCR with other PCR can be versatile in target amplification and signal hybridization, making a twofold accumulation to be detected sensitively. The versatility of PCR and HCR, as well as other enzymatic isothermal technologies, is evident from their ability to combine with other techniques for a more sensitive detection, which can evolve into highly sensitive and portable detection biosensing devices (Chaps. 10 and 13).

Acknowledgments This work is supported by the Ministry of Science and Technology of Beijing (XX2014B069). Many thanks to Chenguang Wang, for his kind help in manuscript conception and preparation.

References

1. Waddington CH. Molecular biology or ultrastructural biology? Nature. 1961;190:184.
2. Erlich HA. PCR technology. Principles and applications for DNA amplification. New York: Stockton Press; 1989.
3. Watson J. The double helix. Paris: Hachette; 2012.
4. Watson JD, Crick FHC. Molecular structure of nucleic acids. Nature. 1953;171(4356):737–8.
5. Bell SD, Méchali M, DePamphilis ML. DNA replication. Cold Spring Harbor: Cold Spring Harbor Laboratory Press; 2013.

6. Cox LS. Molecular themes in DNA replication. London: Royal Society of Chemistry; 2009.
7. Yakovchuk P, Protozanova E, Frank-Kamenetskii MD. Base-stacking and base-pairing contributions into thermal stability of the DNA double helix. Nucleic Acids Res. 2006;34 (2):564–74.
8. Olby RC. The path to the double helix: the discovery of DNA. New York: Courier Corporation; 1974.
9. Hogan M, LeGrange J, Austin B. Dependence of DNA helix flexibility on base composition. Nature. 1982;304(5928):752–4.
10. Keller GH, Manak MM. DNA probes: background, applications, procedures. Basingstoke: Macmillan Press Ltd; 1993.
11. Hardiman G. Microarray innovations: technology and experimentation. Boca Raton: CRC Press; 2009.
12. Crick FHC, Watson JD. The complementary structure of deoxyribonucleic acid. In: Proceedings of the Royal Society of London A: mathematical, physical and engineering sciences. The Royal Society; 1954. p. 80–96.
13. Breslauer KJ, Frank R, Blöcker H, Marky LA. Predicting DNA duplex stability from the base sequence. Proc Natl Acad Sci. 1986;83(11):3746–50.
14. Bansal M. DNA structure: revisiting the Watson-Crick double helix. Curr Sci. 2003;85 (11):1556–63.
15. Williams MC, Wenner JR, Rouzina I, Bloomfield VA. Effect of pH on the overstretching transition of double-stranded DNA: evidence of force-induced DNA melting. Biophys J. 2001;80(2):874–81.
16. Kleppe K, Ohtsuka E, Kleppe R, Molineux I, Khorana HG. Studies on polynucleotides: XCVI. Repair replication of short synthetic DNA's as catalyzed by DNA polymerases. J Mol Biol. 1971;56(2):341–61.
17. Panjkovich A, Melo F. Comparison of different melting temperature calculation methods for short DNA sequences. Bioinformatics. 2005;21(6):711–22.
18. SantaLucia Jr J, Hicks D. The thermodynamics of DNA structural motifs. Annu Rev Biophys Biomol Struct. 2004;33:415–40.
19. Sinden RR. DNA structure and function. 1st ed. St. Louis: Elsevier; 2012.
20. Bloom LB. DNA replication, repair, and recombination. In: Introduction to bioinformatics. New York: Springer; 2003. p. 75–91.
21. McCulloch SD, Kunkel TA. The fidelity of DNA synthesis by eukaryotic replicative and translesion synthesis polymerases. Cell Res. 2008;18(1):148–61.
22. Frick DN, Richardson CC. DNA primases. Annu Rev Biochem. 2001;70(1):39–80.
23. Gellman SH. Introduction: molecular recognition. Chem Rev. 1997;97(5):1231–2.
24. Bielawski C, Chen Y-S. A modular approach to constructing multi-site receptors for isophthalic acid. Chem Commun. 1998;12:1313–4.
25. Lockett MR, Lange H, Breiten B, Heroux A, Sherman W, Rappoport D, Yau PO, Snyder PW, Whitesides GM. The binding of benzoarylsulfonamide ligands to human carbonic anhydrase is insensitive to formal fluorination of the ligand. Angew Chem. 2013;125(30):7868–71.
26. Breiten B, Lockett MR, Sherman W, Fujita S, Al-Sayah M, Lange H, Bowers CM, Heroux A, Krilov G, Whitesides GM. Water networks contribute to enthalpy/entropy compensation in protein–ligand binding. J Am Chem Soc. 2013;135(41):15579–84.
27. Jeffrey GA, Saenger W. Hydrogen bonding in biological structures. Berlin: Springer; 2012.
28. Shinkai S, Ikeda M, Sugasaki A, Takeuchi M. Positive allosteric systems designed on dynamic supramolecular scaffolds: toward switching and amplification of guest affinity and selectivity. Acc Chem Res. 2001;34(6):494–503.
29. Cocolin L, Rajkovic A, Rantsiou K, Uyttendaele M. The challenge of merging food safety diagnostic needs with quantitative PCR platforms. Trends Food Sci Technol. 2011;22:S30–8.
30. Mullis KB, Erlich HA, Gelfand DH, Horn G, Saiki RK. Reacting nucleic acid with oligonucleotide primer in presence of catalytic enzyme DNA polymerase; polymerase chain reaction patent. Google Patents. 1990.

31. Saiki RK, Scharf S, Faloona F, Mullis KB, Horn GT, Erlich HA, Arnheim N. Enzymatic amplification of beta-globin genomic sequences and restriction site analysis for diagnosis of sickle cell anemia. Science. 1985;230(4732):1350–4.
32. Saiki RK, Gelfand DH, Stoffel S, Scharf SJ, Higuchi R, Horn GT, Mullis KB, Erlich HA. Primer-directed enzymatic amplification of DNA with a thermostable DNA polymerase. Science. 1988;239(4839):487–91.
33. Mullis KB, Faloona FA. Specific synthesis of DNA in vitro via a polymerase-catalyzed chain reaction. Methods Enzymol. 1987;155:335.
34. Mullis K. Nobel lecture: the polymerase chain reaction. The Royal Swedish Academy of Sciences Nobel Prize: Chemistry The Royal Swedish Academy of Sciences, Sweden; 1993. p. 8.
35. Weissensteiner T, Nolan T, Bustin SA, Griffin HG, Griffin A. PCR technology: current innovations. Boca Raton: CRC Press; 2003.
36. Green MR, Sambrook J. Molecular cloning: a laboratory manual, vol. 1. New York: Cold Spring Harbor Laboratory Press; 2012.
37. Sharkey DJ, Scalice ER, Christy KG, Atwood SM, Daiss JL. Antibodies as thermolabile switches: high temperature triggering for the polymerase chain reaction. Nat Biotechnol. 1994;12(5):506–9.
38. Rychlik W, Spencer WJ, Rhoads RE. Optimization of the annealing temperature for DNA amplification in vitro. Nucleic Acids Res. 1990;18(21):6409–12.
39. Mitsuhashi M. Technical report: part 1. Basic requirements for designing optimal oligonucleotide probe sequences. J Clin Lab Anal. 1996;10(5):277–84.
40. Lawyer FC, Stoffel S, Saiki RK, Chang S-Y, Landre PA, Abramson RD, Gelfand DH. High-level expression, purification, and enzymatic characterization of full-length Thermus aquaticus DNA polymerase and a truncated form deficient in 5′ to 3′ exonuclease activity. Genome Res. 1993;2(4):275–87.
41. Chien A, Edgar DB, Trela JM. Deoxyribonucleic acid polymerase from the extreme thermophile Thermus aquaticus. J Bacteriol. 1976;127(3):1550–7.
42. Blin N, Stafford DW. A general method for isolation of high molecular weight DNA from eukaryotes. Nucleic Acids Res. 1976;3(9):2303–8.
43. Mullis KB, Faloona FA, Scharf SJ, Saiki RK, Horn GT, Erlich H. Specific enzymatic amplification of DNA in vitro: the polymerase chain reaction. Biotechnol Ser. 1992;24:17–27.
44. Saiki RK, Bugawan TL, Horn GT, Mullis KB, Erlich HA. Analysis of enzymatically amplified β-globin and HLA-DQα DNA with allele-specific oligonucleotide probes. Nature. 1986;324(6093):163–6.
45. van Pelt-Verkuil E, van Belkum A, Hays JP. Deoxynucleotide triphosphates and buffer components. In: Principles and technical aspects of PCR amplification. Dordrecht: Springer; 2008. p. 91–101.
46. Wolffs P, Grage H, Hagberg O, Rådström P. Impact of DNA polymerases and their buffer systems on quantitative real-time PCR. J Clin Microbiol. 2004;42(1):408–11.
47. Bachmann B, Lüke W, Hunsmann G. Improvement of PCR amplified DNA sequencing with the aid of detergents. Nucleic Acids Res. 1990;18(5):1309.
48. Hopfer RL, Walden P, Setterquist S, Highsmith WE. Detection and differentiation of fungi in clinical specimens using polymerase chain reaction (PCR) amplification and restriction enzyme analysis. Med Mycol. 1993;31(1):65–75.
49. Frackman S, Kobs G, Simpson D, Storts D. Betaine and DMSO: enhancing agents for PCR. Promega Notes. 1998;65(27–29):27–9.
50. Chase JW, Williams KR. Single-stranded DNA binding proteins required for DNA replication. Annu Rev Biochem. 1986;55(1):103–36.
51. Henke W, Herdel K, Jung K, Schnorr D, Loening SA. Betaine improves the PCR amplification of GC-rich DNA sequences. Nucleic Acids Res. 1997;25(19):3957–8.

52. Nagai M, Yoshida A, Sato N. Additive effects of bovine serum albumin, dithiothreitol and glycerol on PCR. IUBMB Life. 1998;44(1):157–63.
53. Park YH, Kohel R. Effect of concentration of MgCl2 on random-amplified DNA polymorphism. BioTechniques. 1994;16(4):652–6.
54. Ely JJ, Reeves-Daniel A, Campbell ML, Kohler S, Stone WH. Influence of magnesium ion concentration and PCR amplification conditions on cross-species PCR. Biotechniques. 1998;25(1):38–40, 42.
55. Birch DE, Kolmodin L, Wong J, Zangenberg GA, Zoccoli MA. Simplified hot start PCR. Nature. 1996;381:445–6.
56. Peirson SN, Butler JN. Quantitative polymerase chain reaction. In: Circadian rhythms. New York: Springer; 2007. p. 349–62.
57. Carr AC, Moore SD, Lucia A. Robust quantification of polymerase chain reactions using global fitting. PLoS ONE. 2012;7(5):e37640.
58. Schefe JH, Lehmann KE, Buschmann IR, Unger T, Funke-Kaiser H. Quantitative real-time RT-PCR data analysis: current concepts and the novel "gene expression's C T difference" formula. J Mol Med. 2006;84(11):901–10.
59. Heid CA, Stevens J, Livak KJ, Williams PM. Real time quantitative PCR. Genome Res. 1996;6(10):986–94.
60. Tichopad A, Dilger M, Schwarz G, Pfaffl MW. Standardized determination of real-time PCR efficiency from a single reaction set-up. Nucleic Acids Res. 2003;31(20):e122.
61. Ruijter JM, Ramakers C, Hoogaars WMH, Karlen Y, Bakker O, Van den Hoff MJB, Moorman AFM. Amplification efficiency: linking baseline and bias in the analysis of quantitative PCR data. Nucleic Acids Res. 2009;37(6):e45.
62. Helling RB, Goodman HM, Boyer HW. Analysis of endonuclease R·EcoRI fragments of DNA from lambdoid bacteriophages and other viruses by agarose-gel electrophoresis. J Virol. 1974;14(5):1235–44.
63. Jeppson JO, Laurell CB, Franzen B. Agarose gel electrophoresis. Clin Chem. 1979;25 (4):629–38.
64. Livak KJ, Flood SJ, Marmaro J, Giusti W, Deetz K. Oligonucleotides with fluorescent dyes at opposite ends provide a quenched probe system useful for detecting PCR product and nucleic acid hybridization. Genome Res. 1995;4(6):357–62.
65. Tyagi S, Kramer FR. Molecular beacons: probes that fluoresce upon hybridization. Nat Biotechnol. 1996;14(3):303–8.
66. Wittwer CT, Herrmann MG, Moss AA, Rasmussen RP. Continuous fluorescence monitoring of rapid cycle DNA amplification. Biotechniques. 1997;22(1):130–9.
67. Rychlik W, Rhoads RE. A computer program for choosing optimal oligonucleotides for filter hybridization, sequencing and in vitro amplification of DNA. Nucleic Acids Res. 1989;17 (21):8543–51.
68. Lowe T, Sharefkin J, Yang SQ, Dieffenbach CW. A computer program for selection of oligonucleotide primers for polymerase chain reactions. Nucleic Acids Res. 1990;18 (7):1757–61.
69. Tamura T, Holbrook SR, Kim S-H. A Macintosh computer program for designing DNA sequences that code for specific peptides and proteins. Biotechniques. 1991;10(6):782–4.
70. Lucas K, Busch M, Mössinger S, Thompson JA. An improved microcomputer program for finding gene-or gene family-specific oligonucleotides suitable as primers for polymerase chain reactions or as probes. Comput Appl Biosci CABIOS. 1991;7(4):525–9.
71. Mitsuhashi M. Technical report: part 2. Basic requirements for designing optimal PCR primers. J Clin Lab Anal. 1996;10(5):285–93.
72. McConlogue L, Brow MAD, Innis MA. Structure-independent DNA amplification by PCR using 7-deaza-2′-deoxyguanosine. Nucleic Acids Res. 1988;16(20):9869.
73. Saiki RK. The design and optimization of the PCR. In: Erlich HA, editor. PCR technology: principles and applications for DNA amplification'. New York: Stockton; 1989. p. 7–16. http://www.publish.csiro.au/journals/asb

74. Brooks EM, Sheflin LG, Spaulding SW. Secondary structure in the 3′UTR of EGF and the choice of reverse transcriptases affect the detection of message diversity by RT-PCR. Biotechniques. 1995;19(5):806–12, 814–5.
75. McPherson MJ, Møller SG. PCR. BIOS Scientific Publishers Ltd. Oxford; 2000.
76. Dieffenbach CW, Lowe TM, Dveksler GS. General concepts for PCR primer design. PCR Methods Appl. 1993;3(3):S30–7.
77. Watkins NE, SantaLucia J. Nearest-neighbor thermodynamics of deoxyinosine pairs in DNA duplexes. Nucleic Acids Res. 2005;33(19):6258–67.
78. SantaLucia J. A unified view of polymer, dumbbell, and oligonucleotide DNA nearest-neighbor thermodynamics. Proc Natl Acad Sci. 1998;95(4):1460–5.
79. Xu W, Wang C, Zhu P, Guo T, Xu Y, Huang K, Luo Y. Real-time quantitative nicking endonuclease-mediated isothermal amplification with small molecular beacons. Analyst. 2016;141:2542–52.
80. Koehler RT, Peyret N. Effects of DNA secondary structure on oligonucleotide probe binding efficiency. Comput Biol Chem. 2005;29(6):393–7.
81. Bonnet G, Tyagi S, Libchaber A, Kramer FR. Thermodynamic basis of the enhanced specificity of structured DNA probes. Proc Natl Acad Sci. 1999;96(11):6171–6.
82. Stephenson FH. Calculations for molecular biology and biotechnology: a guide to mathematics in the laboratory. Boston: Academic; 2010.
83. Innis MA, Gelfand DH, Sninsky JJ. PCR strategies. San Diego: Academic; 1995.
84. Letowski J, Brousseau R, Masson L. Designing better probes: effect of probe size, mismatch position and number on hybridization in DNA oligonucleotide microarrays. J Microbiol Methods. 2004;57(2):269–78.
85. Mathews DH, Burkard ME, Freier SM, Wyatt JR, Turner DH. Predicting oligonucleotide affinity to nucleic acid targets. RNA. 1999;5(11):1458–69.
86. Peyret N. Prediction of nucleic acid hybridization: parameters and algorithms. Detroit: Wayne State University; 2000.
87. Nazarenko I, Pires R, Lowe B, Obaidy M, Rashtchian A. Effect of primary and secondary structure of oligodeoxyribonucleotides on the fluorescent properties of conjugated dyes. Nucleic Acids Res. 2002;30(9):2089–195.
88. Nazarenko I, Lowe B, Darfler M, Ikonomi P, Schuster D, Rashtchian A. Multiplex quantitative PCR using self-quenched primers labeled with a single fluorophore. Nucleic Acids Res. 2002;30(9):e37.
89. Chun J-Y, Kim K-J, Hwang I-T, Kim Y-J, Lee D-H, Lee I-K, Kim J-K. Dual priming oligonucleotide system for the multiplex detection of respiratory viruses and SNP genotyping of CYP2C19 gene. Nucleic Acids Res. 2007;35(6):e40.
90. Obernosterer G, Martinez J, Alenius M. Locked nucleic acid-based in situ detection of microRNAs in mouse tissue sections. Nat Protoc. 2007;2(6):1508–14.
91. Whitcombe D, Theaker J, Guy SP, Brown T, Little S. Detection of PCR products using self-probing amplicons and fluorescence. Nat Biotechnol. 1999;17(8):804–7.
92. Moser DA, Neuberger EWI, Simon P. A quick one-tube nested PCR-protocol for EPO transgene detection. Drug Test Anal. 2012;4(11):870–5.
93. Don R, Cox P, Wainwright B, Baker K, Mattick J. ‘Touchdown’ PCR to circumvent spurious priming during gene amplification. Nucleic Acids Res. 1991;19(14):4008.
94. Korbie DJ, Mattick JS. Touchdown PCR for increased specificity and sensitivity in PCR amplification. Nat Protoc. 2008;3(9):1452–6.
95. Wilson IG. Inhibition and facilitation of nucleic acid amplification. Appl Environ Microbiol. 1997;63(10):3741.
96. Liu-Stratton Y, Roy S, Sen CK. DNA microarray technology in nutraceutical and food safety. Toxicol Lett. 2004;150(1):29–42.
97. Kostrzynska M, Bachand A. Application of DNA microarray technology for detection, identification, and characterization of food-borne pathogens. Can J Microbiol. 2006;52 (1):1–8.

98. Rasooly A, Herold KE. Food microbial pathogen detection and analysis using DNA microarray technologies. Foodborne Pathog Dis. 2008;5(4):531–50.
99. Borucki MK, Krug MJ, Muraoka WT, Call DR. Discrimination among Listeria monocytogenes isolates using a mixed genome DNA microarray. Vet Microbiol. 2003;92 (4):351–62.
100. Kato H, Saito K, Kimura T. A perspective on DNA microarray technology in food and nutritional science. Curr Opin Clin Nutr Metab Care. 2005;8(5):516–22.
101. Donhauser SC, Niessner R, Seidel M. Sensitive quantification of Escherichia coli O157: H7, Salmonella enterica, and Campylobacter jejuni by combining stopped polymerase chain reaction with chemiluminescence flow-through DNA microarray analysis. Anal Chem. 2011;83(8):3153–60.
102. Ping-Ping Z, Zhang J-Z, Yuan-Hai YOU, Yong-Ning WU. Detection of genetically modified crops by combination of multiplex PCR and low-density DNA microarray. Biomed Environ Sci. 2008;21(1):53–62.
103. González SF, Krug MJ, Nielsen ME, Santos Y, Call DR. Simultaneous detection of marine fish pathogens by using multiplex PCR and a DNA microarray. J Clin Microbiol. 2004;42 (4):1414–9.
104. Dirks RM, Pierce NA. Triggered amplification by hybridization chain reaction. Proc Natl Acad Sci U S A. 2004;101(43):15275–8.
105. Chemeris DA, Nikonorov YM, Vakhitov VA. Real-time hybridization chain reaction. Doklady biochemistry and biophysics. 2008;419(1):53–55.
106. Vincent M, Xu Y, Kong H. Helicase-dependent isothermal DNA amplification. EMBO Rep. 2004;5(8):795–800.
107. Dahl F, Banér J, Gullberg M, Mendel-Hartvig M, Landegren U, Nilsson M. Circle-to-circle amplification for precise and sensitive DNA analysis. Proc Natl Acad Sci U S A. 2004;101 (13):4548–53.
108. Niu S, Jiang Y, Zhang S. Fluorescence detection for DNA using hybridization chain reaction with enzyme-amplification. Chem Commun. 2010;46(18):3089–91.
109. Wiedmann M, Wilson WJ, Czajka J, Luo J, Barany F, Batt CA. Ligase chain reaction (LCR)-overview and applications. PCR Methods Appl. 1994;3(4):S51–64.
110. Dong J, Cui X, Deng Y, Tang Z. Amplified detection of nucleic acid by G-quadruplex based hybridization chain reaction. Biosens Bioelectron. 2012;38(1):258–63.
111. Evanko D. Hybridization chain reaction. Nat Methods. 2004;1(3):186–7.
112. Huang F, Xu P, Liang H. Using dual-polarization interferometry to study surface-initiated DNA hybridization chain reactions in real time. Biosens Bioelectron. 2014;51:317–23.
113. Huang J, Wu Y, Chen Y, Zhu Z, Yang X, Yang CJ, Wang K, Tan W. Pyrene-excimer probes based on the hybridization chain reaction for the detection of nucleic acids in complex biological fluids. Angew Chem Int Ed. 2011;50(2):401–4.
114. Ikbal J, Lim GS, Gao Z. The hybridization chain reaction in the development of ultrasensitive nucleic acid assays. TrAC Trends Anal Chem. 2015;64:86–99.
115. Sen D, Gilbert W. Formation of parallel four-stranded complexes by guanine-rich motifs in DNA and its implications for meiosis. Nature. 1988;334:364–6.
116. Green SJ, Lubrich D, Turberfield AJ. DNA hairpins: fuel for autonomous DNA devices. Biophys J. 2006;91(8):2966–75.
117. Wang D, Tang W, Wu X, Wang X, Chen G, Chen Q, Li N, Liu F. Highly selective detection of single-nucleotide polymorphisms using a quartz crystal microbalance biosensor based on the toehold-mediated strand displacement reaction. Anal Chem. 2012;84(16):7008–14.

Chapter 4
Reference Materials: A Golden Criterion in Nucleic Acid Identification

Abstract Identification of nucleic acids needs certificated positive samples for quality control, and reference materials (RMs) are developing to be received by most researchers in molecular biology. However, uniformity and disharmonization exist in the production and regulation of reference materials between different organizations. The objective of this chapter is to review and clarify the definition, attributes and applications, and common regulation process of reference materials in the context of nucleic acid identification.

Keywords Reference materials • Classification • CRMs • Biological assessment • Quantification

4.1 Introduction: Definitions and Classification of Reference Materials

Metrology is developed by means of production-mediated increasing accumulation and prosperity of national economy, which aims to fulfill the human thirst for justice and righteousness in an accepted way and to link measured results from a patient sample to a commonly acceptable reference [1]. Officials have deduced a series of measurement standards since ancient times, such as length measured through finger, hand, or arms, which is documented by Codex Hammurabi [2]. Similar incidents happened in the Chinese Qin Dynasty after unification, during which the country regulated the weight and length measurements to simplify the business market [3]. Reference materials that are similar to our recent consciousness trace back to the Industrial Age at the beginning of the twentieth century. The American Foundry Association developed standard iron samples to improve the productivity and quality. In 1906, the first regulation and corresponding reference materials were issued by the National Bureau of Standards (NBS) [4]. These reference materials were soon spread to Europe and Japan. From that time, reference materials started to be a golden criterion in many industries.

In addition to industrial development, the world was also coming through a scientific explosion, namely, analytic and bioanalytical technologies, which acted as an impetus for the production of reference materials. From measurement of acidity to evaluate physical chemistry properties to the standard cholesterol as a

W. Xu, *Functional Nucleic Acids Detection in Food Safety*,
DOI 10.1007/978-981-10-1618-9_4

qualitative control, an all-dimensional permeation of reference materials into scientific realm occurred, facilitating more stable and validated experimental data, as well as scientific progress [5, 6].

Reference materials (RMs) are indispensable in any analytical laboratory. RMs are applicable for the calibration of an instrument, scientific research represented by stable-performed experimental substances, or an official analytical procedure operated by a standard bureau. However, not all measurements need reference materials, and not all the substances can be processed into reference materials. Therefore, we have to differentiate between various tasks and select proper reference materials.

4.1.1 Definition of Reference Materials

The definitions of the basic terms for reference materials were formulated for the first time in 1992, when the International Organization for Standardization (ISO) put forward documents on recommended terms and definitions that should be assigned to them when used in connection with reference materials, with particular attention to terms that are used in reference material certificates and corresponding certification reports [7]. A list of terms is regulated to describe the nature of these materials officially: reference materials, certified reference materials, materials for quality control, laboratory reference materials, etc. Proper use and a commonly false understanding of terminology related to various types of reference materials are currently under critical discussion.

Based on the reports of the ISO/Committee on Referenced Materials (REMCO), the definitions of reference materials are as follows:

- *Reference material (RM).* A general definition of a material or substance of which one or more properties are sufficiently well established to be used for the calibration of an apparatus, for the assessment of a measurement method, and for the verification of one or more measurements in other materials.
- *Certified reference materials (CRMs).* An RM of which one or more characteristics are certified by procedures universally accepted as the more reliable for precision and accuracy. This material is equipped with an official certificate by public organization or local organization with high competitiveness. The differences between RMs and CRMs are listed below. There has been recent interest in factors which influence the quality of these materials, including particle size distribution and stability. Rauret has provided additional data to prove extractable contents of trace metals in sewage sludge amended soil CRM that was certificated in the last century for its EDTA [8]. The additional work mainly focused on the long-term stability of the material and the applicability of an improved version of the BCR three-step sequential extraction procedure on the sewage sludge amended soil (CRM 483) [9, 10].
- *Standard reference material (SRM).* This term refers to the CRMs produced and distributed by the National Institute of Standards and Technology (NIST), a

Table 4.1 Differences of the certified reference material and reference material

	Reference material	Certified reference material
Material characteristics	Well distributed, stability	Well distributed, stability
Additional study	–	With reliable measure of a fixed value
Requested data	Description of uniformity and stability	Description of uniformity and stability, traceability of value of a quantity, operating instruction manual
Main use	Consistency evaluation of methods or the accuracy of lab, methods development	Calibration, precise control, method validation quality control and assessment

government agency of the USA. In particular, the International Organization for Standardization (ISO), in its most recent publications, gives all the correct definitions in this field [11]. Standard Reference Material 720 (Al_2O_3) has been offered by the NBS Office of Standard Reference Materials since the 1970s as a heat capacity and enthalpy standard certified in the temperature range 273.15–2,250 K [12]. Human mitochondrial DNA (mtDNA) is also made into SRM so that mtDNA analysis for its medical diagnosis, mutant study, and gene expression research can be referred. SRM results will also provide other evidence if unknown mtDNA gene or region is amplified correctly, which in turn helps improve this SRM [10, 13].

- *Matrix reference material (MRM).* A material that is characterized by a usually low analyte content level, which is used mainly in testing and validation of analytic procedures and in calibration of measuring instruments. The matrix reference materials contain primary reference materials, secondary reference materials, laboratory reference materials, and materials for quality control [10, 14].
- *Primary reference material (PRM).* A certified reference material of the highest metrological quality (measuring coherence with SI units). Such material is approved after strict validation by national and regional metrological institutes, international organizations, and even foreign countries [15].
- *Secondary reference material (SecRM).* A material for which the value of the characteristic parameter is determined by comparison with the appropriate value for the primary reference material of the same type [16].
- *Laboratory reference material (LRM).* LRM is a reference material developed mainly by laboratories for scientific study in the laboratory or interlaboratory validation, which may not be certified. LRMs are that materials to fulfill the internal quality control between laboratories in the absence of CRMs [17] (Table 4.1).

4.1.2 Institutions to Make Reference Materials

France has contributed to the globalization and application of RMs. The Bureau national de métrologie (BNM) set up a reference material database, called COMAR, which is now charged by Laboratoire national de métrologie et d'essais (LNE) [18]. The system is recognized and used all over the world and thereby deemed by the International Organization of Standardization (ISO) as the recommended managing system in RM development and application [19].

ISO/REMCO, the ISO Committee on Reference Materials, carries out and encourages a broad international effort for the harmonization and promotion of reference materials, as well as their production and their application. The vision of ISO/REMCO is to be the international center of excellence with respect to issues relating to reference materials. The committee develops and maintains a series of ISO guides related to the definitions of reference materials (ISO 30: 2015), as well as contents of certificates, labels and accompanying documentation (ISO 31: 2015), good practice in using reference materials (ISO 33: 2015), general requirements for the competence of reference material producers (ISO 34: 2009), examples of reference materials for qualitative properties (ISO/TR 79: 2015), general and statistical principles for certification (ISO 35: 2006), guidance for the in-house preparation of quality control materials (QCMs) (ISO 80: 2014), keywords used for RM categorization (ISO/TR 10989: 2009), and global distribution of reference materials (ISO/TR 11773: 2013).

An institute was founded in 1957 under the Treaties of Rome and started operation in 1960 under the name of the Central Bureau for Nuclear Measurements (CBNM). In 1992–1994, CBNM obtained the title "Reference materials and reference measurements." The word "nuclear" disappeared from the title, and "reference materials" replaced it. Additionally, at the end of 1992, the Commission authorized the change of the name into Institute for Reference Materials and Measurements (IRMM), located in Geel, Belgium. It is now one of the seven institutes of the Joint Research Centre (JRC), a Directorate-General of the European Commission (EC), which promotes a common and reliable European measurement system in support of European union policies (https://ec.europa.eu/jrc/institutes/irmm/). IRMM was then able to produce new CRMs in sufficient quantities, and the quality of the materials could be significantly increased by employing modern technologies (e.g., jet milling, conditioning, and bottling under controlled atmosphere). IRMM has become an innovative reference material producer with worldwide recognition. It offers 75 nuclear and 700 different nonnuclear high-quality CRMs for calibration, method validation, and quality control in a variety of areas such as food and feed analysis, clinical chemistry, environmental analysis, nuclear security and safeguards, engineered materials, biotechnologies, and nanotechnologies. In addition, the distribution of CRMs to customers worldwide was increased by the nomination of "authorized distributors," operating under strict control of IRMM with respect to their storage and distribution conditions. In 1997, a collaboration agreement was signed with Fluka Inc. to produce joint reference materials for the detection of genetically modified organisms (GMOs). This led to the first GMO CRMs worldwide.

The National Bureau of Standards (NBS) was founded in 1901 and is now developed to National Institute of Standards and Technology (NIST) in the USA (https://www.nist.gov/). NIST promotes US innovation and industrial competitiveness by advancing measurement science, standards, and technology in ways that enhance economic security and improve our quality of life. As part of the mission, NIST supplies industry, academia, government, and other users with over 1300 standard reference materials (SRMs), making them the biggest RM supplier worldwide. The materials are certified as having specific characteristics or component content, used as calibration standards for measuring equipment and procedures, quality control benchmarks for industrial processes, and experimental control samples. Other US institutions that offer RMs are US Geological Survey (USGS) for rock measurement, US Environmental Protection Agency (USEPA) for environmental quality control, and US Atomic Energy Commission (USAEC) for engineering materials.

There are other agencies in influential countries to ensure the national measurement calibration, such as the Physikalisch-Technische Bundesanstalt (PTB) in Germany, National Measurement Institute (NMI) in Australia, National Metrology Institute of Japan (NMIJ), and National Institute of Metrology, P. R. China.

4.1.3 Categories of Reference Materials

RMs first occurred in the industrial realm, mainly focused on chemical substances [20, 21], minerals [22, 23], isotopes [24–26], metallic and nonmetallic elements [27–30], etc. Currently, molecular RMs are widely applied in the laboratories. In addition to nucleic acid RMs, other types of RMs are worth mentioning.

The need for a "stable enzyme preparation of known activity" in protein analysis was noted by Moss in 1971 [31]. Rej reported on aspartate aminotransferase from human erythrocytes as a reference material [32]. Fasce later stated the specifications that enzyme materials should be required to meet [33]. Since then, other research groups have prepared purified enzyme materials and used them in the interlaboratory evaluation of reference methods or to establish intermethod relationships, e.g., for aspartate aminotransferase [34], alanine aminotransferase [35], lactate dehydrogenase [36], amylase [20], creatine kinase [37], alkaline phosphatase [38–40], cystatin C [41], and acid phosphatase [42, 43]. Specifications and functions of reference materials have also been described by Horder and Rej [44]. Pharmaceutical enzymology encountered similar problems in establishing reference materials. International standardization procedures for pharmaceutical enzymes and their assays have been in common use since 1960. These achievements were made possible by the Enzyme Commission of the Federation Internationale Pharmaceutique, whose methods, units, and enzyme reference materials for trypsin, chymotrypsin, and pancreatin (lipase, protease, and α-amylase) were adopted by the European Pharmacopoeia.

Standard Reference Material (SRM) 1950 Metabolites in Human Plasma was developed in collaboration with the National Institutes of Health (NIH) in response to a recognized need for a reference material to support metabolomics technology development. SRM 1950 consists of a human plasma pool collected from healthy adults. Concentrations of approximately 100 analytes have been determined in this material, including both metabolites and environmental contaminants. Isotope-dilution mass spectrometry approaches for quantification were used whenever possible. This SRM is the first reference material developed specifically for metabolomics research. RMs can be used for validating analytical methods and evaluating the quality of data. NIST SRMs have typically been developed with a particular application in mind. For example, SRM 967a Creatinine in Frozen Human Serum is directed toward laboratories performing measurements of creatinine in serum. Because metabolomics studies encompass such a wide range of interests including toxicology, nutrition, and drug development, as well as a number of different analytical platforms, it is nearly impossible to match the sample matrixes and metabolite profiles encountered by each end user of a potential reference material. An alternative approach is to develop a reference material that is as broadly applicable as possible. NIH and a panel of metabolomics investigators provided guidance to NIST in designing SRM 1950 Metabolites in Human Plasma and in selecting target analytes for value assignment.

Nucleic acids need to be quantified and measured in a reasonable way in the field of molecular biology. Most of the nucleic acids used as reference materials are plasmids, which have a genetically stable formation and advantages for polymerase chain reaction (PCR) analysis. Genomes, RNAs, and modified nucleic acids are all applicable reference materials in the field of molecular diagnose, which will be detailed later.

4.2 Procedures for the Production of Reference Materials

The production of reference materials is a difficult and time-consuming process. The preparation process and the characteristics of a reference material consist of various stages [45, 46].

4.2.1 Estimation of the Demand for Reference Material

This is the fundamental first step in the production of a new RM. It is very important that the future RM be profit yielding (evaluation of potential demand for the produced RM) and that there be demand for such material in the analytical environment.

4.2.2 Planning the Production of Reference Material

Market survey on reference material is the prerequisite for the production and therefore needs to be carefully planned. Detailed information should be obtained for scientific research on reference materials, including what is needed, how long the shelf life is, and what price is acceptable [10]. Normally, the planning should last 1–2 years.

4.2.3 Quest and Processing of Reference Material

Detailed qualification of the source of acquisition of materials with the required properties and values is not easy and in the initial search stage may require of the producer some consultation with extraneous persons during the initial search phase. Normally, the supply of raw material depends on the ultimate amount of reference material, as well as the experimental and processing cost. Usually, from 100 to 200 kg of the ordered raw material is used; after processing, some 1000–5000 completed samples of the reference material are produced [10].

Reference material is provided with various types of format, some of which even regulate the concrete content. Therefore, the processing of reference material should be carefully manipulated. The raw material was firstly cleaned and thoroughly mixed, followed by the separation and homogeneity procedure. RMs with certain content must be mixed thoroughly. For liquid RMs, the concentration needs to be controlled, as low concentration makes it difficult for storage.

4.2.4 Homogeneity Testing

Homogeneity testing relates to the validity of the assumption that the proportion of components of RM are the same in the RM product. It is important that the RM packaged into bottles or vials be subjected to tests that ensure that it is homogeneous and that the reference value (CCRM) measured in part of the packages be comparable to that of another sample and that it lies within the admissible uncertainty threshold [10, 47].

To test the homogeneity of RM, samples should be selected randomly, each of which must be measured and evaluated to ensure the quality. The selected samples are also tested for reproducibility, with researchers in different experimental locations involved in the testing process. Moreover, the amount of samples should be large enough, to enhance the credibility at a previously established level of significance [48].

4.2.5 *Stability Testing*

A feasible reference material is one that remains stable under the influence of deposition, storage, transportation, and use [48]. The tests are carried out at different temperatures, times, and storage conditions [10, 49].

A stable RM is resistant to light, humidity, and temperature at the exact time and even after the holding period. Commonly, stability tests require that the samples be put in a freezer at a temperature of −20 °C, in a refrigerator at 4 °C, at an ambient temperature of 20 °C, and in a drier at 40 °C and subsequently analyzed after a period of 0, 1, 3, 6, and 12 months [10].

The medium for RM storage also plays a major role in the preservation of function of RM. The appropriate container provides a comfortable environment to prevent from drying, lighting, and cooling. For example, RM of genetically modified organisms (GMOs) should be kept at −20 °C, which is free from light in the sealed tube.

4.3 Reference Material at the Molecular Level

Biological tests can be highly predictive for the future health of an individual and can be carried out at any stage of life even in the embryo before implantation. The results of such tests are relevant to healthy people as well as those showing symptoms of an unhealthy condition and may also have important implications for the relatives of the person tested. Given the rapid translation from research into clinical practice, research laboratories play a valuable role in service provision worldwide. The genotype established by a single laboratory test is usually not repeated and represents a permanent component of the medical record of the patient [50]. Consequently, molecular testing requires a high level of data reliability based on proper quality assurance of testing laboratories as well as of their measurement and testing procedures. As the field develops, the need for appropriate reference materials that are required to establish this reliability becomes increasingly urgent.

Molecular classification of references materials has its own characteristics. At present, the qualitative research object mainly includes nucleic acids, protein, cells, and microorganisms. Nucleic acids, proteins, and microorganisms are currently the focus of the international biometric research. Nucleic acid measurement is given priority to with content and sequence measurement; protein measurement mainly includes the content, structure of the enzyme activity, immune activity, and metering; microorganism content measurement gives priority to with identification and quantitative research; molecular-level standard components can then be divided into physical and biological characteristics of standard substances. Component of biological reference materials mainly includes agricultural products, mainly for content analysis of food or deep-processed products, as well as analysis of poisonous and harmful substances. In the scope of this chapter, we only discuss about nucleic acid reference materials.

4.3.1 *Development and Application of Standard Material at the Molecular Level*

Molecular biological reference materials enable biological assays to have biometric designs that are reasonable, reliable, and, most important, correct. It is, therefore, necessary to understand the nature of biological reference materials [51].

As early as 1890, research was dedicated to diphtheria toxin and reliable therapy [52]. After antitoxins were produced in the serum of vaccinated animals, the antitoxin was found to protect normal animals from death of an attack by diphtheria toxin. After this discovery, Europeans began to produce antitoxin to protect human beings from diphtheria. Paul Ehrlich, who was awarded the Nobel Prize in 1908, then thought a standard method to evaluate a batch of antitoxins, seen as the antitoxin effect to compare the difference of each batch of sample titer [53]. Theoretically, a standard unit titer should remain unchanged for a long period of time if it is stored at freeze-dried low temperature away from light. These achievements laid the foundation for biological reference standards. Ehrlich officially put forward the diphtheria antitoxin standard and method to evaluate the antitoxin serum content in 1914 [54]. The standard test achieved consistent results from three laboratories (Washington, Franco, and Copenhagen) in 1922. Therefore, the Ehrlich theory on diphtheria antitoxin serum was formally identified as the first international standard material [55].

Although the concept of biometering was born in less than 10 years ago, the use of biological reference materials can be traced back to the 1960s, when the environmental and biological reference material cabbage powder was prepared for the first time in 1964 by an American. The nucleic acid reference material was first developed to measure hepatitis C virus by the World Health Organization (WHO) [56]. Currently, nucleic acid reference materials are dominantly applied in the area of viral diagnostics.

4.3.2 *Production of Reference Materials at the Molecular Level*

Reference materials at the molecular level are produced based on the basic theory (Sect. 4.2), with some alternations in the performance validation. Nucleic acid identification is especially important to ensure the overall content and target quality in the system. We now use production of genetically modified organisms (GMOs) production as an example of the production of reference materials at the molecular level.

The Institute for Reference Materials and Measurements (IRMM; Geel, Belgium) has produced the first powdered genetically modified organism (GMO) CRMs in cooperation with the Institute for Health and Consumer Protection (Ispra, Italy). Until now, different weight percentages in the range of 0–5 % for four GMOs in Europe were produced and certified: Bt (*Bacillus thuringiensis*)-11 and Bt-176 maize, Roundup Ready® soybean, etc.

Characterization of the base material is the first step in production and is especially important for GMO CRMs. The production of powdered GMO CRMs and the methods used for production control are described. Thorough control of homogeneity and stability is essential for certification of reference materials and to ensure validity of the certificate for each bottle of a batch throughout a defined shelf life. Because production of reference materials and their maintenance are very labor- and cost-intensive tasks, the usefulness of new types of GMO CRMs must be carefully estimated.

In the case of biological and environmental materials, the concentration of an analyte very often differs in the different particle size fractions of a given material. For the genetically modified maize sample, a good example is the DNA content of ground maize powder as different parts of the maize kernel, such as the embryo, endosperm, and seed coat, have different hardness, structures, and DNA contents.

In the production of powdered reference materials, the particle size distribution of the sample is a crucial question as was reported by Kramer and coworkers. If the particle size distribution has two peaks or the size of the particles is larger than 100–150 m, the sample can be separated during transport. Smaller particles (below 5–10 m) can be charged, and the sampling from the bottle becomes more complicated; therefore, small particles should not exceed a certain ratio. The ideal particle size distribution has only one peak and shows a maximum between 10 and 100 m. The third generation of genetically modified (GMO) CRMs is prepared using a dry mixing technique in which the necessary amount of the milled non-GMO and GMO maize kernels are mixed with each other.

After measuring the particle size distribution and the concentration of the analyte in the different particle size fractions of a candidate CRM, the relative standard deviation versus sample amount function can be calculated by the mathematical model developed (Table 4.2).

4.3.3 *Classification and Application of Reference Materials of Nucleic Acids*

Reference materials of nucleic acids are mainly plasmids, nucleic acids, and other forms of nucleic acids, to best satisfy the molecular diagnostics demands, such as

Table 4.2 Advantages and disadvantages of different forms of GMO reference materials

Type	Seed	Power		Leaf DNA	Plasmid
		Concentration gradients	Pure		
Advantages	Covered a whole range of measure; used for the detection method of DNA and protein; rapid detection; CRM	Covered a whole range of measure; used for the detection method of DNA and protein; rapid detection; CRM; relative quantification	Used for high concentration samples; absolute quantification; as a detection of the protein; CRM	Suitable for on-site rapid detection and high oil content crops; fewer need for seeds; stable; 100 % purity	High production levels at a low cost; qualitative analysis; do not depend on the GMO plant
Disadvantages	Spread easily; imports and exports; preservation	Large demand for material; complex production; narrow application scope	Large demand for material; do not suit for absolute quantitation	Do not cover a wide range of measure; need for planting place and large number of DNA extraction	Need correction; do not cover a whole range of measure; easy to pollute

qualitative and quantitative identification, gene expression, biological composition, etc.

4.3.3.1 Reference Material for Qualitative and Quantitative Analysis

The use of proper reference materials may enable verification and calibration of various analytical biases in measurements of biological properties and will, therefore, facilitate accurate and reliable measurements. The type of reference material that is required to perform a proper analysis of the sample of interest depends on the analytical problem. An analysis that is directed to qualitative properties of a sample (such as the biological identity of a sample component) needs a RM that allows comparison of this qualitative property in the sample and in the RM during the application of the same measurement or testing procedure. A typical example would be a RM consisting of plasmid DNA with a well-defined base pair sequence used for the quality assurance of the DNA sequencing procedure of an unknown genomic DNA fragment. As the synthetic DNA RMs lack the complexity of the human genomic DNA and are sometimes designed to work with a dedicated platform, they might not perform identically to patient DNA. Therefore, these materials have to be tested in ring trials, proficiency studies, or EQA schemes so that their fitness for purpose is demonstrated. On the other hand, most of the steps in quantitative measurements have to be calibrated because of the lack of completely known mathematical equations to calculate the relation between the targeted quantity in the sample and the measurement signal. Such calibration materials are almost indispensable for biochemical measurements.

Quantitative measurement is necessary for several reasons, one of the most important being to comply with labeling regulations that allow consumers to make an informed choice about their use of a product. Two new labeling regulations for GM crops have been instated: EC/49/2000, which stipulates a 1 % threshold of GM material in products otherwise derived from identity-preserved sources (e.g., organic, non-GMO grain), and EC/50/2000, which requires labeling of flavors and additives (the latter was previously exempt from labeling) [57]. A consequence of the declaration of a minimum threshold of 1 % was the need to progress from a qualitative detection of the transgenic species by using an appropriately validated screening system to more complex quantitative procedures. Following the qualitative detection of a GMO, manufacturers would be required to label the product or, in the case of identity-preserved sources, determine whether the amount of GMOs present was above or below the 1 % threshold. The implication is that the method used to quantitative the level of GMO should be as accurate as possible. The effects of the regulations are such that all aspects of the product require detailed analyses involving a large number of reference materials.

4.3.3.2 Reference Material for Biological Composition

The standard material of biological compositions in reference to the biological matrix materials determines the various component compositions and content in the characteristic value of standard substances. These include inorganic ion composition and content, biological composition and content of small molecules and macromolecules, composition and content of cells, microbial composition and content, etc.

The standard materials of biological compositions determine value with chemical analysis and physical analysis methods. The standard material of biological composition is one of the most widely used and most developed standard materials.

4.3.3.3 Reference Material for Biological Activity

The reference material for biological activity refers to determine biological matrix materials and characteristic values of the components of standard substances, including bacteria, viruses, cells, enzymes, and antibodies that have the ability to reproduce or catalytic ability of biological activity quantity standard material. The amount of biological activity as a method for determining the characteristic value of standard material is used more in biology analysis application [58, 59].

4.3.3.4 Reference Materials of Biological Construction

The reference material for biological construction determines the biological matrix materials and its composition structure characteristic value of standard substances, including bacteria, viruses, proteins, and nucleic acids. Structural characteristics of biological structures of standard materials are determined by physical and chemical analysis methods for confirmation.

Through continuous efforts and development over 30 years, biological products and standard substance formulate the obvious increase in the number of varieties, preparation program analysis scope, strong technical support, increasing precision, and yield. However, in the past 10 years, new products have constantly emerged due to the rapid development of biotechnology and information technology. New technology promotion and application, such as DNA recombinant products instead of part of the traditional products, gene amplification technology instead of traditional products and the safety of the blood product technology as the virus detection of auxiliary or supplement, and molecular biology technology gradually replaces animal testing evaluation of product safety and quality. These trends reflect the new historical stage that biological standardization has entered. With the demand for the shift of original methods to biological composition analysis and the continuing development of biological ingredient standard materials, the standard material of biological activity and construction has become a focus of research in the recent years.

4.4 Current Situation of Reference Material and Its Application in Various Fields

4.4.1 Microorganisms

In microbiology, the standard methodology commonly used for bacterial detection in environmental, clinical, and food samples is based on culture isolation on specific media and confirmation of presumptive colonies by biochemical and/or immunological tests. In this sense, DNA technology offers, among other advantages, higher specificity and more rapid analytical results than culture techniques.

However, because molecular methods are usually developed in academic and research fields, most of the DNA-based detection methods are not included in the current legislation and can be considered more as scientific approaches than as analytical tools. This lack of legislation and normalization makes the transition difficult from conventional to molecular methods in microbiology laboratories. Therefore, there is an urgent need for parallel validation and subsequent standardization of these DNA-based analytical techniques [60].

The validation of a new method requires a comparison of the performance of standard samples to that of standard reference methodologies or reference values.

The standard samples can be prepared using reference materials (RMs), certified reference materials (CRMs), or spiked samples. Apart from method validation, RMs are very useful for internal quality assurance purposes, verification of methods accuracy and precision, process control, proficiency testing schemes (PTS), and certification of products or commercial kits [61].

Consequently, the use of reference materials can help laboratories accomplish ISO 17025 requirements in terms of method validation and both internal and external quality control systems [62].

At present, a range of RMs for application in microbiology is commercially available [36, 63, 64]. Nevertheless, owing to the increasing implementation of molecular biology techniques in diagnostic laboratories, there are many parameters for which an appropriate RM has not been developed as the availability of RMs for qPCR is quite limited.

4.4.2 Molecular Biology

The Standard Reference Materials Program at the US National Institute of Standards and Technology(NIST) has three human DNA standard reference materials (SRM 2390, SRM 2391a, and SRM 2392) currently available [20]. Both the DNA profiling SRM 2390 and the polymerase chain reaction (PCR)-based DNA profiling SRM 2391a are intended for use in forensic and paternity identifications, for instructional law enforcement, or for nonclinical research purposes and are not intended for clinical diagnostics. The mitochondrial DNA (mtDNA) SRM 2392 is to provide standardization and quality control when performing PCR and sequencing any segment or the entire 16,569 base pairs that comprise human mitochondrial DNA. SRM 2392 is designed for use in the forensic, medical, and toxicological communities for human identification, disease diagnosis, or mutation detection.

4.4.3 Genetically Modified Organisms

GM organisms (GMOs) with new traits have been developed by the transfer of exogenous genes that originated from another species into the genomes of host species. GM crops have been approved for commercialization and have been widely planted for more than 20 years [65]. Although the number of new GM crops and their planting areas have constantly increased over the past two decades, the general public has also had constant concerns about their potential food safety and environmental risks. Several international organizations and countries have issued series of regulations, guidelines, and standards to strengthen the administration, labeling, and monitoring of GMOs [66]. To execute GMO labeling regulations, analytical techniques and methods targeting exogenous proteins or nucleic acids (such as ELISA, conventional PCR, and real-time PCR) have been developed and applied [67, 68]. Whatever method is used, appropriate CRMs should be used for quality control purposes and as calibrators in the determination of qualitative and quantitative results [69]. Two types of CRMs have been developed for GMO analysis: matrix-based and DNA-based CRMs [70]. The matrix-based CRMs are produced by blending GM and non-GM seed or ground seed powders in desired ratios [71]. Matrix-based CRMs are more similar to practical GM food/feed samples in physicochemical characteristics than other types of CRMs, and they can be used as calibrators in either DNA-based or protein-based analytical methods. Therefore, matrix based CRMs are primarily used in GMO analysis. In some matrix-based CRMs, small seeds, such as canola, are used directly as a CRM with certified purity and homozygosis [72]. The DNA-based CRMs consist of two types: genomic DNA-based and plasmid DNA-based CRMs. Genomic DNA-based CRM is genomic DNA extracted from leaves of certified homozygous GM plants [72]; plasmid DNA-based CRMs are a new type of CRMs that are developed by the integration of sequences of exogenous genes and/or endogenous reference genes into one plasmid and have certified genome copy number ratios [73].

GM CRMs are essential for ensuring the accuracy of GM content measurement. Matrix-based GM CRMs are quite different from the conventional chemical CRMs because the quantification of GM contents is conducted by the indirect measurement of DNA, and the actual quantification result is the genome copy number ratio besides the mass fraction ratio. The GM and non-GM materials used to produce matrix-based CRMs should be prepared under very strict requirements. The purity of both GM material and its recipient non-GM materials should be strictly verified. To date, approximately 100 CRMs targeting 58 GM events, mainly GM maize, soybean, cotton, and canola, have been developed and commercialized [74]. Compared with the large number of GM events developed (nearly 400), the number of CRMs that has been developed is seriously low, and more GM CRMs need to be developed soon.

4.4.4 Gene Methylation

The use of proper reference materials may enable verification and calibration of various analytical biases in measurements of biological properties and therefore will facilitate accurate and reliable measurements. Hannah Yu suggested a calibration concept employing template-type reference materials to facilitate accurate measurement of gene methylation [75]. The proposed reference materials are characterized by two distinct properties. First, they include a pair of bisulfite-converted sequences that represent exactly 0 and 100 % methylation status of a gene. Second, they are in discrete plasmid constructs that could be easily quantified and mixed as working standards. Yu demonstrated that various biases, either from PCR or post-PCR measurements, can be precisely verified using the reference materials. Additionally, those biases were successfully calibrated leading to accurate and consistent measurements of DNA methylation, independent of target genes, PCR conditions, and post-PCR analytical platforms.

4.4.5 Genetics

Reference materials (RMs) available for the diagnosis of genetic disorders are scarce [13, 76]. Business concerns limit the production of commercially produced RMs to the most frequently performed tests. So far the only commercial material is for cystic fibrosis testing (FDA approved) and is produced by Maine Molecular Quality Controls. Several certified RMs suitable for in vitro diagnostic (IVD) use from noncommercial sources also exist. The US National Institute of Standards and Technology has produced a standard reference material panel for fragile X syndrome testing, consisting of size-defined FMR1 nucleotide repeat PCR products. The Institute for Reference Materials and Measurements (IRMM) has made some plasmid-based reference materials for the G20210A variant in the prothrombin gene. At the National Institute for Biological Standards and Control (NIBSC), we have produced several panels of reference materials which have been established by the World Health Organization (WHO) which are considered "higher-order" materials and are suitable for in vitro diagnostic use. They are not intended as routine "run controls," but as materials to occasionally test reagents, equipment, and staff, as well as to act as calibrants for in-house local controls.

The clinical application of pharmacogenomics warrants the production and use of appropriate reference materials. Given these verity, often lethal, of the adverse reactions, the need for reference materials is particularly strong. This need is just beginning to be met, with Paragon Dx, now offering FDA-approved human genomic DNA quality controls for CYP2D6, CYP2C9, and VKORC1 diagnostic testing.

Currently, genetic tests tend to be designed and performed as gene-specific assays, i.e., the methodology is specifically tailored to the gene in question and the nature of the mutations/variants within it. Each assay therefore ought to have specific RMs to act as suitable controls and calibrators. The number of GRMs required is therefore very large. The development, however, of new massively high throughput sequencing technologies and microarray technologies promises to introduce generic assays into genetic testing, so that all except a handful of unusual genes/mutations/variants can be assayed using a single method. Thus, in the future, it is likely that the need for a large number of very different GRMs will diminish. The advent of these relatively generic assays, however, may pose new validation and quality control problems. Reference materials may need to be tailored to specific platforms, making cross-platform comparisons difficult. Additionally, because these technologies analyze genomic DNA directly rather than amplified forms, relatively large amounts of DNA are required hence the provision of sufficient quantities for regular use of reference materials may pose unique challenges. However, until these new technologies enter routine clinical diagnostics, it is important to continue to develop new GRMs for existing assays to facilitate assay validation, determination of assay sensitivity, test accuracy, and comparability of laboratory performance worldwide.

4.4.6 *Viruses*

NAT-based diagnostic techniques have a crucial role in the ongoing Ebola virus outbreak in Western Africa. Reference materials are needed to assess the validity of the assays used, to compare results across assays, and to provide guidance to the regulatory agencies in the evaluation of new assays. It is crucially important that Ebola virus NAT reference materials standardize and control the entire process from the extraction to the final amplification and detection reaction. Giada Mattiuzzo found the development of safe, noninfectious, stable reference materials for Ebola virus NAT-based assays [77]. This has been achieved by incorporating Ebola virus RNA into HIV-1-like particles. These chimeric particles, resembling spherical HIV-1 particles on the outside, have an internal core containing Ebola virus genes as anti-sense RNA within the two HIV-1 LTRs. The lack of any viral envelope protein and HIV-1 structural genes renders this material noninfectious and unable to replicate. To further increase the safety of these preparations, the cloned Ebola virus gene sequences were designed to lack the start codon and contain random stop codons, ensuring that full-length Ebola virus protein could not be produced. The plasmids used to generate the particles were fully sequenced, and the data are available in the GenBank database to allow end users to check the suitability of the reference material for their assay.

4.5 Prospects

The accuracy and precision of measurements form the basis of accurate application in industry, commerce, and medicine all over the world. Harmonization of the measurement system can be achieved by enforcing various standards and directions. In many cases, varying measurement results are obtained, even when laboratories use the same research method. To avoid such situations, it is necessary to use quality control systems, personnel training, a good management structure of laboratories, validation of methods (e.g., control diagrams), and evaluation by external quality control (e.g., participation in interlaboratory comparisons) and, above all, by the use of CRMs. In contrast, new types of reference materials attract researchers' attention. It would be useful for the scientific community to have access to well-characterized materials that may be appropriate in the short term as test materials for hazard assessment and, in the longer term, could be used for benchmarking toxic effects. Nanomaterials appear to be the profound reference materials in human and food toxicology [78–80].

Acknowledgments This work is supported by the National Science and Technology Major Project (2016ZX08012-001). Many thanks to Wenying Tian and Chenguang Wang for their kind help in manuscript conception and preparation.

References

1. Vesper HW, Thienpont LM. Traceability in laboratory medicine. Clin Chem. 2009;55 (6):1067–75.
2. Sarton G. Ancient science through the golden age of Greece. Courier Corporation; 2012.
3. Hollihan-Elliot S. Ancient history of China. Mason Crest; 2014.
4. Cochrane RC. Measures for progress: a history of the National Bureau of Standards, vol 275. National Bureau of Standards, US Department of Commerce; 1966.
5. van Lubeck H. Why not replace pH and pOH by just one real acidity grade, AG? J Chem Educ. 1999;76(7):892.
6. Larsson A, Palm M, Hansson LO, Axelsson O. Reference values for clinical chemistry tests during normal pregnancy. BJOG Int J Obstet Gynaecol. 2008;115(7):874–81.
7. Guide I. 30 (1992) Terms and definitions used in connection with reference materials. ISO, Geneva; 1992.
8. Rauret G, López-Sánchez J-F, Sahuquillo A, Barahona E, Lachica M, Ure AM, Davidson C, Gomez A, Lück D, Bacon J. Application of a modified BCR sequential extraction (three-step) procedure for the determination of extractable trace metal contents in a sewage sludge amended soil reference material (CRM 483), complemented by a three-year stability study of acetic acid and EDTA extractable metal content. J Environ Monit. 2000;2(3):228–33.
9. Sahuquillo A, Lopez-Sanchez J, Rubio R, Rauret G, Thomas R, Davidson C, Ure A. Use of a certified reference material for extractable trace metals to assess sources of uncertainty in the BCR three-stage sequential extraction procedure. Anal Chim Acta. 1999;382(3):317–27.
10. Kupiec K, Konieczka P, Namieśnik J. Prospects for the production, research and utilization of reference materials. Crit Rev Anal Chem. 2009;39(4):311–22.
11. Verkouteren RM, Klinedinst DB. Standard reference materials®. NIST Spec Publ. 2004;260:149.
12. Ditmars D, Ishihara S, Chang S, Bernstein G, West E. Enthalpy and heat-capacity standard reference material: synthetic sapphire (a-Al2O3) from 10 to 2250 K. J Res Natl Bur Stand. 1982;87(2):159–63.
13. Levin BC, Cheng H, Reeder DJ. A human mitochondrial DNA standard reference material for quality control in forensic identification, medical diagnosis, and mutation detection. Genomics. 1999;55(2):135–46.
14. Sharpless KE, Colbert JC, Greenberg RR, Schantz MM, Welch MJ. Recent developments in food-matrix Reference Materials at NIST. Fresenius J Anal Chem. 2001;370(2–3):275–8.
15. Blirup-Jensen S, Grubb A, Lindström V, Schmidt C, Althaus H. Standardization of Cystatin C: development of primary and secondary reference preparations. Scand J Clin Lab Invest. 2008;68(sup241):67–70.
16. Tate JR, Berg K, Couderc R, Dati F, Kostner GM, Marcovina SM, Rifai N, Sakurabayashi I, Steinmetz A. International Federation of Clinical Chemistry and Laboratory Medicine (IFCC) Standardization Project for the Measurement of Lipoprotein (a). Phase 2: selection and properties of a proposed secondary reference material for lipoprotein (a). Clin Chem Lab Med. 1999;37(10):949–58.
17. Thorpe S, Koenders A, Group ER. Standardization of steroid receptor assays in human breast cancer—III. Selection of reference material for intra-and inter-laboratory quality control. Eur J Cancer Clin Oncol. 1986;22(8):939–44.

18. Klich H, Caliste J. COMAR—database for certified reference materials. Fresenius' Z Anal Chem. 1988;332(6):552–5.
19. Steiger T, Pradel R. COMAR 2—the internet database for certified reference materials. Anal Bioanal Chem. 2004;378(5):1145–6.
20. Stoeppler M, Wolf WR, Jenks PJ. Reference materials for chemical analysis: certification, availability and proper usage. Wiley; 2008.
21. Quevauviller P. Reference materials: an inquiry into their use and prospects in Europe. TrAC Trends Anal Chem. 1999;18(2):76–85.
22. Hammes K, Schmidt MW, Smernik RJ, Currie LA, Ball WP, Nguyen TH, Louchouarn P, Houel S, Gustafsson Ö, Elmquist M. Comparison of quantification methods to measure fire-derived (black/elemental) carbon in soils and sediments using reference materials from soil, water, sediment and the atmosphere. Global Biogeochem Cycles. 2007;21(3):GB3016.
23. Dulski P. Reference materials for geochemical studies: new analytical data by ICP-MS and critical discussion of reference values. Geostand Newslett. 2001;25(1):87–125.
24. Coplen TB, Kendall C, Hopple J. Comparison of stable isotope reference samples. Nature. 1983;302:236–8.
25. Reynolds BC, Aggarwal J, André L, Baxter D, Beucher C, Brzezinski MA, Engström E, Georg RB, Land M, Leng MJ. An inter-laboratory comparison of Si isotope reference materials. J Anal Atom Spectrom. 2007;22(5):561–8.
26. Richter S, Alonso A, De Bolle W, Kühn H, Verbruggen A, Wellum R, Taylor P. -Re-certification of a series of uranium isotope reference materials: IRMM-183, IRMM-184, IRMM-185, IRMM-186 and IRMM-187. Int J Mass Spectrom. 2005;247(1):37–9.
27. Steinnes E, Rühling Å, Lippo H, Mäkinen A. Reference materials for large-scale metal deposition surveys. Accred Qual Assur. 1997;2(5):243–9.
28. Quevauviller P, Rauret G, López-Sánchez J-F, Rubio R, Ure A, Muntau H. Certification of trace metal extractable contents in a sediment reference material (CRM 601) following a three-step sequential extraction procedure. Sci Total Environ. 1997;205(2):223–34.
29. Eaton DF. Reference materials for fluorescence measurement. Pure Appl Chem. 1988;60 (7):1107–14.
30. Emteborg H, Bordin G. Speciation of organic and inorganic selenium in a biological certified reference material based on microbore ion-exchange chromatography coupled to inductively coupled plasma atomic emission spectrometry via a direct injection nebulizer or coupled to electrothermal atomic absorption spectrometry. Analyst. 1998;123(2):245–53.
31. Moss D. Accuracy, precision, and quality control of enzyme assays. J Clin Pathol Suppl (Ass Clin Path). 1970;4:22.
32. Rej R, Vanderlinde RE, Fasce CF. An l-aspartate: 2-oxoglutarate aminotransferase reference material from human erythrocytes: preparation and characterization. Clin Chem. 1972;18 (4):374–83.
33. Fasce CF, Rej R, Copeland WH, Vanderlinde RE. A discussion of enzyme reference materials: applications and specifications. Clin Chem. 1973;19(1):5–9.
34. Burtis CA, Sampson EJ, Bayse DD, McKneally SS, Whitner VS. An interlaboratory study of measurement of aspartate aminotransferase activity with use of purified enzyme materials. Clin Chem. 1978;24(6):916–26.
35. Schiele F, Muller J, Colinet E, Siest G, Arzoglou P, Brettschneider H, Calam D, Ceriotti F, Ferard G, Frei J. Interlaboratory study of the IFCC method for alanine aminotransferase performed with use of a partly purified reference material. Clin Chem. 1992;38(12):2365–71.
36. Philipp WJ, van Iwaarden P, Schimmel H, Meeus N, Kollmorgen N. Development of reference materials for microbiological analysis. Accred Qual Assur. 2007;12(3–4):134–8.
37. Mathieu M, Steghens J, Hørder M, Moss D, Colinet E, Profilis C. The certification of the catalytic concentration of creatine kinase BB in a reconstituted lyophilized material(CRM 299). EUR(Luxembourg); 1992.
38. Duncan P, McKneally S, MacNeil M, Fast D, Bayse D. Development of a reference material for alkaline phosphatase. Clin Chem. 1984;30(1):93–7.

39. Tietz N, Rinker A, Burtis C, Duncan P, Ervin K, Ewen L, Hørder M, Mathieu M, Petitclerc C, Grisley D. Transferability studies for the AACC reference method and the IFCC method for measurement of alkaline phosphatase activity. Clin Chem. 1984;30(5):704–6.
40. Schiele F, Muller J, Colinet E, Siest G. Certification of an enzyme reference material for alkaline phosphatase (CRM 371). Clin Biochem. 1991;24(2):159–68.
41. Grubb A, Blirup-Jensen S, Lindström V, Schmidt C, Althaus H, Zegers I. First certified reference material for cystatin C in human serum ERM-DA471/IFCC. Clin Chem Lab Med. 2010;48(11):1619–21.
42. Duncan PH, Van Etten RL, MacNeil ML, Shaw LM. Development of a stable reference material for prostatic acid phosphatase. Clin Chem. 1984;30(8):1327–31.
43. Francis J, Moss D, Colinet E, Calam D, Bullock D. A reference preparation of human prostatic acid phosphatase: purification, characterization and field trials. Ann Clin Biochem: Int J Biochem Med. 1992;29(2):176–83.
44. Hørder M, Rej R, Goldberg D, Werner M. Requirements and functions of reference materials for enzymes. In: Progress in clinical enzymology, vol 2. Masson Paris; 1983. p. 29–34.
45. Quevauviller P. Requirements for production and use of certified reference materials for speciation analysis: a European Commission perspective. Spectrochim Acta B: At Spectrosc. 1998;53(9):1261–79.
46. Campbell PJ. International biological standards and reference preparations: II. Procedures used for the production of biological standards and reference preparations. J Biol Stand. 1974;2 (4):259–67.
47. Pauwels J, Lamberty A, Schimmel H. Homogeneity testing of reference materials. Accred Qual Assur. 1998;3(2):51–5.
48. Linsinger TP, Pauwels J, van der Veen AM, Schimmel H, Lamberty A. Homogeneity and stability of reference materials. Accred Qual Assur. 2001;6(1):20–5.
49. Lamberty A, Schimmel H, Pauwels J. The study of the stability of reference materials by isochronous measurements. Fresenius J Anal Chem. 1998;360(3–4):359–61.
50. Wolf WR, Chemistry FoA, Meeting SS. Biological reference materials: availability, uses, and need for validation of nutrient measurement. Wiley; 1985.
51. Minoia C, Sabbioni E, Apostoli P, Pietra R, Pozzoli L, Gallorini M, Nicolaou G, Alessio L, Capodaglio E. Trace element reference values in tissues from inhabitants of the European Community I. A study of 46 elements in urine, blood and serum of Italian subjects. Sci Total Environ. 1990;95:89–105.
52. von Behring E. Serum therapy in therapeutics and medical science. Nobel Lectures, Physiology or Medicine 1921; 1901.
53. Kasten FH. Paul Ehrlich: pathfinder in cell biology 1. Chronicle of His life and accomplishments in immunology, cancer research, and chemotherapy 1: Paul Ehrlich's recipe for success: "Patience, Ability, Money and Luck". Biotech Histochem. 1996;71(1):2–37.
54. Silverstein AM. Paul Ehrlich's passion: the origins of his receptor immunology. Cell Immunol. 1999;194(2):213–21.
55. Kaufmann SH. Immunology's foundation: the 100-year anniversary of the Nobel Prize to Paul Ehrlich and Elie Metchnikoff. Nat Immunol. 2008;9(7):705–12.
56. Miller WG, Myers GL, Gantzer ML, Kahn SE, Schönbrunner ER, Thienpont LM, Bunk DM, Christenson RH, Eckfeldt JH, Lo SF. Roadmap for harmonization of clinical laboratory measurement procedures. Clin Chem. 2011;57(8):1108–17.
57. Einspanier R. Quantifying genetically modified material in food: searching for a reliable certification. Eur Food Res Technol. 2001;213(6):415–6.
58. Sabbioni E, Minoia C, Pietra R, Fortaner S, Gallorini M, Saltelli A. Trace element reference values in tissues from inhabitants of the European Community. II. Examples of strategy adopted and trace element analysis of blood, lymph nodes and cerebrospinal fluid of Italian subjects. Sci Total Environ. 1992;120(1):39–61.

59. Muramatsu Y, Parr R. Survey of currently available reference materials for use in connection with the determination of trace elements in biological and environmental materials. International Atomic Energy Agency Vienna; 1985.
60. Primrose S, Saunders GC, Parkes HC. Analytical molecular biology: quality and validation, vol. 8. Cambridge: Royal Society of Chemistry; 1999.
61. King B. The selection and use of reference materials. A basic guide for laboratories and accreditation bodies, EEE/RM/062rev3; 2002.
62. Commission IE. General Requirements for the Competence of Testing and Calibration Laboratories: Prescriptions Generales Concernant la Competence Des Laboratoires D'etalonnages Et D'essais. International Organization of Standardization; 1999.
63. Janning B, Mooijman K, Havelaar A. Development, production and certification of microbiological reference materials. Fresenius J Anal Chem. 1995;352(1–2):240–5.
64. Mooijman K, Havelaar A. Application of microbiological (certified) reference materials. Water Sci Technol. 1997;35(11):467–70.
65. James C. Global status of commercialized biotech/GM crops. Ithaca: International Service for the Acquisition of Agri-biotech Applications (ISAAA); 2010.
66. Zhang D, Guo J. The development and standardization of testing methods for genetically modified organisms and their derived products F. J Integr Plant Biol. 2011;53(7):539–51.
67. Li X, Pan L, Li J, Zhang Q, Zhang S, Lv R, Yang L. Establishment and application of event-specific polymerase chain reaction methods for two genetically modified soybean events, A2704-12 and A5547-127. J Agric Food Chem. 2011;59(24):13188–94.
68. Zhu X, Chen L, Shen P, Jia J, Zhang D, Yang L. High sensitive detection of Cry1Ab protein using a quantum dot-based fluorescence-linked immunosorbent assay. J Agric Food Chem. 2011;59(6):2184–9.
69. Trapmann S, Schimmel H, Kramer GN, Eede G, Pauwels J. Production of certified reference materials for the detection of genetically modified organisms. J AOAC Int. 2002;85(3):775–9.
70. Pi L, Li X, Cao Y, Wang C, Pan L, Yang L. Development and application of a multi-targeting reference plasmid as calibrator for analysis of five genetically modified soybean events. Anal Bioanal Chem. 2015;407(10):2877–86.
71. Broothaerts W, Corbisier P, Emons H, Emteborg H, Linsinger TP, Trapmann S. Development of a certified reference material for genetically modified potato with altered starch composition. J Agric Food Chem. 2007;55(12):4728–34.
72. Jiang Y, Yang H, Quan S, Liu Y, Shen P, Yang L. Development of certified matrix-based reference material of genetically modified rice event TT51-1 for real-time PCR quantification. Anal Bioanal Chem. 2015;407(22):6731–9.
73. Zhang H, Yang L, Guo J, Li X, Jiang L, Zhang D. Development of one novel multiple-target plasmid for duplex quantitative PCR analysis of roundup ready soybean. J Agric Food Chem. 2008;56(14):5514–20.
74. Dong W, Yang L, Shen K, Kim B, Kleter GA, Marvin HJ, Guo R, Liang W, Zhang D. GMDD: a database of GMO detection methods. BMC Bioinforma. 2008;9(1):260.
75. Yu H, Hahn Y, Yang I. Reference materials for calibration of analytical biases in quantification of DNA methylation. PLoS One. 2015;10(9):e0137006.
76. Yoshinaga J, Nakama A, Morita M, Edmonds JS. Fish otolith reference material for quality assurance of chemical analyses. Mar Chem. 2000;69(1):91–7.
77. Mattiuzzo G, Ashall J, Doris KS, MacLellan-Gibson K, Nicolson C, Wilkinson DE, Harvey R, Almond N, Anderson R, Efstathiou S. Development of lentivirus-based reference materials for Ebola virus nucleic acid amplification technology-based assays. PLoS One. 2015;10(11):e0142751.
78. Stone V, Nowack B, Baun A, van den Brink N, von der Kammer F, Dusinska M, Handy R, Hankin S, Hassellöv M, Joner E. Nanomaterials for environmental studies: classification, reference material issues, and strategies for physico-chemical characterisation. Sci Total Environ. 2010;408(7):1745–54.

79. Li XH, Long WG, Li QL, Liu Y, Zheng YF, Yang YH, Chamberlain KR, Wan DF, Guo CH, Wang XC. Penglai zircon megacrysts: a potential new working reference material for microbeam determination of Hf–O isotopes and U–Pb age. Geostand Geoanalytical Res. 2010;34 (2):117–34.
80. Nelson BC, Petersen EJ, Marquis BJ, Atha DH, Elliott JT, Cleveland D, Watson SS, Tseng I-H, Dillon A, Theodore M. NIST gold nanoparticle reference materials do not induce oxidative DNA damage. Nanotoxicology. 2013;7(1):21–9.

Chapter 5
Reference Gene: In-Species Universality Versus Between-Species Uniquity

Abstract Reference gene is widely used in species identification. According to gene expression level, it can be classified into two categories: the reference gene on the genome and on the transcriptome. Reference genes that related to the animals, plants, microorganisms, and genetically modified (GM) crops have been reported. This chapter has detailed and introduced the reference gene from definition, classification, identification methods, and application. The future research advances have been stated in the end.

Keywords Reference gene • Classification • Identification methods • Application

5.1 Definition

The reference gene is commonly used in the molecular biology, such as gene expression analysis, species identification, and quantitative detection. Usually, the reference gene is an internal reaction control, which has the sequences different from the other target. For the different use, the reference gene has the different definitions.

For the transcriptome expression analysis, simply, the reference gene is a gene which is expressed stably in the cell or organism. It must meet several requirements: the most important one is the expression level of the reference gene which is stable and cannot be affected by the experimental factors. Meanwhile, it should show minimal variability on its expression between different tissues and physiological states of the organism. Reference gene must in turn demonstrate the variability resulting from imperfections of the technology used and preparatory procedures—this ensures that any variation in the amount of genetic material will relate to the same extent as the object of research and control. It seems that the perfect fulfillment of these conditions is the basic metabolism genes (called housekeeping genes, HKGs), which by definition are involved in processes essential for the survival of cells, must be expressed in a stable and nonregulated constant level, and in fact were first to be examined as reference genes. This was always questioned, even at the time of forming the assumption since many of them participate not only in basic metabolic processes and what seems to be perfect in one experiment does not guarantee its functionality in another. What sounds simple and obvious forces in

W. Xu, *Functional Nucleic Acids Detection in Food Safety*,
DOI 10.1007/978-981-10-1618-9_5

turn an individual and complex approach to each experiment and the necessity of their careful choice, along with validation which, to some extent, was and still is being deficient in many papers.

For the genome analysis, the reference gene, which is also called endogenous reference gene, is a conserved DNA sequence with species specificity and copy number constant and doesn't show any allelic variation. The important characters of the endogenous reference gene are the species specificity and the low and constant copy number. For the species identification and the quantitative detection, the intraspecific nonspecificity and the interspecific specificity are critical. In the PCR reaction, the primer of the endogenous reference gene is usually designed in the exon region. The primers that are designed based on the intron region have many disadvantages, such as many homologous sequences, inestimable variation, and different amplification results between different cultivars. Hence, in order to guarantee the equal amplification, the primers of the endogenous reference gene are always designed based on the exon region.

5.2 The Classifications of the Reference Genes

5.2.1 The Reference Gene on the Genome

Classified by the expression level, the endogenous reference gene can be divided into two categories, the reference gene on the genome and on the transcriptome expression level. The traits of the endogenous reference gene are showed as follows:

1. Species specific, including interspecies specificity and intraspecies stability. Interspecies specificity means that the endogenous reference gene exhibits low homology among different species, and the products can only be amplified in the target species through polymerase chain reaction (PCR) amplification. Intraspecies stability means the endogenous reference gene has high homology in different varieties.
2. Constant copy number. The endogenous reference gene has the constant copy number on the genomes with the same species and has no copy number variation such as repeat, missing, insertion, etc.
3. No allelic variation among cultivars.

For different species, including plant, animal, and microorganism, the methods of selecting endogenous reference gene on the genome are different. For the microorganism, it is the most difficult.

5.2.1.1 The Reference Gene of the Plant

For the plant, the genome is large, and, usually, the endogenous reference gene is selected based on the genome of the plant. At present, in the main crops, such as soybeans, corn, canola, cotton, rice, and wheat, more than one reference gene has been developed. For example, the reference gene *lectin* [1–5] is the endogenous reference gene for soybean, and *β-actin* [6]and *hsp* [7] are also the reference genes of soybean; the *hmga* [8], *zSSIIb* [9], *10 kDa zein* [10], *Adhl* [11], and *Ivr1* [2, 12] are the reference genes for maize; the *CruA* [13, 14], *BnAccg8* [13, 15], *FatA* [13, 16], *HMG-I/Y* [13, 17], and *PEP* are for rapeseed; the *Cotton-ppi-PPF* [18], *ACP1* [19, 20], *Sad1* [21–23], and *SAH7* [24] are for cotton; the *RBE4* [25], *gos9* [26, 27], *PLD* [28, 29], and *SPS* [29] are for rice; the *acc1* [30, 31], *waxy-D1* [32], and *PKABA1* [33] are for wheat; the *Lhcb2* is for peach [34]; the LAT52 is for tomato [35]; and the *papain* is for papaya [36]. The detailed reference gene information is listed in Table 5.1.

Table 5.1 The reported reference genes of main genetically modified crops

Specific species	Gene	GenBank	Copy number
Soybean	*Lectin*	K00821	1
	β-Actin	GMU60506	
Maize	*Hmga*	AJ131373	1
	zSSIIb	X04050	1
	10 kDa zein	U16123	2
	Adhl	X07535	1
	Ivr1	AF019297	1
Rapeseed	*CruA*	X14555	2~3
	BnAccg8	X77576	4
	FatA	AJ294419	
	HMG-I/Y	AF127919	
	PEP	D13987	
Cotton	*Cotton-ppi-PPF*	EF445628	1
	ACP1	GHU48777 1	2
	Sad1	AJ132636 2	
	SAH7	AY117065	
Rice	*RBE4*	EF055878	1
	gos9	X519090	1~2
	PLD	AB001919	1
	SPS	U33175	
Wheat	*acc1*	AF029895	1
	waxy-D1	AF113844	1–2
	PKABA1	M94726	1
Peach	*Lhcb2*	EF127291.1	1
Tomato	*LAT52*	No. 19263	1
Papaya	Papain	M15203	2

5.2.1.2 The Reference Gene of the Animal

For the animal, the mitochondrial genes are commonly used as the target genes for the animal species identification, e.g., the *cytb* gene in pork [37], the *ND2* gene in donkeys [38], the *Myostatin* gene in sheep [39], *D-loop* gene in chickens [40], and the *Cox* II gene in beef [41].

In the studies of Xu et al. [42, 43], a conserved DNA sequence of chicken, cattle, sheep, pig, and horse mitochondrial cytochrome *b* gene is selected as the species-specific sequence. On the mitochondrial cytochrome *b* gene, there is an identical fragment of sequence in the five different species, and meanwhile, the sequence variations also exist on this gene. The common single primer (CSP) is designed on the identical conserved DNA sequence, the other primers are on the variations regions, and the CSP-M-PCR method is developed. Based on the different lengths of the products, the five species can be detected.

The chromosomal gene *Actb* is selected as endogenous reference gene of chicken species for the first time by Xiang et al. Because the copy numbers of mitochondrial genes are high, they are useful for the qualitative detection but unsuitable for quantitative analyses. Hence, it is important to develop a quick, precise, and specific-species quantitative technology for the detection of animal species.

5.2.2 The Reference Gene on the Transcriptome

In the study of transcriptome gene expression, a common expression housekeeping gene is essential for accurately evaluating the gene expression level. Housekeeping gene can be used as a reference gene to standardize the total RNA amount of unknown sample [44]. The expression levels of the reference gene in different materials are stable but others are not [45]. If the expression of the housekeeping gene which is used as the reference gene has been changed, it may have a significant influence on the accuracy of the standardization of the target gene [46]. The expression level of the reference gene should not only be stable but also be close to the target gene. Ideally the housekeeping gene should not be regulated or influenced by the experimental procedure. Gene transcriptome studies in different tissues showed it is necessary that a reference gene should exhibit constant RNA transcription in all tissues. According to the central dogma, the genetic information is transferred from DNA to RNA and then transferred from RNA to protein. In different species, the expression levels of the different proteins are not constant and consistent. The gene which regulates the protein that has constant expression amount is selected as the reference gene on the transcriptome level, and it can be used as a normalizer to evaluate other genes. The commonly used housekeeping genes that are chosen as the reference gene are albumin (for hepatocytes) [47], β-actins, γ-actins [48], cyclophilin [49], G3PDH [50], α-tubulins, β-tubulins [48], hypoxanthine phosphoribosyltransferase (HPRT) [51], L32 for other cell types

[52], or 18S and 28S rRNA [53]. GAPDH gene is an enzyme with 37 kDa that catalyzes the sixth step of glycolysis and thus serves to break down glucose for energy and carbon molecules and often stably and constitutively expressed at high levels in most tissues and cells. GAPDH is overexpressed in multiple human cancers, such as cutaneous melanoma, and its expression is positively correlated with the tumor growth [54, 55]. The glycolytic and antiapoptotic functions contribute to proliferation and protection of tumor cells, promoting tumorigenesis. Besides, GAPDH has been implicated in several neurodegenerative diseases and disorders, largely through interactions with other proteins that are specific to disease or disorder.

ACTB is one of the six different actin isoforms which have been identified in humans. *ACTB* is one of the two nonmuscle cytoskeletal actins. Actins are highly conserved proteins [56, 57] that are involved in cell motility, structure, and integrity. Alpha actins are major constituents of the contractile apparatus.

18S rRNA is a part of the ribosomal RNA. The "S" in the 18S represents for the Svedberg units. 18S rRNA is a component of the small eukaryotic ribosomal subunit (40S). 18S rRNA is the structural RNA for the small component of eukaryotic cytoplasmic ribosomes and thus one of the basic components of all eukaryotic cells. It is the eukaryotic nuclear homologue of 16S ribosomal RNA in prokaryotes and mitochondria.

28S rRNA is the structural RNA for the large component or large subunit (LSU) of eukaryotic cytoplasmic ribosomes and thus one of the basic components of all eukaryotic cells. It is the eukaryotic nuclear homologue of the prokaryotic 23S ribosomal RNA.

Many other reference genes have been investigated, including peptidylprolyl isomerase A (*PPIA*), glucose 6-phosphate dehydrogenase (*G6PDH*), TATA-box-binding protein (*TBP*), β2-microglobulin (β2M), porphobilinogen deaminase (*PBGD*), and the ribosomal protein L13 (*L13*). The details of the reference genes are listed in Table 5.2.

Table 5.2 The reference gene of RNA

Reference gene	Function of the gene	Reported to be used in identification for	Reference
GAPDH (*G3PDH*)	Important glycolytic pathway	Tissues and cells	[50]
Albumin	Intravascular osmotic pressure great contributor		[47]
Cyclophil	Involved in cellular protein folding and protein interactions		[49]
β-, γ-Actins	Essential for the structure and kinetics of the cytoskeleton	Cell motility, structure, and integrity	[48]
α-, β-Tubulins	Essential for the structure and kinetics of the cytoskeleton		[48]
L32	Ribosomal protein		[52]
18S rRNA	Ribosomal subunits		[53]
28S rRNA			

MicroRNAs (miRNAs) are 22-nucleotide, short, noncoding RNAs that are thought to regulate gene expression through sequence-specific base pairing with target mRNAs. Hundreds of microRNAs have been identified in worms, flies, fish, frogs, mammals, and flowering plants using molecular cloning and bioinformatics prediction strategies [58]. MicroRNAs play an important role in various kinds of physiological processes such as differentiation, growth and homeostasis balance, etc. For an exact comparison of miRNA in different samples, it is crucial to choose the reference genes. There are four criteria [59] used to select candidate normalize miRNAs in the tissues of interest: (1) the miRNA must be highly expressed in most, if not all, of the samples; (2) the miRNA must be consistently expressed, as measured by the modified Z-score; (3) only one representative from a given miRNA family should be considered; and (4) the miRNA must be a target of a commercially available qRT-PCR assay at the time of the work. The reference genes are generally expressed in all kinds of cells and necessary to the biological processing. 5S rRNA [60] and U6 snRNA [61] (RNU6B, 45 nt) are the most commonly used in relative quantitative RT-PCR. 5S rRNA is an approximately 120-nucleotide-long ribosomal RNA molecule with a mass of 40 kDa. It is a structural and functional component of the large subunit of the ribosome in all domains of life (bacteria, archaea, and eukaryotes), with the exception of mitochondrial ribosomes of fungi and animals. U6 is expressed consistent in different tissues and widely used as tissue cell relative quantitative RT-PCR reference gene. And U6 snRNA is the noncoding snRNA component of U6 snRNP (small nuclear ribonucleoprotein), an RNA–protein complex that combines with other snRNPs, unmodified pre-mRNA, and various other proteins to assemble a spliceosome, a large RNA–protein molecular complex upon which splicing of pre-mRNA occurs. Splicing, or the removal of introns, is a major aspect of posttranscriptional modification and takes place only in the nucleus of eukaryotes. The lengths of the SnRNA and SnoRNA are about 200 bp, which are similar with that of miRNA, and have high expressed levels in varieties of tissue cells. They are not involved in the regulation of the miRNA pathways, and the design method is the same as miRNA. Thus, many researchers considered SnRNA and SnoRNA as the good reference genes. Commonly used SnoRNA reference genes in human tissue include U6, RNU48, RNU44, U47, and RNU6B. But U6 isn't as the reference gene in the serum. In recent years, many studies have reported that there were new reference genes under different conditions. Although generally recognized that the miRNA reference genes should choose the same genus of RNA [62], now there are rarely reports about the miRNAs reference gene or unified standard. Some researchers pointed out that *miR-26a*, *miR-345*, *miR-425*, and *miR-454* were identified as having expression profiles closest to the global mean. From a panel of six miRNAs (*let-7a*, *miR-16*, *miR-26a*, *miR-345*, *miR-425*, and *miR-454*) and two small nucleolar RNA genes (*RNU48* and *Z30*), *miR-16* and *miR-345* were identified as the most stably expressed reference genes in tumor and normal colorectal tissue [63]. On animal reference gene studies, researchers identified five miRNAs, including miR-146a, miR-16, miR-195, miR-30e, and miR-744, to be stably expressed in all strains, which could serve as mouse serum miRNA endogenous references for single assay experiments [64]. On pig miRNA reference gene research, the stability of ten miRNAs (Ssc-let-7a, Ssc-miR-103, Ssc-miR-17-3p, Hsa-miR-25, Hsa-miR-93, Ssc-miR-106a, Ssc-miR-191, Ssc-miR-16, Ssc-miR-26a, and Ssc-miR-17-5p) was investigated by reverse transcription-quantitative polymerase chain reaction (RT-qPCR) in different tissues (skeletal muscle, kidney, liver, ovary, and uterus) and in different pig breeds (Iberian, Landrace, Large White, Meishan, and Vietnamese) as variation factors. All ten miRNAs analysis had good stability values and, therefore, can be used as reference miRNAs. When all tissues were considered, miR-93 was the most stable miRNA. Dividing dataset by tissues, let-7a was the most stable in skeletal muscle and ovary, miR-17-5p in kidney, miR-26a in liver, and miR-103 in uterus. The use of five reference miRNAs (miR-93, miR-25, miR-106a, miR-17-5p, and miR-26a) in multi-tissue experimental designs and the use of three reference miRNAs as the optimal number in single-tissue studies (let-7a, miR-17-5p, and miR-25 in skeletal muscle; miR-17-5p, miR-93, and miR-26a in kidney; miR-26a, miR-103, and let-7a in liver; let-7a, miR-25, and miR-106a in ovary; and miR-103, let-7a, and miR-93 in uterus) are suggested [65].

Many reports suggest that the use of multiple reference gene combination is more accurate than using one single reference gene. Schaefer conducted experiments on cancer tissue samples [66]; miR-130-b was used as reference gene in prostate cancer tissues, and the geometric mean of miR-130-b and RNU6-2 as reference genes is of great advantage. WotSchofSky [67] suggested that in miRNA reference gene in clear cell renal carcinoma study, if only one type of miRNA reference gene was used as normalizer, the miR-28 was the best choice, but the combination of miR-28 and miR-103 or miR-28, miR-103, and miR-106 might be better. Various miRNA reference gene selection results of tissues indicated that miR-191 was the most stably expressed miRNA in a variety of normal human tissues and miR-103 was the most stably expressed in tumor tissues. The details of microRNA reference genes are listed in Table 5.3.

5.2.3 Microorganism Reference Gene Identification

To correct the variation of microorganisms, a number of strategies have been proposed, including normalizing against total RNA, genomic DNA, or incorporation of an introduced RNA molecule [68]. Although the latter method, particularly, shows promise, the generally accepted method of normalization for qPCR involves the use of endogenous control genes [69]. Reference genes can limit errors resulting from experimental variation as they are subject to the same experimental procedures as the target genes [70]. For the genome of microorganism, the copy

Table 5.3 The reference gene of MicroRNA

MiRNA reference gene	Function of gene	Reported to be used in identification for	Reference
5s rRNA	Most kinds of tissue cells	Tissue cell	[17]
U6 snRNA	Most kinds of tissue cells	Tissue cell	[18]
SnRNA	Most kinds of tissue cells	Tissue cell	[18]
SnoRNA	Most kinds of tissue cells	Human tissue	[19]
miR-146a, miR-16,miR-195, miR-30e, and miR-744	Mouse serum miRNA	Mouse	[20]
miR-93	Pig stable miRNA	Pig	[21]
miR-16 and miR-345	Tumor and normal colorectal tissue	Human	[20]
miR-130-b and RNU6-2	Prostate cancer tissues	Cancer tissues	[23]
miR-191	Human tissue	Human	[24]
miR-103	Tumor tissues	Human	[24]
miR-28	Clear cell renal carcinoma	Human cell renal carcinoma	[24] [24]

number of the target gene is usually unstable, and with multiple copy number. Due to the evolutionary relationship, the whole sequence of the reference gene is difficult to select. Hence, for the microorganism, the conserved sequence either on DNA or RNA is always used as the reference gene.

In the case of bacteria, a number of housekeeping genes for bacterial identification, such as 16S rRNA, gyrB, rpoA, rpoB, rpoC, rpoD, etc. [71], can be used as reference genes. The most widely utilized reference gene is 16S rRNA, which has been regarded as the most useful region in bacterial DNA as reference not only to correct the quantitative error but also to study the polygenetic relationship for all kinds of bacteria on our planet. 16S rRNA contains conserved, variable, and hypervariable regions [72]. As a consequence, in most cases, we can find suitable regions as reference genes for bacteria of the same species, genus, or family. On the other hand, there is a fatal shortcoming of 16S rRNA when used as a reference gene—the copy number of this gene is variable in different species. Thus, when using 16S rRNA as a reference gene for a taxon containing one bacterial species, for example, a reference gene for *Lactobacillus* genus, if different species in this taxon have different 16S rRNA copy numbers, the results will be inaccurate and confusing.

Besides the 16S rRNA gene, which is regarded as a universal and preferred reference gene for any kind of bacteria, many other genes have been develop to be reference gene of specific taxa. The examples include *rpoA* for *Enterococcus* sp. and lactic acid bacteria; recN for *Streptococcus*, *E. coli*, *Neisseria gonorrhoeae*, and *Amycolatopsis*; *pheS* for *Enterococcus* sp. [73]; *GapA* for *Escherichia coli* and

Table 5.4 Some recently found reference gene of microorganisms

Targeted gene sequence	Function of the gene/product of the gene	Reported to be used in identification for	Reference
pheS	Phenylalanyl-tRNA synthase	*Enterococcus* sp., *Lactobacillus*	[73]
gapA	Production of D-glyceraldehyde-3-phosphate dehydrogenase	*E. coli*, *Mycoplasma gallisepticum*	[74, 75]
luxABE	Production of luciferase enzyme that catalyzes light-emitting reaction	*Aliivibrio logei*, *Aliivibrio salmonicida*	[76]
dnaK	Initiation of DNA replication	*Ensifer* sp., *Sinorhizobium*, *Nitrosomonas europaea*	[77]
pyrH	uridylate kinase	*Vibrio vulnificus*, *Vibrio* sp.	[78]
	UMP phosphorylation		
ectBC	Ectoine: diaminobutyric acid aminotransferase	*Halomonas variabilis*, *Halomonadaceae*	[79]

Mycoplasma gallisepticum [74, 75]; *luxABE* gene for *Aliivibrio logei*, *Aliivibrio salmonicida* [76], and *Photobacterium mandapamensis*; *dnaK* gene for *Ensifer* sp., *Sinorhizobium* [77], and *Nitrosomonas europaea*; PyrH for *Vibrio vulnificus* and *Vibrio* sp. [78]; ectBC for *Halomonas variabilis* and *Halomonadaceae* [79]; and *dnaA* for *Mycobacterium* sp., *Rhizobium meliloti*, and *Lactobacillus* (Table 5.4).

In the case of fungi, rDNA consists of the SSU 18S, internal transcribed spacer (ITS), ITS1 + 5.8S + ITS2, and LSU 25-28S regions. ITS has been considered to be the best fungal barcode for identification purposes and can also be used as a reference gene for fungi. Compared with 16S rRNA in bacteria, the major disadvantage of ITS region is the absence of registered fungal DNA sequences in the DNA data libraries, which makes the design of ITS reference genes and pre-analysis of the specificity and sensitivity of designed reference genes in fungi harder than in 16S rRNA of bacteria [80].

5.3 The Universal Endogenous Reference Gene

Based on the species-specific gene mentioned above, there still is the endogenous reference gene for the universal applications. The universal endogenous reference gene is a fragment of conserved sequence, which is existed common in many species.

As reported, the gene *cob* is selected as the universal endogenous reference gene for the plant, the cytochrome b for the animal, 16S rDNA for bacteria, and the ITS1/ITS4 for fungi.

For the molecular detection, there are many experiment objectives. Excepting the species identifications, the basic DNA quality detection is always needed. Hence, it is unnecessary to use the species reference gene, and the universal endogenous reference gene can satisfy the demand. The use of universal endogenous reference gene would not only save the cost but also realize the identification of the large group of the category.

5.4 Methods for Identifying the Reference Gene

Though the reported genes are known to present in all cells, their gene expression levels may vary depending on the cellular functions and the corresponding environmental conditions. It is a well-accepted fact that these reference genes are not very resistant to the changes in their environment. They are susceptible, sometimes very easily, to any variations in temperature, stress, and other attributes that may be affecting the environment. Thus, the use and reliability of the experimental results involving these reference genes in the past must be validated with considerable interest. There is a need to validate these genes for the specific experiments as the behavior of each gene may change based on the experiments, environment conditions, and the accuracy of manual operation.

There are several methods in the identification of reference genes. Internal control genes (or reference genes) can be validated using regression analysis [81]. The methods used for identifying reference genes are listed as follows:

- Reverse transcription PCR (RT-PCR), which is a variant of PCR, has been used extensively for many years [82–84] for the identification of suitable reference genes. In this technique, RNA sequence is reverse transcribed into complementary DNA with the help of the reverse transcriptase enzyme. The cDNA is then amplified using traditional or real-time PCR. In a recent study [85], RPS4, UBQ, and eEF1A1 genes were found to be useful in larvae studies using quantitative PCR. In the past, this technique has been used without any normalization of the genes, which could indicate an increased risk of inaccurate experimental results. Thus, there is a need to focus on normalization strategies and validate the reference genes which are used in experiments [68].
- Sequential analysis of gene expression (SAGE) is a genetic sequence profiling technique [86] used for obtaining high-quality, accurate, and quantitative analysis of gene sequences in the given datasets. One of the main methods requires data-mining the microarray datasets for highly expressed and relatively constant transcripts [87] as done in a study involving SAGE [88]. SAGE is a technique used for getting a complete analysis of the gene expression patterns in the given gene datasets. This is conducted by isolating unique sequence tags from given mRNA sequences and then concatenating these sequences into long DNA sequences [89]. This technique has the following steps [86, 89]: the mRNA of the target sample is first extracted and isolated; for each mRNA sequence, a section is removed at a pre-specified position; all these obtained sections of mRNA sequences are put together to form a lengthy chain-like sequence; these chain-like sequences are cloned into vectors; these vectors are sequenced using

automatic DNA sequencing technologies; the number of sequence tags is counted using computer algorithms and supported by software. In another study [87], SAGE-based records were analyzed, and then the specificity of reference genes was evaluated using quantitative PCR. This study suggested the SAGE is a useful method for the normalization of housekeeping gene.

- Microarray data analysis is a technique which is used to uniquely identify the reference gene [90, 91]. In a separate study [92], microarray data analysis was used for normalizing gene expression levels of the canine osteoarthritic joint tissue, and then the reference genes were identified using three different algorithms. New reference genes [92] were identified using the traditionally normalized microarray data and were found to be more stable than those found using reverse transcription-quantitative polymerase chain reaction (RT-qPCR). A similar method was used to identify reference genes for breast cancer studies [93] in which three microarray datasets of breast tumors were normalized and matched with their normal counterparts. ACTB (beta actin) and succinate dehydrogenase (SDH) were found to be the most stable reference genes in this study involving quantitative qPCR.
- Statistical algorithm-based software such as geNorm and NormFinder [94] are also used to normalize reference genes. In a separate study, while analyzing the expression levels in hemocytes of *Ostrea edulis*, the reference genes for the same were identified by analyzing the RNA sequences [95] using PCR, in pools of hemocytes that were in touch with the parasite and the hemocyte alone. The abovementioned software was used to normalize the gene expression levels in the study. This is just one of the several experiments in which the abovementioned algorithms have played an important role for improving the appropriateness of the housekeeping genes.
- In the study for ovarian tissues, 20 reference genes were identified from 52 samples [96] involving nonmalignant and malignant carcinogenic cells. One-way analysis of variance (ANOVA) method was used to study the gene changes. Next, geNorm [62] and NormFinder [94] were used for further validation. There are some drawbacks of using RT-PCR when compared to other methods. It is time-consuming [97], and the results are not always reliable as ethidium bromide (which is used as a staining material for detection of PCR products) and have a low sensitivity. RT-PCR has an increased cross contamination risk of the samples under consideration here. At the same time, there had been reported instances of contamination of cDNA transcripts with genomic pseudogenes [98] or degradation of RNA [99], which may arise from RNA preparation [100]. This will significantly increase the detection error, as the detection of the PCR products requires the post-amplification processing of the samples. However, there had been a sample processing steps to prevent genomic pseudogene contamination [101] but these will increase the number of steps in sample preparation and validation using mitochondrial DNA [102]. Moreover, the specificity of the assay is determined by the primers and which can give false-positive results. Another issue is that it is semi- or even low quantitative technique, whereas the amplicon (it is a piece of DNA which is formed as the

product of amplification events) can be visualized only after the amplification ends. Real-time PCR is the most sensitive method and can discriminate closely related mRNAs. It is technically simple but it is difficult to get truly quantitative results using conventional PCR. Northern blotting and ribonuclease protection assays (RPAs) require no amplification, while in situ hybridization is qualitative in nature. Northern blotting and ribonuclease protection assays (RPAs) provide good results, but consume more RNA than their counterparts. PCR considers amplification of the DNA; thus, it is more sensitive. Real-time PCR gives more quantitative results and is also easier and convenient to use [97, 103]. SAGE and microarrays are different due to the need for prior knowledge (in the latter) of the gene sequences of the samples under consideration here [104]. Microarrays need a uniform standard for platform fabrication, assay protocols, and analysis. They have large datasets and need complex statistical algorithms for processing them with high accuracy. A significant issue, besides normalization of data and the reduction of dimensionality of data, is the sequencing requirement in SAGE. It is comparatively easy and more reliable to search sequences from SAGE databases (which store experimental data using the SAGE technique) stored at different places and then perform Northern blots on them. However, microarrays have a higher degree of random and systematic errors, due to which a comparison of experimental data from different places (experimental sources) is difficult.
- Basic Local Alignment Search Tool (BLAST) method. In the principle of choosing the reference gene, it should show the lowest homology between the reference gene and that of other species. On the dataset of the National Center for Biotechnology Information (NCBI), all the targets of the organisms that have been sequenced are submitted onto NCBI. In selecting the reference gene, all the sequences related to the target species are collected, and the potential candidates are conducted with BLAST; the gene with the lowest homology is selected. And then, the copy number is validated.

At the development, there are two commonly used methods to detect the copy number, the Southern blot and digital PCR.

Southern blot is a method used in molecular biology for detection of a specific DNA sequence in DNA samples, which is regarded as the gold standard for the determination of the copy numbers of target genes. Restriction endonucleases are used to digest the DNA strands into smaller fragments. The DNA fragments are then electrophoresed on an agarose gel to separate them by size. The specific probe and primers are designed based on the sequence of the candidate. Southern blotting combines transfer of electrophoresis-separated DNA fragments to a filter membrane and subsequent fragment detection by probe hybridization. According to the result of the bands of the Southern blot, the copy number of the reference gene is validated.

Digital PCR is a novel absolute quantification technique; in evaluating the copy number of the candidate, a housekeeping gene with known and stable copy number is needed. Based on the principle, the following equations are obtained:

$$N(\text{endogenous gene}) = -\ln[(N - X)/N] \times N \quad (5.1)$$

$$N(\text{event-specific gene})304 = -\ln[(N - Y)/N] \times N \quad (5.2)$$

For the two equations above, N (endogenous gene) and N (event-specific gene) are the estimated copy number of the endogenous and event-specific genes for each panel, respectively. N is the total number of wells, and X and Y are the wells that are positive for the endogenous and event-specific genes, respectively.

5.5 Application

5.5.1 Ingredient Source Identification

Endogenous reference gene has important application value in the identification of composition source. Some people are allergic to certain ingredients; hence, it is a meaningful issue to detect allergenic ingredients. Peanuts and sesames in oil crops are one of common allergen ingredients; there are some relevant reports of allergen detections by endogenous reference genes. The peanut endogenous gene *Arah* 3 has been used to detect peanut ingredient in peanut products such as peanut butter, peanut cake, etc. [105]. *Ses i* 1 gene has been used to detect sesame ingredient in mixed products [106]. Kerstin Schoringhumer and Margit Cichna-Mark also used *Ses i* 1 gene to determine sesame in several crackers, salty snacks, biscuits, tahini sesame paste, and sesame oil [107].

In the PCR test of genetically modified organisms (GMOs) and products, they cannot be distinguished directly in the appearance, such as solid powder, cake or liquid oil, soy sauce, jam, etc. Testing the source of the ingredient and then making the specificity detection and analysis is an effective method, which is mainly depending on the PCR amplification. Amplification of such reference sequences will allow the detection of DNA from the species of interest in food samples and at the same time assay for the quality of the extracted DNA, providing a means to quantify the amount of GMO in the processed food sample [108]. If the further event specificity of the GMO is wanted, according to the identified species, the screen of the approved events can be time and labor saving.

5.5.2 The Adulteration Identification

With the development of the product processing technology, many original simple materials are processed into convenient and suitable products for all kinds of people, such as edible oil, cake, food, and industrial raw materials. These products cannot be distinguished easily as their appearances are similar. Hence, in order to get high profits, many small retailers on the market often sell seconds at best quality prices or make adulteration directly. This behavior severely disrupts the normal market order and endangers the interests and health of consumers. Identifying the composition of the products is with great significance. There are many reports about detecting soybean protein powder, chocolate drink, baby rice flour, and soybean ingredients in cooking oil by employing the soybean reference gene lectin [109]. By traditional PCR, the detection sensitivity was 0.5 %, and through triple-nested PCR, the detection sensitivity can reach to 0.005 %. Hellebrand successfully detect the rapeseed ingredients in edible oil of the with rapeseed PEP gene [110].

5.5.3 DNA Quality Evaluation in PCR

In the food species detection, the testing samples are divided into plant raw materials, primary processed products, and deep-processed products. The DNA preparation and purification of the raw materials and primary processed products may be relatively easy, but still certain species inhibitors of PCR reaction, such as peanut phenol in peanut DNA, would still be contained in the extracted DNA, which can severely inhibit the subsequent PCR reaction. Although the DNA quality evaluation can be completed by spectrophotometer method, agarose gel electrophoresis detection, and fluorimetry, it is difficult to judge whether the inhibiting factor such as peanut phenol is contained in prepared DNA solution.

In addition, for deeply processing plant products, due to the processing usually use high temperature and high pressure, the genomic DNA may degrade seriously. It makes the preparation and purification of DNA very difficult, especially to some oil crops. The DNA quality is directly related to the accuracy of PCR analysis results; in order to increase the analysis reliability of the PCR system, the amplifications of the reference genes as positive controls are necessary. Hence, the results can evaluate whether the prepared DNA samples is suitable for PCR reaction or not.

5.5.4 Quantification of the Target Materials in the Mixture

Researchers have developed several methods of quantitative PCR and RT-PCR. One approach measures PCR product quantity in the log phase of the reaction before the plateau. This method requires that each sample has equal input amounts of nucleic acid and that each sample under analysis amplifies with identical efficiency up to the point of quantitative analysis. A gene sequence contained in all samples at relatively constant quantities, such as reference gene, can be used for sample amplification efficiency normalization [111]. Using conventional methods of PCR detection and quantification, it is extremely laborious to assure that all samples are analyzed during the log phase of the reaction (for both the target gene and the normalization gene). Another method, quantitative competitive (QC)-PCR, has been developed and is used widely for PCR quantitation. QC-PCR relies on the inclusion of an internal control competitor in each reaction. The efficiency of each reaction is normalized to the internal competitor. A known amount of internal competitor can be added to each sample. To obtain relative quantitation, the unknown target PCR product is compared with the known competitor PCR product. Both the two methods need to amplify exogenous DNA sequence and reference genes at the same time.

Standard curve method is based on the linear relationship between Ct values of each template and initial copy number logarithm of the template. Utilize the known initial copy number standard to draw the standard curve of exogenous genes and reference gene, respectively. Hence, only to obtain exogenous genes and reference gene amplification Ct value of the unknown samples, according to standard curve of the reference gene and the exogenous genes, the amount of the target DNA and the total amount of DNA can be calculated. The ratio of target DNA amount and the total DNA amount is the content of the target material in sample. This method is very perfect for quantifying the percentage of the GMO in the common materials and also widely used. Now, reference genes with species specificity, constant copy number, no allele variation, and high sensitivity in many species have been developed and have been applied to the quantitative detection of the target materials in the products.

5.6 Prospect

The incorrect use of normalized genes would have a profound effect on the conclusions drawn from studies of gene expression. The searching and application of multiple reference genes often cost much. The publishing of erroneous results would result in the future research being predicated on the incorrect or biased data. Given that many such studies are performed in the field of medical research, the consequences could also have negative impacts on drug development and human health. Hence, comparing with those, we consider that the high cost of using the multiple reference genes is insignificant. Furthermore, a reference gene is validated stable in a certain condition, which does not mean it has the same stability in other species and tissue/cell type when used in other laboratories. We advocate that the researchers should routinely test the stability of the potential reference genes that are use as endogenous controls before conducting any gene expression study.

Additionally, novel available methods and analytical software for normalizing qPCR data [112, 113] should be investigated in appropriate experiment. However, we advocate that the studies of the reference gene stability should be continuously conducted, as they provide a useful starting point for the species identification method. What's more is it can conduct not only the gene expression study but also the identification of the species and adulteration, making of the rapid detection kit and instrument, DNA or RNA quality control, and so on. Hence, the study of the endogenous reference gene makes important significance in molecular biology and food safety detection.

Acknowledgments This work is supported by National Science and Technology Major Project (2016ZX08012-004). Many thanks to Ying Shang and Wenjin Xiang, for their kindly help in manuscript conception and preparation.

References

1. Heid CA, Stevens J, Livak KJ, Williams PM. Real time quantitative PCR. Genome Res. 1996;6(10):986–94.
2. Hurst CD, Knight A, Bruce IJ. PCR detection of genetically modified soya and maize in foodstuffs. Mol Breed. 1999;5(6):579–86.
3. Vollenhofer S, Burg K, Schmidt J, Kroath H. Genetically modified organisms in food-screening and specific detection by polymerase chain reaction. J Agr Food Chem. 1999;47 (12):5038–43.
4. Wurz A, Bluth A, Zeltz P, Pfeifer C, Willmund R. Quantitative analysis of genetically modified organisms (GMO) in processed food by PCR-based methods. Food Control. 1999;10(6):385–9.
5. Costa J, Mafra I, Amaral JS, Oliveira MBPP. Detection of genetically modified soybean DNA in refined vegetable oils. Eur Food Res Technol. 2010;230(6):915–23.
6. Guo LH, Qiu B, Chi YW, Chen GN. Using multiple PCR and CE with chemiluminescence detection for simultaneous qualitative and quantitative analysis of genetically modified organism. Electrophoresis. 2008;29(18):3801–9.
7. Yang LT, Chen JX, Huang C, Liu YH, Jia SR, Pan LW, Zhang DB. Validation of a cotton-specific gene, Sad1, used as an endogenous reference gene in qualitative and real-time quantitative PCR detection of transgenic cottons. Plant Cell Rep. 2005;24(4):237–45.
8. Heide BR, Dromtorp SM, Rudi K, Heir E, Holck AL. Determination of eight genetically modified maize events by quantitative, multiplex PCR and fluorescence capillary gel electrophoresis. Eur Food Res Technol. 2008;227(4):1125–37.
9. Kim JH, Kim SY, Lee H, Kim YR, Kim HY. An event-specific DNA microarray to identify genetically modified organisms in processed foods. J Agr Food Chem. 2010;58(10):6018–26.
10. Chang KH, Mestdagh P, Vandesompele J, Kerin MJ, Miller N. MicroRNA expression profiling to identify and validate reference genes for relative quantification in colorectal cancer. BMC Cancer. 2010. doi:10.1186/1471-2407-10-173.
11. Hernandez M, Duplan MN, Berthier G, Vaitilingom M, Hauser W, Freyer R, Pla M, Bertheau Y. Development and comparison of four real-time polymerase chain reaction systems for specific detection and quantification of Zea mays L. J Agr Food Chem. 2004;52(15):4632–7.
12. Hupfer C, Hotzel H, Sachse K, Moreano F, Engel KH. PCR-based quantification of genetically modified Bt maize: single-competitive versus dual-competitive approach. Eur Food Res Technol. 2000;212(1):95–9.
13. Huang HL, Cheng F, Wang RA, Zhang DB, Yang LT. Evaluation of four endogenous reference genes and their real-time PCR assays for common wheat quantification in GMOs detection. PloS ONE. 2013;8(9):e75850.
14. Van Ooyen AJJ, Dekker P, Huang M, Olsthoorn MMA, Jacobs DI, Colussi PA, Taron CH. Heterologous protein production in the yeast Kluyveromyces lactis. Fems Yeast Res. 2006;6(3):381–92.
15. Pan LW, Yuen P, Lin L, Garcia EJ. Flip chip electrical interconnection by selective electroplating and bonding. Microsyst Technol. 2003;10(1):7–10.
16. Akiyama H, Makiyama D, Nakamura K, Sasaki N, Minegishi Y, Mano J, Kitta K, Ozeki Y, Teshima R. A novel detection system for the genetically modified canola (Brassica rapa) line RT73. Anal Chem. 2010;82(23):9909–16.

17. Scarth R, Tang J. Modification of Brassica oil using conventional and transgenic approaches. Crop Sci. 2006;46(3):1225–36.
18. Chaouachi M, Giancola S, Romaniuk M, Laval V, Bertheau Y, Brunel D. A strategy for designing multi-taxa specific reference gene systems. example of application – ppi phosphofructokinase (ppi-PPF) used for the detection and quantification of three taxa: maize (Zea mays), cotton (Gossypium hirsutum) and rice (Oryza sativa). J Agric Food Chem. 2007;55 (20):8003–10.
19. Castillo AR, Gallardo MR, Maciel M, Giordano JM, Conti GA, Gaggiotti MC, Quaino O, Gianni C, Hartnell GF. Effects of feeding rations with genetically modified whole cottonseed to lactating Holstein cows. J Dairy Sci. 2004;87(6):1778–85.
20. Jennings JC, Whetsell AJ, Nicholas NR, Sweeney BM, Klaften MB, Kays SB, Hartnell GF, Lirette RP, Glenn KC. Determining whether transgenic or endogenous plant DNA is detectable in dairy milk or beef organs. Bulletin Of the International Dairy Federation No 383/2003:41–46.
21. Guan QF, Wang XM, Teng D, Yang YL, Tian F, Yin QQ, Wang JH. Construction of a standard reference plasmid for detecting GM cottonseed meal. Appl Biochem Biotechnol. 2011;165(1):24–34.
22. Elenis DS, Kalogianni DP, Glynou K, Ioannou PC, Christopoulos TK. Advances in molecular techniques for the detection and quantification of genetically modified organisms. Anal Bioanal Chem. 2008;392(3):347–54.
23. Wang XM, Teng D, Xi D, Guan QF, Wang JH. Construction of a reference plasmid containing ten targets for the detection of genetically modified crops. Plasmid. 2013;69 (1):108–13.
24. Baeumler S, Wulff D, Tagliani L, Song P. A real-time quantitative PCR detection method specific to widestrike transgenic cotton (event 281-24-236/3006-210-23). J Agr Food Chem. 2006;54(18):6527–34.
25. Zhang L, Wu G, Wu YH, Cao YL, Xiao L, Lu CM. The gene MT3-B can differentiate palm oil from other oil samples. J Agr Food Chem. 2009;57(16):7227–32.
26. Wahler D, Schauser L, Bendiek J, Grohmann L. Next-generation sequencing as a tool for detailed molecular characterisation of genomic insertions and flanking regions in genetically modified plants: a pilot study using a rice event unauthorised in the EU. Food Anal Method. 2013;6(6):1718–27.
27. Wang XH, Jiang DM, Yang DC. Fast-tracking determination of homozygous transgenic lines and transgene stacking using a reliable quantitative real-time PCR assay. Appl Biochem Biotechnol. 2015;175(2):996–1006. doi:10.1007/s12010-014-1322-3.
28. Wang C, Jiang LX, Rao J, Liu YN, Yang LT, Zhang DB. Evaluation of four genes in rice for their suitability as endogenous reference standards in quantitative PCR. J Agr Food Chem. 2010;58(22):11543–7.
29. Takabatake R, Onishi M, Futo S, Minegishi Y, Noguchi A, Nakamura K, Kondo K, Teshima R, Mano J, Kitta K. Comparison of the specificity, stability, and PCR efficiency of six rice endogenous sequences for detection analyses of genetically modified rice. Food Control. 2015;50:949–55.
30. Hernandez M, Esteve T, Pla M. Real-time polymerase chain reaction based assays for quantitative detection of barley, rice, sunflower, and wheat. J Agric Food Chem. 2005;53 (18):7003–9.
31. Wei J, Li F, Guo J, Li X, Xu J, Wu G, Zhang D, Yang L. Collaborative ring trial of the papaya endogenous reference gene and its polymerase chain reaction assays for genetically modified organism analysis. J Agric Food Chem. 2013;61(47):11363–70.
32. Vautrin S, Zhang D. Real-time polymerase chain reaction assay for endogenous reference gene for specific detection and quantification of common wheat-derived DNA (Triticum aestivum L.). J AOAC Int. 2007;90(3):794–801.

33. Kim JH, Park SB, Roh HJ, Park S, Shin MK, Moon GI, Hong JH, Kim HY. A simplified and accurate detection of the genetically modified wheat MON71800 with one calibrator plasmid. Food Chem. 2015;176:1–6.
34. Shang Y, Zhu PY, Huang KL, Liu WH, Tian WY, Luo YB, Xu WT. A peach (Prunus persica)-specific gene, Lhcb2, used as an endogenous reference gene for qualitative and real-time quantitative PCR to detect fruit products. LWT-Food Sci Technol. 2014;55(1):218–23.
35. Yang LT, Pan AH, Jia JW, Ding JY, Chen JX, Cheng H, Zhang CM, Zhang DB. Validation of a tomato-specific gene, LAT52, used as an endogenous reference gene in qualitative and real-time quantitative PCR detection of transgenic tomatoes[J]. J Agric Food Chem. 2005;53 (2):183–90.
36. Xu WT, Bai WB, Guo F, Luo YB, Yuan YF, Huang KL. A papaya-specific gene, papain, used as an endogenous reference gene in qualitative and real-time quantitative PCR detection of transgenic papayas. Eur Food Res Technol. 2008;228(2):301–9.
37. Murugaiah C, Noor ZM, Mastakim M, Bilung LM, Selamat J, Radu S. Meat species identification and Halal authentication analysis using mitochondrial DNA. Meat Sci. 2009;83(1):57–61.
38. Kesmen Z, Gulluce A, Sahin F, Yetim H. Identification of meat species by TaqMan-based real-time PCR assay. Meat Sci. 2009;82(4):444–9.
39. Ibeagha-Awemu EM, Kgwatalala P, Zhao X. A critical analysis of production-associated DNA polymorphisms in the genes of cattle, goat, sheep, and pig. Mamm Genome. 2008;19 (9):591–617.
40. Mane BG, Mendiratta SK, Tiwari AK. Polymerase chain reaction assay for identification of chicken in meat and meat products. Food Chem. 2009;116(3):806–10.
41. Verkaar ELC, Nijman IJ, Boutaga K, Lenstra JA. Differentiation of cattle species in beef by PCR-RFLP of mitochondrial and satellite DNA. Meat Sci. 2002;60(4):365–9.
42. Bai WB, Xu WT, Huang KL, Yuan YF, Cao SC, Luo YB. A novel common primer multiplex PCR (CP-M-PCR) method for the simultaneous detection of meat species. Food Control. 2009;20(4):366–70.
43. Xu WT, Bai WB, Luo YB, Yuan YF, Zhang W, Guo X, Huan KL. A novel common single primer multiplex polymerase chain reaction (CSP-M-PCR) method for the identification of animal species in minced meat. J Sci Food Agric. 2008;88(15):2631–7.
44. Thellin O, Zorzi W, Lakaye B, De Borman B, Coumans B, Hennen G, Grisar T, Igout A, Heinen E. Housekeeping genes as internal standards: use and limits. J Biotechnol. 1999;75 (2–3):291–5.
45. Iskandar HM, Simpson RS, Casu RE, Bonnett GD, Maclean DJ, Manners JM. Comparison of reference genes for quantitative real-time polymerase chain reaction analysis of gene expression. Plant Mol Biol Rep. 2004;22(4):325–37.
46. Nicot N, Hausman JF, Hoffmann L, Evers D. Housekeeping gene selection for real-time RT-PCR normalization in potato during biotic and abiotic stress. J Exp Bot. 2005;56 (421):2907–14.
47. Goldsworthy SM, Goldsworthy TL, Sprankle CS, Butterworth BE. Variation in expression of genes used for normalization of northern blots after induction of cell-proliferation. Cell Prolif. 1993;26(6):511–8.
48. Choi JK, Holtzer S, Chacko SA, Lin ZX, Hoffman RK, Holtzer H. Phorbol Esters selectively and reversibly inhibit a subset of myofibrillar genes responsible for the ongoing differentiation program of chick skeletal myotubes. Mol Cell Biol. 1991;11(9):4473–82.
49. Bjarnason R, Wickelgren R, Hermansson M, Hammarqvist F, Carlsson B, Carlsson LMS. Growth hormone treatment prevents the decrease in insulin-like growth factor I gene expression in patients undergoing abdominal surgery. J Clin Endocr Metab. 1998;83(5):1566–72.
50. Petersen BH, Rapaport R, Henry DP, Huseman C, Moore WV. Effect of treatment with biosynthetic human growth-hormone (Gh) on peripheral-blood lymphocyte populations and function in growth hormone-deficient children. J Clin Endocr Metab. 1990;70(6):1756–60.

51. Marten NW, Burke EJ, Hayden JM, Straus DS. Effect of amino-acid limitation on the expression of 19 genes in rat hepatoma-cells. FASEB J. 1994;8(8):538–44.
52. Lemay S, Mao CC, Singh AK. Cytokine gene expression in the MRL/lpr model of lupus nephritis. Kidney Int. 1996;50(1):85–93.
53. Finnegan MC, Goepel JR, Hancock BW, Goyns MH. Investigation of the expression of housekeeping genes in non-Hodgkin's lymphoma. Leuk Lymphoma. 1993;10(4–5):387–93.
54. Ramos D, Pellin-Carcelen A, Agusti J, Murgui A, Jorda E, Pellin A, Monteagudo C. Deregulation of glyceraldehyde-3-phosphate dehydrogenase expression during tumor progression of human cutaneous melanoma. Anticancer Res. 2015;35(1):439–44.
55. Wang D, Moothart DR, Lowy DR, Qian X. The expression of glyceraldehyde-3-phosphate dehydrogenase associated cell cycle (GACC) genes correlates with cancer stage and poor survival in patients with solid tumors. PLoS ONE. 2013;8(4), e61262.
56. Gunning PW, Ghoshdastider U, Whitaker S, Popp D, Robinson RC. The evolution of compositionally and functionally distinct actin filaments. J Cell Sci. 2015;128(11):2009–19.
57. Hanukoglu I, Tanese N, Fuchs E. Complementary DNA sequence of a human cytoplasmic actin. Interspecies divergence of 3′ non-coding regions. J Mol Biol. 1983;163(4):673–8.
58. Alvarez-Garcia I, Miska EA. MicroRNA functions in animal development and human disease. Development. 2005;132(21):4653–62.
59. Peltier HJ, Latham GJ. Normalization of microRNA expression levels in quantitative RT-PCR assays: identification of suitable reference RNA targets in normal and cancerous human solid tissues. RNA. 2008;14(5):844–52.
60. Lardizabal MN, Nocito AL, Daniele SM, Ornella LA, Palatnik JF, Veggi LM. Reference genes for real-time PCR quantification of MicroRNAs and messenger RNAs in rat models of hepatotoxicity. PloS ONE. 2012;7(5):e36323.
61. Markou A, Tsaroucha EG, Kaklamanis L, Fotinou M, Georgoulias V, Lianidou ES. Prognostic value of mature microRNA-21 and microRNA-205 overexpression in non-small cell lung cancer by quantitative real-time RT-PCR. Clin Chem. 2008;54 (10):1696–704.
62. Vandesompele J, De Preter K, Pattyn F, Poppe B, Van Roy N, De Paepe A, Speleman F. Accurate normalization of real-time quantitative RT-PCR data by geometric averaging of multiple internal control genes. Genom Biol. 2002;3(7).
63. Chang KH, Mestdagh P, Vandesompele J, Kerin MJ, Miller N. MicroRNA expression profiling to identify and validate reference genes for relative quantification in colorectal cancer. BMC Cancer. 2010;10:173.
64. Mi QS, Weiland M, Qi RQ, Gao XH, Poisson LM, Zhou L. Identification of mouse serum miRNA endogenous references by global gene expression profiles. PloS ONE. 2012;7(2): e31278.
65. Timoneda O, Balcells I, Cordoba S, Castello A, Sanchez A. Determination of reference microRNAs for relative quantification in porcine tissues. PloS ONE. 2012;7(9):e44413.
66. Schaefer A, Jung M, Miller K, Lein M, Kristiansen G, Erbersdobler A, Jung K. Suitable reference genes for relative quantification of miRNA expression in prostate cancer. Exp Mol Med. 2010;42(11):749–58.
67. Wotschofsky Z, Meyer HA, Jung M, Fendler A, Wagner I, Stephan C, Busch J, Erbersdobler A, Disch AC, Mollenkopf HJ, Jung K. Reference genes for the relative quantification of microRNAs in renal cell carcinomas and their metastases. Anal Biochem. 2011;417(2):233–41.
68. Huggett J, Dheda K, Bustin S, Zumla A. Real-time RT-PCR normalisation; strategies and considerations. Genes Immunol. 2005;6(4):279–84.
69. De Jonge HJM, Fehrmann RSN, de Bont ESJM, Hofstra RMW, Gerbens F, Kamps WA, de Vries EGE, van der Zee AGJ, Meerman GJT, ter Elst A. Evidence based selection of housekeeping genes. PloS ONE. 2007;2(9):e898.
70. Radonic A, Thulke S, Mackay IM, Landt O, Siegert W, Nitsche A. Guideline to reference gene selection for quantitative real-time PCR. Biochem Bioph Res Co. 2004;313(4):856–62.

71. Glazunova OO, Raoult D, Roux V. Partial sequence comparison of the rpoB, sodA, groEL and gyrB genes within the genus Streptococcus. Int J Syst Evol Micrbiol. 2009;59:2317–22.
72. Clarridge JE. Impact of 16S rRNA gene sequence analysis for identification of bacteria on clinical microbiology and infectious diseases. Clin Microbiol Rev. 2004;17(4):840–5.
73. Naser SM, Thompson FL, Hoste B, Gevers D, Dawyndt P, Vancanneyt M, Swings J. Application of multilocus sequence analysis (MLSA) for rapid identification of Enterococcus species based on rpoA and pheS genes. Microbiol-Sgm. 2005;151:2141–50.
74. Tenaillon O, Skurnik D, Picard B, Denamur E. The population genetics of commensal Escherichia coli. Nat Rev Microbiol. 2010;8(3):207–17.
75. Szczepanek SM, Tulman ER, Gorton TS, Liao X, Lu Z, Zinski J, Aziz F, Frasca S, Kutish GF, Geary SJ. Comparative genomic analyses of attenuated strains of mycoplasma gallisepticum. Infect Immun. 2010;78(4):1760–71.
76. Manukhov IV, Khrul'nova SA, Baranova A, Zavilgelsky GB. Comparative analysis of the lux Operons in Aliivibrio logei KCh1 (a Kamchatka Isolate) and Aliivibrio salmonicida. J Bacteriol. 2011;193(15):3998–4001.
77. Martens M, Dawyndt P, Coopman R, Gillis M, De Vos P, Willems A. Advantages of multilocus sequence analysis for taxonomic studies: a case study using 10 housekeeping genes in the genus Ensifer (including former Sinorhizobium). Int J Syst Evol Microbiol. 2008;58:200–14.
78. Wilson B, Muirhead A, Bazanella M, Huete-Stauffer C, Vezzulli L, Bourne DG. An improved detection and quantification method for the coral pathogen vibrio coralliilyticus. PloS ONE. 2013;8(12):e81800.
79. De la Haba RR, Marquez MC, Papke RT, Ventosa A. Multilocus sequence analysis of the family Halomonadaceae. Int J Syst Evol Microbiol. 2012;62:520–38.
80. Rogall T, Wolters J, Flohr T, Bottger EC. Towards a phylogeny and definition of species at the molecular-level within the genus mycobacterium. Int J Syst Bacteriol. 1990;40 (4):323–30.
81. Dhar AK, Bowers RM, Licon KS, Veazey G, Read B. Validation of reference genes for quantitative measurement of immune gene expression in shrimp. Mol Immunol. 2009;46 (8–9):1688–95.
82. Gerard CJ, Andrejka LM, Macina RA. Mitochondrial ATP synthase 6 as an endogenous control in the quantitative RT-PCR analysis of clinical cancer samples. Mol Diagn. 2000;5 (1):39–46.
83. Savli H, Karadenizli A, Kolayli F, Gundes S, Ozbek U, Vahaboglu H. Expression stability of six housekeeping genes: a proposal for resistance gene quantification studies of Pseudomonas aeruginosa by real-time quantitative RT-PCR. J Med Microbiol. 2003;52(5):403–8.
84. Wu YY, Rees JL. Variation in epidermal housekeeping gene expression in different pathological states. Acta Derm Venereol. 2000;80(1):2–3.
85. Infante C, Matsuoka MP, Asensio E, Canavate JP, Reith M, Manchado M. Selection of housekeeping genes for gene expression studies in larvae from flatfish using real-time PCR. BMC Mol Biol. 2008;9:28.
86. Hu M, Polyak K. Serial analysis of gene expression. Nat Protoc. 2006;1(4):1743–60.
87. Chari R, Lonergan KM, Pikor LA, Coe BP, Zhu CQ, Chan THW, MacAulay CE, Tsao MS, Lam S, Ng RT, Lam WL. A sequence-based approach to identify reference genes for gene expression analysis. BMC Med Genomics. 2010;3:32.
88. Velculescu VE, Zhang L, Vogelstein B, Kinzler KW. Serial analysis of gene-expression. Science. 1995;270(5235):484–7.
89. Yamamoto M, Wakatsuki T, Hada A, Ryo A. Use of serial analysis of gene expression (SAGE) technology. J Immunol Methods. 2001;250(1–2):45–66.
90. Chia CY, Lim CWX, Leong WT, Ling MHT. High expression stability of Microtubule Affinity Regulating Kinase 3 (MARK3) makes it a reliable reference gene. Iubmb Life. 2010;62(3):200–3.

91. Too IH, Ling MH. Signal peptidase complex subunit 1 and hydroxyacyl-CoA dehydrogenase beta subunit are suitable reference genes in human lungs. ISRN Bioinform. 2012;2012:790452.
92. Maccoux LJ, Clements DN, Salway F, Day PJ. Identification of new reference genes for the normalisation of canine osteoarthritic joint tissue transcripts from microarray data. BMC Mol Biol. 2007;8:62. doi:10.1186/1471-2199-8-62.
93. Gur-Dedeoglu B, Konu O, Bozkurt B, Ergul G, Seckin S, Yulug IG. Identification of endogenous reference genes for qRT-PCR analysis in normal matched breast tumor tissues. Oncol Res. 2009;17(8):353–65.
94. Andersen CL, Jensen JL, Orntoft TF. Normalization of real-time quantitative reverse transcription-PCR data: a model-based variance estimation approach to identify genes suited for normalization, applied to bladder and colon cancer data sets. Cancer Res. 2004;64 (15):5245–50.
95. Morga B, Arzul I, Faury N, Renault T. Identification of genes from flat oyster Ostrea edulis as suitable housekeeping genes for quantitative real time PCR. Fish Shellfish Immunol. 2010;29 (6):937–45.
96. Fu J, Bian LH, Zhao L, Dong ZH, Gao X, Luan HF, Sun YJ, Song HF. Identification of genes for normalization of quantitative real-time PCR data in ovarian tissues. Acta Bioch Bioph Sin. 2010;42(8):568–74.
97. Dundas J, Ling M. Reference genes for measuring mRNA expression. Theor Biosci. 2012;131(4):215–23.
98. Smith RD, Ogden CW, Penny MA. Exclusive amplification of cDNA template (EXACT) RT-PCR to avoid amplifying contaminating genomic pseudogenes. Biotechniques. 2001;31 (4):776–8.
99. Yang LA, Takuno S, Waters ER, Gaut BS. Lowly expressed genes in arabidopsis thaliana bear the signature of possible pseudogenization by promoter degradation. Mol Biol Evol. 2011;28(3):1193–203.
100. Pinto FL, Thapper A, Sontheim W, Lindblad P. Analysis of current and alternative phenol based RNA extraction methodologies for cyanobacteria. BMC Mol Biol. 2009;10(1):1–8.
101. Lehmann MH, Weber J, Gastmann O, Sigusch HH. Pseudogene-free amplification of human GAPDH cDNA. Biotechniques. 2002;33(4):766–9.
102. Malik AN, Shahni R, Rodriguez-De-Ledesma A, Laftah A, Cunningham P. Mitochondrial DNA as a non-invasive biomarker: accurate quantification using real time quantitative PCR without co-amplification of pseudogenes and dilution bias. Biochem Bioph Res Co. 2011;412 (1):1–7.
103. Van Guilder HD, Vrana KE, Freeman WM. Twenty-five years of quantitative PCR for gene expression analysis. Biotechniques. 2008;44(5):619–26.
104. Patino WD, Mian OY, Hwang PM. Serial analysis of gene expression – technical considerations and applications to cardiovascular biology. Circ Res. 2002;91(7):565–9.
105. Scaravelli E, Brohee M, Marchelli R, van Hengel AJ. Development of three real-time PCR assays to detect peanut allergen residue in processed food products. Eur Food Res Technol. 2008;227(3):857–69.
106. Schoringhumer K, Redl G, Cichna-Markl M. Development and validation of a duplex real-time PCR method to simultaneously detect potentially allergenic sesame and hazelnut in food. J Agr Food Chem. 2009;57(6):2126–34.
107. Schoringhumer K, Cichna-Markl M. Development of a real-time PCR method to detect potentially allergenic sesame (Sesamum indicum) in food. J Agr Food Chem. 2007;55 (26):10540–7.
108. Hernandez M, Rio A, Esteve T, Prat S, Pla M. A rapeseed-specific gene, Acetyl-CoA carboxylase, can be used as a reference for qualitative and real-time quantitative PCR detection of transgenes from mixed food samples. J Agr Food Chem. 2001;49(8):3622–7.
109. Zhang MH, Gao XJ, Yu YB, Ao JX, Qin J, Yao YH, Li QZ. Detection of Roundup Ready soy in highly processed products by triplex nested PCR. Food Control. 2007;18(10):1277–81.

110. Hellebrand M, Nagy M, Morsel JT. Determination of DNA traces in rapeseed oil. Z Lebensm Unters FA. 1998;206(4):237–42.
111. Bustin SA, Benes V, Nolan T, Pfaffl MW. Quantitative real-time RT-PCR – a perspective. J Mol Endocrinol. 2005;34(3):597–601.
112. Feuer R, Vlaic S, Arlt J, Sawodny O, Dahmen U, Zanger UM, Thomas M. LEMming: a linear error model to normalize parallel quantitative real-time PCR (qPCR) data as an alternative to reference gene based methods. PloS ONE. 2015;10(9):e0135852.
113. Ballester M, Cordon R, Folch JM (2013) DAG expression: high-throughput gene expression analysis of real-time PCR data using standard curves for relative quantification. PloS ONE. 2013;8(11):e80385.

Chapter 6
PCR-Based Technologies for Identifying Unknown Gene Sequences

Abstract Genome walking is a basic molecular biology technique for obtaining unknown sequences. In biology, unknown sequences are usually related to gene regulation, diseases, undiscovered functions of genes, and food safety. The first genome walking method described, inverse PCR, has been established for 28 years. Despite the shortcomings of I-PCR, it has permitted the identification of several previously unknown sequences. To overcome the drawbacks of I-PCR and enhance its efficiency, and in conjunction with recent advances in molecular biology, many genome walking methods have been developed. Essentially, these methods may be classified into the two following categories: those with or those without genome digestion involving restriction enzymes in the first step. According to the features of specific genome walking techniques, these methods are suitable for different types of templates, such as long or short genome walking, single or multiple insertion identification, or simple or complex DNA structures. As the world begins to place greater importance on food security, the future of genome walking is promising, especially in identifying the unintended consequences of GM foods.

Keywords Genome walking • Unknown sequences • Flanking sequences • PCR • DNA sequencing

6.1 Introduction

Unknown gene mining is very important to biology and the human species. Its primary applications can be summarized as follows: (1) obtain the regulatory genes for humans, animals, and plants based on known genes or molecular markers to study the expression regulation of structural genes; (2) obtain the non-conserved region of the new species to assemble complete gene sequences; (3) identify insertion sites of transposons or T-DNA and exogenous genes in transgenic technology; (4) overlap artificial chromosomes (e.g., PAC, YAC, BAC); (5) construct map-based cloning of overlaps; and (6) discover new genes related to diseases [1–5].

Genome walking (chromosome walking) is a molecular procedure for the direct identification of unknown nucleotide sequences from purified genomes. Several genome walking methods have been developed in the last 30 years, with continuous

W. Xu, *Functional Nucleic Acids Detection in Food Safety*,
DOI 10.1007/978-981-10-1618-9_6

improvements added to the first basic strategies, including the recent coupling with next-generation sequencing technologies. According to the basic principle of genome walking, these methods can be classified into two groups. The first category mainly depends on the genomic library, which is suitable for the long length of walking. This type of method makes possible the acquisition of contigs in longer contiguous extents that represent a specific chromosome. The other category is based on the polymerase chain reaction (PCR) as the main technique, and the distance of walking is relatively short. The procedures used with this type of method are simple, which is especially suitable when the sequence flanking the unknown region is clear. For modern biology, the sequences around unknown genes are always known; therefore, genome walking methods based on PCR amplification are the major techniques employed for exploring unknown genes.

The traditional genome walking methods, including inverse PCR, TAIL-PCR, cassette-ligation PCR, and so on, are relatively old technologies. The success rate of these methods is relatively low for several reasons, including the length of the genome, the cleavage sites of the restriction endonuclease which are not presented uniformly, coding and noncoding regions, nonspecific amplification, and so on. Subsequently, novel methods, which overcome the aforementioned disadvantages, have been developed, including A-T linker PCR, loop-linker PCR, and randomly broken fragment PCR with 5′ end-directed adaptor. If the GC content is rich in the unknown region, the defects of these methods become gradually clear. With the development of biotechnology, the cost of sequencing has become significantly reduced to obtain comprehensive molecular information, and next-generation sequencing techniques are evolving rapidly to help identify the flanking sequences of known genes.

6.2 Traditional Genome Walking Methods

6.2.1 Inverse PCR

Inverse PCR (I-PCR) is the basic and earliest genome walking method based on conventional PCR that was developed by Ochman [6]. I-PCR has primers that orient in the reverse direction of their usual orientation. Additionally, the amplification direction of I-PCR is opposite with the common PCR method, hence its named for I-PCR. During the experiment, the restriction enzyme with no cutting site within the known sequence is selected to digest the genomic DNA. Then, the target sequence with a cohesive end is self-looped, and a couple of reverse primers are used for the PCR reaction. The amplified products contain the unknown sequence outside the two primers, which can be analyzed after the vector clone and sequencing steps.

The success of I-PCR depends mainly on the two following factors: one is the choice of the restriction enzyme, and the other is the self-looping rate of the digested fragments. Currently, completed genomes that have been sequenced are few, which has limited the amount of available data for reference when choosing restriction enzymes. Meanwhile, for species with large and complex genomes, the digestion rate cannot be guaranteed, which affects the success rate of I-PCR.

Five model organisms (i.e., *Oryza sativa*, *Escherichia coli* O157:H7 EDL933, *Saccharomyces cerevisiae*, *Homo sapiens* chromosome 1, and *Arabidopsis thaliana*) and 13 common restriction enzymes (BamH I, Bgl I, Bgl II, EcoR I, EcoR V, Hind III, Kpn I, Not I, Pst I, Sac I, Sal I, Xba I, and Xho I) were studied. It was found that most lengths of the digested fragments are focused between 0 and 499 bp. However, the self-looping rate of 500 bp DNA fragments is far less than 1,000–3,000 bp fragments [7]. The above results provide a rationalization of the lower success rate of I-PCR.

Because I-PCR is a straightforward and low-cost method, it is applied in analyzing unknown sequences, such as to obtain the flanking sequence of genetically modified (GM) crops [8, 9], clone puroindoline gene promoters of wheat [10], separate the T-DNA insertions in the *Arabidopsis* genome [11], to name a few. Variations of I-PCR include long-range inverse PCR (LR-IPCR) [12] and bridged inverted PCR [13].

6.2.2 Thermal Asymmetric Interlaced PCR

Liu et al. developed thermal asymmetric interlaced PCR (TAIL-PCR), which is a method based on thermal asymmetric PCR [14]. TAIL-PCR is a semi-specific PCR reaction that refers to a reaction where the unknown sequence is amplified by a series of nested sequence-specific primers and a short arbitrary primer. Because the annealing temperatures of the two primers are different, changing the annealing temperature during the reaction process can control the amplification of specific and nonspecific products.

In TAIL-PCR, the nested specific primers are longer than the arbitrary primers, and the annealing temperatures are also higher. The long primers can anneal with a specific sequence at both high and low temperatures; however, the short arbitrary primers can only anneal with the unknown sequence at low temperature. Target genes are effectively amplified by alternating high and low annealing temperatures. Under the alternated PCR conditions, the nonspecific amplification can be effectively reduced; meanwhile, more than two nested fragments can be produced. Terauchi et al. [5] and Yang et al. [15] isolated the promoter sequences of yams and the flanking sequence of MON863 by TAIL-PCR, respectively. Chaves et al. detected minus strand hepatitis A virus RNA by TAIL-PCR [16], and Chen et al. cloned the PM H+-ATPase gene promoter from wheat genomic DNA [17].

Because of the randomness of the arbitrary primer, the annealing site of the reverse specific primer is limited. For an individual unknown region, positive specific amplicons are rarely acquired, even when different arbitrary primers are used. Hence, establishing the TAIL-PCR conditions requires refinement and a series of consecutive reactions to obtain a specific product. Meanwhile, the nonspecific amplification is another disadvantage of TAIL-PCR, as several reactions are required to obtain the target amplicons.

6.2.3 Cassette-Ligation PCR

In 1992, Isegawa developed cassette-ligation-mediated PCR (CL-PCR) [18]. The first step of this method is identical to I-PCR, that is, the template must be digested by multiple restriction enzymes. Double-stranded DNA with a short length is termed the cassette, which has a cohesive end that can ligate with the digested fragment. The specific primer is designed based on the known sequence, and the other primer is designed based on the cassette sequence. Although the entire digested fragment contains the cassette sequence, only the ligation product that contains the target of the known sequence can be amplified successively. Hence, the amplicons contain three parts of sequence, as follows: the known sequence, the unknown sequence, and the cassette sequence. Sequencing from the cassette primer provides information to design a new primer for the next walking step. Through multiple walking steps, a length of unknown sequence as long as 6 kb can be obtained.

Nthangeni cloned complete nucleotide sequences encoding functional lipolytic genes from *B. licheniformis* and *Bacillus pumilus* [19]. Further, Tamori detected 15 cases of HCC infected with HBV by CL-PCR [20].

The principles of ligation-mediated PCR (LM-PCR) [21] and T linker PCR [22] are similar to CL-PCR, with the function of the adaptor being identical to the cassette. The general drawbacks of these methods are the self-ligation between the cassettes or adaptors, the low ligation rates between the cassettes and digested fragments, the nonspecific amplifications, and the choice of the restriction enzyme. Many improvements were made to CL-PCR based on these traditional methods; meanwhile, some genome walking with novel principles were also developed.

6.3 Novel Genome Walking Methods

6.3.1 A-T Linker PCR

Developed in 2012, A-T linker adaptor PCR has integrated both LM-PCR and T linker PCR [23]. It was first used to detect the T-DNA insertion sites of *Arabidopsis*.

The A-T adaptor includes two parts, long- and short-strand primers. The long-strand primer consists of 40–50 bp and has a high GC content at the 3′ end, which ensures its strong binding with the short-strand primer. The formation of hairpin structures should have low energy and should be avoided at the 3′ end because they could interfere with the binding of the short-strand primer. The short-strand primer consists of 8–15 bp and can generate a T overhang when binding with the long strand. Its 5′ end is phosphorylated, which can increase the efficiency of ligation with the digested fragments. Moreover, an amino C6 moiety was added at the 3′ end as an NH_2 group, on the one hand to prevent the extension of the short-strand primer and on the other hand to avoid generating nonspecific products. The two adaptor primers with high melting temperature were designed according to the sequence of the long-strand adaptor.

The principle of the A-T linker adaptor PCR method is as follows: (1) The genome is digested by restriction enzymes, with the digested fragments containing 3′ overhangs, 5′ overhangs, or blunt ends. The first step of A-T linker PCR is the digestion of the genome, which is identical to I-PCR. Hence, the selection of the restriction enzymes is the most important factor in this method. No restriction site should occur in the region between the specific primer binding site and the border of the known sequence, and the distance between the nearest restriction site and the edge of the border of the known sequence should be appropriate for the PCR amplification. (2) The digested fragments were modified using Taq DNA polymerase or via simple extension to produce an A overhang. (3) The A-T linker adaptor was ligated onto the 3′ end of the target fragment by a T4 DNA ligase. (4) Two rounds of nested PCR were conducted using the specific and adaptor primers to obtain the amplicons that contained the unknown target sequence, and the specificity and sensitivity were effectively increased. After the secondary PCR, the amplification products were analyzed on agarose gel and retrieved with a purification kit, and the products were directly sequenced. This method was shown to successfully detect the T-DNA insertion of 18 *Arabidopsis* mutants, and the insertion sites and flanking sequences were obtained, even in multiple insertions.

The novelty of the approach focuses on the following two points: the addition of an A at the 3′ ends of the restriction fragments and the modification of the adapter. On one hand, the digested DNA was modified at the 3′ end by the utilization of the terminal, non-template extension capacity of Taq DNA polymerases. With this modification, all types of restriction enzyme sites would be converted to an A overhang. This transformation could increase the ligation rate of digested DNA with the adapter by preventing digested DNA from self-ligating and permitting the use of a single T adapter for any restriction enzyme. On the other hand, PO4 and NH_2 groups were added to the 5′ and 3′ ends of the short adapter, respectively. The PO_4 group ensured that the A-T adapter ligated efficiently onto the 3′ end of the genomic DNA, and the NH_2 group blocked elongation at the 3′ end of the short adapter primer. In addition, the adapter does not contain an AP1 primer binding site, and the short adapter primer cannot extend to create an AP1 complementary sequence. Therefore, in the primary PCR, the extension of the AP1 primer must depend on the initial synthesis of the sequence from the specific primer, SP1. For these reasons, the amplification of "adapter–nontarget DNA–adapter" structures does not occur, and only specific products generated from AP1, the adapter primer, and SP1, the specific primer, could be amplified [5]. In conclusion, the addition of both the NH_2 and PO_4 groups is pivotal for efficient amplification. The simultaneous presence of a T overhang in the adapter and an A overhang in the digested DNA enhances specific ligation in this method and prevents digested DNA–DNA or adapter–adapter ligation, as is observed in conventional LM-PCR.

This method overcomes the disadvantage of CL-PCR mentioned above, such as effectively avoiding the self-ligation between the adaptors by appropriate modifications. Moreover, the use of a common adaptor for all restriction enzymes makes the method more flexible and cost-saving. The novel A-T linker adaptor PCR method is versatile, with high specificity, flexibility, and efficiency, and can be used in high-throughput work. This method is also an efficient approach to isolate mutated genes in the field of functional genome research and to isolate flanking sequences for further research in transgenic plants.

6.3.2 *Loop-Linker PCR*

Loop-linker PCR was first developed in 2012 by Trinh to obtain the flanking sequences of genetically modified (GM) crops [24]. The stem-loop, or hairpin loop, is a specific nucleotide structure that can be formed when two regions on the same strand are complementary. Its function is identical to the double stranded adaptor. Several genome walking methods including Vectorette PCR [25], boomerang PCR [26], panhandle PCR [27], and self-formed adaptor PCR [28] have adopted the loop-linker structure.

In this novel method, the loop-linker adaptor is a 52 bp single-strand oligonucleotide with two complementary regions at its 5′ and 3′ ends. At the 5′ end of the sequence, an additional 5 bp oligonucleotide (5′ GATCT-3′ or 5′-AATTT-3′) is used to generate an enzymatic protruding end site and a thymine nick site when the adapter is ligated with restricted DNA. This ligation site includes *Bam*HI (/GATC) or *Eco*RI (/AATT) protruding ends and one thymine that generates a nick when the adapter is ligated with restricted DNA. To ensure the stability of the stem-loop structure, complementary sequences of approximately 8–12 bp were designed. The annealing of the formed-adapter primer can generate either self-complementary or trans-complementary products. To guarantee the efficient ligation between the adaptor and digested DNA, the 5′ end of the adaptor primer is modified with a -PO_4 group.

The nick position in the adapter is beneficial to prevent adapter self-ligation and to form a site for the extension of Taq DNA polymerase. In the PCR process, the initial annealing step in the primary PCR is a key factor that determines the specificity and efficiency of this method. Through extension by Taq DNA polymerase, each ligation product (i.e., adapter-digested DNA-adapter) will form two extension products with panhandle structures. These structures are more stable than the primer-template hybrid and will therefore suppress exponential amplification except from ligation products that contain the specific primer binding site.

As with several genome walking methods, the first step of this method involves DNA digestion. To increase the utilization rate of the linker adapter, enzymes that generate the same protruding ends were selected. The fragments for manipulation could be truncated by using double digestion with more than one restriction enzyme. This step increases the likelihood that the resulting fragments will have a suitable size for manipulation. The 5′ overhang restriction enzymes that can generate the similar binding sites are adopted, and then the loop-linker adaptor is ligated onto the 3′ end of the digested fragment. The ligation products are extended using Taq DNA polymerase to generate a DNA strain with a blunt end. Two rounds of nested PCR are conducted; the first round of nested PCR is performed to amplify the extended fragment using the outer primers, a sequence-specific (SP1) primer and a loop-linker adaptor (LAP1) primer, and the second round of nested PCR is performed to amplify the product of the primary PCR using the inner primers SP2 and LAP2. In loop-linker PCR, the adapter primers LAP1 and LAP2 were designed according to the sequence of the stem-loop structures. These primers were designed to be identical with and have no binding site in the adapter; thus, the extension of LAP1 depends on the initial synthesis of the complementary sequence in the stem-loop region by the specific primer SP1 after the first PCR cycle. The binding site in the adapter primer is generated only after extension, and a high and stringent annealing temperature (Tm) is used for the thermal cycling, effectively limiting nonspecific background amplification [16]. Moreover, similar to boomerang PCR, the specific primers SP1 and SP2 can be used to complete the PCR amplification on their own. However, the efficiency of this pattern is low because the length of the products is double that of the conventional LM-PCR method and the complex structure is generated in the thermal process. By using the LAP1 and LAP2 primers, one half of the PCR product lengths are significantly reduced; accordingly, the efficiency of the amplification is enhanced. Similar to the LM-PCR method, loop-linker PCR only weakly amplifies desired products and suppresses nonspecific amplification in a single round of PCR. Thus, the use of the second round of nested PCR is necessary because it can increase the probability of amplifying the desired flanking sequence and suppressing nonspecific amplification.

To achieve the desired amplification products, the following three aspects should be noted: (1) the formation of the adaptor primer, (2) the concentrations of the pairs of primers, and (3) the choice of the restriction enzyme. Meanwhile, an isocaudamer that generates the same protruding ends could be selected to increase the utilization rate of the linker adaptor.

Similar to the adaptors used in Vectorette PCR, the loop-linker adaptors have a mismatching region in the stem-loop structure that ensures efficient amplification by inhibiting nonspecific amplification. Moreover, the adaptor is designed to form a nick site when it is ligated with digested DNA. This modification could prevent ligation between the adaptors and thus increase the ligation rate. In addition, in the initial extension step of the primary PCR, the loop-linker adaptor extension will generate stem-loop structures from ligation products that can decrease nonspecific amplification and enhance the success rate of this approach.

6.3.3 Randomly Broken Fragment PCR with 5′ End-Directed Adaptor

Randomly broken fragment PCR (RBF-PCR) with 5′ end-directed adaptor was developed by Xu and Shang in 2013 [29]. The first step of this method fragments genomic DNA with the help of an ultrasonic processor and not restriction enzymes, thus overcoming the disadvantages inherent with the use of restriction enzymes.

The 5′ end-directed adaptor was developed in this method. The RBF-PCR adaptors consist of 33 bp that are partially (in 11 bp) reverse complemented with each other. When the fragments are ligated to the adaptors, the adaptors in the 5′ ends of the ligation products are different from those in the 3′ ends (Fig. 6.1). Because the adaptors are not completely reverse complemented, the adaptor primers are designed. The sequence of AP1 is identical with adaptor-1, and the sequence of AP2 is reverse complemented with adaptor-2. There would be four types of amplification styles during the PCR when different pairs of primers are used. After verification, the specificity of GW can be enhanced when we use AP1 with SP, which is also one of the advantages of this new directed adaptor.

The 5′ end-directed adaptor is similar with the Y-adaptors, and Y-adaptors are used in many next-generation sequencing technologies and other methods related to sequencing. The sequences of the Y-adaptors used in next-generation sequencing are unknown and may be protected by patents. In this study, the adaptor sequences are publically available. Meanwhile, the function of the adaptors in this study is directed randomly, which is different from the Y-adaptors.

An ultrasonic processor randomly fragments genomic DNA. Ideally, the length of the broken fragments is moderate. If the length is too short, it is difficult to conduct subsequent modifications, and the genome walking sequence will be short. Conversely, if the length is too long, the subsequent PCR amplifications will be negatively affected. A series of modifications are required to provide the fragments with blunt ends, and an A overhang is generated. Then, the 5′ end-directed adaptor, which can lead to the T/A-specific ligation, are ligated onto both of the 5′ and 3′ ends of the fragments. The self-ligation between the adaptors is effectively avoided by the T overhang on the 3′ end of the adaptor. The partially reverse complementary has been shown to enhance the PCR specificity by real-time PCR and further increase the efficiency of obtaining the unknown sequences. The primer designed on the sequence of the adaptor (AP) is chosen as the fixed primer, and the specific semi-nested primers are the specific primers (SP) designed in the region of the known sequence. Finally, the unknown sequence can be obtained through semi-nested PCR, vector cloning, and sequencing.

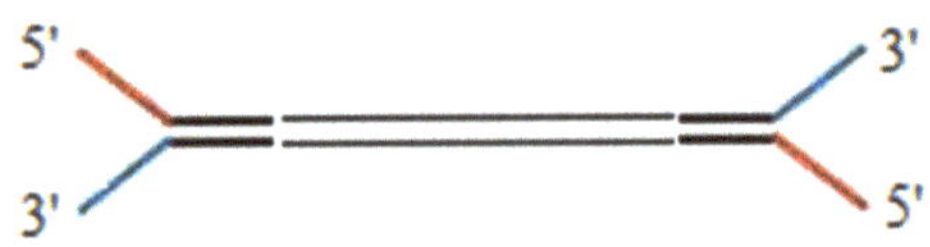

Fig. 6.1 The scheme of the ligation of the adaptor with the DNA fragment

As a novel method, the feasibility of RBF-PCR has been identified. The optimizations were conducted using a series of real-time PCR, including the choice of the adaptor, its specificity, detection limit, identity of amplification between AP/SP and SP/SP, and the ratio of AP to SP. RBF-PCR was applied to obtain the flanking sequence of genetically modified (GM) maize. In this case, the experimental results provided a 588 bp left flanking sequence and a 364 bp right flanking sequence for GM LY038.

This method has been shown to be a very simple but accurate gene walking method that includes the following advantages: (1) all of its technologies are easy to operate, such as end blunting, adaptor ligating, and vector cloning; (2) the genome is broken randomly without dependence on restriction enzymes; (3) a pair of partially reverse complemented adaptors can lead to T/A-specific ligation in the ends of the fragments, and the self-ligation between the adaptors is avoided effectively by a T overhang in the 3′ end; (4) PCR can be conducted through combining different adaptor primers with specific primers; and (5) through cloning the PCR products of interest into vectors and sequencing, the unknown sequence can be efficiently obtained. Directed against specific sequences, especially those with complex secondary or tertiary structures, the digestion of the genome may fail because of restriction endonuclease limitations. RBF-PCR can satisfy the need of genome walking and would function well for the identification of multiplex insertion sites.

6.3.4 Genome Walking with Combined Single-Strand DNA Circularization and Rolling-Circle Amplification

In 2011, Vijay et al. developed a PCR-dependent genome walking protocol that combines single-strand DNA circularization and rolling-circle amplification [30].

Rolling-circle amplification (RCA) is an isothermal amplification method that can replicate circularized oligonucleotide probes with either linear or geometric kinetics. This method generates 10^9 or more copies of each circle in 1.5 h during the DNA strand displacement amplification. Based on its high amplification efficiency, the application of RCA is useful, despite its disadvantages of low reproducibility and poor detection limit as a genome walking method.

In the first step of this method, a phosphorylated primer that is designed based on the known sequence is used to direct a single-stranded extension. Then, the single-stranded amplified products are circularized using CircLigase, which is a thermostable ATP-dependent ligase that catalyzes intramolecular ligation (i.e., circularization) of ssDNA templates. The circularized fragment is taken as the initial template of RCA, with Phi29 DNA polymerase, which initiates the isothermal amplification. Linear DNA amplicons are generated following RCA, which then serve as templates for subsequent I-PCR. The nested primers were designed based on the known sequences, and after two rounds of nested PCR, vector cloning, and sequencing, the unknown sequence was obtained.

This method involves the combination of isothermal amplification with genome walking. The initial template that contains the unknown sequence is enriched, which consequently enhances the success rate of I-PCR. The proposed methodology does not entail laborious testing of restriction enzymes or adaptor ligation. Meanwhile, the length of the known sequence requires only 100–150 bp, which is far less than other genome walking methods.

6.3.5 Identification of Comprehensive Molecular Characterization Based on Next-Generation Sequencing

Molecular characterization includes the flanking sequencing of a target of interest, the copy number, the location in the chromosome, among other details. The basic method to obtain a molecular characterization is genome walking, and the flanking sequence is regarded as the unknown gene. All the methods mentioned previously are appropriate either for one or low-copy insertions or for those with long experimental times. Especially for identifying multiplex unknown insertion sites, such as a GM crop generated by microparticle bombardment, the existing genome walking methods are prone to be inaccurate and incomplete and possess low efficiency.

Next-generation sequencing (NGS) was first applied in 2012 for molecular characterization [31]. With the importance of developing fast and low-cost sequencing technology and the development of informatics, NGS has made possible the sequencing of thousands of genomes. In addition, various computational methods have been developed. The NGS platforms have been applied in many molecular biology fields, such as gene deletions, insertions, inversions, and duplications.

As an example, we consider the molecular characterization of a GM crop. The strategy of the NGS method is to produce DNA sequence fragments that comprehensively cover the entire genome of the GM crop and use bioinformatic tools to analyze these sequencing data. The basic principle compares separately the sequencing reads with the entire genome of its wild type and the exogenous transformation plasmids. The sequencing read, which contains both the exogenous plasmid sequence and the native genomic DNA in the wild type, is the potential insertion site. Primers are designed according to the candidates identified in the reads, the wild type and the GM crop are used as templates, and the PCR amplifications are then conducted. An insertion site is verified if a positive result is shown in the GM crop that is found in the wild type.

In a study involving the application of this method to GM soybean MON17903, the insertion of a single copy of T-DNA was detected at a single genomic locus. An identical result was obtained using Southern blot analysis. Additionally, the NGS method is capable of detecting complex cases of multiple T-DNAs and detailed sequence structure, such as DNA rearrangement and deletion.

By applying similar protocols, the comprehensive molecular characteristics of two GM rice species TT51-1 and T1c-19 were revealed based on paired-end resequencing. Comparing the results with PCR and Southern blotting, the additional unintended insertions were obtained [32]. This method provides an opportunity to identify and characterize such GM events, especially for complex and multiple unknown genes mining. Moreover, this method has the potential to be an effective detection technique solely based on the sequences. Currently, except for obtaining the molecular characterization of GMOs, this technique has been used to identify transposon insertions in maize, obtain long novel sequence insertions and multiple types of genetic variations in the human genome, determine large transcript sequences in *Sesamum indicum*, identify T-DNA insertion loci in activation-tagged *Arabidopsis thaliana*, and sequence the chloroplast genome [33–36].

To sum up, there are primarily two strategies to identify the location of unknown genes, as follows:

1. Sequence alignment with an exogenous insertion vector and then with the entire genome sequence of the same species. In this strategy, the whole-genome sequence of the same species must be complete and published. Further, the strains between the sample and the reference with a known genome sequence are better identical. Take *Zea mays* as an example, it is reported that the DNA content of the unreplicated haploid genome ranges from 2292 to 2716 Mbp/1C (a nearly 20 % difference).

   During the first read alignments, the reads are matched with the sequence of the exogenous insertion vector, leading to three possible situations, as shown in Fig. 6.2.

   Situations A and B contain a potential insertion site of the unknown gene. For A, the insertion site is laid on the partially matched read; for B, the insertion site is laid between the two reads in one pair. The verification of the insertion site is usually conducted by PCR.
2. Direct sequence alignment with the whole-genome sequence of the same species. Regarding the sequencing reads that were aligned with the whole sequence reference genome, if the reads of the sample and its wild type directly match the whole-genome sequence of the same species, then differences would exist. First, the sequencing reads of the sample and its wild type were separately

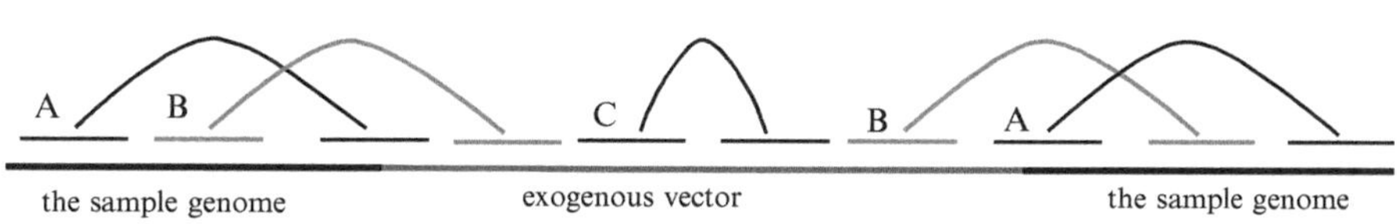

Fig. 6.2 The read alignment results with an exogenous insertion vector. (**a**) a partial sequence of one read in a pair matches the exogenous vector sequence; (**b**) the entire sequence on one read in a pair matches the exogenous vector sequence; (**c**) both of the two reads match the exogenous vector sequence

mapped to the whole-genome sequence; the unmatched reads were collected as set 1 and set 2. Then, sets 1 and 2 were compared, and the potential insertion sites were identified. This strategy is without the help of the exogenous insertion vector sequence, and the analysis procedure is simple. Theoretically, if the genetic backgrounds of the genomes were similar, this method would be appropriate for analyzing the insertion sites, especially when no information is available for detecting unknown GM organisms.

There are multiple advantages to using NGS, such as simplicity and high reproducibility, a decrease in complex lab procedures, and independence from the radioactive ^{32}P-labeled Southern blot technique. Therefore, this method offers a distinct advantage for the supervision and inspection of notable organisms, especially those relevant for food safety and the unintended consequences of GMOs.

In summary, according to whether restriction enzymes are required, genome walking techniques can be classified into two categories. The first category requires the use of restriction enzymes to initially fragment the genome, including I-PCR, CL-PCR, Vectorette PCR, single-specific-primer PCR, template-blocking PCR, RCA-GIP, straight-walk ligation-mediated genome walking [37], A-T linker PCR, loop-linker PCR, and affinity-based genome walking [38]. Compared with traditional methods, the improvements of these methods mainly focus on enhancing the efficiency of ligation between the adaptor and digested fragment or inhibiting nonspecific amplifications. However, if the genome is large or does not have an appropriate restriction endonuclease cutting site, these methods do not work effectively.

The second category of genome walking techniques does not use restriction enzymes and involves primer- and extension-based methods. This category includes site-finding PCR, TAIL-PCR, semi-random primer PCR, and two-step gene walking PCR. The generation of nonspecific DNA products is the main problem with this category, and especially TAIL-PCR, where the specific product is seldom observed following gel electrophoresis. Poor reproducibility is another problem with this category because of the randomness of the primer. RBF-PCR is also included in this category. However, according to the results, RBF-PCR can apparently circumvent the abovementioned drawbacks in an effective manner. The specific categories of the genome walking methods are listed in Table 6.1.

6.4 Prospect

With the development of the aforementioned techniques, genome walking methods are becoming easier, faster, and more accurate. The enormous amount of data obtained by these methods indicates that genome walking, with its large range of applicability, multiplicity of strategies, and recent developments, will continue to

Table 6.1 Categories of genome walking methods

Basic technique	Category		Name
PCR	With restriction enzyme	Inverse PCR	Inverse PCR
		On the base of adaptor or cassette	Cassette-ligation PCR
			Template-blocking PCR
			Vectorette PCR
			Capture PCR
			A-T linker PCR
			Loop-linker PCR
			Single-specific-primer PCR
			RCA-GIP
			Straight-walk ligation-mediated genome walking
	Without restriction enzyme	On the base of extension	TAIL-PCR
			Site-finding PCR
		On the base of primer	Semi-random primer PCR
			Two-step gene walking PCR
			Self-formed adaptor PCR
			Lariat-dependent nested PCR
Next-generation sequencing	The library construction		

have much to offer for the rapid identification of unknown sequences in several fields of genomic research.

Acknowledgments This work is supported by the National Science and Technology Major Project (2016ZX08012-001). Many thanks to Ying Shang, for her kindly help in manuscript conception and preparation.

References

1. Yuanxin Y, Chengcai A, Li L, Jiayu G, Guihong T, Zhangliang C. T-linker-specific ligation PCR (T-linker PCR): an advanced PCR technique for chromosome walking or for isolation of tagged DNA ends. Nucleic Acids Res. 2003;31(12):e68–68.
2. Wahl GM, Lewis KA, Rulz JC, Rothenberg B, Zhao J, Evans GA. Cosmid vectors for rapid genomic walking, restriction mapping, and gene transfer. Proc Natl Acad Sci. 1987;84 (8):2160–4.
3. Berriman M, Ghedin E, Hertz-Fowler C, Blandin G, Renauld H, Bartholomeu DC, Lennard NJ, Caler E, Hamlin NE, Haas B. The genome of the African trypanosome Trypanosoma brucei. Science. 2005;309(5733):416–22.
4. Acevedo JP, Reyes F, Parra LP, Salazar O, Andrews BA, Asenjo JA. Cloning of complete genes for novel hydrolytic enzymes from Antarctic sea water bacteria by use of an improved genome walking technique. J Biotechnol. 2008;133(3):277–86.

5. Terauchi R, Kahl G. Rapid isolation of promoter sequences by TAIL-PCR: the 5′-flanking regions of Pal and Pgi genes from yams (Dioscorea). Mol Gen Genet MGG. 2000;263 (3):554–60.
6. Ochman H, Gerber AS, Hartl DL. Genetic applications of an inverse polymerase chain reaction. Genetics. 1988;120(3):621–3.
7. Shang Y, Zhang N, Zhu P, Luo Y, Huang K, Tian W, Xu W. Restriction enzyme cutting site distribution regularity for DNA looping technology. Gene. 2014;534(2):222–8.
8. Zhang N, Xu W, Bai W, Zhai Z, Luo Y, Yan X, He J, Huang K. Event-specific qualitative and quantitative PCR detection of LY038 maize in mixed samples. Food Control. 2011;22 (8):1287–95.
9. Yang R, Xu W, Luo Y, Guo F, Lu Y, Huang K. Event-specific qualitative and quantitative PCR detection of roundup ready event GT73 based on the 3′-integration junction. Plant Cell Rep. 2007;26(10):1821–31.
10. Digeon J-f, Guiderdoni E, Alary R, Michaux-Ferriere N, Joudrier P, Gautier M-f. Cloning of a wheat puroindoline gene promoter by IPCR and analysis of promoter regions required for tissue-specific expression in transgenic rice seeds. Plant Mol Biol. 1999;39(6):1101–12.
11. Benkel BF, Fong Y. Long range-inverse PCR (LR-IPCR): extending the useful range of inverse PCR. Genet Anal: Biomol Eng. 1996;13(5):123–7.
12. Ponce MR, Quesada V, Micol JL. Rapid discrimination of sequences flanking and within T-DNA insertions in the Arabidopsis genome. Plant J. 1998;14(4):497–501.
13. Kohda T, Taira K. A simple and efficient method to determine the terminal sequences of restriction fragments containing known sequences. DNA Res. 2000;7(2):151–5.
14. Liu Y-G, Whittier RF. Thermal asymmetric interlaced PCR: automatable amplification and sequencing of insert end fragments from P1 and YAC clones for chromosome walking. Genomics. 1995;25(3):674–81.
15. Yang L, Xu S, Pan A, Yin C, Zhang K, Wang Z, Zhou Z, Zhang D. Event specific qualitative and quantitative polymerase chain reaction detection of genetically modified MON863 maize based on the 5′-transgene integration sequence. J Agric Food Chem. 2005;53(24):9312–8.
16. Chaves RL, Graff J, Normann A, Flehmig B. Specific detection of minus strand hepatitis A virus RNA by Tail-PCR following reverse transcription. Nucleic Acids Res. 1994;22 (10):1919.
17. Chen J-Y, Shen X-l, Xin Y-R, Chen X-J. Application of the modified TAIL-PCR technique in cloning of PM H~+-ATPase gene promoter from wheat genomic DNA [J]. J Henan Agric Univ. 2008;1:000.
18. Isegawa Y, Sheng J, Sokawa Y, Yamanishi K, Nakagomi O, Ueda S. Selective amplification of cDNA sequence from total RNA by cassette-ligation mediated polymerase chain reaction (PCR): application to sequencing 6 5 kb genome segment of hantavirus strain B-1. Mol Cell Probes. 1992;6(6):467–75.
19. Nthangeni MB, Ramagoma F, Tlou MG, Litthauer D. Development of a versatile cassette for directional genome walking using cassette ligation-mediated PCR and its application in the cloning of complete lipolytic genes from Bacillus species. J Microbiol Methods. 2005;61 (2):225–34.
20. Tamori A, Yamanishi Y, Kawashima S, Kanehisa M, Enomoto M, Tanaka H, Kubo S, Shiomi S, Nishiguchi S. Alteration of gene expression in human hepatocellular carcinoma with integrated hepatitis B virus DNA. Clin Cancer Res. 2005;11(16):5821–6.
21. Pfeifer GP, Steigerwald SD, Mueller PR, Wold B, Riggs AD. Genomic sequencing and methylation analysis by ligation mediated PCR. Science. 1989;246(4931):810–3.
22. Tonooka Y, Fujishima M. Comparison and critical evaluation of PCR-mediated methods to walk along the sequence of genomic DNA. Appl Microbiol Biotechnol. 2009;85(1):37–43.
23. Trinh Q, Xu W, Shi H, Luo Y, Huang K. An AT linker adapter polymerase chain reaction method for chromosome walking without restriction site cloning bias. Anal Biochem. 2012;425(1):62–7.

24. Trinh Q, Shi H, Xu W, Hao J, Luo Y, Huang K. Loop-linker PCR: an advanced PCR technique for genome walking. IUBMB Life. 2012;64(10):841–5.
25. Arnold C, Hodgson I. Vectorette PCR: a novel approach to genomic walking. Genome Res. 1991;1(1):39–42.
26. Hengen PN. Vectorette, splinkerette and boomerang DNA amplification. Trends Biochem Sci. 1995;20(9):372–3.
27. Riley J, Butler R, Ogilvie D, Finniear R, Jenner D, Powell S, Anand R, Smith J, Markham A. A novel, rapid method for the isolation of terminal sequences from yeast artificial chromosome (YAC) clones. Nucleic Acids Res. 1990;18(10):2887–90.
28. Wang S, He J, Cui Z, Li S. Self-formed adaptor PCR: a simple and efficient method for chromosome walking. Appl Environ Microbiol. 2007;73(15):5048–51.
29. Xu W, Shang Y, Zhu P, Zhai Z, He J, Huang K, Luo Y. Randomly broken fragment PCR with 5′ end-directed adaptor for genome walking. Scientific reports 3; 2013.
30. Gadkar VJ, Filion M. A novel method to perform genomic walks using a combination of single strand DNA circularization and rolling circle amplification. J Microbiol Methods. 2011;87 (1):38–43.
31. Kovalic D, Garnaat C, Guo L, Yan Y, Groat J, Silvanovich A, Ralston L, Huang M, Tian Q, Christian A. The use of next generation sequencing and junction sequence analysis bioinformatics to achieve molecular characterization of crops improved through modern biotechnology. Plant Genome. 2012;5(3):149–63.
32. Yang L, Wang C, Holst-Jensen A, Morisset D, Lin Y, Zhang D. Characterization of GM events by insert knowledge adapted re-sequencing approaches. Scientific reports 3; 2013.
33. Williams-Carrier R, Stiffler N, Belcher S, Kroeger T, Stern DB, Monde RA, Coalter R, Barkan A. Use of Illumina sequencing to identify transposon insertions underlying mutant phenotypes in high-copy Mutator lines of maize. Plant J. 2010;63(1):167–77.
34. Studer A, Zhao Q, Ross-Ibarra J, Doebley J. Identification of a functional transposon insertion in the maize domestication gene tb1. Nat Genet. 2011;43(11):1160–3.
35. Hajirasouliha I, Hormozdiari F, Alkan C, Kidd JM, Birol I, Eichler EE, Sahinalp SC. Detection and characterization of novel sequence insertions using paired-end next-generation sequencing. Bioinformatics. 2010;26(10):1277–83.
36. Polko JK, Temanni M-R, van Zanten M, van Workum W, Iburg S, Pierik R, Voesenek LA, Peeters AJ. Illumina sequencing technology as a method of identifying T-DNA insertion loci in activation-tagged Arabidopsis thaliana plants. Mol Plant. 2012;5(4):948–50.
37. Leoni C, Volpicella M, De Leo F, Gallerani R, Ceci LR. Genome walking in eukaryotes. FEBS J. 2011;278(21):3953–77.
38. Thirulogachandar V, Pandey P, Vaishnavi C, Reddy MK. An affinity-based genome walking method to find transgene integration loci in transgenic genome. Anal Biochem. 2011;416 (2):196–201.

Chapter 7
A Developed Accurate Digital PCR Detection Technology in Food Safety

Abstract With the development of the quantitative PCR detection methodologies, different types of PCR techniques have been developed, such as real-time quantitative PCR, competitive real-time PCR, etc. However, all the above techniques only achieve quantitative detection by standard curves and are considered labor-intensive experiments. Based on this background, digital PCR (dPCR), an absolute quantitative detection method, was developed in the 1990s and could achieve quantitative detection without the generation of standard curves. The idea of digital PCR was first raised by Kinzler and Vogelstein in the 1990s based on the limited separation of the original PCR reaction. Digital PCR has the advantage of being highly sensitive, specific, and well stable and having absolute quantitation compared to the original real-time PCR. Over nearly three decades, digital PCR has achieved wide usage in different areas, such as clinical diagnostics, GMO detections, biosensor detections, etc. In this chapter, the background of the development of digital PCR, the concept of digital PCR, and the use of digital PCR in different areas are reviewed in detail. By reading this chapter, the reader could have a better comprehension of all the aspects of digital PCR.

Keywords Digital PCR • Absolute quantification • Viral detection • Diagnostic detection • GMO detection

7.1 Background of the Development of Digital PCR

The quantitative detection of all types of samples heavily relies on quantitative polymerase chain reaction (qPCR) to detect and quantify the target sequence. In the past 20 years, based on nucleic acid chemistry, real-time PCR has revolutionized the diagnosis and became the gold standard for the quantification and detection of various samples [1], such as clinical samples, samples of genetically modified organisms, bacterial samples, and others. During quantitative PCR, DNA is amplified until it generates a signal that is embedded in a dye or a sequence-specific fluorescent probe. The period threshold is defined as the number of signals required to achieve the desired level of amplification cycles and the number of DNA molecules for the standard curve based on the presence of the original [2].

W. Xu, *Functional Nucleic Acids Detection in Food Safety*,
DOI 10.1007/978-981-10-1618-9_7

Although quantitative PCR has led to significant progress in diagnosis, the technology has obvious limitations. Quantification is based on the standard curve, which requires careful calibration and the same source material. In addition, the signal selection threshold may be adjusted by the operator and is introduced into the main analysis. Due to changes in the structure of the standard curve and potential differences, differences in subjective analysis between commercial kits and standardized protocols can be enormous. Furthermore, even in a trained laboratory, the variation coefficient of any single measurement may be 20–30 % or more of the lower template copy number [3, 4].

Real-time PCR method for the determination of genetically modified, real-time quantitative polymerase chain reaction (quantitative PCR) is the most widely accepted quantitative methods to detect and identify genetically modified organisms. While it is true that this technology can have excellent sensitivity and specificity, along with high reproducibility and accuracy, it is necessary to understand qPCR analysis through many and divergent protocols, as well as different instruments, enzymes, liquid buffers, and nonidentical goals [5].

For the reliable quantitative analysis of GMOs, a laboratory requires appropriate organization and quality management system, and several key points should be considered in the analysis of the extraction process, such as sampling, sample preparation, and DNA. The number of new GMOs is increasing rapidly, and laboratories need to introduce new methods within their scope of certification in a timely manner. To promote a unified Europe-wide recognition of a flexible scope, technical guidance documents are currently limited to quantified GMO laboratories and focus on flexibility related to products, activities, and programs at different levels of analysis [6]. The content of genetically modified organisms (i.e., quantization control) and additional independent samples can be analyzed in parallel with the unknown to confirm the appropriate analytical methods and calculation results and to generate control charts.

The analysis of each step may introduce a certain degree of uncertainty to the final data and the interpretation. In such measurements, the uncertainty of the test data can be used together with internal quality control data and guidance documents to describe the uncertainty of measurements in GMO testing laboratory estimates [7]. Measurement uncertainty and analysis of the results with appropriately estimated uncertainty are crucial to determine the compliance of test samples. The uncertainty value obtained by subtracting the expanded, reported GM content uncertainty has been proposed as an assessment of compliance. For a test sample not to conform to the requirements, the concentration of the analyte should be above the legal threshold without any reasonable doubt.

To address the drawbacks of real-time PCR that are discussed above, digital PCR was developed and can achieve absolute quantitation by meeting all the different demands of detection. Digital PCR (dPCR) is expected to make up for some deficiencies of qPCR by transforming the exponential, analog nature of the PCR into a digital and linear signal [8].

Originally, emulsion PCR was regarded as the main precursor of digital PCR and an improved version of the original PCR process. Emulsion PCR uses a large number of droplets to separate the reaction into many small volumes [9–14]. By separating the original amplification volume and producing a water-in-oil (W/O) structure, this method balances amplification efficiency with avoiding substrate effects caused by PCR tubes [11]. Thus, emulsion PCR is more suitable for the amplification of samples with low-abundance targets or complex substrates. Moreover, because of the generation of a W/O structure, the amplification mixture and templates are also divided. The distribution of materials follows a Poisson distribution [15]. By combining a flow instrument with fluorescence detection, the fluorescence of each droplet can be detected. After the droplets are read, positive and negative droplets can be recognized in a final heat map. The copy number of the target sequence can be calculated using the Poisson distribution by considering the ratio of positive to negative droplets. Emulsion PCR has developed rapidly since it was first introduced. Different applications, such as the amplification of gene libraries [11, 12, 16] and single-molecule detection [10, 17], have been devised to take advantage of the balanced amplification efficiency of the droplets. Because the division of targets follows the Poisson distribution, quantitative detection can be achieved using fluorescence detection in flow instruments. This quantitative detection method can be regarded as an important alternative to quantifying targets using real-time PCR and standard curves.

7.2 The Basic Principle of Digital PCR

"Digital PCR" was proposed in 1999 by Kinzler and Vogelstein [8], although the conceptual frameworks of limiting the dilution of DNA and single-molecule detections had been reported previously [18, 19]. The origin and principles of digital PCR have been extensively reviewed in other studies [8, 20–22], but the concept remains relatively poorly understood, and we briefly review the principles here.

The success of digital PCR depends on to detection of the target locus of individual molecules. The sample is greatly diluted and aliquoted into a number of droplets; some aliquots are "positive" with at least one target molecule, while others are not. The number of positive aliquots is determined by PCR.

If the sample is sufficiently diluted, only a few aliquots will be positive, and each of these positive aliquots may be assumed to contain only a single target molecule. In this case, the process is equivalent to direct and simple molecule counting, representing the "digital" component of "digital PCR."

However, more often, the sample is not considerably diluted. In this case, many, but not all, droplets will be considered positive, and some of the positive aliquots will probably (from the perspective of the uninformed experimenter) contain two, three, or more target molecules. Therefore, a simple calculation may underestimate the true number of molecules. The number of molecules can be estimated using confidence intervals and the Poisson equation, $A = -\log_e(1 - P)$, wherein A is the average number of molecules per aliquot and *P* is the proportion of positive aliquots.

In this way, it is easy to calculate the absolute abundance of a target sequence in a sample [8, 19]. More commonly, the abundance of a target sequence is compared with that of a reference sequence analysis to determine the relative abundance of the target [8, 23]. The reference sequence is usually chosen such that the abundance is known, e.g., a sequence that is present in two copies per diploid cell.

In addition to the accurate quantification of target sequences, digital PCR can also be used to identify rare variants, such as those that arise from DNA mutations and exist in only a few isolated cells. Digital PCR experiments can be designed to yield amplified results from aliquots containing a single or a few template molecules. In aliquots containing a rare variant, the probe is much more common with respect to the variant than it would be in most DNA amplification processes [24, 25]. In practice, rare variant detection requires a non-template control verification and well-designed experimental program; however, digital PCR offers notable potential sensitivity for detecting and quantifying rare variants.

The precision (reproducibility) of digital PCR-based quantitation and its capacity to detect very rare variants depend on the total number of aliquots that are interrogated; precision and sensitivity increase as more aliquots are analyzed.

The degree of sample dilution is also important. If the sample is too dilute, then very few aliquots will be positive, and the data will not be reliable. If the sample is not diluted enough, then all of the aliquots will be positive; however, quantitative information could be obtained. Many authors have studied linear responses to a variety of platforms for the determination of DNA concentration [26–28]. A recent study used a droplet digital PCR (ddPCR) system and 20,000 partition and reaction droplets; a linear response to DNA concentration was obtained from 0.16 to 99.6 % saturation [27]. However, the influence of random events increased copy number estimation due to the relative uncertainty in the dynamic range, especially at the lower limit of the range. A study using microfluidic-based methods has made similar observations [28].

Uniform partition volume is also regarded as a key factor in estimation accuracy. The number of partitions becomes particularly important when it is over 1000 [27].

In contrast with real-time PCR, in digital PCR, if the efficiency of the reaction reaches a threshold, then any product present will be detected. As a result, whether a response is more effective than another may not be important, as long as both are fully amplified. This feature increases the likelihood of biomarker detection for the purpose of primer design.

7.3 The Commercialization of Digital PCR

The process of single-molecule amplification must take place in a highly stable environment to transform the powerful potential of dPCR into efficient platforms. The choice of platform depends mainly on the trade-offs between precision, throughput, and system and assay costs.

The first generation of dPCR platforms was based on chambers containing microfluidic channels, such as those made by Fluidigm [29], and plates with hydrophilic and hydrophobic surfaces, such as those produced by Life Technologies [30]. The target molecule is monitored in real-time conditions; thus, false-positive reactions could be inferred from the amplification curve of each reaction. With the demand for higher sensitivity and precision, platforms providing many more digital partitions were needed. Hence, companies have launched systems where each dPCR reaction takes place in aqueous droplets in oil coupled with end-point PCR readings. Although these platforms cannot perform real-time PCR measurements, the immense number of uniformly sized reaction partitions led to a dramatic extension of the dynamic range. Twenty thousand reaction droplets could be generated per reaction by the platform developed by Bio-Rad Laboratories, which provides quantitative measurement across four orders of magnitude [25, 27]. Furthermore, millions of reaction partitions could be handled by the system produced by RainDance Technologies, further expanding the dynamic range [31]. The BEAMing digital PCR technology (beads, emulsion, amplification, and magnetics) provided by Sysmex Inostics [32, 33] clonally amplifies nucleic acids in the presence of magnetic particles and quantifies them using flow cytometry.

7.3.1 The BioMark System by Fluidigm

The BioMark system for chamber-based digital PCR was developed by Fluidigm in the 2000s. This platform features microfluidic channels within the chambers that separate the partitions of this platform. With the BioMark system, the fluorescence signal is collected in real time to generate the final amplification curve of the target sequence. Thus, both the amplification curves and the results could be collected according to the final heat maps. Compared with other digital PCR platforms, this chamber-based digital PCR platform by Fluidigm is easier to operate and requires fewer liquid transfer and amplification curve generation steps. With these above advantages, the BioMark chamber-based digital PCR platform has facilitated many advances in many applications.

7.3.2 The QX100/200 System by Bio-Rad

The Bio-Rad QX100/200 digital PCR platform is regarded as the earliest of all droplet digital PCR platforms. This platform mainly consists two parts, the droplet generation facilities and the droplet reading platforms. PCR amplification can be achieved by many commercially available PCR platforms. However, compared with other platforms, this droplet digital PCR platform developed by Bio-Rad was more cost efficient.

7.3.3 *The RainDrop System by RainDance*

The RainDrop droplet digital PCR system was developed recently by RainDance. It is the latest generation of droplet digital PCR systems and can be regarded as an improved version of the Bio-Rad platforms. Compared with the QX platform from Bio-Rad, the RainDrop can generate many more droplets, i.e., 2,000,000 compared with the 20,000 generated by the QX platform. However, the cost per reaction is more expensive than that of the QX platform. With the largest number of generated droplets, the RainDrop platform can achieve the widest dynamic range among all the digital PCR platforms.

7.4 Comparing Digital PCR with Real-Time PCR

Digital PCR is considered to be an improved version of real-time PCR. Therefore, digital PCR has surpassed real-time PCR in many areas. In this section, we compare the advantages and disadvantages of digital PCR and real-time PCR.

7.4.1 *dPCR Exhibits Superior Precision to That of qPCR*

Typical clinical virus PCR experiments can be highly imprecise, especially at lower template copy numbers. Even the coefficients of variation (CVs, inaccurate measurements) of highly specialized laboratories may be 20–30 % or higher [3, 4, 34, 35].

One advantage of dPCR over qPCR in practical detection is the potential for increased precision both within and between runs. This feature has been demonstrated by many studies comparing digital and real-time PCR. In a study by Hindson, ddPCR consistently displayed lower variation than qPCR for a set of synthetic miRNAs. The mean ddPCR CVs were 37–86 % than those of qPCR for overall variation and 48–72 % lower for PCR-specific variation [36].

The increased precision provided by ddPCR can not only be an absolute measure to improve the reproducibility of the data from a single weight but may also be a determination of the ratio. One application enabled by this increase in the proportion of diagnostic accuracy is a ddPCR assay to identify chromosomally integrated human herpesvirus 6 (ciHHV-6), which can be transmitted through the germ line and is present in approximately 1 % of the population [37, 38].

In addition to increased precision, or repeatability, some data suggest that ddPCR may also be more reproducible (i.e., has less day-to-day variation). A comparison of ddPCR and qPCR miRNA quantitation determined that ddPCR exhibited a day-to-day reproducibility greater by a factor of 7. Similarly, a comparison of ddPCR and qPCR assays for CMV showed lower interassay (between-run) CVs for ddPCR than for qPCR across multiple dilutions of a highly positive patient sample down to 150 copies/mL.

7.4.2 dPCR May Standardize Quantitation

dPCR has numerous internal and cross-use clinical uses, and laboratory standardization has increasingly developed its potential. Interoperability between laboratories provides meaningful load values, and research in different testing laboratories has established reference viral load treatments for the management of key issues [39].

7.4.3 dPCR and qPCR Have Comparable Sensitivity

Although recent work has shown the advantages of dPCR accuracy over that of quantitative PCR, the sensitivity of comparative data is not so clear. To date, no studies have shown dPCR to be more or less sensitive than quantitative PCR when the same amount input of DNA is measured. In comparing ddPCR research and quantitative PCR for detecting cytomegalovirus, we emphasize the fact that any amplified nucleic acid sensitivity is mainly determined by the input amount or concentration of the reaction. Two laboratories that developed a cytomegalovirus detection method performed a comparison showing that quantitative PCR had sensitivity tenfold higher than that of ddPCR. However, in this study, four times the volume of the DNA used in the ddPCR measurement (5 μl) was used in the quantitative PCR (20 μl), which accounts for much of the difference between the limits of detection [40].

One way to maximize PCR sensitivity is to "merge" holes or droplets and analyze them as one metawell, thereby increasing the DNA template per a certain sample amount. This method can be used in quantitative PCR, but not in dPCR, because each reaction would be counted as a positive or negative digital output, rather than analog output [41].

7.4.4 dPCR Is Resistant to Inhibition

dPCR also offers advantages over qPCR such as increased tolerance to -PCR-inhibitory substances [41, 42]. In the clinical setting, Dingle et al. demonstrated that ddPCR was more refractory to inhibition than qPCR by testing substances such as heparin and SDS, which are common inhibitors to DNA extraction of clinical samples [43]. Greater than a half log increase in IC_{50} was observed for ddPCR compared with qPCR when heparin and SDS were spiked into CMV assays.

7.4.5 dPCR Is Less Affected by Target Sequence Variability

Quantitation by ddPCR may be less affected than qPCR by incorrect sequences for the same reason; partial inhibition is less of a problem in ddPCR than in qPCR. When the template DNA and primers or probe sets that do not match among amplified positive droplets are not completely abolished, they are still detected, although at a lower amplitude (represented by a lower reaction efficiency). These lower-amplitude drops are still counted as positive, resulting in accurate quantification. However, in quantitative PCR, amplification efficiency is low; efficient response using an incorrect standard curve Ct value corresponding to the value will lead to misquantification.

7.4.6 Current Drawbacks of dPCR over qPCR

dPCR is a promising new commercial amplification method that has several potential advantages over using qPCR to molecularly diagnose viruses. However, dPCR is still an emerging technology that has only recently been applied for actual detection. Therefore, several problems with dPCR remain to be overcome as laboratories gain experience with this technology, and companies can still make improvements to their systems for this high-throughput molecular method.

1. While dPCR bypasses the need for a traditional standard curve, the lack of a standard curve for quantitation does not automatically guarantee better accuracy. The accurate definition of a gold standard or calibration material is still required to ensure interoperability between molecular diagnostic laboratories.
2. Performance differences of dPCR systems, with a particular restriction on the DNA input and microfluidic functions of the system, define different dynamic ranges and sensitivity levels. This may cause inconvenient analysis of data from different dPCR platforms.
3. dPCR systems have a linear dynamic range logs lower than that of qPCR systems, depending on the number of partitions utilized in each particular system.
4. Currently, many dPCR systems are "open systems" that risk the pollution of the operation and carry-over from run to run or from transfers. For example, in Bio-Rad's droplet systems, the greatest need is to mix and transfer to a droplet-forming box and then transfer it to a 96-well plate that had been sealed prior to the reaction, thermal cycling, and pipetting.
5. dPCR is currently more expensive than qPCR.

7.5 New Developments of Digital PCR

7.5.1 *The Development of Digital PCR for Clinical Samples*

7.5.1.1 The Development of dPCR for the Detection of Circulating Nucleic Acids

Rare Variant Detection

A tumor-derived trace associated with CNA has been needed as an extremely sensitive detection method for the early stages of tumor formation and release into the bloodstream. A great benefit of cancer research has been the ability to use dPCR to trace scarce genetic material. By valid nucleic acid partition and reaction partition counts from the thousands to the millions, one can detect the presence of tumor-derived mutants or variants at extremely low concentrations. dPCR analysis capability allows the testing of every 100,000 wild-type sequences for single-nucleotide mutations [44].

Gene Copy Number Variation Analysis

Copy number variations (CNVs) are regarded as the most prevalent cause of structural DNA variability that is responsible for human genomic diversity [45]. At the molecular level, CNVs are composed of gene amplifications or deletions, and genetic variability is often observed in the process of tumorigenesis [46].

Potential dPCR techniques have not been widely used for exploring copy number variation in circulating nucleic acids in plasma. However, previous studies have shown that dPCR would be an ideal molecular tool in this research field [47–49]. Investigations on breast cancer cell lines have shown superior precision and sensitivity of dPCR compared with qPCR in the detection of subtle differences in HER2 gene copy number variation [47]. Quantitative real-time PCR was used to reveal the presence of HER2 gene amplification in the plasma of patients with breast cancer [49], thus providing an opportunity to refine the detection using the digital single-molecule counting strategy [50]. The use of nonspecific, double-stranded binding dye in dPCR similarly provides a highly sensitive and specific quantification of CNVs in a multiplex format simply by adjusting the size of the targeted region [51].

Gene Expression Analysis

Several studies have reported that miRNAs have a high potential to be blood-based biomarkers in cancer detection [52], as well as in prenatal testing [53–55]. Ma et al. reported the use of dPCR as a potential tool for the quantitative assessment of miRNA in the plasma of lung cancer patients by observing significantly higher copy numbers of miRNA compared to controls [56]. Similarly, encouraging results were observed in a study focused on the measurement of circulating lncRNA expression in human serum [57]. Nevertheless, as stated in the Minimum Information for Publication of Quantitative Digital PCR Experiments (dMIQE) guidelines, the quantitative analysis of RNA using RT-dPCR still requires careful consideration of the experimental design and final data reporting [58].

Analysis of Methylation Loci

Digital PCR has been used to quantitatively evaluate the methylation of specific loci in the plasma of patients with different cancers, such as colorectal cancer [59] and breast cancer [60]. Li et al. developed a methyl-BEAMing technology that increases the ability to detect curable, early-stage colorectal cancers with the same precision as currently used next-generation sequencing platforms [59]. Additionally, a method called microfluidic digital MethyLight has been successfully tested for cancer-specific DNA hypermethylation events present in the plasma of breast cancer patients [60].

7.5.1.2 The Development of dPCR for Viral Detection

Absolute Quantitation

Digital polymerase chain reaction provides a sensitive method for the direct measurement of viral nucleic acids, providing an absolute number of copies per mL/mL without standard curves. White et al. [61] utilized the Fluidigm detection system for the quantification of GB virus type C (GBV-C), which is regarded as a mysterious and viral RNA HIV-1 infection. GBV-C infection in patients with HIV-1 has been shown to lead to a reduction in the progression of AIDS [62].

Another comparison of viral qPCR and dPCR was taken by Henrich et al. [63] on HIV-1 quantitation. They found that the dilution of human DNA CCR5 amplified HIV-1 or standard droplet number PCR (mRNA, Bio-Rad) matched the expected nominal copy number. When they ran the same patient specimens, they found that mRNA and qPCR had similar sensitivities, but the number of copies of the formula was 10–40 % less than that of qPCR and DNA. Although the reasons for this gap are speculative, the authors provide a quantitative analysis of the errors that can be measured by the standard DNA concentration spectrophotometric method [63].

A study by Kiss et al. [64] has been validated by the RainDance dPCR platform for sensitive, absolute quantification. The system uses millions of tiny liquid droplets to produce millions of separate, positive or negative PCR outputs. They detected an adenovirus at levels as low as 1 template molecule/167 droplets, or 92 molecules/μL [64].

Absolute virus quantification can be accomplished not only by commercially available testing systems but also by a simple system that is suitable for medical points of care and resource-limited environments. Shen et al. [65] used a SlipChip system to demonstrate that the absolute quantitation of HIV and hepatitis C virus (HCV) can be achieved over a large dynamic range. They used a rotating and sliding micro platform and manipulated liquid samples on the microliter to picoliter scale through varying the relative motion of the plates without the need for a complex control system. They validated their detection with viral RNA from two HIV patients. Their measured values were well correlated with clinical standards, achieving a dynamic range of threefold (0.5 $\log_{10}$) resolution from 1.7×10^2 to 2.0×10^7 molecules/mL and a low detection limit of 40 molecules/mL. They have also validated a multiplex SlipChip with a 5-plex panel to detect HIV and HCV simultaneously along with a negative and positive control and a dynamic range of 1.8×10^3 to 1.2×10^7 molecules/mL.

Clinical diagnosis using dPCR virus detection is still in the initial stages. However, these preliminary studies show that rapid and sensitive detection is possible for potential clinical applications, such as the accurate quantitation of viral loads in patient samples.

Rare Mutant Detection

Another important application of viral diagnostic testing, which has been the most studied so far in the field of oncology, is the detection of a rare point mutation in a wild-type sequence. Pekin et al. used a RainDance dPCR platform for a sensitive and quantitative KRAS mutation detection, which is the most common among human cancer types [31]. In tumor diagnosis, the somatic mutations of the tumor gene are used as specific biomarkers to distinguish the difference between cancer cells and normal cells. Current qPCR TaqMan assays and pyrosequencing can identify less than 1/10 % mutant genes in a nonmutated DNA background [31]. Pekin has quantified the KRAS gene at codon 12 and 13 mutations in genomic DNA from several different human cell lines and was able to detect 1 of the mutations in a background of 200,000 wild-type KRAS genes (0.005 % isolates) by analyzing 106 drops [31].

Another example of dPCR is the detection of low abundance of epidermal growth factor receptor (EGFR) in tumor tissue and in plasma by using rare mutation detection [66, 67]. Epidermal growth factor receptor tyrosine kinase inhibitors are involved in the development of some lung cancer types. In response to these inhibitors, the mutant epidermal growth factor receptor kinase domain is activated. In early studies, direct sequencing was usually used, but this technique can only detect the total genetic content of a mutant sequence larger than 30 % of the total DNA [66]. By dPCR testing, they were able to identify the sequence of mutations that could not be detected by conventional sequencing methods. In these samples, the sequence of mutations constituted 2–14 % of the total DNA. Similarly, Wang et al. utilized dPCR (Fluidigm platform) to detect and quantify the effects of drugs (0.02–9.26 %) on low-abundance EGFR mutations of sensitized tumor DNA.

This study used a powerful tool for identifying the low-abundance mutant alleles in a background of high-abundance wild-type sequences. Although these data are concentrated in the diagnosis of tumor, these principles could be translated to virological diagnosis applications of mutations in the detection of low-abundance sequences, such as those mediating antiviral properties and significantly affecting treatment outcome.

7.5.1.3 The Development of dPCR in Personalized Medicine

Pharmacogenetics

In addition to somatic mutation detection, there is evidence that some germline mutations can profoundly affect individual response to some treatments [68, 69]. In some cases, these are point mutations or single-nucleotide polymorphisms that may affect drug metabolism; in other cases, copy number variation at specific loci may be predictive of how individuals are treated with a drug. Improved digital PCR accuracy could become useful in the future, when more data appear in the pharmacological context of CNVs in the reproductive system.

Gene Expression Analysis

Using the method of gene expression analysis in theory and in practice is very simple using digital polymerase chain reaction technology. Dynamic range analysis requires variable expression genes and reference/housekeeping genes that are more suitable for quantitation in comparison with common high-expression transcription loci. The dynamic ranges provided by new digital polymerase chain reaction platforms can support this application, as well as the selection of reference sites.

A potential function of digital polymerase chain reaction analysis may be for transcriptions from single cells or small cell numbers. For the reasons discussed above, digital polymerase chain reaction is particularly suitable when the sample material is limited. The transcriptional analysis of single cells has recently been reported in the literature, especially for cells with large RNA variant levels; the use of qPCR precludes the preamplifier bias possibility present with many steps [70]. With respect to dynamic range, most digital polymerase chain reaction platforms (assuming good experimental designs and reference transcripts are chosen) will have enough partition reactions to provide accurate absolute quantification of specific transcription levels in single cells. However, it remains to be seen whether single-cell expression analysis can be used to develop clinical biomarkers, even if it is optimized.

Recently, great interest has developed in biological noncoding RNA (ncRNA), and microRNAs, in particular, have been extensively researched in the context of human pathology [71, 72]. Microfluidic systems have been successfully used for miRNA analysis in pre-amplified RT-PCR protocols [73]. In principle, microRNA analysis using digital PCR is feasible, as it has been used to quantify other transcripts and to validate and analyze the potential of the discovery platform (NGS) method.

Methylation-Specific Digital PCR

Changes in the epigenetic regulation of gene expression occur in multiple disease states, especially in cancer. Over the years, there has been interest in the use of this molecular biomarker for cancer prediction and prognosis. Progress has been disappointing to some extent, as the US Food and Drug Administration has not approved diagnosis applications, although there is evidence to support the use of specific methylation sites as markers of some diseases [74]. Methylation-specific polymerase chain reaction has been the cornerstone of these efforts. The digital polymerase chain reaction protocol has been optimized for methylation analysis, and it may be better than some standard techniques [75]. Examples include the use of the digital protocol to quantitatively assess the methylation state of DNA in colorectal cancer samples and in plasma samples from breast cancer patients [60]. Other researchers have developed a digital protocol that allows single-molecule capture and assessment of methylation and histone modification [76].

7.5.2 The Development of dPCR in GMO Content Detection

Although dPCR has been used for many different applications, mainly in the determination of absolute copy number [25, 26], it has been shown to be particularly useful for detecting rare and low copy number targets [28] and to determine copy number variations [47, 77].

Inhibitory substances existing in the background of sample matrices could cause differences in amplification efficiency between samples and reference materials. However, when dPCR is used as an end-point measurement, its tolerance to inhibitors is reduced, as often observed in qPCR sample bias [7]. The experimental formula has been shown to affect the individual microreactions, but after analyzing the amplitude of the plot and setting the threshold, all of these responses are positive, so it does not affect the final quantification [43].

Another very important advantage of dPCR is the easier operation over assays of the qPCR platform, which makes the implementation of the dPCR platforms in laboratories easier compared with other approaches.

The importance of reference materials is associated with the reduced use for detection in absolute quantitative analyses that do not depend on the availability of reference materials [7]. For example, for quantitative qPCR data conflicts, the sample and standard material inequality can be easily resolved by testing. At present, although not widely applied, this technology has been able to improve the comparability of the CRM replication/copy data. The use of cdPCR in the CRM copy number ratio has been shown to illustrate the suitability of this measurement [28, 78].

All of the abovementioned advantages contribute to the cost-effectiveness of the dPCR methods, especially ddPCR. The qPCR experimental apparatus can be quite extensive for relative quantification when it is required to include a reference material for several dilutions, such as the dilution of each sample by at least a factor of two. Regarding comparison of 96-hole reaction plates, in the same time frame, with the number of mRNA samples processed three times, the price of each sample is 30 % lower than that of the quantitative mRNA.

ddPCR has also been shown to be suitable for routine use in control laboratories, especially when addressing large numbers of samples [79]. When comparing ddPCR and cdPCR price performance, the cdPCR machine is about two times the price of ddPCR the machine, and the array used for cdPCR is relatively expensive.

Reusability could be used to improve price of the detection performance. Some detection systems can be reused from two to ten times. Multiplex amplification was dependent on the use of different labeled probes, with up to five different fluorescent groups for cdPCR and two different fluorescent differential displays for PCR. In ddPCR, additional multiplexing is possible by using different primer/probe concentrations in the same reaction index up to 10 [80]. Recent advances in ddPCR include the possible use of DNA in combination with dye chemistry, which also allows multiplexing [81].

7.6 Future Prospects of dPCR

7.6.1 Future Prospects in Viral Detection

Although it might not be feasible or advantageous to replace qPCR with dPCR in molecular virology laboratories in the next 5 years due to the low throughput and high cost of dPCR, dPCR can be utilized for standardizing calibration materials within and between laboratories. CMV is the first virus to have a digitally quantified standard material, and more virus standard materials will be developed as the dPCR technology gains traction in the field. Even within a single laboratory, the maintenance of standards requires considerable time and monetary resources. dPCR offers a reliable method for maintaining standard values over time, reducing problems associated with lot-to-lot variability and revealing potential quality control issues.

Within the next few years, dPCR systems will become more automated, making them more attractive for use in routine diagnostic applications by increasing their throughput and providing "closed systems" with less potential for error or contamination from manual handling. As these systems become more advanced, the sensitivity of current dPCR assays might also increase. The sensitivity of these systems is mainly limited by the amount of DNA input they can handle, which can be improved by offering more concentrated reaction master mixes or increasing the number of partitions utilized.

7.6.2 Future Prospects for Clinical Samples

dPCR is a cost-effective alternative to the currently used next-generation sequencing platforms. However, it is only suitable for analyses in which knowledge about the mutation is available and thus relies on customized assay designs. This technology could be used in combination with next-generation sequencing platforms to determine the required input material by accurately quantifying DNA libraries. Sequencing technologies can also benefit from separate heterogeneous templates generated by dPCR for detecting heterogeneous methylation patterns in cancer research. Currently, dPCR is one of the most powerful methods for accurately quantifying small amounts of CNAs in plasma. It is therefore expected that more research in this area will enable the development of this technology for use in a wide spectrum of clinical scenarios in which the high precision and sensitivity offered by dPCR is of the highest priority.

7.6.3 Future Prospects in GMO Detection

In the field of genetically modified organisms, this technology is the preferred detection method. Because this technology is employed in laboratories worldwide, it must be used correctly in routine analyses, and all possible sources of bias, such as the amount of nucleic acids extracted, the presence of inhibitors, and the quality reference material, must be considered and controlled. Considerable effort has been devoted to understanding and critically evaluating this technology.

However, due to the increasing development and approval of genetically modified products worldwide, the current qPCR method is no longer fully applicable for their detection and quantification. It is expected that in the near future, new methods for absolute quantification, better sequencing technology, and more generated data will enable the detection and quantification of genetically modified organisms worldwide. Ideally, several biological detections and multiple target screening systems for individual samples will be combined with simultaneous quantification techniques in future transgenic methods [82–84].

On the other hand, in practical GMO detection by digital PCR, the availability and stability of dPCR are more important than its sensitivity. Therefore, the number of steps before the main digital PCR step should be minimized. Accordingly, we have explored many methodologies to simplify practical GMO detection by digital PCR, which is the main motivation for developing digital PCR. We developed digital PCR methods that do not require pretreatment steps for qualitative screening, which have been published [6], and for quantitative detection, which are under review.

Acknowledgments This work is supported by the National Science and Technology Major Project (2016ZX08012-001). Many thanks to Pengyu Zhu for his kindly help in manuscript conception and preparation.

References

1. Mackay IM, Arden KE, Nitsche A. Real-time PCR in virology. Nucleic Acids Res. 2002;30:1292–305. doi:10.1093/nar/30.6.1292.
2. Bustin S, Benes V, Garson J. The MIQE guidelines: minimum information for publication of quantitative real-time PCR experiments. Clin Chem. 2009;55:611–22. doi:10.1373/clinchem.2008.112797.

3. Cook L, Atienza EE, Bagabag A, et al. Comparison of methods for extraction of viral DNA from cellular specimens. Diagn Microbiol Infect Dis. 2009;64:37–42. doi:10.1016/j.diagmicrobio.2009.01.003.
4. Lai KK-Y, Cook L, Wendt S, et al. Evaluation of real-time PCR versus PCR with liquid-phase hybridization for detection of enterovirus RNA in cerebrospinal fluid. J Clin Microbiol. 2003;41:3133–41. doi:10.1128/JCM.41.7.3133-3141.2003.
5. Kodama T, Kurosawa Y, Kitta K, et al. Tendency for interlaboratory precision in the GMO analysis method based on real-time PCR. J AOAC Int. 2010;93:734–49.
6. Fu W, Zhu P, Wang C, et al. A highly sensitive and specific method for the screening detection of genetically modified organisms based on digital PCR without pretreatment. Sci Rep. 2015;5:12715.
7. Burns MJ, Burrell AM, Foy CA. The applicability of digital PCR for the assessment of detection limits in GMO analysis. Eur Food Res Technol. 2010;231:353–62. doi:10.1007/s00217-010-1277-8.
8. Vogelstein B, Kinzler KW. Digital PCR. Proc Natl Acad Sci U S A. 1999;96:9236–41. doi:10.1073/pnas.96.16.9236.
9. Ge Q, Liu Z, Bai Y, et al. Emulsion PCR-based method to detect Y chromosome microdeletions. Anal Biochem. 2007;367:173–8.
10. Zhu Z, Jenkins G, Zhang W, et al. Single-molecule emulsion PCR in microfluidic droplets. Anal Bioanal Chem. 2012;403:2127–43.
11. Williams R, Peisajovich SG, Miller OJ, et al. Amplification of complex gene libraries by emulsion PCR. Nat Methods. 2006;3:545–50.
12. Shao K, Ding W, Wang F, et al. Emulsion PCR: a high efficient way of PCR amplification of random DNA libraries in aptamer selection. PLoS ONE. 2011;6, e24910.
13. Corbisier P, Bhat S, Partis L, et al. Absolute quantification of genetically modified MON810 maize (Zea mays L.) by digital polymerase chain reaction. Anal Bioanal Chem. 2010;396:2143–50.
14. Demeke T, Gräfenhan T, Holigroski M, et al. Assessment of droplet digital PCR for absolute quantification of genetically engineered OXY235 canola and DP305423 soybean samples. Food Control. 2014;46:470–4.
15. Morisset D, Štebih D, Milavec M, et al. Quantitative analysis of food and feed samples with droplet digital PCR. PLoS ONE. 2013;8, e62583.
16. Hori M, Fukano H, Suzuki Y. Uniform amplification of multiple DNAs by emulsion PCR. Biochem Biophys Res Commun. 2007;352:323–8.
17. Nakano M. Single-molecule PCR using water-in-oil emulsion. J Biotechnol. 2003;102:117–24.
18. Dear PH, Cook PR. Happy mapping: linkage mapping using a physical analogue of meiosis. Nucleic Acids Res. 1993;21:13–20. doi:10.1093/nar/21.1.13.
19. Sykes PJ, Neoh SH, Brisco MJ, et al. Quantitation of targets for PCR by use of limiting dilution. Biotechniques. 1992;13:444–9.
20. Pohl G, Shih I-M. Principle and applications of digital PCR. Expert Rev Mol Diagn. 2004;4:41–7. doi:10.1586/14737159.4.1.41.
21. McCaughan F, Dear PH. Single-molecule genomics. J Pathol. 2010;220:297–306. doi:10.1002/path.2647.
22. Bhat S, Herrmann J, Armishaw P, et al. Single molecule detection in nanofluidic digital array enables accurate measurement of DNA copy number. Anal Bioanal Chem. 2009;394:457–67. doi:10.1007/s00216-009-2729-5.
23. Daser A, Thangavelu M, Pannell R, et al. Interrogation of genomes by molecular copy-number counting (MCC). Nat Methods. 2006;3:447–53. doi:10.1038/nmeth880.
24. Zhang B, Li M, McDonald T, et al. Microenvironmental protection of CML stem and progenitor cells from tyrosine kinase inhibitors through N-cadherin and Wnt-β-catenin signaling. Blood. 2013;121:1824–38. doi:10.1182/blood-2012-02-412890.

25. Hindson BJ, Ness KD, Masquelier DA, et al. High-throughput droplet digital PCR system for absolute quantitation of DNA copy number. Anal Chem. 2011;83:8604–10. doi:10.1021/ac202028g.
26. Sanders R, Huggett JF, Bushell CA, et al. Evaluation of digital PCR for absolute DNA quantification. Anal Chem. 2011;83:6474–84. doi:10.1021/ac103230c.
27. Pinheiro LB, Coleman VA, Hindson CM, et al. Evaluation of a droplet digital polymerase chain reaction format for DNA copy number quantification. Anal Chem. 2012;84:1003–11. doi:10.1021/ac202578x.
28. Bhat S, Herrmann J, Armishaw P, et al. Single molecule detection in nanofluidic digital array enables accurate measurement of DNA copy number. Anal Bioanal Chem. 2009;394:457–67.
29. Spurgeon SL, Jones RC, Ramakrishnan R. High throughput gene expression measurement with real time PCR in a microfluidic dynamic array. PLoS ONE. 2008;3, e1662. doi:10.1371/journal.pone.0001662.
30. Morrison T, Hurley J, Garcia J, et al. Nanoliter high throughput quantitative PCR. Nucleic Acids Res. 2006;34, e123. doi:10.1093/nar/gkl639.
31. Pekin D, Skhiri Y, Baret J-C, et al. Quantitative and sensitive detection of rare mutations using droplet-based microfluidics. Lab Chip. 2011;11:2156–66. doi:10.1039/c1lc20128j.
32. Diehl F, Li M, He Y, et al. BEAMing: single-molecule PCR on microparticles in water-in-oil emulsions. Nat Methods. 2006;3:551–9. doi:10.1038/nmeth898.
33. Dressman D, Yan H, Traverso G, et al. Transforming single DNA molecules into fluorescent magnetic particles for detection and enumeration of genetic variations. Proc Natl Acad Sci U S A. 2003;100:8817–22. doi:10.1073/pnas.1133470100.
34. Hayden RT, Yan X, Wick MT, et al. Factors contributing to variability of quantitative viral PCR results in proficiency testing samples: a multivariate analysis. J Clin Microbiol. 2012;50:337–45. doi:10.1128/JCM.01287-11.
35. Hayden RT, Hokanson KM, Pounds SB, et al. Multicenter comparison of different real-time PCR assays for quantitative detection of Epstein-Barr virus. J Clin Microbiol. 2008;46:157–63. doi:10.1128/JCM.01252-07.
36. Hindson CM, Chevillet JR, Briggs HA, et al. Absolute quantification by droplet digital PCR versus analog real-time PCR. Nat Methods. 2013;10:1003–5. doi:10.1038/nmeth.2633.
37. Flamand L, Komaroff AL, Arbuckle JH, et al. Review, part 1: human herpesvirus-6-basic biology, diagnostic testing, and antiviral efficacy. J Med Virol. 2010;82:1560–8. doi:10.1002/jmv.21839.
38. Arbuckle JH, Medveczky MM, Luka J, et al. The latent human herpesvirus-6A genome specifically integrates in telomeres of human chromosomes in vivo and in vitro. Proc Natl Acad Sci U S A. 2010;107:5563–8. doi:10.1073/pnas.0913586107.
39. Hirsch HH, Lautenschlager I, Pinsky BA, et al. An international multicenter performance analysis of cytomegalovirus load tests. Clin Infect Dis. 2012;56:367–73. doi:10.1093/cid/cis900.
40. Hayden RT, Gu Z, Ingersoll J, et al. Comparison of droplet digital PCR to real-time PCR for quantitative detection of cytomegalovirus. J Clin Microbiol. 2013;51:540–6.
41. Strain MC, Lada SM, Luong T, et al. Highly precise measurement of HIV DNA by droplet digital PCR. PLoS ONE. 2013;8:2–9. doi:10.1371/journal.pone.0055943.
42. Hoshino T, Inagaki F. Molecular quantification of environmental DNA using microfluidics and digital PCR. Syst Appl Microbiol. 2012;35:390–5. doi:10.1016/j.syapm.2012.06.006.
43. Dingle TC, Sedlak RH, Linda C, Jerome KR. Tolerance of droplet-digital PCR vs real-time quantitative PCR to inhibitory substances. Clin Chem. 2013;59:1668–9.
44. Heyries KA, Tropini C, Vaninsberghe M, et al. Megapixel digital PCR. Nat Methods. 2011;8:649–51. doi:10.1038/nmeth.1640.
45. Conrad DF, Pinto D, Redon R, et al. Origins and functional impact of copy number variation in the human genome. Nature. 2010;464:704–12. doi:10.1038/nature08516.
46. Berger MF, Lawrence MS, Demichelis F, et al. The genomic complexity of primary human prostate cancer. Nature. 2011;470:214–20. doi:10.1038/nature09744.

47. Whale AS, Huggett JF, Cowen S, et al. Comparison of microfluidic digital PCR and conventional quantitative PCR for measuring copy number variation. Nucleic Acids Res. 2012;40, e82. doi:10.1093/nar/gks203.
48. Petriv OI, Heyries KA, VanInsberghe M, et al. Methods for multiplex template sampling in digital PCR assays. PLoS ONE. 2014;9:1–8. doi:10.1371/journal.pone.0098341.
49. Page K, Hava N, Ward B, et al. Detection of HER2 amplification in circulating free DNA in patients with breast cancer. Br J Cancer. 2011;104:1342–8. doi:10.1038/bjc.2011.89.
50. Gevensleben H, Garcia-Murillas I, Graeser MK, et al. Noninvasive detection of HER2 amplification with plasma DNA digital PCR. Clin Cancer Res. 2013;19:3276–84.
51. Miotke L, Lau BT, Rumma RT, Ji HP. High sensitivity detection and quantitation of DNA copy number and single nucleotide variants with single color droplet digital PCR. Anal Chem. 2014;86:2618–24. doi:10.1021/ac403843j.
52. Mitchell PS, Parkin RK, Kroh EM, et al. Circulating microRNAs as stable blood-based markers for cancer detection. Proc Natl Acad Sci U S A. 2008;105:10513–8. doi:10.1073/pnas.0804549105.
53. Mouillet J-F, Chu T, Hubel CA, et al. The levels of hypoxia-regulated microRNAs in plasma of pregnant women with fetal growth restriction. Placenta. 2010;31:781–4. doi:10.1016/j.placenta.2010.07.001.
54. Wu L, Zhou H, Lin H, et al. Circulating microRNAs are elevated in plasma from severe preeclamptic pregnancies. Reproduction. 2012;143:389–97. doi:10.1530/REP-11-0304.
55. Chim SSC, Shing TKF, Hung ECW, et al. Detection and characterization of placental MicroRNAs in maternal plasma. Clin Chem. 2008;54:482–90. doi:10.1373/clinchem.2007.097972.
56. Ma J, Li N, Guarnera M, Jiang F. Quantification of plasma miRNAs by digital PCR for cancer diagnosis. Biomark Insights. 2013;8:127–36. doi:10.4137/BMImi.Ss13154.
57. Takahashi K, Yan IK, Kim C, et al. Analysis of extracellular RNA by digital PCR. Front Oncol. 2014;4:2–7.
58. Huggett JF, Foy CA, Benes V, et al. The digital MIQE guidelines: minimum information for publication of quantitative digital PCR experiments. Clin Chem. 2013;59:892–902. doi:10.1373/clinchem.2013.206375.
59. Li M, Chen W-D, Papadopoulos N, et al. Sensitive digital quantification of DNA methylation in clinical samples. Nat Biotechnol. 2009;27:858–63. doi:10.1038/nbt.1559.
60. Weisenberger DJ, Trinh BN, Campan M, et al. DNA methylation analysis by digital bisulfite genomic sequencing and digital MethyLight. Nucleic Acids Res. 2008;36:4689–98. doi:10.1093/nar/gkn455.
61. White RA, Quake SR, Curr K. Digital PCR provides absolute quantitation of viral load for an occult RNA virus. J Virol Methods. 2012;179:45–50. doi:10.1016/j.jviromet.2011.09.017.
62. Bhattarai N, Stapleton JT. GB virus C: the good boy virus? Trends Microbiol. 2012;20:124–30. doi:10.1016/j.tim.2012.01.004.
63. Henrich TJ, Gallien S, Li JZ, et al. Low-level detection and quantitation of cellular HIV-1 DNA and 2-LTR circles using droplet digital PCR. J Virol Methods. 2012;186:68–72. doi:10.1016/j.jviromet.2012.08.019.
64. Kiss MM, Ortoleva-Donnelly L, Beer NR, et al. High-throughput quantitative polymerase chain reaction in picoliter droplets. Anal Chem. 2008;80:8975–81.
65. Shen F, Davydova EK, Du W, et al. Digital isothermal quantification of nucleic acids via simultaneous chemical initiation of recombinase polymerase amplification reactions on SlipChip. Anal Chem. 2011;83:3533–40. doi:10.1021/ac200247e.
66. Yung TKF, Chan KCA, Mok TSK, et al. Single-molecule detection of epidermal growth factor receptor mutations in plasma by microfluidics digital PCR in non-small cell lung cancer patients. Clin Cancer Res. 2009;15:2076–84. doi:10.1158/1078-0432.CCR-08-2622.
67. Wang J, Ramakrishnan R, Tang Z, et al. Quantifying EGFR alterations in the lung cancer genome with nanofluidic digital PCR arrays. Clin Chem. 2010;56:623–32. doi:10.1373/clinchem.2009.134973.

68. Wang L, McLeod HL, Weinshilboum RM. Genomics and drug response. N Engl J Med. 2011;364:1144–53. doi:10.1056/NEJMra1010600.Genomics.
69. Ouahchi K, Lindeman N, Lee C. Copy number variants and pharmacogenomics. Pharmacogenomics. 2006;7:25–9. doi:10.2217/14622416.7.1.25.
70. Reiter M, Kirchner B, Müller H, et al. Quantification noise in single cell experiments. Nucleic Acids Res. 2011;39:1–12. doi:10.1093/nar/gkr505.
71. Rottiers V, Näär AM. MicroRNAs in metabolism and metabolic disorders. Nat Rev Mol Cell Biol. 2012;13:239–50. doi:10.1038/nrm3313.
72. Nair VS, Maeda LS, Ioannidis JPA. Clinical outcome prediction by MicroRNAs in human cancer: a systematic review. J Natl Cancer Inst. 2012;104:528–40. doi:10.1093/jnci/djs027.
73. Petriv OI, Kuchenbauer F, Delaney AD, et al. Comprehensive microRNA expression profiling of the hematopoietic hierarchy. Proc Natl Acad Sci U S A. 2010;107:15443–8. doi:10.1073/pnas.1009320107.
74. Baylin SB, Jones PA. A decade of exploring the cancer epigenome – biological and translational implications. Nat Rev Cancer. 2011;11:726–34. doi:10.1038/nrc3130.
75. Mikeska T, Candiloro IL, Dobrovic A. The implications of heterogeneous DNA methylation for the accurate quantification of methylation. Epigenomics. 2010;2:561–73. doi:10.2217/epi.10.32.
76. Cipriany BR, Zhao R, Murphy PJ, et al. Single molecule epigenetic analysis in a nanofluidic channel. Anal Chem. 2010;82:2480–7.
77. Weaver S, Dube S, Mir A, et al. Taking qPCR to a higher level: analysis of CNV reveals the power of high throughput qPCR to enhance quantitative resolution. Methods. 2010;50:271–6. doi:10.1016/j.ymeth.2010.01.003.
78. Corbisier P, Bhat S, Partis L, et al. Absolute quantification of genetically modified MON810 maize (Zea mays L.) by digital polymerase chain reaction. Anal Bioanal Chem. 2010;396:2143–50. doi:10.1007/s00216-009-3200-3.
79. Morisset D, Štebih D, Milavec M, et al. Quantitative analysis of food and feed samples with droplet digital PCR. PLoS ONE. 2013;8, e62583. doi:10.1371/journal.pone.0062583.
80. Baker M. Digital PCR hits its stride. Nat Methods. 2012;9:541–4. doi:10.1038/nmeth.2027.
81. McDermott GP, Do D, Litterst CM, et al. Multiplexed target detection using DNA-binding dye chemistry in droplet digital PCR. Anal Chem. 2013;85:11619–27. doi:10.1021/ac403061n.
82. Waiblinger H-U, Ernst B, Anderson A, Pietsch K. Validation and collaborative study of a P35S and T-nos duplex real-time PCR screening method to detect genetically modified organisms in food products. Eur Food Res Technol. 2008;226:1221–8. doi:10.1007/s00217-007-0748-z.
83. Kluga L, Folloni S, van den Bulcke M, et al. Applicability of the "Real-Time PCR-Based Ready-to-Use Multi-Target Analytical System for GMO Detection" in processed maize matrices. Eur Food Res Technol. 2012;234:109–18. doi:10.1007/s00217-011-1615-5.
84. Shao N, Jiang SM, Zhang M, et al. MACRO: a combined microchip-PCR and microarray system for high-throughput monitoring of genetically modified organisms. Anal Chem. 2014;86:1269–76. doi:10.1021/ac403630a.

Chapter 8
Development of Accurate Nucleic Acid Detection Technology for Target Quantification

Abstract As the requirement of detection, qualitative detection can no longer fulfill the current needs. In the fields of disease diagnosis, food adulteration, and clinical medicine and the detection of pathogenic bacteria and viruses, quantitative detection is crucial. From early real-time quantitative PCR to digital PCR, quantification has progressed from relative to absolute, and the instrumentation is growing much more automated. No matter how developed qualitative detection becomes, the fluorescent probe and dye are the key factors in achieving quantification. Hence, in the development of quantitative detection, various probes and fluorescent dyes have been created to enhance the specificity, improve the detection limit, and decrease the background interference. This review mainly describes the basic principle of each commonly used quantitative method and the advantages and weaknesses.

Keywords Quantification • Competitive PCR • Physicochemical • qPCR • Digital PCR

8.1 Introduction

In the development of biotechnology, methods based on molecular biology with high accuracy, efficiency and detection limits, short detection periods, and simple operation have been widely used for various purposes, such as: (1) the detection of harmful or pathogenic factors, (2) the identification of the food adulteration, (3) medicolegal expertise, (4) disease diagnosis, (5) the detection of genetically modified organisms, and so on [1–7].

Based on genetics and the flow direction of the biological information, DNA, RNA, proteins, and metabolites can all be used as target materials in molecular biological detection.

In nature, RNases are widespread and highly stable, and RNA is vulnerable to pollution and degradation by RNase [8]. The extraction of proteins and metabolites is always complex and easily influenced by environmental factors. Meanwhile, DNA carries the genetic information of an organism, and the information content of DNA is greater than the information content of protein due to the degeneracy of the genetic code going from DNA to protein [9]. In light of the advantages of DNA,

W. Xu, *Functional Nucleic Acids Detection in Food Safety*,
DOI 10.1007/978-981-10-1618-9_8

which is stable, easy to obtain, and sensitive, PCR-based techniques are widely used to detect target DNA sequences.

In the food safety field, there are many limitation standards for harmful factors, such as pathogenic bacteria, mycotoxins, heavy metals, pesticide residues, and so on. In recent years, many food safety incidents have occurred, mostly related to pathogenic bacteria exceeding these limits. In 2008, at least 1220 people in 42 states in the USA were attacked by a disease caused by salmonella. This incident involved the worst salmonella disease in 10 years and might have been caused by polluted food. In 2011, the "toxic cucumber" issue occurred first in Germany, and the culprit was enterohemorrhagic *Escherichia coli* (EHEC). During that time, all of Europe suffered a panic. In addition, genetically modified crops have been approved for food and feed in many countries. Many countries have established their own thresholds for the content of genetically modified crops, such as 0.9 % in the EU; 3 % in Korea; 5 % in Taiwan; 1 % in Australia, New Zealand, and Brazil; and 5 % in Japan [10].

Qualitative detection not only affects people's health but is also relevant to national policies. With the increasing demands for detection and labeling policies, the traditional qualitative detection cannot address the current situation and requirements. Thus, accurate nucleic acid detection technologies for target quantification were developed.

8.2 Physicochemical Nucleic Acid Quantification

Due to the special structure and properties of nucleic acids, the methods that directly quantify the concentration or amount of nucleic acid are based on physicochemical techniques.

8.2.1 Spectrophotometric Method

The bases that compose nucleic acids have certain ultraviolet absorption features, with the maximum absorption wavelength between 250 and 270 nm (adenine, 260.5 nm; cytosine, 267 nm; guanine, 276 nm; thymine, 264.5 nm; uracil, 259 nm). The maximum absorption wavelength of nucleic acids is at approximately 260 nm, and the absorption low is at 230 nm. These physical properties provide a basis for determining nucleic acid concentration. At the UV wavelength of 260 nm, 1OD (optical density) is equivalent to 50 μg/mL double-stranded DNA, 40 μg/mL of single-stranded DNA or RNA, or 20 μg/mL of a single oligonucleotide. Based on the absorbance, the concentration of the nucleic acid samples can be quantified.

The spectrophotometric method cannot only determine the concentration of nucleic acid but also estimate the purity through determining the UV absorption ratio at 260 nm and 280 nm (A260/A280). The ratio for pure DNA is 1.8 and for

RNA is 2.0. If the ratio of DNA is higher than 1.8, there is residual RNA in the solution. If there is residual phenol or protein, the OD ratio might be decreased. Ultraviolet spectrophotometry is only appropriate for measuring nucleic acid solutions with a concentration greater than 0.25 μg/mL, and the loading quantity of the sample is large, ranging from 50 to 100 μL.

As the technique developed, portable and minitype instruments such as the NanoDrop, made by Thermo Scientific, and the SmartSpec Plus, made by Bio-Rad, have been sold on the marketplace. The loading quantity of the sample is small, requiring only 0.5–2 μL, and the detection limit is as low as 2 ng/μL, which is appropriate for precious and rare samples.

8.2.2 *Fluorospectrophotometry*

For dilute nucleic acid solutions, fluorospectrophotometry is adopted. The commonly used fluorescent dyes include ethidium bromide (EB), EvaGreen, LC Green, and so on. When the fluorescent dye embeds into the minor groove of the DNA, the fluorescent signal is enhanced. The fluorescence intensity is proportional to the DNA concentration.

There are two ways to estimate the DNA concentration using fluorescent dyes. The first is agarose gel electrophoresis: EB is added into the agarose gel, only a 5–10 μL DNA sample is needed, and 0.05–0.1 g DNA can be observed by the naked eye. This method provides a relatively rough estimate of the DNA concentration and requires a standard control sample with known content.

The second is to measure the fluorescence intensity. An instrument such as the Modulus™ made by Turner Biosystems can detect the fluorescence. A sample with known concentration is used as the standard, usually λ calf thymus DNA, and then based on the relationship between the fluorescent signal and the DNA concentration, a standard curve is established. The concentration of the unknown sample can be obtained through calculation based on the standard curve. The NanoDrop 2000, developed by Thermo Scientific, is the most widely used ultramicro nucleic acid/protein analyzer based on the principle of fluorospectrophotometry. As little as 0.5 μL DNA is needed, and the detection range is from 2 ng/μl (dsDNA) to 15,000 ng/μL with a brief detection period (<5 s).

The reference material mentioned above should have the following characteristics: a stable, uniform, and determinate characteristic value, which can be used in instrument calibration, evaluation of the measurement method, or valuing the substances or materials. There are three kinds of reference materials: the matrix material with the known concentration, the DNA standard reference, and the protein standard reference. In the detection of GM organisms, a special reference material is employed: the constructed plasmid containing the target fragment. The matrix material of the GM organism is difficult to obtain, whereas the plasmid is easy to construct and quantify.

8.2.3 Chromatographic and Mass-Spectrographic Techniques

8.2.3.1 High-Performance Liquid Chromatography

Based on the high separation efficiency, good selectivity, and simultaneous separation and detection functions, high-performance liquid chromatography (HPLC) has been applied as a strong method in the separation and analysis of biological macromolecules (such as nucleic acid and protein). There are six separation modes in HPLC: size-exclusion chromatography, anion-exchange chromatography, mixed-mode chromatography, reversed-phase chromatography, ion-pair reversed-phase chromatography, and affinity chromatography [11]. Among the abovementioned modes, ion-pair reversed-phase chromatography (IR-RP-HPLC) is acknowledged as the ideal nucleic acid detection method, due to its high resolution and wide range of analyzed fragment lengths, including single- and double-stranded DNA. Usually, through choosing the appropriate chromatographic column, using a standard substance with known concentration, and optimizing the mobile phase solutions, the samples can be quantified accurately.

Some studies have reported that not only the DNA concentration but also the PCR products can be effectively quantified and separated by HPLC [12–14]. The commonly used columns in HPLC are TSK gel DEAE-NPR, 3.5×4.6 mm and 4.6×7.5 mm. Using different conditions, such as pH, recovery temperature, and flow rate, different lengths of fragments can be analyzed. For instance, the flow rate of 0.5–0.75 mL/min is suitable for fragments <3000 bp, and 1.0 mL/min is suitable for fragments of 100–25,000 bp. Currently, accurate quantification and purification for the new synthetic probes and primers always employ HPLC. Based on similar principles, capillary electrophoresis, Southern blot, and LC-MS are also reported as methods to quantify nucleic acids and PCR products.

8.2.3.2 Denaturing High-Performance Liquid Chromatography

Denaturing high-performance liquid chromatography (DHPLC) is also an effective instrument for the analysis of PCR sequences and is widely used in DNA sequence variation to separate the PCR amplicons. DHPLC can discriminate between very short amplicons based on sequence differences, even if the fragments are of identical length and the amplicons differ only in a small number of nucleotides. The advantages of DHPLC are as follows: (1) the ability to view the PCR product in a semiquantitative manner by comparing peak heights and overall profiles, which allows the researcher to monitor reproducibility and bias within and between samples, (2) effective analysis for mixed-template reactions with low DNA concentrations, and (3) the automation of the procedure, the short analysis time, and the ease of fragment collection [15].

8.2.3.3 MALDI-TOF Mass Spectrometry

MALDI-TOF mass spectrometry has been widely used in the analysis of proteins, peptides, and nucleic acids. In DNA sequence analysis, the detection of SNPs (single-nucleotide polymorphisms) is frequently used.

Analyzing nucleic acids by MALDI-TOF/MS has several advantages. First, it is fast; the ionization, separation by size, and detection of nucleic acids take milliseconds to complete. Second, the results are absolute, being based on the intrinsic property of the mass-to-charge ratio (m/z). Third, the complete automation of all steps, from sample preparation through the acquisition and processing of the data, is feasible, giving MALDI-TOF MS great potential for high-throughput nucleic acid analysis applications [16].

Based on the principles of SNP detection, MALDI-TOF/MS has been developed for the quantitative, semiquantitative detection and separation of multiple PCR products. To create the detection system, a competitor that differs only in one base part is employed, which is the same principle as in competitive PCR. First, in the PCR reaction, both the target template and its corresponding competitor are placed in a single tube. Second, single-base extension is conducted. Third, MALDI-TOF/MS detection is performed: because only one base difference exists between the target and the competitor, and the mass-to-charge ratios (m/z) are also different, the target can be identified [17, 18]. Hence, based on the different lengths of the targets and the different extended bases, multiplex detection can be realized. Meanwhile, because the concentration of the competitor is known, the amount of the corresponding target can be calculated. A mature technique and analysis software using this principle have been established already, named MassARRAY, and been widely used in the gene expression analysis. MALDI-TOF/MS has many advantages, such as sensitivity, specificity, and quantitative results, which make it feasible for application to DNA detection.

8.3 Competitive PCR for Quantitative Detection

Competitive PCR is a quantitative method that requires the use of an internal reference and elimination of the differences between the tubes or samples and then can achieve accurate quantification. As in common real-time quantitative PCR, the quantification is conducted in the exponential phase and depends on the initial sample quantity. Due to the exponential nature of PCR, even small and unpredictable biases in amplification efficiency during the early cycles can lead to large differences in the final product yield. This problem is especially important when beginning with small amounts of target sequences [19]. In contrast, the quantification of competitive PCR is realized at the reaction end point, not affected by initial sample concentration or the total final amplification yield.

After amplification, the PCR products of the target and competitor can be recognized by a difference in length or by the presence of a point mutation that creates a novel restriction site, or they can be resolved by temperature gradient gel electrophoresis or distinguished by real-time PCR.

In the competitive PCR approach, the two templates (target and competitor) have the same primer annealing sequences and are subject to the same predictable and unpredictable variables that might affect the PCR, and thus, they amplify at the same rate. The amplicons of the target and the competitor can be represented by the following formula:

$$X_{\text{unknown}}(1+E)^n = Y_{\text{unknown}} \tag{8.1}$$

$$X_{\text{reference}}(1+E)^n = Y_{\text{reference}} \tag{8.2}$$

X represents the initial copy number of the sample.
Y represents the final copy number of the sample.
E represents the amplification efficiency.
n represents the number of the cycle.

Because the target and competitor are nearly identical in competitive PCR, the amplification efficiency E is regarded as the same. Hence, the relationship can be expressed as follows:

$$\frac{X_{\text{reference}}}{X_{\text{unknown}}} = \frac{Y_{\text{reference}}}{Y_{\text{unknown}}} \tag{8.3}$$

When the final amplicons of the two templates are the same, the initial copy number of the unknown sample is the same as the initial copy number of the reference.

Competitive PCR is a powerful technique for the absolute quantification of nucleic acids. It has been applied in the quantitative detection of GMOs and pathogenic bacteria in food [20, 21].

Nonetheless, it suffers from a few drawbacks that somewhat limit its application. In particular, the technique can be time-consuming because of the requirement of constructing the competitor, the necessity of assembling multiple competitive reactions for a single determination, and, most notably, the need for a post-PCR electrophoresis-based detection and analysis step. This requirement turns out to be particularly cumbersome for the analysis of a large number of samples. Moreover, the commonly used detection techniques, such as agarose or acrylamide gel electrophoresis, have low dynamic range and low throughput and are ineffective for dealing with heterodimers (competitor/target hybrids) [19].

8.4 Quantitative PCR

In 1993, the real-time fluorescent quantitative PCR technique was first reported by Higuchi et al. [22], and the first fully automated real-time PCR instrument was produced by the ABI Company (USA) in 1996. The so-called fluorescent quantitative detection uses a fluorophore in the PCR reaction. The amplification is monitored in real time through the accumulation of the fluorescent signal, and the generated standard curve is ultimately used for quantitative analysis of the unknown sample. Compared with traditional qualitative PCR, this real-time PCR has made the leap from qualitative to quantitative, and it has stronger specificity, higher automation, and effective solution to the problem of PCR contamination.

In the earliest quantitative detection, the nucleic acid dye EB was added to the PCR reaction. Because the fluorescence increased as the products accumulated, at the end of the amplification, the fluorescent signal was detected to identify whether the PCR had occurred. This approach is called "end point" analysis. The traditional quantitative PCR methods, such as the endo-reference method and PCR-ELISA, are also end point detection methods, which always have poor reproducibility.

In the work of Higuchi et al., a video camera (cooled CCD camera) was used to monitor the accumulation of the fluorescent signal of EB in real time throughout the PCR reaction. According to the principles of PCR, fluorescent signal acquisition was conducted during each annealing or extension phase. By analyzing all the collected images using software, the intensity of the fluorescence at each PCR cycle is obtained, and based on these data, the initial amplification curve is drawn, and kinetic PCR analysis can be performed. This study achieved real-time monitoring with the help of a camera. Although it was not automated, it laid the essential groundwork for the development of the automated real-time PCR instrument.

Due to the toxicity and specificity of EB, this kind of fluorescence is seldom used in real-time PCR. Currently, the basic principle of real-time fluorescent quantitative PCR is based on the fluorescence resonance energy transfer (FRET) phenomenon. FRET is the physical process by which energy is transferred non-radiatively from an excited molecular chromophore (the donor) to another chromophore (the acceptor) by means of intermolecular long-range dipole-dipole coupling. The essential requirements for effective transfer over distances from 10 to 100 Å are that the fluorescence spectrum of the donor and the absorbance spectrum of the acceptor overlap adequately and that both the quantum yield of the donor and the absorption coefficient of the acceptor are sufficiently high. In addition, for the dipole-dipole vectorial interaction to occur, either the transition dipoles of D and A must be oriented favorably relative to each other or one (or both) must have a certain degree of rapid rotational freedom; this latter condition is usually satisfied for chromophores attached to biomolecules in solution. The transfer is non-radiative; that is, the donor does not actually emit a photon and the acceptor does not absorb a photon [23]. The development of FRET has led to quantitative improvements, and hence, real-time PCR has been widely used in the quantitative detection of bacteria, viruses, and GMOs [24–26].

8.4.1 The Hot Start Reaction in Real-Time PCR

Two types of undesired DNA synthesis can occur during PCR setup: mispriming on less than specific sites in the template and/or the formation of a so-called primer dimer, where the primers are extended by serving each other as a template. Each of these events, occurring in seconds at temperatures below the cycling conditions, can compromise the success, yield, and/or specificity of PCR, particularly for amplifications involving high cycle numbers or high GC content of the amplicon [27]. To prevent the nonspecific amplification or the primer dimer, the hot start is employed in real-time PCR, delaying the start of PCR amplification until the reaction system is heated to the desired temperature.

There are two ways to achieve the hot start of real-time PCR. The first way is to separate the compositions from each other, such as by a wax barrier in the mixture. The primers, Mg^{2+}, and dNTP are below the wax film, and the rest of the reagents including DNA polymerase are above the wax. During the PCR cycle, the wax film is melted as the mixture is heated, and amplification then begins. The second way is using hot start DNA polymerase. Anti-Taq antibodies reduce the DNA polymerase activity but, being thermolabile, release the enzyme at PCR cycling temperatures to achieve a hot start [28]. Some enzyme preparations, such as AmpliTaq Gold (PerkinElmer and Roche Molecular Systems) and hot start Taq (Qiagen), are chemically inactivated and can be reactivated by heating in a special preincubation step [29].

8.4.2 Nonspecific Dye-Binding Real-Time PCR

A fluorescent dye, such as SBYR GREEN, Eva GREEN, or LC GREEN, can combine with the DNA minor groove and release a single fluorophore, which operates by the same principle as ethidium bromide (EB) during DNA agarose electrophoresis. The fluorescence intensity increases with amplification. The use of fluorescent dye decreases the detection cost; however, because the dye can insert into any double-stranded DNA, including nonspecific amplicons and primer dimers, the actual result is somewhat affected.

8.4.3 Fluorescent-Labeled Probe in Real-Time PCR

The use of fluorescent probes in real-time quantitative PCR mainly involves the TaqMan probe, MGB probe, self-quenched probes with single fluorophores, double hybridized probes, and molecular beacons [30].

8.4.3.1 TaqMan Probe

The TaqMan probe is the most widely used hydrolysis probe. The reporter group, such as FAM or VIC, is labeled at the 5′ end of the probe, and a fluorescence quenching group, such as TAMRA, is tagged at the 3′ end of the probe. While the probe maintains its integration, the quenching group absorbs the emission fluorescence of the reporter group. When the probe hybridizes with the template, there is no fluorescent signal, and when the amplification begins, the probe is hydrolyzed by Taq DNA polymerase, the reporter energy cannot be absorbed, and a fluorescent signal can be detected. Therefore, both the fluorescent signal and the target fragment increase exponentially with the PCR cycle number. The signal strength is proportional to the copy number of the DNA template.

8.4.3.2 TaqMan MGB Probe

The TaqMan MGB probe is another kind of hydrolysis probe and is an improvement based on the TaqMan probe. The MGB molecule (a minor groove binder) is added at the 3′ end of the probe and can bind to the minor groove part in the double-stranded DNA. The MGB probe enhances the Tm value of the probe, allowing it to distinguish a mismatch of one base. If there is only one mismatched base, one signal can be detected. Moreover, the labeled quenching group is a nonfluorescent substance, so during the reaction, there is no background fluorescence. The better quenching effect, labeling flexibility, improved signal-to-noise ratio of the fluorescent signal, and higher annealing temperature increase the recognition rate of SNP and give the TaqMan MGB wide potential applications.

8.4.3.3 Double Hybridized Probes

The use of double hybridized probes is a quantitative PCR technique developed by Roche Company, which is also based on FRET. The reporter and quenching probes are separately labeled on the 5′ and 3′ ends of two probes. The two probes can bind to the same template at two adjacent sites (1–5 bp), and FRET can occur during the hybridization. The method is characterized by high quenching efficiency; however, as two probes are bound to the same template, the amplification efficiency is affected. In addition, due to the synthesis of two long probes, the cost is relatively high.

8.4.3.4 Self-Quenched Probe with Single Fluorophore

This kind of primer is usually labeled with only one fluorescent group. Guanosine, an efficient electron donor, was demonstrated to have a significant effect on fluorescence. The proximity of guanosine in the complementary strand to the fluorescent group upon hybridization can decrease the fluorescence intensity. Through labeling with a single fluorescent group at the 5′ end of the oligonucleotides, the fluorescent signal can be decreased upon hybridization, which can provide a basis for a cost-effective DNA quantitative detection method. However, the amount of signal generated by this decrease is small compared to the signal generated by FRET-labeled probes, which might affect the sensitivity of the detection [31].

8.4.3.5 LUX (Light Upon eXtension) Primer

In the following study, the fluorogenic primer is designed to be "self-quenched" until it is incorporated into a double-stranded PCR product, whereupon its fluorescence increases, i.e., "dequenched." The fluorogenic primer is called a LUX (Light Upon eXtension) primer, which is also a self-quenched probe. In the quantitative PCR reaction, the counterpart primer used in conjunction with the fluorogenic primer is a standard, unlabeled oligonucleotide. The LUX primer design is based on studies that demonstrate the effects of the primary and secondary structure of oligonucleotides on the emission properties of a conjugated fluorescent group. The design factors are largely based on the necessity of having guanosine bases in the primary sequence near the conjugated fluorescent group. The fluorescent group may be any of various commonly used dyes such as FAM and JOE. Furthermore, the LUX primers have a non-sequence-specific 5′ tail that is complementary to the 3′ end of the primer. The 5′ tail enables the primer to assume a hairpin conformation at temperatures below the melting point of the hairpin, and it also effects the fluorescence of the LUX primer [32].

8.4.3.6 Molecular Beacons (Hybridization Probes)

A molecular beacon is a double-labeled stem-loop oligonucleotide probe with a hairpin structure. The sequence of the loop, whose length is 15–30 bp, can hybridize with the complementary target sequence. The reporter and quenching groups are separately labeled on the 3′ and 5′ ends. The stem part, 5–7 bp long, cannot match the target. When the molecular beacon hybridizes with the target in the reaction system, the hairpin structure is destroyed, and FRET can occur. The most commonly used fluorescent groups are FAM and Texas Red.

8.4.3.7 Locked Nucleic Acids (LNA Probes)

The locked nucleic acid probe is a nucleic acid chemical modification technique developed in recent years. A locked structure is formed by using a methylene bridge to connect the O2 and C4 of the ribose. Due to the use of LNA in the conventional

TaqMan probe, the affinity between the probe and the target sequence is greatly enhanced, and the Tm value of the probe is increased, which can make the TaqMan probe much shorter and thus highly suitable for the detection of short DNA segments. Hence, the detection sensitivity is improved.

8.4.3.8 Complex Probe

The complex probe combines the advantages of molecular beacons and double hybridized probes. The fluorescent probe and quenching probe are synthesized first. The length of the fluorescent probe is 25 bp, and the fluorescent group is labeled on its 5′ end; the quenching probe is approximately 15 bp, and the quenching group is on its 3′ end. These two probes can hybridize with each other, and no fluorescent signal is detected. In PCR amplification, when the target templates are present in the reaction, the specific probe can hybridize with the template first during the annealing stage, and the probe pair is separated. Hence, the fluorescent signal can be detected, and quantitative detection is achieved.

8.4.3.9 AmpliSensor Detection System

The AmpliSensor detection system is a quantitative PCR system developed by BioTronics. Its basic principle is to conduct an asymmetric amplification; in the second round of amplification, the fluorescent and quenching groups are separately labeled at the ends of the complementary strands of the double-stranded signal primers. When the primers are in a double-stranded structure, the fluorescent signal can be detected. During the PCR amplification, the two groups are separated by the denaturation, and in the annealing stage, the signal primers are hybridized with the asymmetric amplification product prior to the complementary probes, and fluorescence is no longer produced. The AmpliSensor detection system uses asymmetric amplification and double-stranded primers, and the specificity and sensitivity are high, similar to the qualities of competitive PCR [33].

8.4.3.10 Simple Probe

The simple probe was developed by Roche Company. Only one reporter fluorescent group is labeled on the 5′end of the probe. There is a "linker" structure between the fluorescent group and the 5′ end of the target-specific sequence on the probe. During PCR amplification, when the probe hybridizes with the target template, the structure of the "linker" is changed, and the fluorescent signal emerges. The benefit of this kind of probe is the absence of the quenching group, which prevents interference by the background fluorescence.

8.4.4 Content of the Data Analysis in Real-Time Quantitative PCR

Because PCR amplification is an exponential amplification process, there are some indexes that appear during data analysis that indicate the specificity or sensitivity of the amplification.

8.4.4.1 Threshold

The threshold is an artificially set value on the fluorescent amplification curve. It can be set at any position during the phase of exponential amplification. Generally, the fluorescence threshold is ten times the standard deviation of the fluorescent signal of the baseline (background).

The amplification kinetics curves of the different tubes are expected to be parallel during the reaction, and the exact value of the threshold level should therefore not be critical. Different instrument software programs use different methods and algorithms to select the threshold.

8.4.4.2 Ct Value

The number of PCR cycles to reach the threshold is called the Ct value. The Ct value is mainly determined by the initial template concentration in the amplification reaction system. If the initial template concentration is high, only a small number of amplification cycles will accumulate enough product to be higher than the background fluorescent signal; therefore, the Ct value is small. In contrast, if the initial concentration of the template is low, more amplification cycles are needed to exceed the background fluorescent signal, producing a large Ct value. The relationship between the concentration of the template and the Ct value is the basis for establishing quantitative detection.

8.4.4.3 Melting Curve and Melting Temperature

The melting curve is the degradation degree curve of the DNA double helix structure with increasing the temperature. The temperature at which half of the total DNA double helix structure has been degraded is called the melting temperature (Tm). Tm values vary with different DNA sequences and are proportional to the GC content. Only real-time PCR using fluorescent dye can generate the melting curve. During the melting curve, the specificity of the amplification primer can be identified, and the different amplicons can be distinguished.

8.4.4.4 Amplification Efficiency

Using the principles of PCR, when the reaction is conducted for one cycle, the amount of the template is doubled. Then, if the reaction has proceeded for N cycles, the amount of the template C_N theoretically equals $2^N \times C_0$ (C_0 is the initial amount of the template). At this point, the theoretical amplification efficiency is 100 %, which is 1. However, in reality, due to factors such as the enzyme, dNTPs, primers, and template, the amplification efficiency is always less than 100 %.

In quantitative PCR analysis, calculation of the starting concentrations for analysis requires an estimate of the efficiency, the setting of a fluorescence threshold, and the determination of the Ct value, which is the fractional cycle number required to reach this threshold. The efficiency value in quantitative PCR analysis was commonly assumed to be a constant derived from a standard curve (the slope of the standard curve reflects the amplification efficiency) or calculated as the mean efficiency per amplicon [34].

In the standard curve, if the horizontal axis shows the initial concentration of the template ($\log_{10}$), and the vertical axis shows the Ct value, then the amplification efficiency E is as follows:

$$E = 10^{-\frac{1}{k}} - 1 \ (k \, \text{is the slope of the standard curve}) \tag{8.4}$$

If the horizontal axis shows the initial Ct value, and the vertical axis shows the concentration of the template (log10), then the amplification efficiency E is as follows:

$$E = 10^{-k} - 1 (k \, \text{is the slope of the standard curve}) \tag{8.5}$$

Actually, the value of the amplification efficiency (E) is approximately 0.8–1.2. The closer the E value is to 1, the better the quantitative PCR reaction is.

8.4.4.5 The Reproducibility

The reproducibility of quantitative PCR is an important parameter in evaluating the results. Usually, in one experiment, 96 parallels are conducted using the same template, and fine differences occur among the Ct values (Fig. 8.1). The standard deviation (SD) and the variable coefficient (VC) are the two indexes that reflect the reproducibility.

During quantitative PCR, many factors can affect the reproducibility:

1. The amplification efficiency. Optimizing the reaction conditions can allow the reaction system to achieve the best amplification efficiency.
2. The initial concentration of the target gene. The sample should have an initial concentration on the same order of magnitudes as the template, or multiple parallel experiments should be performed.

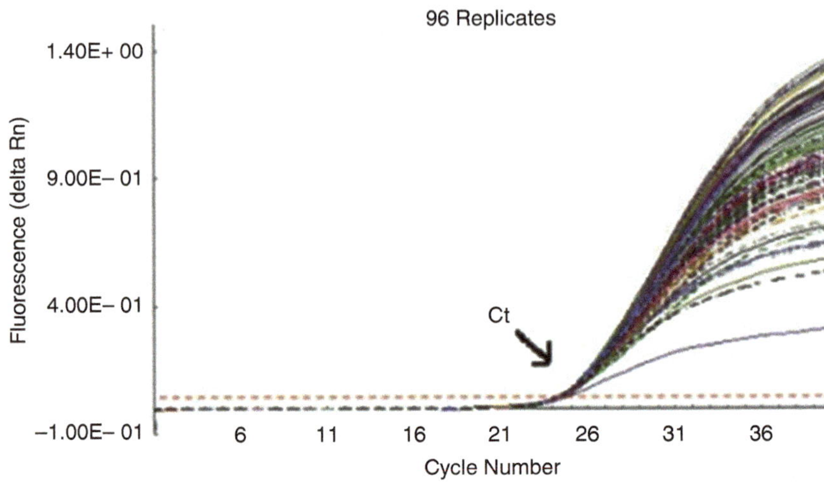

Fig. 8.1 The reproducibility of the same template in 96 amplifications

3. The standard curve. In drawing the standard curve, the concentrations of the standard reference with five gradient dilutions are always employed, and thus, the potential amount of the target gene to be tested can be covered. The ideal standard reference should have high homology compared with the target, such as plasmid DNA or in vitro synthesized or transcribed RNA.

8.4.4.6 Cooperative Verification

In quantitative PCR detection, cooperative verification is a key step in identifying the "plasmid reference material," which is an artificial plasmid containing the target fragment. In particular, in the quantitative detection of genetically modified organisms, the plasmid reference material is widely used to replace the actual genetically modified standard material. After constructing the vector, methodological validations are conducted, including the purity of the plasmid, the quantitative detection standard curve, and reproducibility. Then, cooperative verification is needed.

1. Six or more qualified laboratories are chosen to lead this cooperative verification. The principle of choosing the laboratory is confirmed by ISO/IEC 17025 or the Ministry of Agriculture CNAS-CL01:2006 or by the ability verification plan or by passing a blind test.
2. Alternative study. The chosen laboratories should conduct alternative studies, including the verification of the amplification efficiency, linearly dependent coefficient, and intercept. The verification follows the internationally accredited operational guidelines and analysis results, including ISO 5725-2: 1994. Accuracy (trueness and precision) of measurement methods and results—Part2: Basic

method for the determination of repeatability and reproducibility of a standard measurement method and protocol for the design, conduct, and interpretation of method performance studies [35, 36].

3. The results of the cooperative verification. All the detection results among the laboratories should obey a normal distribution with no outliers.

8.4.5 *Absolute Quantification Detection*

Absolute quantitative PCR is performed based on the use of reference standards. There are three ways to prepare the reference standard: (1) chemically synthesized genes, (2) serially diluted PCR products, and (3) the vector cloned with the PCR products; through extracting the plasmid and measuring its concentration, the copy number can be calculated. The plasmid DNA and the transcriptional RNA in vitro are usually used as the reference standards in absolute quantification [37]. The reference standard is used with different concentrations on the abscissa, while the ordinate is the Ct values obtained for the standard curve for quantitative analysis. Based on the Ct value, the copy number of the unknown sample can be calculated using the standard curve.

Completing the standard curve is crucial. For RNA samples, the standard is to use transcriptional RNA in vitro. Although the preparation and storage of the sample are easy if cDNA or PCR products are used as the standard, these materials cannot indicate the efficiency of transcription in the samples, and thus, the initial copy number is affected.

8.4.6 *Relative Quantification Detection*

Relative quantification refers to measuring a target gene as well as the endogenous housekeeping gene at the same time. Because a housekeeping gene has constant expression across several tissues, its amount can be used as a standard. By comparing the expression differences between the housekeeping gene and the target, relative quantitative detection is realized.

Housekeeping genes, including GAPDH, β2-microglobulin, and rRNA [38], are mainly used for copy number comparison and to make sure there is no influence on the PCR amplification. The copy number of the unknown sample obtained by this method is relative to the reference in terms of quantity; therefore, the standard curve is easy to draw [39].

8.4.6.1 Basic Formula Calculations and Mathematical Relationships in Quantitative PCR

In quantitative PCR, the Ct value is proportional to the initial concentration of the template. The mathematical relationship can be represented by the formula below:

$$R_n = R_B + X_0(1 + E)^n R_S \tag{8.6}$$

R_n represents the total fluorescent signal of the nth thermal cycle.
R_B represents the native signal.
R_S represents the unit signal intensity.
X_0 represents the initial copy number of the DNA.
E represents the amplification efficiency.
When the amplification reaches the Ct cycle,

$$R_T = R_B + X_0(1 + E)^{\mathrm{CT}} R_S \tag{8.7}$$

$$\lg(R_T - R_B) = \lg\mathrm{X}_0 + C_T \lg(1 + E) + \lg R_s \tag{8.8}$$

$$C_T \lg(1 + E) = -\lg X_0 + \lg(R_T - R_B) - \lg R_s \tag{8.9}$$

$$C_T = -\frac{\lg X_0}{\lg(1 + E)} + \frac{\lg(R_T - R_B) - \lg R_s}{\lg(1 + E)} \tag{8.10}$$

Because E, R_B, R_S, and R_T are constants, the last formula above can be represented as follows:

$$C_T = -k \lg\mathrm{X}_0 + b \text{ (a linear equation)} \tag{8.11}$$

Meanwhile, based on the above linear equation, the standard curve is formed using serially diluted reference samples. The abscissa is the log concentration of the initial template, and the ordinate is the Ct value.

In ideal conditions, the amplification efficiency E is regarded as 1, and the slope of the standard curve is −3.3219280. Normally, the slope ranges from −3.5 to −4.2. The formula (I) can be rewritten as follows:

$$R_T = R_B + X_0 2^{\mathrm{CT}} R_S \tag{8.12}$$

$$2^{\mathrm{CT}} = \frac{R_T - R_B}{X_0\, R_S} = \frac{K}{X_0} \tag{8.13}$$

$$C_T = \log_2 \frac{K}{X_0} \tag{8.14}$$

Another analysis method is $2^{-\Delta\Delta\mathrm{Ct}}$ method. The principle of the $2^{-\Delta\Delta\mathrm{Ct}}$ method is to assume that the amplification efficiencies of the target genes and reference are 100 % [40].

$$\Delta\text{Ct}\ (test) = \text{Ct}(\text{target}, test) - \text{Ct}(\text{ref}, test) \tag{8.15}$$

$$\Delta\text{Ct}(control) = \text{Ct}(\text{target}, control) - \text{Ct}(\text{ref}, control) \tag{8.16}$$

The *test* represents the test or treatment group, and the *control* represents the control group. Target represents the interested target gene, and ref represents the housekeeping gene.

The concentration of the unknown sample can be simplified to:

$$\Delta\Delta\text{Ct} = \Delta\text{Ct}(test) - \Delta\text{Ct}(control) \tag{8.17}$$

$$2^{-\Delta\Delta\text{Ct}} = \text{amount of the target} \tag{8.18}$$

The advantage of this method is that the amplification efficiency is not calculated: thus, it is easy to perform with high efficiency. It is also unnecessary to calculate the standard curves of the target and housekeeping genes in the control group each time. This analysis method is one of the mainstream methods currently used for relative quantitative analysis. However, the disadvantage is that the accuracy is relatively low, especially when inhibitors are present in the sample, which can cause the actual amplification efficiency to be significantly less than the ideal amplification efficiency (100 %), resulting in a larger error in the analysis results.

8.4.6.2 The Double-Standard Curve Method

The principle of the double-standard curve method is based on the amplification of a series of gradient dilutions of standard templates (the concentrations are unknown), obtaining the Ct values of each sample. Then, the Ct values are taken as the ordinate and the logarithm of the initial concentration of the template ($\lg X_0$) as the abscissa. Linear regression analysis is performed, and the amplification efficiency is calculated based on the slope. Finally, the amplification efficiency of the target gene and the reference gene can be analyzed. The formula can be represented as follows:

$$R = \frac{{E_{\text{target}}}^{\Delta\text{Ct}_{\text{target}}(control\text{-}test)}}{{E_{\text{ref}}}^{\Delta\text{Ct}_{\text{ref}}(control\text{-}test)}} \tag{8.19}$$

The advantage of this method is its accuracy, and the *E* values of the target and reference genes are not the same, but multiple standard curves must be built each time. Hence, the workload is high and the efficiency low. It is currently only used for the accurate study of quantitative analysis.

Many kinds of real-time quantitative PCR methods are due to its widespread use in modern quantitative detection, especially in food safety.

- *Pathogenic bacteria detection*. Bhagwat applied real-time PCR to detect *Salmonella* bacteria in food and feed samples for quantitative microbial risk assessment [41]. A protocol for the simultaneous detection of *Escherichia coli* O157:H7, *Listeria monocytogenes*, and *Salmonella* strains was also devised by Bhagwat et al. [42]. The detection limit in artificially contaminated water was 1–10 cells/ml for *E. coli* O157:H7 and 1000 cells/ml for *S. typhimurium* and *L. monocytogenes*.

- *Meat species identification.* Dooley et al. developed species-specific real-time PCR (TaqMan) assays for detecting beef, pork, lamb, chicken, and turkey [43]. It was possible to detect each species when spiked in any other species at a 0.5 % level in the raw meat, and the experimentally determined limits for beef, lamb, and turkey were below 0.1 %. Fajardo et al. used SYBR Green real-time PCR to successfully detect red deer, fallow deer, and roe deer DNA in meat mixtures, and the quantification of the target cervid DNAs was in the range 0.1–0.8 % [44].
- *Fruit species identification.* Xu et al. selected the papaya-specific gene, *papain*, as an endogenous reference gene in the qualitative and real-time quantitative PCR detection of papayas. In real-time quantitative PCR analysis, the detection limit was as low as 10 pg of DNA [45]. Shang Y et al. chose a peach-specific gene, *Lhcb2*, as an endogenous reference gene for the real-time quantitative detection of fruit products that contain peach components, and the detection limit of quantitative PCR was as low as 5 pg of DNA [2].
- *Identification of GM food.* GM crops including soybean, maize, rice, cotton, and canola can be specifically quantitatively detected, and the detection limit can be used to satisfy the GM-labeling policy or the limitation standards. To summarize, real-time PCR is an efficient, accurate, simple, and cost-saving quantitative detection technique in the field of food safety.

8.4.6.3 Multiplex Quantification Combined with Multivariate Statistics

The combination of real-time PCR with multivariate statistical analysis was first developed by Xu et al. in 2009 for the accurate quantification of stacked GM maize Bt11 × GA21 in a 100 % GM maize mixture (Bt11×GA21, Bt11 and GA21) [46].

Based on the construction specificities of the three GM maizes, the following mathematical relations can be constructed for the four genes (*IVR*, *NOS*, *Bt11event*, *GA21 event*):

$$\begin{cases} X+Y+Z=100\% & \text{(a)} \\ X+Y=C_{\text{Bt11}}/C_{\text{ivr}} & \text{(b)} \\ 3X+2Y+Z=C_{\text{NOS}}/C_{\text{ivr}} & \text{(d)} \end{cases} \qquad \begin{cases} X+Y+Z=100\% & \text{(a)} \\ X+Z=C_{\text{GA21}}/C_{\text{ivr}} & \text{(c)} \\ 3X+2Y+Z=C_{\text{NOS}}/C_{\text{ivr}} & \text{(d)} \end{cases}$$

$$\begin{cases} X+Y+Z=100\% & \text{(a)} \\ X+Y=C_{\text{Bt11}}/C_{\text{ivr}} & \text{(b)} \\ X+Z=C_{\text{GA21}}/C_{\text{ivr}} & \text{(c)} \end{cases}$$

C_{Bt11}, copy number of Bt11 event-specific sequence
C_{GA21}, copy number of GA21 event-specific sequence

C_{NOS}, copy number of NOS gene
C_{IVR}, copy number of IVR gene
X, content of maize Bt11 × GA21 in 100 % GM mixture
Y, content of maize Bt11 in 100 % GM mixture
Z, content of maize GA21 in 100 % GM mixture

When the three kinds of GM maize were mixed together, the percentage of each GM maize satisfied the four equations above; any three of these equations could compose a three-linear-equation system; and the coefficient was determined by the gene composition of the three GM maize materials.

During the quantification, the copy number of the four target genes could be obtained first by real-time PCR as the known data X, Y, and Z in the equations. The contents of each GM maize in the mixture are taken as the three unknown variables of ternary linear equation sets, and the average values of the solutions from the three equations are taken as the accurate contents of each GM maize in the mixture.

In these calculations, the detection limit was as low as 0.5 %; when the content of stacked maize was lower than the detection limit, the test result was far from the true value. The traditional quantitative method of detecting stacked GM organisms is to apply individual kernel analysis, which always suffers from lower sensitivity and the disparity of different primers.

In this novel method, the detection limit is lower than the GM-labeling limit of the European Union and satisfies the requirements of all the countries for the labeling test. The test results were proven to be close to the real value. This approach solves the problem of detecting stacked maize in a GM mixture and may find wide application in the detection of other GM maize crops, other stacked GM plants like soybean and rice, and even food products containing stacked events.

8.5 Digital PCR

Digital PCR (dPCR) was first developed by Vogelstein and Kinzler to extend the applications of conventional PCR in 1999. It differs from traditional real-time quantitative PCR in that the direct counting of the target molecules without any calibration or external standard of reference allows dPCR to achieve absolute quantification. The dPCR technique transforms the exponential data of traditional PCR into a digital signal.

In 1992, Sykes et al. reported a method of quantitating the total number of initial targets present in a sample using limiting dilution, PCR, and Poisson statistics, and the concept of digital PCR was proposed [47]. Kalinina et al. used a glass capillary to conduct the PCR reaction in the volume of a microliter, collecting the signal of single DNA molecules on the fluorescent probes, followed by the development of single-molecule quantitative techniques [48]. In 1998, the water-in-oil emulsion for PCR was developed. During the simulation of the reaction needed to complete life activities, Tavvfik et al. proposed emulsion PCR [49]. An emulsifier was added to the PCR system, and through separating the microparticles, the amplifications could be conducted in the micro "room" without affecting each other.

Based on the abovementioned techniques, dPCR was developed in 1999 [50]. This experimental approach involves two components. First, the DNA to be analyzed is diluted into multi-well plates with one template molecule per two wells (on average), and PCR is performed in optimal conditions designed to amplify a single copy of the PCR template. The amplicons are hybridized with fluorescence probes, such as molecular beacons, that allow the detection of sequence-specific products using different fluorescent groups. Thus, digital PCR is employed to directly count, one by one, the number of each of the two (paternal/maternal or wild type/mutant) alleles in the samples. Second, several statistical analyses, including Bayesian-type likelihood methods, can be applied to measure the strength of the evidence for the allele distribution being different from normal. This approach supplies a rigorous statistical basis to analyze the allelic status and is expected to provide more reliable information than was heretofore possible in allelic studies of tissue or body fluid samples. Thus, digital PCR transforms the exponential and analog signals of conventional PCR to linear and digital signals.

There are several advantages of dPCR compared with other quantitative methods. First, compared with microsatellite markers, the PCR products derived from the two SNP alleles at every locus are the same size, and therefore, their analysis is not biased by the preferential DNA degradation of larger alleles. Second, the digital PCR approach, which amplifies single-allele templates in the PCR reaction, can precisely determine the number of alleles examined in each experiment. Accordingly, SNP genotyping is digital, involving the detection of the presence or absence of a specific allele, rather than analog, like microsatellite genotyping, which measures the length of microsatellites. Third, a statistical method such as SPRT can be employed to determine whether AI is present in the background DNA. However, there are some disadvantages that cannot be ignored. The instrument for dPCR is expensive, and the conditions of the reaction must always be carefully optimized.

Because dPCR can achieve absolute quantification, it has wide application prospects, such as mutational analysis, assessment of allelic imbalance in tissue, cancer detection in body fluid DNA or in stool DNA, quantification of the gene expression of specific alleles, measurement of copy number, and detection of GMOs and metal ions.

8.6 Third-Generation Sequencing

Third-generation sequencing technology has been developed to avoid the deviation caused by amplification and allows the direct sequencing of a single DNA molecule. Three technology platforms comprise this third generation. Two achieve single-molecule sequencing by incorporating and detecting fluorescently labeled nucleotides: Helicos Genetic Analysis System and Pacific Biosciences single-molecule real-time (SMRT) technology. The third, Oxford Nanopore's nanopore sequencing, is likely to be the cheapest of the three. "Nanopore sequencing has the advantage that it does not require any labeling of the DNA, no expensive fluorescent reagents or really expensive CCD [charge-coupled device] cameras to record from optical chips," says Hagan Bayley from Oxford Nanopore Technologies [51].

Third-generation sequencing has been widely used in research fields of life sciences, such as proteomics, genomics, whole-cell electron microscopy, neuroscience and protein biochemistry, transcriptome sequencing, and methylation. In genomics, especially in research involving whole-genome sequencing and resequencing, and metagenome sequencing, third-generation sequencing has greatly promoted whole-genome sequencing projects for non-model species, and more and more organism genomes have been published [52, 53].

Third-generation sequencing has many advantages. (1) The cost is dramatically reduced. The cost of whole-genome sequence might decrease to a thousand dollars by using third-generation sequencing. (2) The sequencing reads are longer. In NGS, the length of the sequencing reads ranges from 100 bp to approximately 500 bp. For third-generation sequencing, the longest sequencing length can reach 40,000 bp, which makes the mapping of the reads much more accurate. This technique may quickly replace NGS in evacuating unknown gene sequences due to this advantage. In addition, third-generation sequencing can be used in accurate DNA detection based on its low cost [54, 55].

However, there are many obstacles to the applications of third-generation sequencing: the quality of the data needs to be enhanced, the construction of the genome library needs to be simplified, and the data analysis and information platform are not yet perfected. Still, third-generation sequencing has great potential and advantages in bioinformatics and gene excavation.

8.7 Prospects

Quantitative PCR allows the sensitive, accurate, and specific detection of nucleic acids. Since its development, real-time PCR has revolutionized the field of molecular diagnostics and is being used in a rapidly expanding number of applications with the development of biotechnology.

Acknowledgments This work is supported by National Science and Technology Major Project (2016ZX08012-004). Many thanks to Ying Shang, for her kindly help in manuscript conception and preparation.

References

1. Yuan Y, Xu W, Zhai Z, Shi H, Luo Y, Chen Z, Huang K. Universal primer-multiplex PCR approach for simultaneous detection of Escherichia coli, Listeria monocytogenes, and Salmonella spp. in food samples. J Food Sci. 2009;74(8):M446–52.
2. Shang Y, Zhu P, Huang K, Liu W, Tian W, Luo Y, Xu W. A peach (Prunus persica)-specific gene, Lhcb2, used as an endogenous reference gene for qualitative and real-time quantitative PCR to detect fruit products. LWT-Food Sci Technol. 2014;55(1):218–23.
3. Xu W, Bai W, Luo Y, Yuan Y, Zhang W, Guo X, Huang K. A novel common single primer multiplex polymerase chain reaction (CSP-M-PCR) method for the identification of animal species in minced meat. J Sci Food Agric. 2008;88(15):2631–7.
4. Huang J, Wang X, Hao B, Chen Y, Liu H, Quan L, Tang D, Sheng L, Li M, Huang E. Genetic variants in KCNE1, KCNQ1, and NOS1AP in sudden unexplained death during daily activities in Chinese Han population. J Forensic Sci. 2015;60(2):351–6.
5. Van der Velden V, Hochhaus A, Cazzaniga G, Szczepanski T, Gabert J, Van Dongen J. Detection of minimal residual disease in hematologic malignancies by real-time quantitative PCR: principles, approaches, and laboratory aspects. Leukemia. 2003;17(6):1013–34.
6. Gachet E, Martin G, Vigneau F, Meyer G. Detection of genetically modified organisms (GMOs) by PCR: a brief review of methodologies available. Trends Food Sci Technol. 1998;9(11):380–8.
7. Huang S, Xu Y, Yan X, Shang Y, Zhu P, Tian W, Xu W. Development and application of a quantitative loop-mediated isothermal amplification method for detecting genetically modified maize MON863. J Sci Food Agric. 2015;95(2):253–9.
8. Kay S, Van den Eede G. The limits of GMO detection. Nat Biotechnol. 2001;19(5):405.
9. Wolf C, Scherzinger M, Wurz A, Pauli U, Hübner P, Lüthy J. Detection of cauliflower mosaic virus by the polymerase chain reaction: testing of food components for false-positive 35S-promoter screening results. Eur Food Res Technol. 2000;210(5):367–72.
10. Song Q, Wei G, Zhou G. Analysis of genetically modified organisms by pyrosequencing on a portable photodiode-based bioluminescence sequencer. Food Chem. 2014;154:78–83.
11. Huber CG. Biopolymer chromatography. In: Encyclopedia of Analytical Chemistry. Chichester: Wiley; 2000.
12. Dehainault C, Lauge A, Caux-Moncoutier V, Pages-Berhouet S, Doz F, Desjardins L, Couturier J, Gauthier-Villars M, Stoppa-Lyonnet D, Houdayer C. Multiplex PCR/liquid chromatography assay for detection of gene rearrangements: application to RB1 gene. Nucleic Acids Res. 2004;32(18):e139.
13. Palma F, Potenza L, Amicucci A, Fiorani M, Labella D, Di Biase S, Stocchi V. HPLC and CE analysis of PCR products: a comparative study. J Liq Chromatogr Relat Technol. 1998;21(10):1527–40.
14. Wilde C, Schmidt G, Moellenhoff E, Chatterjee B, Nakatani S, Tsuda T, Yamasaki Y, Moriyama H, Watanabe H, Kato Y. Quantification of gene expression by combining competitive RT-PCR with HPLC analysis. LC GC EUROPE. 2003. p. 17–9.
15. Epp LS, Stoof-Leichsenring KR, Trauth MH, Tiedemann R. Molecular profiling of diatom assemblages in tropical lake sediments using taxon-specific PCR and Denaturing High-Performance Liquid Chromatography (PCR-DHPLC). Mol Ecol Resour. 2011;11(5):842–53.
16. Griffin TJ, Smith LM. Single-nucleotide polymorphism analysis by MALDI–TOF mass spectrometry. Trends Biotechnol. 2000;18(2):77–84.
17. Yang H, Yang K, Khafagi A, Tang Y, Carey T, Opipari A, Lieberman R, Oeth P, Lancaster W, Klinger H. Sensitive detection of human papillomavirus in cervical, head/neck, and schistosomiasis-associated bladder malignancies. Proc Natl Acad Sci U S A. 2005;102(21):7683–8.
18. Pomerantz MM, Beckwith CA, Regan MM, Wyman SK, Petrovics G, Chen Y, Hawksworth DJ, Schumacher FR, Mucci L, Penney KL. Evaluation of the 8q24 prostate cancer risk locus and MYC expression. Cancer Res. 2009;69(13):5568–74.

19. Zentilin L, Giacca M. Competitive PCR for precise nucleic acid quantification. Nat Protoc. 2007;2(9):2092–104.
20. Hübner P, Studer E, Lüthy J. Quantitative competitive PCR for the detection of genetically modified organisms in food. Food Control. 1999;10(6):353–8.
21. Choi WS, Hong C-H. Rapid enumeration of Listeria monocytogenes in milk using competitive PCR. Int J Food Microbiol. 2003;84(1):79–85.
22. Higuchi R, Fockler C, Dollinger G, Watson R. Kinetic PCR analysis: real-time monitoring of DNA amplification reactions. Biotechnology. 1993;11:1026–30.
23. Clegg RM. Fluorescence resonance energy transfer. Curr Opin Biotechnol. 1995;6(1):103–10.
24. Kimura H, Morita M, Yabuta Y, Kuzushima K, Kato K, Kojima S, Matsuyama T, Morishima T. Quantitative analysis of Epstein-Barr virus load by using a real-time PCR assay. J Clin Microbiol. 1999;37(1):132–6.
25. Matsuki T, Watanabe K, Fujimoto J, Takada T, Tanaka R. Use of 16S rRNA gene-targeted group-specific primers for real-time PCR analysis of predominant bacteria in human feces. Appl Environ Microbiol. 2004;70(12):7220–8.
26. Berdal KG, Holst-Jensen A. Roundup Ready® soybean event-specific real-time quantitative PCR assay and estimation of the practical detection and quantification limits in GMO analyses. Eur Food Res Technol. 2001;213(6):432–8.
27. Hébert B, Bergeron J, Potworowski EF, Tijssen P. Increased PCR sensitivity by using paraffin wax as a reaction mix overlay. Mol Cell Probes. 1993;7(3):249–52.
28. Kellogg D, Rybalkin I, Chen S, Mukhamedova N, Vlasik T, Siebert P, Chenchik A. TaqStart Antibody: "hot start" PCR facilitated by a neutralizing monoclonal antibody directed against Taq DNA polymerase. BioTechniques. 1994;16(6):1134–7.
29. Moretti T, Koons B, Budowle B. Enhancement of PCR amplification yield and specificity using AmpliTaq Gold DNA polymerase. Biotechniques. 1998;25(4):716–22.
30. Ginzinger DG. Gene quantification using real-time quantitative PCR: an emerging technology hits the mainstream. Exp Hematol. 2002;30(6):503–12.
31. Nazarenko I, Lowe B, Darfler M, Ikonomi P, Schuster D, Rashtchian A. Multiplex quantitative PCR using self-quenched primers labeled with a single fluorophore. Nucleic Acids Res. 2002;30(9):e37.
32. Lowe B, Avila HA, Bloom FR, Gleeson M, Kusser W. Quantitation of gene expression in neural precursors by reverse-transcription polymerase chain reaction using self-quenched, fluorogenic primers. Anal Biochem. 2003;315(1):95–105.
33. Wang C, Wu KY, Wang H-T. Quantitative PCR using the AmpliSensor assay. PCR Primer: a laboratory manual. Cold Spring Harbor: Cold Spring Harbor Laboratory Press; 1995. p. 193–202.
34. Ruijter J, Ramakers C, Hoogaars W, Karlen Y, Bakker O, Van den Hoff M, Moorman A. Amplification efficiency: linking baseline and bias in the analysis of quantitative PCR data. Nucleic Acids Res. 2009;37(6):e45.
35. Iso I. Accuracy (trueness and precision) of measurement methods and results-Part 6: use in practice of accuracy values. Geneva: International Organization for Standardization; 1994. p. 5725–6.
36. Horwitz W. Protocol for the design, conduct and interpretation of method-performance studies: revised 1994 (Technical Report). Pure Appl Chem. 1990;62(6):1193–208.
37. Zimmermann B, Holzgreve W, Wenzel F, Hahn S. Novel real-time quantitative PCR test for trisomy 21. Clin Chem. 2002;48(2):362–3.
38. Galbiati S, Smid M, Gambini D, Ferrari A, Restagno G, Viora E, Campogrande M, Bastonero S, Pagliano M, Calza S. Fetal DNA detection in maternal plasma throughout gestation. Hum Genet. 2005;117(2–3):243–8.
39. Wadowsky RM, Laus S, Green M, Webber SA, Rowe D. Measurement of Epstein-Barr virus DNA loads in whole blood and plasma by TaqMan PCR and in peripheral blood lymphocytes by competitive PCR. J Clin Microbiol. 2003;41(11):5245–9.

40. Kang TS, Korber DR, Tanaka T. Contributions of citrate in redox potential maintenance and ATP production: metabolic pathways and their regulation in Lactobacillus panis PM1. Appl Microbiol Biotechnol. 2013;97(19):8693–703.
41. Bhagwat AA. Rapid detection of Salmonella from vegetable rinse-water using real-time PCR. Food Microbiol. 2004;21(1):73–8.
42. Bhagwat AA. Simultaneous detection of Escherichia coli O157: H7, Listeria monocytogenes and Salmonella strains by real-time PCR. Int J Food Microbiol. 2003;84(2):217–24.
43. Dooley JJ, Paine KE, Garrett SD, Brown HM. Detection of meat species using TaqMan real-time PCR assays. Meat Sci. 2004;68(3):431–8.
44. Fajardo V, González I, Martín I, Rojas M, Hernández PE, García T, Martín R. Real-time PCR for detection and quantification of red deer (Cervus elaphus), fallow deer (Dama dama), and roe deer (Capreolus capreolus) in meat mixtures. Meat Sci. 2008;79(2):289–98.
45. Xu W, Bai W, Guo F, Luo Y, Yuan Y, Huang K. A papaya-specific gene, papain, used as an endogenous reference gene in qualitative and real-time quantitative PCR detection of transgenic papayas. Eur Food Res Technol. 2008;228(2):301–9.
46. Xu W, Yuan Y, Luo Y, Bai W, Zhang C, Huang K. Event-specific detection of stacked genetically modified maize Bt11× GA21 by UP-M-PCR and Real-Time PCR. J Agric Food Chem. 2008;57(2):395–402.
47. Sykes P, Neoh S, Brisco M, Hughes E, Condon J, Morley A. Quantitation of targets for PCR by use of limiting dilution. Biotechniques. 1992;13(3):444–9.
48. Kalinina O, Lebedeva I, Brown J, Silver J. Nanoliter scale PCR with TaqMan detection. Nucleic Acids Res. 1997;25(10):1999–2004.
49. Tavvfik DS, Griffiths AD. Man-made cell-like compartments for molecular evolution. Nat Biotechnol. 1998;16(7):652–6.
50. Pohl G, Shih I-M. Principle and applications of digital PCR. Expert Rev Mol Diagn. 2004;4 (1):41–7.
51. Rusk N. Cheap third-generation sequencing. Nat Methods. 2009;6(4):244.
52. Li R, Fan W, Tian G, Zhu H, He L, Cai J, Huang Q, Cai Q, Li B, Bai Y. The sequence and de novo assembly of the giant panda genome. Nature. 2010;463(7279):311–7.
53. Huang S, Li R, Zhang Z, Li L, Gu X, Fan W, Lucas WJ, Wang X, Xie B, Ni P. The genome of the cucumber, Cucumis sativus L. Nat Genet. 2009;41(12):1275–81.
54. Leoni C, Volpicella M, De Leo F, Gallerani R, Ceci LR. Genome walking in eukaryotes. FEBS J. 2011;278(21):3953–77.
55. Thirulogachandar V, Pandey P, Vaishnavi C, Reddy MK. An affinity-based genome walking method to find transgene integration loci in transgenic genome. Anal Biochem. 2011;416 (2):196–201.

Chapter 9
Recent Progress in High-Throughput Detection Technology for Food Safety

Abstract Nucleic acid detection (specifically, DNA and RNA analysis) is of great importance to clinical diagnosis, forensic analysis, biological and biomedical fields, and food safety. Because of the increasing number of targets to be detected, traditional single detection techniques cannot fulfill the necessary requirements. Hence, multiple and high-throughput detection methods have been developed. These technologies can save time and reduce costs; moreover, the validation of their results is simpler and more intuitive. This review mainly describes the basic principle as well as the advantages and applications of each commonly used high-throughput detection method.

Keywords Multiplex PCR • Universal primer • High-throughput detection • DNA chips • Pyrosequencing

9.1 Introduction

With growing detection demands and the development of new techniques, common single PCR (polymerase chain reaction) methods can no longer satisfy current requirements. To reduce costs and shorten detection periods, multiplex PCR detection techniques have emerged at precisely the right moment. A method of this kind can achieve not only the simultaneous detection of multiple targets but also high-throughput detection.

Multiplex PCR has the same specificity and sensitivity of single PCR, offers quick and economic operation, and shows a great deal of flexibility in the design of the primers and PCR reaction conditions.

As a highly efficient and economic technology, multiplex PCR has been widely used in scientific research, in the medical field, in business, and elsewhere, for applications such as pathogenic microorganism detection, gene expression, whole-genome sequencing, forensic identification, and infectious disease diagnosis. In recent years, multiplex PCR has also become the core technology for high-throughput SNP classification research [1–5]. With the increasing variety of unsafe factors in food products, such as pathogenic bacteria, adulteration, the addition of nonfood substances, and the abuse of food additives, multiple detection techniques have brought considerable convenience to regulation organizations.

W. Xu, *Functional Nucleic Acids Detection in Food Safety*,
DOI 10.1007/978-981-10-1618-9_9

9.2 Techniques That Achieve the Simultaneous Detection of Multiple Targets

9.2.1 Common Multiplex PCR

Multiplex PCR (M-PCR) is a variant of traditional PCR, and its principle is the same as that of traditional single PCR, except that more than one pair of primers is required in the reaction system. These primers can specifically combine with their corresponding DNA templates, allowing more than one DNA fragment to be simultaneously amplified in one reaction [6].

In the context of food safety and labeling, we consider the detection of GMOs as an example. With the development of GM crops, an increasing number of exogenous genes must be detected and an increasing capability for event specificity in detection is needed to prevent the import of unauthorized GM crops. Hence, one of the most common applications of multiplex PCR is GMO detection. It has been reported that GM maize Mon810 and NK603 could be simultaneously detected via multiplex PCR with a detection limit lower than 0.5 % [7]. A multiplex PCR system that could simultaneously detect GM maize MON810, Bt11, Bt 176, and GA21 and GM soybean RRS was built [8] with nine kinds of primer pairs to detect relatively stable transgenic components from GM soybeans, corn, and canola; the detection limit was as low as 0.25 %, and the system could detect 185 samples (soybean, corn, canola, and their processed products) in 1–2 days. All of the research cited above indicates that M-PCR is a fast, reliable, sensitive, accurate, specific, simple, and low-cost technology for detecting multiple targets.

Although M-PCR possesses many advantages, it also suffers several disadvantages that cannot be ignored: (1) self-inhibition among different sets of primers, (2) low amplification efficiency, and (3) a lack of identical efficiency among different templates. These shortcomings restrict its further development and broad application, especially in high-throughput detection. Even the reported nine-target M-PCR method cannot avoid the disadvantage of poor reproducibility and stability.

9.2.2 Universal Primer Multiplex PCR

To overcome the disadvantages of common M-PCR, especially the self-inhibition among different sets of primers and the low amplification efficiency, a novel technique called Universal Primer Multiplex PCR (UP-M-PCR) with sequencing gel electrophoresis analysis was developed [9].

Universal primer (UP) is a key component in this UP-M-PCR reaction. In the developed system, both the UP and compound-specific primers are added. A compound-specific primer consists of two components: one is the specific sequence that can hybridize with the target, and the other is a common sequence at its 5′ end;

this latter sequence is the same as that of the UP and has no binding site, and the annealing temperature of this common sequence is lower than that of the specific sequence.

During amplification, in the first few cycles, the compound-specific primers bear the primary responsibility for amplifying the target sequences because of their higher annealing temperature, whereas the UP produces nearly no amplification because of its lack of a binding site. Once the compound-specific primers have been used up and amplified products containing the complementary sequence of the UP have accumulated, the UP begins to play a leading role, taking the amplicons as templates and demonstrating its ability to amplify fragments of different targets. Meanwhile, the factors that affect the amplification of UP-M-PCR, such as the concentrations of the compound-specific primers and the universal primer, the *Taq* polymerase, and the annealing temperature, need to be optimized carefully.

The UP-M-PCR method was originally developed by Xu et al. [10], and it was first applied for the detection of stacked GM maize Bt11×GA21. Subsequently, it has been used to simultaneously detect four meat species [11] and three foodborne pathogens [12]. Zhang also used UP-M-PCR combined with capillary electrophoresis with laser-induced fluorescence (CE-LIF) detection to simultaneously detect six GM maize lines [13].

In the above applications, the number of targets is up to 6 at most. However, with further development, the throughput must improve. Because of the presence of multiple amplicons of different lengths in the same reaction, a product separation technique with high resolution, such as polyacrylamide sequencing gel electrophoresis (with a resolution as sensitive as 1 bp), is usually employed. By applying this technique, as many as nine targets in GM crops, including nine commonly used selectable marker and reporter genes, as well as six endogenous genes of six of the most common GM crops, papaya, maize, soybeans, rice, cotton, and canola, have been successfully detected and separated. The detection limit was as low as 0.1 ng for a single type of crop when screening for the presence of GM crops in mixed samples.

With the development of detection instruments, many new methods, such as multiple digital PCR and microemulsion PCR, have been developed based on UP-M-PCR. Through this research, UP-M-PCR has become able to satisfy the demand for verification of different targets in mixed food samples, and it has become widely used in high-throughput detection.

9.2.3 Multiplex Ligation-Dependent Probe Amplification

High specificity is important and difficult to achieve in multiplex PCR. Researchers have made many improvements in an effort to resolve this problem. Multiplex ligation-dependent probe amplification (MLPA) was first developed in 2002 by Schouten et al. [14] to resolve the problems associated with conventional multiplex

PCR, especially the specificity of the amplification. As many as 40 nucleic acid sequences could be relatively quantified using this method.

The MLPA reaction is separated into two steps. The first step is probe ligation. Two ligation probes (LPs), including a forward and a reverse probe, are needed. Each probe always consists of three components: the target-specific sequence, the stuffer sequence, and the common sequence. Both target-specific sequences on the probes can be ligated with each other in the presence of the ligase when hybridizing to the target of interest, and the length of this component usually ranges from 25 to 43 bp. All ligated probes have an identical sequence, called the common sequence, at their 5′ and 3′ ends, permitting simultaneous amplification in a PCR using only one primer pair. The stuffer sequence, with a length from 19 to 370 bp, is located between the target-specific sequence and the common sequence, and its function is to adjust the length of the ligation product, consequently realizing the separation of the products in gel electrophoresis through the creation of different lengths. After the ligation reaction, each probe gives rise to an amplification product of a unique size between 130 and 480 bp.

The second step is common PCR amplification, the principle of which is similar with that of UP-M-PCR. Because the 5′ and 3′ ends of each probe pair contain the same common sequence, the UPs used in this step are designed based on this common sequence.

Because of the small differences among the lengths of the amplicons, the products are always analyzed using capillary electrophoresis (CE). Moreover, the ligation probes are always labeled with fluorescence to permit relative quantitative detection. Since the emergence of this method, it has been widely used in medical diagnostics and clinical applications [15, 16]. Because of the high specificity of MLPA, this method has also now been used to detect multiple GMO events [17–19].

9.2.4 *Single Universal Primer Multiplex Ligation-Dependent Probe Amplification*

The single universal primer multiplex ligation-dependent probe amplification (SUP-MLPA) technique, in which only one universal primer is used to perform multiplex PCR, was developed based on MLPA by Shang et al. in 2013 [20], and it was reported for the detection of multiple GM crops.

In common MLPA, CE and agarose gel electrophoresis are always used as the detection apparatus [21, 22]. However, the resolution of agarose gel is always greater than at least 20–30 bp, which restricts the flexibility of the primer design for common detection. Second, to satisfy the requirements of CE detection, the ligation probes must always be labeled with fluorescent groups. Therefore, this detection method is expensive and is subject to high requirements on the PCR products for which it is suitable. Moreover, the detection apparatus is often incompatible with the high-throughput detection of complex samples. In common MLPA, two universal PCR primers are required to perform the amplification after the ligation reaction, which makes the reaction system more complex.

In view of the shortcomings discussed above, SUP-MLPA combined with sequencing gel electrophoresis (SGE) was developed. In this method, only a single universal primer is used to perform the PCR amplification after the ligation reaction; therefore, the common sequences at the 5′ end of the forward LP and the 3′ end of the reverse LP are reverse complementary to allow the single universal primer to bind with both of them.

A subtle issue arises in the design of ligation probes: when the annealing temperature of the target-specific sequences is higher than that of the common sequences, then during thermal cycling, the target-specific sequences will hybridize with the templates first. By contrast, the common sequences on the two LPs will hybridize with themselves. To achieve successful amplification while preventing hybridization between the LPs, the annealing temperatures of each component of the LPs and the SUP must be carefully and strictly designed. In SUP-MLPA, the LP concentrations, feasibility, and PCR enhancer are all optimized, and the detection limit for the target gene is 0.1 ng of target DNA per reaction, which is comparable to the limits for other M-PCR methods.

In an electrophoresis-based analysis, all amplicons are separated on the gel based on the differences in their lengths; MLPA is no exception. However, limited by the annealing temperature of the target-specific sequences, the region that is suitable for the design of target-specific sequences is short, and the lengths of the target-specific sequences for all probes may be the same; as a result, the lengths of the amplification products in multiplex detection typically have little diversity.

To facilitate separation, a stuffer sequence is added to each LP between the target-specific sequence and the common sequence. To determine whether the different lengths of the stuffer sequences on the LPs would affect the efficiency of the ligation reaction or the amplification, a series of LPs with stuffer sequences of different lengths was designed, and the quantitative PCR results showed that the length of the stuffer sequence did not affect the ligation efficiency. Therefore, the stuffer sequence can be easily adjusted to achieve better separation based on the size of the products, thus increasing the multiplicity of the assay. This new SUP-MLPA technique can be used in all fields related to multiplex PCR and has great potential for and value in application.

To summarize the MLPA and SUP-MLPA methods discussed above, amplification can be performed only when the specific sequences on the ligation probes hybridize with the targets. Hence, the specificity of these two methods is very high. In highly processed products, such as edible oil, the genomes are fragmented into small segments during processing, causing traditional PCR amplification methods to be difficult to perform and to exhibit poor specificity. However, based on the ligation reaction of the probes and the short length of the specific sequences on the ligation probes, MLPA and SUP-MLPA are well suited for the detection of short DNA fragments.

9.2.5 Multiplex PCR Combined with Pyrosequencing

In traditional multiplex PCR, the detection strategy is based solely on the different lengths of the amplicons, and the amplicons are always separated and analyzed via gel electrophoresis. The most important factor in the design of the primers is the size of the products, which can lead to amplification differences and a negative influence between primer pairs.

Moreover, in existing multiplex detection methods, regarding the sequencing gel electrophoresis, skilled operation is necessary and the time required for sequence analysis is long. The resolution of common agarose gel electrophoresis is poor for the separation of multiple amplicons, and ethidium bromide (EB) dye is toxic to the human body. Consequently, analysis of the electrophoresis segregation pattern is time- and labor-consuming, requiring a carefully optimized detection method. Additionally, the nonspecific amplification affects the accuracy of the detection results.

For gene detection, PCR amplification followed by sequencing is the "gold standard" approach. Based on this understanding, a method was developed in which the PCR products are directly subjected to pyrosequencing immediately after amplification. The products are verified solely based on their different sequences.

Pyrosequencing, which is a series of enzyme cascade reactions that is based on the real-time bioluminometric quantification of pyrophosphate (PPi) released during deoxyribonucleoside triphosphate (dNTP) incorporation, has been widely applied for single-nucleotide polymorphism (SNP) typing, microbe genotyping, gene expression analysis, pharmacogenomics, polymorphic analysis of plants, gene methylation analysis, and analysis of GMOs [23–25]. This technology has the advantages of accuracy, a lack of need for labeling with fluorescent groups, and parallel processing; therefore, it is a powerful tool in the field of nucleic acid analysis.

Based on the aforementioned advantages offered by pyrosequencing, a novel multiplex qualitative detection method coupled with UP-M-PCR and pyrosequencing was established. In accordance with the principle of pyrosequencing, the dNTP dispensing order is designed based on the different sequences of the PCR products, and qualitative detection is achieved through observation of the characteristic peak signal. Therefore, when designing the primers, the amplification efficiency is the first factor to be considered rather than the length of the product. The amplification products are directly pyrosequenced, thereby not only realizing a qualitative and quantitative analysis but also incorporating the sensitivity and accuracy of PCR technology. Meanwhile, because of the presence of the same common sequences at the ends of the amplicons by virtue of the UP-M-PCR approach, multiple products can be sequenced using only one sequencing primer. In a verification analysis, the detection limit is as low as 1 % for a single GM maize in a mixture, thereby satisfying the requirements for the labeling policies in most countries.

The need for only one sequencing primer greatly reduces costs and shortens the working hours needed to perform multiplex detection. The UP-M-PCR method can be used for large-scale mixed sample detection within a short time, without relying on the length of the PCR products. This approach enables greater flexibility in primer design and has a wide variety of prospective applications in the fields of molecular biology and food species identification.

9.2.6 Multiplex Digital PCR

As the need for higher throughput in the parallel analysis of multiple targets continues to escalate in the fields of genomics and genetics, and as the need for more efficient sample use grows in medically related fields, the ability to perform and quantify multiple amplifications simultaneously within the same reaction volume has become paramount for both PCR and quantitative PCR.

In quantitative PCR, the ability to multiplex targets using differently colored fluorescent probes is typically limited to fourfold multiplexing by the spectral overlap of the fluorescent groups. Furthermore, multiplexing quantitative PCR assays requires expensive instrumentation and most often lengthy assay development cycles. Digital PCR (dPCR), as mentioned in Sect. 8.3, which is based on the amplification of single-target DNA molecules in many separate reactions, is an attractive alternative to quantitative PCR.

In 2011, Zhong et al. reported a novel strategy for multiplexed digital PCR assays that promises much higher multiplexing than is possible with existing quantitative PCR or dPCR techniques [26]. It is based on the singular nature of amplifications at terminal dilution that arises because only a single target allele is typically present in any one droplet even when multiple primers/probes targeting different alleles are present. This alleviates the complications that otherwise plague simultaneous competing reactions, such as varying arrival times at the exponential stage and unintended interactions between primers. As shown, reactions within microfluidic droplets yield very uniform fluorescence intensities at the end point, and ultimately, the intensity depends on the efficiency of probe hydrolysis. Thus, different reactions with different efficiencies can be discriminated based on end-point fluorescence intensity alone, even if they are labeled with the same color. Furthermore, it has also been shown that the efficiencies can be tuned simply by adjusting the probe concentration, resulting in an easy-to-use and general purpose method for multiplexing.

In multiplex detection, the use of multiple colors increases the number of possible reactions geometrically, rather than linearly as with quantitative PCR, because individual reactions can be labeled with multiple fluorescent groups.

This novel multiple dPCR technique, with its combination of high precision, high accuracy, and unprecedented multiplexing capability as well as high sensitivity, endows dPCR with superior performance compared with real-time quantitative PCR for many important applications, such as copy number variation, expression analysis, and the detection of rare mutants. In food safety detection, digital PCR has widely been used in GMO quantification [27], environmental bacteria monitoring [28], and other applications.

9.3 Chip Detection Techniques for High-Throughput Detection

In addition to multiple detection capabilities, high throughput is another key objective in scientific research. Multiplex detection methods can achieve the detection of multiple targets; however, in terms of high throughput, they are somewhat limited. The throughput is constrained by two main limitations. One is the formation of primer dimers due to uncontrollable interferences among primers. The other is the fact that the identification of multiplex PCR amplicons typically requires a secondary method, such as slab gel, melting curve, microarray hybridization, or capillary electrophoresis, for size separation or sequence verification [29]. To minimize interferences in multiplex PCR, techniques called solid- or liquid-phase PCR on chips have been developed.

9.3.1 Detection Techniques Based on Solid Arrays

Targets are detected either via hybridization to target-specific probes or by the direct extension of immobilized probes during the PCR itself (one-step protocol). The latter technique is referred to as solid-phase PCR (SP-PCR). Among its various applications are genotyping and the identification of SNPs.

9.3.1.1 DNA Microarrays

A DNA microarray, also commonly known as a DNA chip or biochip, is a collection of microscopic DNA spots attached to a solid surface. A key requirement for grafting PCR primers onto substrates in a SP-PCR-compatible manner is that the chemical bonds between the oligonucleotides and the array substrates must withstand thermocycling conditions, such as incubation at temperatures of approximately 95 °C. In addition, the binding chemistry must ensure a free 3′-OH end that is accessible and extendable by a DNA polymerase. At present, several immobilization protocols have been proven suitable for array-based SP-PCR: (1) direct immobilization methods, in which the oligonucleotides are covalently bonded to substrates, such as the use of 1-ethyl-3(3-dimethylaminopropyl)-carbodiimide (EDC) to mediate the linkage of 5′-NH2-modified DNA to hydroxylated substrates [30], and (2) indirect immobilization methods, in which homobifunctional linking molecules such as glutaraldehyde and 1,4-phenylene diisothiocyanate (PDITC) are used to attach oligonucleotides to activated or modified (mostly aminosilanized) surfaces [31, 32].

A universal protocol reported by Hoffmann et al. [33] for grafting PCR primers onto glass, PDMS, COP, COC, or PP for solid-phase PCR has been developed and evaluated. The primers are immobilized in a PCR-compatible way, featuring spots with high homogeneity and integrity. The presented immobilization and SP-PCR protocols may enable the direct integration of DNA microarrays into microfluidic lab-on-a-chip cartridges of various materials for analysis via SP-PCR. In addition to planar microarrays, another interesting application could be to coat the inner surfaces of a chip with PCR primers to recover the generated PCR products in a digital PCR system.

In modern food safety evaluations, molecular diagnostics may play a key role related to gene expression, toxicology and dietary components, genetically modified foods, foodborne pathogens, novel nutraceuticals, and so on. DNA microarray technology offers a new dimension of strength in molecular diagnostics by permitting the simultaneous analysis of large sets of genes for applications such as the following:

1. Assurance of safety of dietary supplements. There is an increasing trend toward dietary supplementation with certain levels of micronutrients, taken in either a single or multi-nutrient format. However, researchers have found that some dietary supplements, especially novel nutraceuticals, may pose potential risks. For instance, β-carotene prevents inducible skin carcinogenesis in animals, whereas a potential genotoxic effect on DNA damage is observed when both vitamin C and vitamin E supplements are provided together. The use of DNA microarray technology offers the opportunity for the simultaneous study of toxin-responsive genes, including those responsible for xenobiotic metabolism, DNA repair, cell growth, and stress response.
2. GM food safety assessment. Spotted DNA microarrays have been employed to evaluate the gene expression patterns of GM tomatoes; the results have revealed reproducible alterations of gene expression patterns outside the ranges of natural variation due to different stages of ripening, although no evidence has been found that genes from GM organisms are transferred to humans or animals that consume them as foods.
3. Detection of foodborne pathogens. DNA microarrays offer the opportunity to establish molecular signatures specific to certain types or subtypes of pathogenic bacteria. Studies have been conducted using DNA microarrays spotted with genetic markers in bacterial genomes that are associated with various pathogens and virulence factors. Chizhikov et al. chose six genes (slt-I, slt-II, eaeA, rfbE, fliC, and ipaH) to detect 15 strains of *Salmonella*, *Shigella*, and *E. coli*. This technology permits not only the simultaneous detection of multiple targets but also the simultaneous, high-throughput screening of a large number of specimens [34].

9.3.1.2 PCR-ELISA

The success of ELISA has motivated the development of an ELISA-based detection method for the solid-phase capture of PCR products. The basic principle of PCR-ELISA is the hybridization of nucleic acids.

In PCR-ELISA, microwells are first packaged using avidin, which can capture probes with biotin labeling at their 3′ ends, allowing the capture probes to become fixed on the microporous plate. Meanwhile, the sequence on the 5′ ends of the capture probes can hybridize with the PCR products, thus forming a solid-phase capture system. Second, during PCR amplification, the primers are labeled with antigens (biotin, digoxin, luciferase, etc.), such that the generated amplification products will contain the corresponding antigens. The capture probes on the microwells become hybridized with the amplification products, thus capturing the targets. The antibodies that can specifically bind with the antigens on the products and are labeled with horseradish peroxidase are added; through the further addition of the substrate and coloration, the detection results can be verified.

Because the PCR-ELISA method can achieve both quantitative and qualitative detection, it has been widely used in food safety assessments, such as for the detection of potential allergens, bacterial toxins, and pathogenic bacteria. Allergen detection is currently a topic of increasing interest for food labeling purposes. Holzhauser et al. applied PCR-ELISA to detect hazelnut proteins and compared the results with those of the sandwich-type ELISA method. The detection limit of PCR-ELISA was found to be <10 ppm in complex food matrixes, and no cross-reactions with non-hazelnut foods were observed, thus reducing the probability of false positives at the trace level [35]. Ge et al. applied PCR-ELISA to detect *E. coli* O157:H7 and other Shiga toxin-producing *E. coli* (STEC) in food. The assay was based on the incorporation of digoxigenin-labeled dUTP and a biotin-labeled primer specific for Shiga toxin genes during PCR amplification. The labeled PCR products were bound to streptavidin-coated wells of a microtiter plate and detected using the ELISA technique. Both the specificity and sensitivity of the PCR-ELISA method were determined, and the detection limit was found to be as low as 0.1–10 CFU, depending on the STEC serotypes and the genotypes of the Shiga toxins. Meanwhile, this PCR-ELISA method was able to detect approximately 10^5 CFU of STEC per gram of ground beef without any culture enrichment [36]. Taboada et al. employed PCR-ELISA to unequivocally authenticate three different kinds of fish, including Atlantic cod (*Gadus morhua*), ling (*Molva molva*), and Alaska pollock (*Gadus chalcogrammus*). A set of primers for *M. molva* and a probe for each species were designed using the *ND4* and *cytochrome b* genes as molecular markers. In no case were false negatives or false positives obtained in any of the analyzed samples, demonstrating the applicability of this method for the analysis of food products [37].

In addition to the advantages described in the examples above, the specificity, accuracy, and sensitivity of PCR-ELISA are all enhanced compared with the gel electrophoresis detection method. However, the disadvantages of this method, such as the fact that it is easy to contaminate and complex to perform, require considerable mitigation; otherwise, they will limit its further application.

9.3.2 Detection Techniques Based on Liquid Arrays

Microfluidic technology for molecular identification is developing rapidly, providing advantages such as high throughput, precise fluid control, and low consumption of reagents and energy.

The first continuous-flow PCR chip was developed in 1998 by Kopp et al. [38]. Compared with conventional PCR devices, the thermal cycling time can be significantly reduced by using a continuous-flow PCR chip because of the very short time in which the PCR mixture passes through the temperature zones (for denaturation and annealing) in the microchannels. Miniaturized PCR and detection devices are portable, perfect for field application. Furthermore, compared with chamber stationary PCR, the temperature control system is much simpler because only different temperature zones are needed, instead of complicated high-accuracy heating and cooling cycle systems as in conventional chamber PCR systems. Rapid thermal cycling of 17 s/cycle can be achieved in a continuous-flow PCR chip because of the facile temperature zone control and fast thermal transmission. Moreover, the amount of PCR product that can be obtained using a continuous-flow-based PCR chip is theoretically unlimited.

9.3.2.1 The Traditional Liquid Chip

The Luminex company developed a multifunctional flow lattice (Luminex 100™) development platform for multifunctional biochips, which is commonly used in the research on immune analysis, nucleic acids, enzymes, receptors, and ligand recognition analysis. The core of this technology is the encoding of tiny polystyrene spheres (5.6 um) with fluorescent staining, followed by the covalent cross-linking of a specific detection probe, antigen, or antibody on each differently colored set of microspheres. For different test applications, the coded microspheres are first mixed with different detection targets, and samples are then added in trace amounts; each target is mixed with a specific cross-linked molecule on the surfaces of the microspheres in the suspension, and as many as 100 different biological reactions can be performed at the same time. The results are analyzed using the Luminex™ analysis software. Because the molecular hybridization or immune response occurs in a suspended solution, the detection speed is fast.

The most common application of the traditional liquid chip is the detection of foodborne pathogens. Wang et al. developed and preliminarily applied a liquid chip test for the rapid detection of *Staphylococcus aureus*; the sensitivity of the developed method for *S. aureus* was 10^3 CFU/ml, with no cross-reactions with other foodborne pathogens. The detection results obtained for 200 food samples using the developed method were found to be essentially consistent with those of the method prescribed by the national standard [39]. Lü et al. developed an assay for simultaneously detecting seven common foodborne pathogens, including *Staphylococcus aureus*, *Salmonella*, *Vibrio parahaemolyticus*, *Listeria monocytogenes*, *E. coli* O157:H7, *Vibrio vulnificus*, and *Campylobacter jejuni*, using an xMAP liquid chip (Luminex) in a single-tube reaction. The developed liquid chip xMAP assay was used to test 140 strains of bacteria and then 56 food samples; no cross-reactions or false signals were observed, and the detection limit was 1–100 pg and 10^5–10^6 CFU/ml [40].

9.3.2.2 Quantum Dots in Liquid Chips

Quantum dots are a new kind of fluorescent labeling material and primarily consist of elements in groups II–VI or III–V. A quantum dot is a stable nanocrystal with a size of 1–100 nm. Compared with organic fluorescent materials, the excitation spectrum of quantum dots is wide, and the distribution is continuous; moreover, the half-peak width of the fluorescence emission spectrum is narrow and symmetrically distributed. In addition, quantum dots also have advantages such as good quantum yield, good stability of resistance to photobleaching, low background noise, and a long fluorescent life span.

Since the first use of quantum dots in the biological field, various quantum dot-based nanoprobes constructed by conjugating proteins, peptides, and DNA with quantum dots have been employed in the field of bionanotechnology. In food safety detection, quantum dots have also been widely used. Chen et al. developed a rapid indirect competitive fluorescence-linked immunosorbent assay (cFLISA) based on quantum dots as the fluorescent marker for the detection of enrofloxacin (ENR) in chicken muscle tissue. The cFLISA method enabled ENR determination in a linear working range of 1–100 ng/mL with a 50 % inhibition value (IC50) of 8.3 ng/mL and a detection limit of 2.5 ng/mL [41]. Yang and Li explored the use of semiconductor quantum dots as fluorescence labels in immunoassays for the simultaneous detection of two species of foodborne pathogenic bacteria, *E. coli* O157:H7 and *Salmonella typhimurium*. Quantum dots of different sizes can be excited with a single wavelength of light, resulting in different emission peaks that can be measured simultaneously. Highly fluorescent semiconductor quantum dots with different emission wavelengths (525 nm and 705 nm) were conjugated to anti-*E. coli* O157 and anti-*Salmonella* antibodies, respectively. The detection limit of this method was 10^4 CFU/mL, and the detection could be completed within 2 h [42]. Moreover, quantum dots can also be applied in the determination of pesticide residues, biotoxins, heavy metals, and other contaminants in food safety evaluations [43].

Compared with traditional electrophoresis methods, microfluidic microarrays possess several advantages, including the following: (1) they enable detection at low DNA concentrations, (2) the hybridization time can be considerably decreased because of the enhanced reaction rate inside a small chamber, and (3) high-throughput analysis can be achieved by printing all necessary probes on the same microarray [44]. Liquid chips can be used in the detection of protein with the same mean detection results as those of ELISA, but liquid chips offer much higher accuracy, sensitivity, and cost and time savings compared with ELISA.

9.3.2.3 Biosensor Chips

A biochip that combines a silicon wafer or glass with living material is called a biosensor. The essence of the biosensor concept is the orderly fixation of addressable recognition molecules on the surface of a substrate; through the labeling of the sample nucleic acids, the targets can hybridize with the molecules on the biosensor. The detailed sequence of the sample can then be obtained by detecting the hybridization signal. Hence, biosensors provide an effective technique for analyzing the sequences and functions of target genes at a large scale and with high throughput. Depending on different research perspectives, biosensors can be classified in many ways (Table 9.1).

In food safety detection, classical hybridization techniques have been complemented by DNA chip technology. DNA biosensors also exploit the hybridization of the target DNA in a sample solution with complementary DNA probes immobilized on the surface of a transducer. The details of various biosensor applications and examples of other kinds of biosensors are listed as followed.

Optical DNA Sensor Chips for Real-Time Detection

A DNA biosensor is a sensor that couples the specificity of a biological detection reaction with a physical transducer that generates an electrical signal proportional

Table 9.1 Biosensor classification

Category	The basis of classification	The name of the biosensor
Sensitive material on the molecular recognition element	Enzyme and substrate	Enzymatic sensor
	Microbial metabolism	Microbial sensor
	Tissue metabolism	Tissue sensor
	Cellular metabolism	Organelle sensor
	Antigen-antibody reaction	Immunosensor
	Nucleic acid hybridization	**DNA biosensor**
Signal transducer	Electrochemical electrode	Electrochemical sensor
	Ion-sensitive field-effect transistor	Ion-sensitive field-effect sensor
	Thermistor	Thermistor sensor
	Piezoelectric crystal	Piezoelectric crystal sensor
	Optoelectronic devices	Electro-optical sensor
	Acoustic device	Acoustic sensor

to the concentration of analyte. In nucleic acid detection, a promising alternative for addressing certain problems in nucleic acid analysis is offered by DNA biosensors.

To enhance the specificity, lower the detection limit, and shorten the detection period, Peter et al. [45] developed an evanescent optical sensor system consisting of a cheap, disposable sensor chip and a benchtop-sized readout device that allows the parallel detection of two fluorophore-labeled targets in real time. Oligonucleotide probes are directly immobilized on the surface of the disposable sensor chip via biotin/neutravidin linkage and hybridize to complementary Cy5-labeled target DNA in the sample. This technique has been successfully applied in the detection of the gene encoding neomycin phosphotransferase II (NPTII), which is present as a marker gene in a variety of genetically modified organisms. After optimization, a fragment with a length of 18 bp was used as a probe, and a 56 bp single-stranded DNA target from NPTII was used as the target. After PCR amplification and detection, a linear calibration curve was obtained. The detection limit of this technique is as low as 0.21 nmol/L, which is lower than that of traditional gel electrophoresis. This novel method offers a route toward a real-time DNA array format with analysis times as short as 1–2 min.

Porous Silicon-Based Biosensors

Porous silicon has been reported as a potential platform for biosensor applications. Compared with traditional planar platforms, porous silicon has many advantages, such as an increased surface area, a high detection sensitivity, and facile synthesis.

Mathew and Alocilja successfully applied a porous silicon-based biosensor for the rapid detection of the pathogen *E. coli*, with a sensitivity of 10^1–10^2 colony-forming units (CFU) [46]. First, a p-type silicon wafer was fabricated as the raw material for the production of porous silicon, using an anodization current to control the pore and surface characteristics. Second, the porous silicon was functionalized into a biosensor via the diffusion and physical adsorption of the dioxetane substrate, as confirmed by a chemiluminescence-based assay with the β-galactosidase enzyme. When the sample contained the pathogen *E. coli*, the light emission of the porous silicon biosensor changed compared with the negative control. Furthermore, the detection time and sensitivity of the porous silicon biosensor were also discussed and compared with those of traditional planar silicon. The detection limit of this novel biosensor is 10^1–10^2 CFU, which is better than that of planar silicon in a chemiluminescence assay (10^3). The developed sensor is sensitive and easily operated, and it has great potential for use in the rapid detection of foodborne pathogenic bacteria.

Surface Plasmon Resonance (SPR) Sensors

Surface plasmon resonance (SPR) plays a dominant role in the real-time interaction sensing of biomolecular binding events. A surface plasmon polariton is a kind of physical optics phenomenon; it is a non-radiative electromagnetic surface wave that propagates in a direction parallel to the interface between a material with negative permittivity and a dielectric material. Because the wave lies on the boundary between the conductor and the external medium (e.g., air, water, or vacuum), the oscillations are very sensitive to any change in this boundary, such as the adsorption of molecules onto the conducting surface [47].

Based on the above principle, an SPR biosensor can detect changes in the index of refraction. SPR biosensors have many advantages, as follows: (1) detection is achieved in real time, enabling the dynamic monitoring of the entire process of biomolecular interaction; (2) there is no need to label the samples, and the activity is maintained; (3) fewer samples are consumed; (4) the sensors are easy to operate with high sensitivity, throughput, and quality; and (5) no sample pretreatment is required [48]. Currently, such sensors have been widely applied in drug screening, clinical diagnosis, food and environmental monitoring, and membrane biology, and they have become a standard biophysical tool in the life sciences and pharmaceutical studies.

In the food/dairy industry, SPR is one of the most widely used technologies for purposes such as quality control, contaminant detection, and concentration analysis. For example, SPR biosensors have been used to measure the levels of nutrients and antibiotics in food as well as the numbers of bacterial and fungal contaminants, the amounts of toxins spread through the air or water, and the levels of pesticides and herbicides. Indyk et al. have employed biosensors of this type to analyze the non-endogenous R2 protein, which can provide automatic detection of the amounts of vitamin B12 in milk, meat, liver, and other kinds of foods [49].

9.4 Prospects

With the increasing application of molecule biotechnology in clinical medicine, food safety, archaeology, medical diagnosis, and other fields, the capabilities of multiple detection and high throughput are the key criteria on which a newly developed method must be evaluated. With the development of novel probes and gene chips, high-throughput detection methods must innovate toward faster, simpler, cheaper, more highly multiplexed, and flexible technologies that can better satisfy growing detection demands.

Acknowledgments This work is supported by the National Science and Technology Major Project (2016ZX08012-001). Many thanks to Ying Shang for her kindly help in manuscript conception and preparation.

References

1. Pinar A, Bozdemir N, Kocagoz T, Alacam R. Rapid detection of bacterial atypical pneumonia agents by multiplex PCR. Cent Eur J Public Health. 2004;12(1):3–5.
2. Ding C, Cantor CR. A high-throughput gene expression analysis technique using competitive PCR and matrix-assisted laser desorption ionization time-of-flight MS. Proc Natl Acad Sci. 2003;100(6):3059–64.

3. Hess C, Denkers F, Ossenkoppele G, Waisfisz Q, McElgunn C, Eldering E, Schouten J, Schuurhuis G. Gene expression profiling of minimal residual disease in acute myeloid leukaemia by novel multiplex-PCR-based method. Leukemia. 2004;18(12):1981–8.
4. Tettelin H, Radune D, Kasif S, Khouri H, Salzberg SL. Optimized multiplex PCR: efficiently closing a whole-genome shotgun sequencing project. Genomics. 1999;62(3):500–7.
5. Inagaki S, Yamamoto Y, Doi Y, Takata T, Ishikawa T, Imabayashi K, Yoshitome K, Miyaishi S, Ishizu H. A new 39-plex analysis method for SNPs including 15 blood group loci. Forensic Sci Int. 2004;144(1):45–57.
6. Wen-Tao X, Wei-Bin B, Yun-Bo L, Yan-Fang Y, Kun-Lun H. Research progress in techniques for detecting genetically modified organisms. Chin J Agric Biotechnol. 2009;6(01):1–9.
7. Huang H-Y, Pan T-M. Detection of genetically modified maize MON810 and NK603 by multiplex and real-time polymerase chain reaction methods. J Agric Food Chem. 2004;52 (11):3264–8.
8. Germini A, Zanetti A, Salati C, Rossi S, Forré C, Schmid S, Marchelli R. Development of a seven-target multiplex PCR for the simultaneous detection of transgenic soybean and maize in feeds and foods. J Agric Food Chem. 2004;52(11):3275–80.
9. Xu W, Zhai Z, Huang K, Zhang N, Yuan Y, Shang Y, Luo Y. A novel universal primer-multiplex-PCR method with sequencing gel electrophoresis analysis. PLoS ONE. 2012;7(1), e22900.
10. Xu W, Yuan Y, Luo Y, Bai W, Zhang C, Huang K. Event-specific detection of stacked genetically modified maize Bt11× GA21 by UP-M-PCR and Real-Time PCR. J Agric Food Chem. 2008;57(2):395–402.
11. Xu W, Bai W, Luo Y, Yuan Y, Zhang W, Guo X, Huang K. A novel common single primer multiplex polymerase chain reaction (CSP-M-PCR) method for the identification of animal species in minced meat. J Sci Food Agric. 2008;88(15):2631–7.
12. Yuan Y, Xu W, Zhai Z, Shi H, Luo Y, Chen Z, Huang K. Universal primer-multiplex PCR approach for simultaneous detection of Escherichia coli, Listeria monocytogenes, and Salmonella spp. in food samples. J Food Sci. 2009;74(8):M446–52.
13. Zhang C, Xu W, Zhai Z, Luo Y, Yan X, Zhang N, Huang K. Universal primer-multiplex-polymerase chain reaction (UP-M-PCR) and capillary electrophoresis–laser-induced fluorescence analysis for the simultaneous detection of six genetically modified maize lines. J Agric Food Chem. 2011;59(10):5188–94.
14. Schouten JP, McElgunn CJ, Waaijer R, Zwijnenburg D, Diepvens F, Pals G. Relative quantification of 40 nucleic acid sequences by multiplex ligation-dependent probe amplification. Nucleic Acids Res. 2002;30(12):e57–7.
15. Véronèse L, Tournilhac O, Combes P, Prie N, Pierre-Eymard E, Guièze R, Veyrat-Masson R, Bay J-O, Vago P, Tchirkov A. Contribution of MLPA to routine diagnostic testing of recurrent genomic aberrations in chronic lymphocytic leukemia. Cancer Genet. 2013;206(1):19–25.
16. Cui J, Azimi M, Adekile AD, Al Awadhi H, Hoppe CC. Detection of anti-Lepore Hb P-Nilotic by multiplex ligation-dependent probe amplification. Hemoglobin. 2012;36(3):276–82.
17. Moreano F, Ehlert A, Busch U, Engel K-H. Ligation-dependent probe amplification for the simultaneous event-specific detection and relative quantification of DNA from two genetically modified organisms. Eur Food Res Technol. 2006;222(5–6):479–85.
18. Holck A, Vaitilingom M, Didierjean L, Rudi K. 5′-Nuclease PCR for quantitative event-specific detection of the genetically modified Mon810 MaisGard maize. Eur Food Res Technol. 2002;214(5):449–54.
19. Ehlert A, Moreano F, Busch U, Engel K-H. Development of a modular system for detection of genetically modified organisms in food based on ligation-dependent probe amplification. Eur Food Res Technol. 2008;227(3):805–12.
20. Shang Y, Zhu P, Xu W, Guo T, Tian W, Luo Y, Huang K. Single universal primer multiplex ligation-dependent probe amplification with sequencing gel electrophoresis analysis. Anal Biochem. 2013;443(2):243–8.

21. Shrestha HK, Hwu K-K, Wang S-J, Liu L-F, Chang M-C. Simultaneous detection of eight genetically modified maize lines using a combination of event-and construct-specific multiplex-PCR technique. J Agric Food Chem. 2008;56(19):8962–8.
22. Lalic T, Vossen RH, Coffa J, Schouten JP, Guc-Scekic M, Radivojevic D, Djurisic M, Breuning MH, White SJ, den Dunnen JT. Deletion and duplication screening in the DMD gene using MLPA. Eur J Hum Genet. 2005;13(11):1231–4.
23. Wu H, Wu W, Chen Z, Wang W, Zhou G, Kajiyama T, Kambara H. Highly sensitive pyrosequencing based on the capture of free adenosine 5′ phosphosulfate with adenosine triphosphate sulfurylase. Anal Chem. 2011;83(9):3600–5.
24. Song Q, Wei G, Zhou G. Analysis of genetically modified organisms by pyrosequencing on a portable photodiode-based bioluminescence sequencer. Food Chem. 2014;154:78–83.
25. Colella S, Shen L, Baggerly K, Issa J, Krahe R. Sensitive and quantitative universal Pyrosequencing™ methylation analysis of CpG sites. Biotechniques. 2003;35(1):146–51.
26. Zhong Q, Bhattacharya S, Kotsopoulos S, Olson J, Taly V, Griffiths AD, Link DR, Larson JW. Multiplex digital PCR: breaking the one target per color barrier of quantitative PCR. Lab Chip. 2011;11(13):2167–74.
27. Burns M, Burrell A, Foy C. The applicability of digital PCR for the assessment of detection limits in GMO analysis. Eur Food Res Technol. 2010;231(3):353–62.
28. Ottesen EA, Hong JW, Quake SR, Leadbetter JR. Microfluidic digital PCR enables multigene analysis of individual environmental bacteria. Science. 2006;314(5804):1464–7.
29. Sun Y, Dhumpa R, Bang DD, Høgberg J, Handberg K, Wolff A. A lab-on-a-chip device for rapid identification of avian influenza viral RNA by solid-phase PCR. Lab Chip. 2011;11 (8):1457–63.
30. Adessi C, Matton G, Ayala G, Turcatti G, Mermod J-J, Mayer P, Kawashima E. Solid phase DNA amplification: characterisation of primer attachment and amplification mechanisms. Nucleic Acids Res. 2000;28(20):e87–7.
31. Kranaster R, Ketzer P, Marx A. Mutant DNA polymerase for improved detection of single-nucleotide variations in microarrayed primer extension. ChemBioChem. 2008;9(5):694–7.
32. Khodakov DA, Zakharova NV, Gryadunov DA, Filatov FP, Zasedatelev AS, Mikhailovich VM. An oligonucleotide microarray for multiplex real-time PCR identification of HIV-1, HBV, and HCV. Biotechniques. 2008;44(2):241.
33. Hoffmann J, Hin S, von Stetten F, Zengerle R, Roth G. Universal protocol for grafting PCR primers onto various lab-on-a-chip substrates for solid-phase PCR. RSC Adv. 2012;2 (9):3885–9.
34. Liu-Stratton Y, Roy S, Sen CK. DNA microarray technology in nutraceutical and food safety. Toxicol Lett. 2004;150(1):29–42.
35. Holzhauser T, Stephan O, Vieths S. Detection of potentially allergenic hazelnut (Corylus avellana) residues in food: a comparative study with DNA PCR-ELISA and protein sandwich-ELISA. J Agric Food Chem. 2002;50(21):5808–15.
36. Ge B, Zhao S, Hall R, Meng J. A PCR–ELISA for detecting Shiga toxin-producing Escherichia coli. Microbes Infect. 2002;4(3):285–90.
37. Taboada L, Sánchez A, Velasco A, Santaclara FJ, Pérez-Martín RI, Sotelo CG. Identification of Atlantic Cod (Gadus morhua), Ling (Molva molva), and Alaska Pollock (Gadus chalcogrammus) by PCR–ELISA Using Duplex PCR. J Agric Food Chem. 2014;62 (24):5699–706.
38. Kopp MU, De Mello AJ, Manz A. Chemical amplification: continuous-flow PCR on a chip. Science. 1998;280(5366):1046–8.
39. Wang Y, Cai Y, Liu T, Ong Z, Wang N, Wang Z. Development and application of liquid chip test for Staphylococcus aureus. Chin J Biol. 2012;25(10):1383–6.
40. Lü D, Shi X, Chen M, Wu P, He L, Li Y, Lin Y, Qiu Y, Hu Q. Development of a xMAP liquid chip assay for the rapid identification of 7 common foodborne pathogens and its application. Wei sheng yan jiu= J Hyg Res. 2012;41(1):96–101.

41. Chen J, Xu F, Jiang H, Hou Y, Rao Q, Guo P, Ding S. A novel quantum dot-based fluoroimmunoassay method for detection of Enrofloxacin residue in chicken muscle tissue. Food Chem. 2009;113(4):1197–201.
42. Yang L, Li Y. Simultaneous detection of Escherichia coli O157: H7 and Salmonella Typhimurium using quantum dots as fluorescence labels. Analyst. 2006;131(3):394–401.
43. Su S, Fan J, Xue B, Yuwen L, Liu X, Pan D, Fan C, Wang L. DNA-conjugated quantum dot nanoprobe for high-sensitivity fluorescent detection of DNA and micro-RNA. ACS Appl Mater Interfaces. 2014;6(2):1152–7.
44. Jiang X, Shao N, Jing W, Tao S, Liu S, Sui G. Microfluidic chip integrating high throughput continuous-flow PCR and DNA hybridization for bacteria analysis. Talanta. 2014;122:246–50.
45. Peter C, Meusel M, Grawe F, Katerkamp A, Cammann K, Börchers T. Optical DNA-sensor chip for real-time detection of hybridization events. Fresenius J Anal Chem. 2001;371 (2):120–7.
46. Mathew FP, Alocilja EC. Porous silicon-based biosensor for pathogen detection. Biosens Bioelectron. 2005;20(8):1656–61.
47. Zeng S, Baillargeat D, Ho H-P, Yong K-T. Nanomaterials enhanced surface plasmon resonance for biological and chemical sensing applications. Chem Soc Rev. 2014;43(10):3426–52.
48. Rich RL, Myszka DG. BIACORE J: a new platform for routine biomolecular interaction analysis†. J Mol Recognit. 2001;14(4):223–8.
49. Indyk HE, Persson BS, Caselunghe MC, Moberg A, Filonzi EL, Woollard DC. Determination of vitamin B12 in milk products and selected foods by optical biosensor protein-binding assay: method comparison. J AOAC Int. 2002;85(1):72–81.

Chapter 10
Detecting Targets Without Thermal Cycling in Food: Isothermal Amplification and Hybridization

Abstract Isothermal technology is an option for point-of-care diagnosis to replace PCR amplification. Recently, this approach has been updated into a simple, rapid, and cost-effective tool. The input can be defined as DNA, RNA, or certain artificial nucleic acids, and the output response can be an amplified signal or catalytic chemical reaction, which allowing biosensing assays on various classes of biomolecules. The isothermal technologies used in the food safety detection can be classified into the following two categories: amplification and hybridization. Each category includes various methods designed to achieve a rapid and sensitive diagnose. Additionally, isothermal methods have made progress in the single cell analysis due to the unbiased, linear nature of the amplification process. This review mainly describes specific isothermal detection techniques to give a better understanding of the method of choice when performing fast and simple point-of-care detection.

Keywords Isothermal amplification • Strand displacement • Replication • Hybridization • Thermal

10.1 Introduction

Nucleic acids in the host are limited to nanogram levels, and amplification is often required as a prerequisite for molecular diagnosis. An ideal amplification and identification method would be able to efficiently amplify target sequences; to differentiate between sequences by amplicon length, especially for single-base changes; to produce a highly sensitive signal with low noise; and to be performed conveniently [1]. In the last 10 years, molecular biology techniques have been increasingly used by researchers as a way to specifically find targets. Nucleic acid amplification techniques can be classified into two groups: thermal cycling and isothermal amplification. The main difference between them is the requirement of a temperature change during the amplification, usually achieved by automatic instrument [2]. In the last 20 years, thermal cycling, normally recalled polymerase chain reaction (PCR), has received unprecedented welcome in the fields of gene expression, target location, clinical diagnosis, and gene cloning [2–5].

W. Xu, *Functional Nucleic Acids Detection in Food Safety*,
DOI 10.1007/978-981-10-1618-9_10

However, limitations of PCR are becoming evident with the increasing progress in molecular biology. The most distinct disadvantage of PCR is the requirements of an automatic thermal cycler to perform and provide temperature changes among the melting temperature, annealing and extension temperature, making it difficult to execute gene analysis in vivo. Furthermore, the efficiency and specificity of exponential target amplification by PCR relies heavily on the few cycles. These factors led to the birth and the subsequent rise of the isothermal amplification.

The isothermal amplification of nucleic acids has emerged as a promising alternative in which rapid and efficient amplification is achieved at a constant temperature without the thermocycling required in PCR. Additionally, isothermal amplification can be performed under constant temperature conditions, or in vivo, which is preferable to conventional PCR. Since the early 1990s, dozens of isothermal nucleic acid amplification techniques employing various amplification mechanisms have been developed [2, 4, 6–8]. Most of these methods possess impressive sensitivity for the detection of nucleic acids, and some have achieved commercial success [9, 10]. More significantly, isothermal amplification techniques have been expanded to detect a wide spectrum of targets, including proteins, cells, small molecules, and ions. The combination of isothermal amplification and digital technology has recently attracted more attention, with methods to prevent contamination, reduce cost, simplify the use of amplification reactions by eliminating the need for laboratory equipment, and provide a compact format for running several reactions in parallel [11].

In this review, two groups of isothermal technologies are discussed: amplification and hybridization. Each will be presented in a brief blueprint on their suitable scope, influential factors, and use in the practical analysis.

10.2 Isothermal Amplification Technologies

For detecting a scarce target in vivo or in vitro, enzymatic amplification is usually the first thought of researchers, as developed by Kary Mullis [12, 13]. This American scientist was seeking a new way to analyze changes (mutations) in DNA when he realized that he had instead invented a method of amplifying any DNA region through repeated cycles of duplication driven by DNA polymerase. Polymerase has since been developed for high catalytic ability, high temperature tolerance, and a particular recognition style, and the corresponding amplification style has been simplified.

Isothermal amplification technologies use a specific DNA polymerase to eliminate thermal cycling. Isothermal technologies are classified into two categories: one is to displace one strand with another new extended primer using a special DNA polymerase, and the other is to make single-stranded DNA copies by the normal method in the living organisms, with sufficient reagents involved. A list will be presented below. To best evaluate the isothermal technologies, several factors must be concerned: template restriction, amplicon length, simplicity, and the preferred detection model. Since the development of the first isothermal detection method, hundreds of applications have been developed to achieve the rapid identification of foodborne pathogens. Table 10.1 below presents a brief summary of isothermal application in food safety.

10.2.1 Strand Displacement Amplification (SDA)

Strand displacement amplification (SDA) was first developed by Walker in 1992 [136]. The amplification mechanism of strand displacement amplification is based on the continuous nicking and displacement process catalyzed by endonuclease and exonuclease-deficient (exo−) DNA polymerase. The basic principle of SDA is as follows: double-stranded DNA and a specific primer bind firstly under the principle of Watson–Crick base pairing after denaturation, forming a duplex that contains a recognition site for the restriction endonuclease enzyme at the 5′ end. An DNA polymerase with exonuclease-deficient activity extends the 3′ side of the template using dGTP, dCTP, dTTP, and dATP[aS], which produces a phosphorothioate recognition site on one strand of the duplex. An endonuclease enzyme cleaves the recognition site on the unphosphorothioated site, leaving a nicking site in the duplex. The DNA polymerase used exhibits high strand displacement activity that can displace the cleavage target strand during the extension (Fig. 10.1a). This cycle of nicking and displacement step cycle is repeated continuously to produce single-stranded complements of the target strand [137].

SDA is an important technology of isothermal amplification that has been in use for almost 20 years. The dNTPs used in SDA include a special deoxy-adenosine 5′-[a-thiotriphosphate (dATP[αS]). The restriction endonuclease only recognizes the site but only cleave the single strand that has no thiophosphate modification [138]. However, this dATP[αS] is difficult to synthesis and has no other sources. Extensive efforts have been made in our group to find and contact dATP[αS]-producing companies all over the world. Pharmacia, the company that made dATP [αS], is no longer producing commercial dATP[αS] since it was merged with Amersham Life Science (UK). Currently, there only exist a limited number of research project on SDA target detection, including *Mycobacterium tuberculosis* [87, 88, 139, 140], *Chlamydia trachomatis* [141], and *Escherichia coli* [89]. These researches are all traced back to the 1990s and have shown no progress since then. Alternative simpler and more universal methods are developed in the interim.

Successful application of SDA without dATP restriction is the whole genome amplification (WGA) [142–144]. Genomes must be further analyzed to guarantee that every region in the sequence possesses equal quantity or that the sequence can be amplified nonspecifically, allowing WGA to amplify the sequence with high efficiency to balance genomic regions. WGA normally uses a random oligonucleotide as a primer and a DNA polymerase with strong strand displacement activity. φ29 DNA polymerase has a 3′→5′ exonuclease activity and high replication fidelity, making it an optimal choice for DNA amplification. Under isothermal

Table 10.1 Isothermal application on the food safety detection fields

Method	Category	Target	Source	LOD	References
LAMP	Pathogens	*Escherichia coli*		10 copies number	[14, 15]
		Salmonella	Eggs	2.8 cfu	[16]
		H5N1 avian influenza virus	Birds	0.1 pfu	[17]
		White spot syndrome virus	Shrimp	1 fg	[18]
		Escherichia coli	Processed food	10 fg	[19]
		Escherichia coli	Eggs		[20]
		Salmonella	Meat	1 pg	[21–24]
		Campylobacter coli	Meat	1.2 cfu	[25]
		Vibrio parahaemolyticus	Oyster, shell fish, shrimp	1 pg or 0.8 cfu/reaction	[26–33]
		Escherichia coli, *Listeria monocytogenes*	Processed food	10 fg	[34]
		Cryptosporidium oocysts	Water	400 fg	[35]
		Listeria monocytogenes	Chicken	2 cfu	[36]
		Listeria monocytogenes	Retail food	6 cfu/reaction	[37]
		Listeria monocytogenes	Chicken, pork, beef	100 cfu/mL	[38]
		Listeria monocytogenes	Milk	186 cfu/mL	[39]
		Fusarium tricinctum	Wheat, barley	0.95 pg	[40, 41]
		Salmonella enterica	Chicken	2 pg	[42]
		Salmonella spp.	Lettuce, parsley, tomato, pepper, milk, egg, yogurt, hamburger, pork, beef, chicken, juice, cake	10^2–10^8 cfu/mL	[23, 43–45]
		Salmonella serovar	Eggs	10^7 cfu/25 mL	[46]
		Salmonella typhimurium	Pork, chicken, vegetables	10 cfu/mL	[47, 48]
		Salmonella serotype D	Chicken	10 cfu/reaction	[49]
		Cronobacter sakazakii	Infant formula	1.2 cfu/100 g	[50, 51]

(continued)

Table 10.1 (continued)

Method	Category	Target	Source	LOD	References
		Shigella spp.	Milk	6–8 cfu/mL	[52]
		Shewanella putrefaciens	Fish	5.4 copy	[53]
		Brucella abortus	Milk	49,000 cfu/mL	[54]
		Staphylococcus aureus	Milk	8–10 cfu/mL	[55]
		Staphylococcus aureus	Pork	1,000 cfu/g	[56]
		Staphylococcus aureus	Bakery	100 fg	[57]
		Clostridium botulinum	Fish, honey	1 cfu/reaction	[58]
		Clostridium perfringens	Meat	3 cfu/g	[59]
		Alicyclobacillus acidoterrestris	Apple juice	22.5 cfu/mL	[24, 60]
		Dekkera bruxellensis	Wine, beer	10 cfu/mL	[61]
	Adulteration	Pork		1 pg	[62, 63]
		Hedyotis diffusa		3 pg	[64]
		Pork	Halal products	0.5 pg	[65]
		Horse		0.1 ng	[66]
	Mycotoxins	Aflatoxins	Peanut, nut and coffee bean	2.4 pg	[67, 68]
		OTA	Grapes	0.1 pg	[69]
		OTA			[70]
	GMOs	TT51-1, KMD, KF6	Rice	0.01 %	[71]
		KMD	Rice	0.5 %	[72]
		TT51-1, KF6	Rice	0.1 %	[73]
		T1c-19, KF6, TT51-1	Rice	5 copies number	[74]
		Huahui No.1	Rice	0.5 %	[75]
		GTS40-3-2, MON89788	Soybean	4 copy	[73, 76]
		GTS40-3-2	Soybean	0.01 %	[77]
		RRS	Soybean	0.5 %	[78]
		59122, T25, MON810, MON863, BT11, TC1507, BT176,	Maize	4 copy	[79]
		MON810	Maize	0.5 %	[80]
		MON89034	Maize	4	[73]
		T25	Maize	0.5 %	[81]

(continued)

Table 10.1 (continued)

Method	Category	Target	Source	LOD	References
		MON863	Maize	4	[82]
		MON89034, MIR604	Maize	25	[83]
		MS8, RF3	Canola	0.01 %	[77]
		B73-6-1	Wheat	6	[84]
		HBsAg, hATIII	Goat	10	[85]
		hLTF/hLALBA	Cattle	10	[43]
SDA	Small moleculars	Cocaine		2 nM	[86]
	Pathogens	*Mycobacterium tuberculosis*			[87]
		Mycobacterium tuberculosis		10 copy	[88]
		Escherichia coli	Clinical samples	4.3 cells/reaction	[89]
NEMA	Pathogens	*Bacillus subtilis*		1 copy	[90]
	MicroRNAs	let-7a	Lung tissue	1 zM	[91]
		miR-141		3 aM	[92]
	Mycotoxins	Ochratoxin A		0.0003 ppb	[93]
RCA	Pathogens	*Listeria monocytogenes*		0.16 pg	[94]
		Listeria monocytogenes		10 aM	[95]
		Salmonella	Chicken		[96]
		Indian cassava mosaic virus	Cassava	1 pg	[97]
		Torque teno virus	Pig		[98]
		Barley dwarf virus, oat dwarf virus	Barley		[99]
		Candida, *Aspergillus*, and *Scedosporium* spp.		10^7 copies	[100]
		Escherichia coli	Human blood	1 copy	[101]
		Vibrio cholerae	Clinical samples	25 ng	[102]
	MicroRNAs	let-7a		0.025 pM	[103]
	Small molecules	Cocaine		0.48 nM	[104]
	Mycotoxins	Ochratoxin A		0.2 ppb	[105]
	Heavy metals	Pb^{2+}		7.8 pM	[106]
EXPAR	Heavy metals	Hg^{2+}		100 pM	[106]
		Pb^{2+}		200 pM	[107]

(continued)

Table 10.1 (continued)

Method	Category	Target	Source	LOD	References
	Pathogens	Influenza A	Chicken	9.4 fM	[108]
HDA	Pathogens	*Escherichia coli*		10 cfu/ reaction	[109]
		Clostridium difficile	Feces	20 copies	[110]
		Staphylococcus aureus		1 ng	[111]
		Helicobacter pylori	Gastric biopsy samples	10 copies	[112]
NASBA	GMOs	MON863, MON810	Maize	0.1 %	[113]
	Pathogens	*Listeria monocytogenes*	Processed food	10 cfu/25 g	[114]
		Listeria monocytogenes		10 cfu/g	[115]
		Aspergillus fumigatus	Blood	1 cfu/ reaction	[116]
		Hepatitis A	Turkey, lettuce	1 pfu/reaction	[117]
		Hepatitis A	Lettuce, blueberries	2 pfu/reaction	[118]
		Noroviruses	Turkey, lettuce		[119]
		Noroviruses	Water		[120]
		Campylobacter jejuni	Beef, eggs, milk, lettuce	2 cfu/10 g	[121]
		Vibrio cholerae	Water	50 cfu/mL	[122]
		Salmonella	Milk, pepper, chicken, shrimp, fish, cake	10 cfu/25 g	[123]
		Salmonella	Eggs		[124]
		Saccharomyces cerevisiae, *Escherichia coli*, *Staphylococcus aureus*		64 cells/L	[125]
		Salmonella enteritidis, *Salmonella typhimurium*		10 cfu/mL	[126]
		Escherichia coli	Water	1 cfu/mL	[127]
		Vibrio cholerae	Water	1 cfu/ 100 mL	[128]
		H5N1 virus	Chicken	10 copies	[129]
RPA	GMOs	BT11, TC1507, MIR604, TT51-1, KF6	Maize, rice	100 copies	[130]

(continued)

Table 10.1 (continued)

Method	Category	Target	Source	LOD	References
	Pathogens	*Staphylococcus aureus*		1 copy	[131]
		Staphylococcus aureus		10 cfu/ reaction	[132]
		Salmonella spp., *Cronobacter* spp., *Fusarium* spp.	Cookie, water, tomato, infant formula	6–13 cfu/ mL	[133]
		Francisella tularensis		20,000 cop-ies/uL	[134]
SMART	Pathogens	*Escherichia coli*		0.1 ng	[135]

conditions, random primers bind to the complementary target and produce branched structures, which can be extended by other primers and eventually form multibranched structures (Fig. 10.1b).

Walker's efforts have been developed into a type of isothermal amplification style to displace the extended sequence and continue amplification. Significant progress has been made toward optimizing the amplification and facilitating point-of-care diagnosis, represented by loop-mediated isothermal amplification (LAMP), nicking endonuclease-mediated isothermal amplification (NEMA), rolling circle amplification (RCA), and exponential amplification reaction (EXPAR).

10.2.1.1 Loop-Mediated Isothermal Amplification (LAMP)

Loop-mediated isothermal amplification (LAMP) is the most common isothermal technology and was first described in 2000 by Notomi [145]. The operation temperature of amplification is 55–65 °C for 40–60 min, making it faster and easier to perform than traditional PCR. LAMP has extremely high amplification efficiency, causing large amount of precipitate for detection. The LAMP method requires six or even more independent sequences to design specific primers and the optional loop primers. Although the large amount of products generated makes contamination, it is still now the prevailing method for isothermal amplification [146].

In the LAMP reaction (Fig. 10.2), two primers in the forward primer set are termed the inner (F1c–F2) and outer (F3) primers. At the reaction temperature, F2 region in the primer binds to the target and initiate extension, while another primer (F3) binds to the same target with different region (F3c). The two primers both extend by the polymerase and the outer primer (F3) will replace the inner primer (F1c–F2). F1 and F1c region in the displaced strand bind together to form a stem-loop structure at the 5′ site of strand. Then the 3′ primers (inner and outer primer) start to hybridize to the strand and extend to form a double stem-loop dumbbell-like

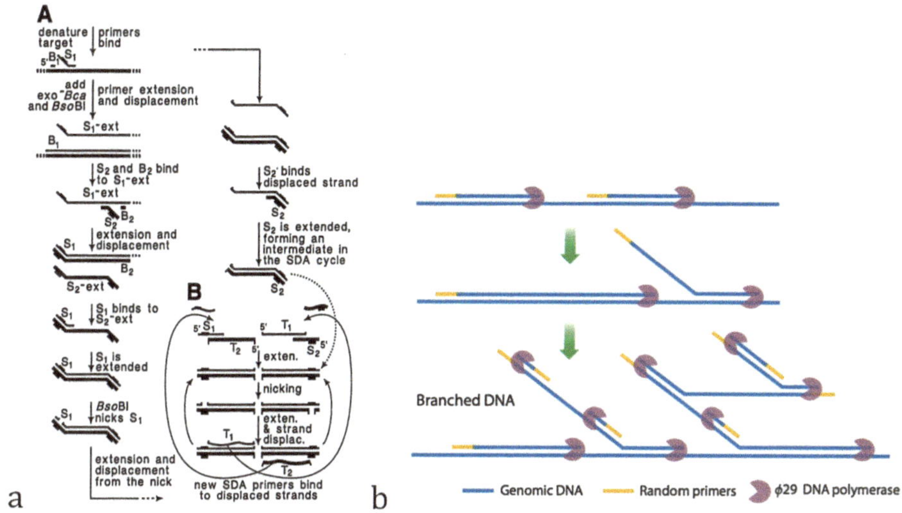

Fig. 10.1 (**a**) Mechanism of SDA process. SDA is performed using an excess of four oligodeoxynucleotide primers (B1, B2, S1, and S2). S1 and S2 contain target binding regions at their 3′ ends and a recognition site (5′ CTCGGG) for BsoBl located 5′ to the target-binding regions (BsoBl recognition sites are designated by the raised boxes). S1 and S2 bind to opposite strands of the target sequence, flanking the region to be amplified. B1 and B2 are simply target-binding sequences (containing no BsoBl recognition sites) and bind at positions 5′ to S1 and S2. S1 and B1 hybridize to one strand of the target upon lowering the temperature to 60 °C. BsoBl and exo-Bca are then added to the sample. At this point the remaining steps in (A) proceed as a single cascade: exo-Bca, which is present in large molar excess over the number of target sequences, simultaneously extends S1 and B1 using dATP, dGTP, TTP, and dCTP S. As S1 is extended, the extension product (S1-ext) is displaced through extension of B1. S1-ext serves as target for binding of S2 and B2. Simultaneous extension of S2 and B2 results in displacement of an S2 extension product (S2-ext). An S1 primer binds to S2-ext and is extended, forming a double-stranded structure with a hemiphosphorothioate BsoBl site at each end. BsoBl nicks the unmodified strands of the hemiphosphorothioate recognition site on S1 leaving intact the modified complementary strand of the BsoBl site (BsoBl nicking can also occur at the hemiphosphorothioate site on the opposite end of the fragment (not shown)). Exo-Bca then extends the 3′ end at the nick and displaces the downstream strand. An S2 primer binds to the displaced strand and is extended, forming an intermediate in the SDA cycle shown in (B) as indicated by the *dashed arrow*. (B) The SDA cycle is where the majority of the amplification occurs. During each round of the cycle, the 3′ end of S1 binds to the 3′ end of the displaced target strand T2, forming a duplex with 5′ overhangs. Likewise, S2 binds to T1, the complement of T2. Exo-Bca extends the recessed 3′ ends of the duplexes producing hemiphosphorothioate recognition sites that are nicked by BsoBl. These nicking and extension/displacement steps cycle continuously [*short upturned arrows* in (B)] because extension at a nick regenerates a nickable BsoBl recognition site. The strand displaced from the S1T2 duplex is identical to T1. Likewise, the displaced strand from the S2T1 duplex is identical to T2. Consequently, target amplification is exponential (Reprinted from Ref. [139], copyright 1996, with permission from Elsevier) (**b**) The WGA diagram. Random primers are used for isothermal amplification with ϕ29 DNA polymerase, which has strong strand displacement activity (Reprinted by permission from Macmillan Publishers Ltd: Ref. [144], copyright 2015)

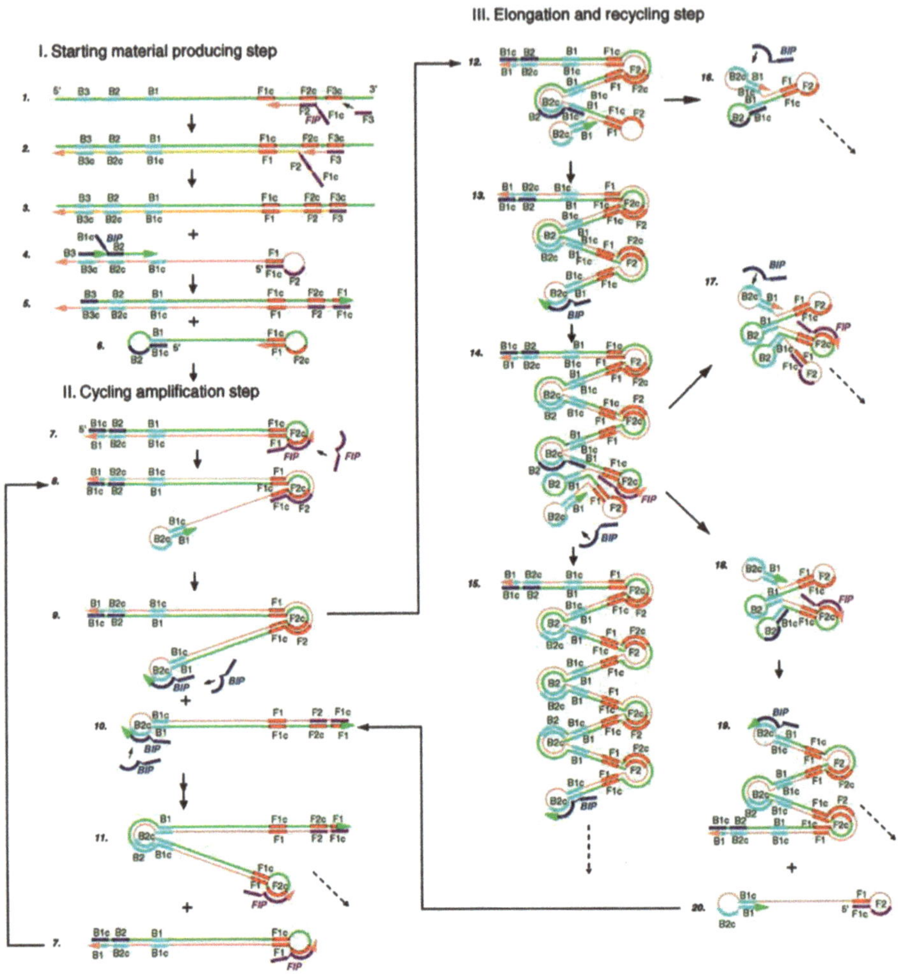

Fig. 10.2 Mechanism of loop-mediated amplification (LAMP). Four probes (F1c-F2, F3, R1c-R2, R3) are used for this method. F1 is complementary with F1c (c stands for complementary sequence) (Reprinted from Ref. [145] with permission from Oxford)

structure. This dumbbell-structured DNA initiates the exponential amplification cycle, and the primer will bind to the corresponding site and displace the extended strand. LAMP produced several types of amplicons with different lengths, whose structure is even similar to cauliflower. LAMP can also be improved through loop primers to accelerate the amplification speed to within 1 h [147].

The LAMP system requires no endonuclease digestion site of endonuclease enzyme but does need a selectable primer binding region. Therefore, primers design work in LAMP is so difficult that only professional design software can meet the requirements (http://primerexplorer.jp/e/). Otherwise, LAMP is flexible in amplifying long sequences in the genome target (up to 500 bp) to satisfy various needs in different fields. LAMP can target either single- or double-stranded sequences, which means that DNA, RNA, and oligonucleotides can all be detected via a LAMP system without complex template pretreatment.

It has also been proven that LAMP can be quantified through turbidity [34, 148–151] or fluorescence [82, 152–154]. Fluorescent detection is similar to conventional real-time quantitative PCR [155]. Turbidity measurement is available because LAMP can produce large amounts of pyrophosphate precipitation. Pyrophosphate ion is released from dNTP during the DNA extension. When a large amount of this pyrophosphate ion is produced, it reacts with magnesium ion in the reaction system to yield a visible precipitate. The accumulation of precipitation in the LAMP reaction can be represented by the following reactions:

$$(\mathrm{DNA})_{n-1} + \mathrm{dNTP} \rightarrow (\mathrm{DNA})_n + \mathrm{P_2O_7}^{4-} \tag{10.1}$$

$$\mathrm{P_2O_7}^{4-} + \mathrm{Mg}^{2+} \rightarrow \mathrm{Mg_2P_2O_7} \tag{10.2}$$

Magnesium ion in the LAMP reaction can also be utilized for target identification. Currently, the colorimetric detection of LAMP has been achieved through addition of metal indicator. Mg^{2+} in the LAMP system can be formed into chelate with specific metal indicator, such as hydroxyl naphthol blue and acid chrome blue K (ACBK). In the presence of target, the pyrophosphate ion reacts with magnesium ion, inhibiting the formation of chelate.

LAMP is currently the most applicable isothermal technology in the world, despite the identification confusion and false-positive interference. It has now been applied in the detection of food pathogens, GMO, virus, and mycotoxins. The table provides a comprehensive summary of reported LAMP applications in the point-of-care field.

10.2.1.2 Nicking Endonuclease-Mediated Isothermal Amplification (NEMA)

Nicking endonuclease is a specific endonuclease that cleaves only one strand of DNA in a double-stranded DNA duplex [156], instead of a special phosphorothioate adenine in the sequence in the SDA reaction. Nicking endonuclease-mediated isothermal amplification (NEMA) was then developed and shown in Fig. 10.3. The target is first amplified with a single strand displaced by another new strand under stable temperature conditions. The duplex is then broken by endonuclease recognition site cleavage. Extension and displacement reactions on the templates produce two types of fragments: one has a recognition site at each end and the other has a recognition site at just one end. Then, the amplification initiated with continous nicking and replacement. During each round of the cycle, the 3′ end of primer 1 (S1) binds to the 3′ end of the displaced target strand T2, forming a duplex with 5′ overhangs. The DNA polymerase extends the 3′ side of the duplexes, producing recognition sites that can be

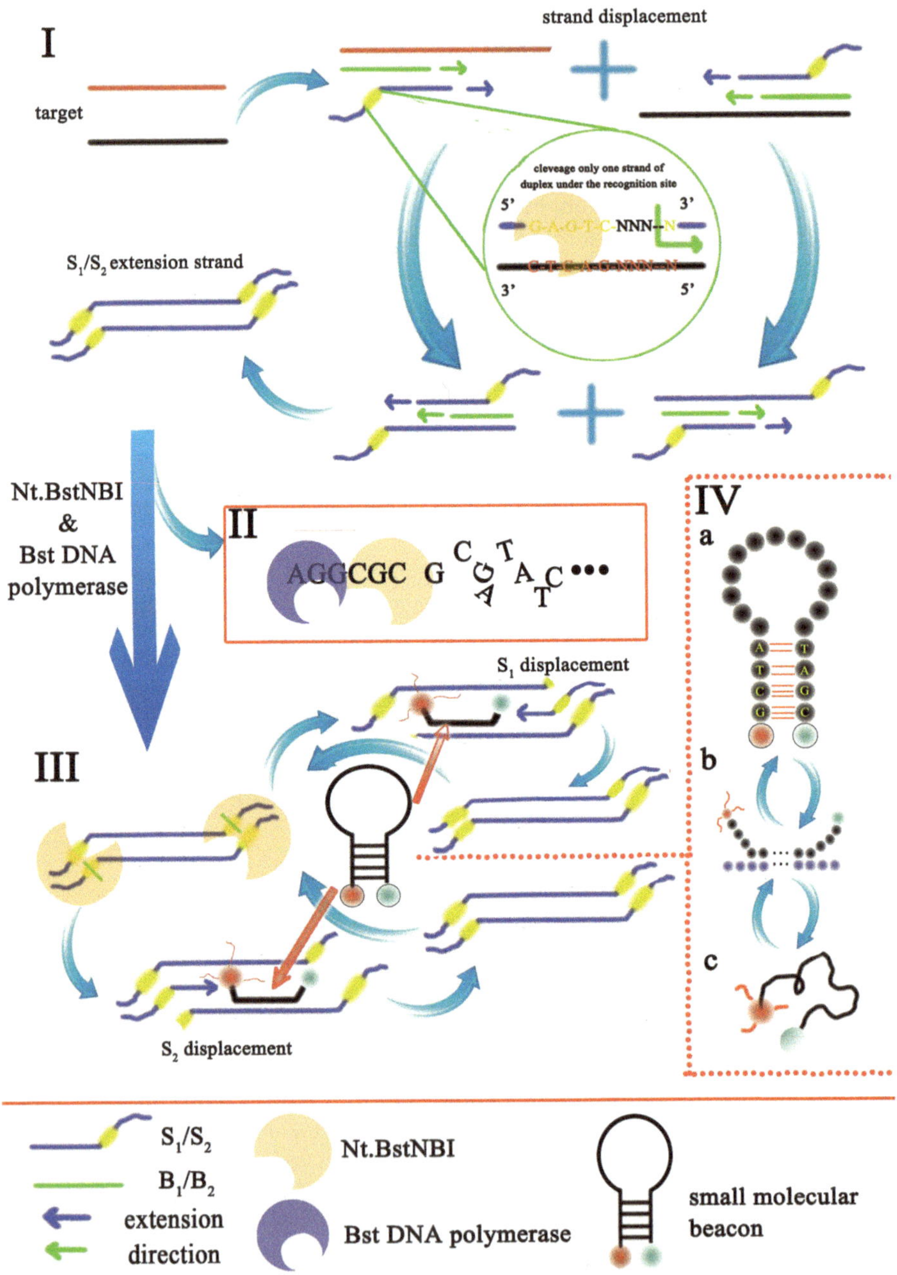

Fig. 10.3 Mechanism of nicking endonuclease-mediated amplification (NEMA). The *blue arrows* indicate all process of amplification (I, II, and III). The duplex target is amplified with S1/S2 and B1/B2 after denaturation, producing a new duplex with two nicking endonuclease recognition sites (step I). Then, the extension strand falls into a cycle. S1/S2 continues to generate a single strand displaced under the activity of Bst DNA polymerase (step III). In this step, one extension strand will produce two identical amplicons, indicating the exponential amplification format. Small molecular beacons bind to complementary target to monitor real-time amplification process. In

recognized by nicking endonuclease. These nicking and displacement steps are continuously repeated to produce large amount of target sequences.

Our group has been researching on NEMA for a long time. A popular nicking endonucleases that often have an optimal incubation temperature from 55 to 65 °C, which can be purchased from New England Biolabs, Inc. Nt.BstNBI should be the best option when DNA polymerase is present in the reaction system. Table 1 shows a clear representation. Otherwise, NEMA can be capable of amplifying long sequences (400–500 bp) that fulfill the needs of common molecular diagnosis with the help of trehalose. Betaine is also added to the system borrowed from LAMP, to balance the incubation temperature of two enzymes. Most importantly, the whole genome becomes target template after our optimizing efforts, guided by the addition of trehalose and betaine.

Without primer and template, dNTPs used in the template amplification show a light but inevitable background. This phenomenon, namely, de novo *synthesis*, is a disadvantage to the reaction, as we can barely discriminate the target from the nonspecific background [157, 158]. The presence of dNTPs in the reaction can automatically generate another strand with palindromic structure and high G-C content in the presence of nicking endonuclease and Bst DNA polymerase, without template. However, many efforts have been made to prevent de novo *synthesis* in the presence of Nt. BstNBI.

NEMA can also be quantified according to the previous work [159, 160]. Our group developed a quantitative nicking endonuclease-mediated isothermal amplification with a small molecular beacons (NEMA-SMB) [161]. This small molecular beacon, which is based on the revised principles of the molecular beacon devising rule, exhibits stem separation and conformational changes at certain temperatures. The whole genome is accurately and quantitatively detected, possessing high specificity and sensitivity (Fig. 10.3).

Although there are few reports of developing the NEMA technique as a functional tool for signal amplification and molecular diagnosis, partly because of its unstable reaction system in amplifying the whole genome, NEMA can still be used in research centers as an inherited SDA alternative.

Fig. 10.3 (continued) addition to this process, dNTPs in the reaction generate another strand with palindromic structure and high G-C content in the presence of nicking endonuclease and Bst DNA polymerase (step II). This phenomenon, as we called de novo synthesis, occurs in all processes of SMB-NEMA amplification. Small molecular beacons in the reaction act in dynamic equilibrium with three main states: completely folded (*a*), bound to target (*b*), and random coil (*c*). The first two is included in the two-state model (step IV) (Reprinted from Ref. [161] by permission of The Royal Society of Chemistry)

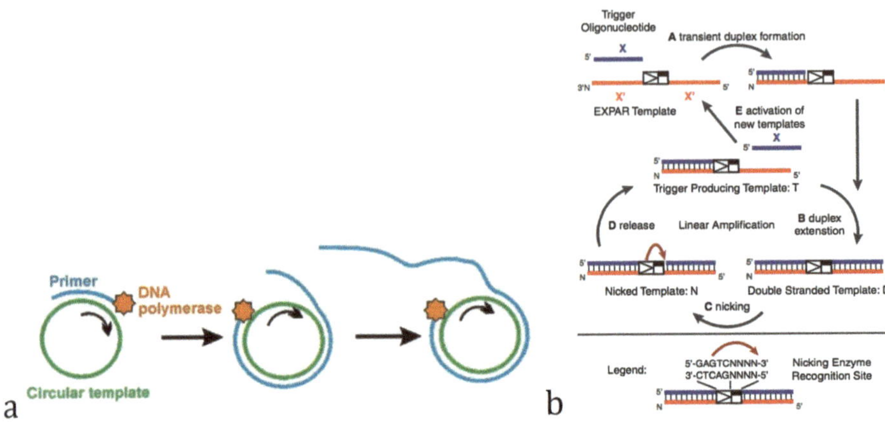

Fig. 10.4 (**a**) The mechanism of rolling circle amplification (RCA). A short DNA primer (*blue curve*) first hybridizes with a circular ssDNA template (*green circle*). A DNA polymerase (*orange octagon*) then adds nucleotides continuously to the primer by copying the circular template (Reproduced from Ref. [162] by permission of John Wiley & Sons Ltd). (**b**) The mechanism of exponential amplification reaction (EXPAR). (A) Trigger X transiently binds to the complementary recognition sequence at the 30 end of the amplification template; (B) the trigger sequence is extended by the DNA polymerase, forming the double-stranded nicking enzyme recognition site 50-GAGTCNNNN-30 on the top strand; (C) the top strand is cleaved through the nicking endonuclease Nt.BstNBI; (D) at the temperature of the reaction (55 C), the newly formed trigger is released from the amplification template. The trigger-producing form of the amplification template reenters the linear amplification cycle, and new trigger oligonucleotides are generated through duplex extension, nicking, and release; (E) the newly formed trigger oligonucleotides activate additional template sequences, giving rise to exponential amplification of the trigger X (Reprinted from Ref. [170] by permission of Oxford)

10.2.1.3 Rolling Circle Amplification (RCA)

It was reported that circular templates can be utilized to produce a long strand that contains multiple copies of target sequence in 1990s. This process was then called rolling circle amplification (RCA) [162, 163]. Initially, RCA was employed to find circle templates in viruses [97, 98, 162]. RCA reaction involves φ29 DNA polymerase extension to continuously prolong the primer that is binding to the circular template and replacement once the primer extends to the initial hybridization site (Fig. 10.4). The single-stranded nature of amplicons in case of linear RCA may be beneficial for subsequent manipulations of these DNAs for detection [8].

Simpler than LAMP and NEMA, RCA involves fewer reagent addition and less primer to design. The circle template is the only limiting factor. Much progress has been achieved since the invention of the padlock, an oligonucleotide that hybridizes to the target leaving a small gap of 6–10 nt. This gap can be filled by a short, allele-specific oligonucleotide with a 5′-phosphate that fits exactly within the gap, completing a stacked duplex structure. Then, the padlock is ligated with DNA ligase to form a closed padlock probe [2]. Another padlock is a circular probe that binds to the end of a short target sequence. The structure can be ligated by a ligation enzyme to make the target circular. This padlock can then act as a primer to initiate RCA, producing a long amplicon with a regular repeated sequence. Once the target can be circularized for RCA, the technique can fulfill the majority of needs in molecular diagnosis, including mutation detection, genotyping, virus diagnosis, toxin identification, and sensitive biosensing [164].

RCA can be quantified, no matter in linear amplification, by allowing one primer to extend and perform continuous cycling, and in exponential amplification, achieved by amplifying and displacing the elongated sequence through the addition of other primers [165]. Multiple electric or electrochemical methods can help to quantify the target amount, which will be detailed in Chap. 13 [163, 166]. The target length of RCA is severely restricted. Short targets (no more than 100 bp) are required for multiple purposes, except in one report [94].

MicroRNA (miRNA) has been widely applied in the field of microRNA detection. MicroRNAs are series of crucial regulators of gene expression. The abnormal expression of miRNAs may contribute to serious diseases, such as cancers, cardiovascular diseases, and neurological disorders. Therefore, the identification of microRNAs is a prerequisite for clinical diagnosis, as well as an indicative tool for pathogen determination. As the length of microRNAs is normally 19–25 nt, an ideal template length for RCA amplification, researchers are more focusing more intently on the RCA identification of microRNAs. Meanwhile, simpler and more sensitive RCA signal amplification can be achieved through microRNA-mediated ligation to form a circular target molecule [167].

10.2.1.4 Exponential Amplification Reaction (EXPAR)

Highly sensitive detection can be achieved by producing and storing large short amplified sequences in a short time, rather than using amplicons as the template in the next ground. The EXPAR was developed in 2003, intended to provide a cycle whose sequence specificity is independent on primer extension and replacement [168]. By arranging a specific nicking endonuclease recognition site and adding DNA polymerase, a cycle of polymerization and subsequent oligonucleotide release is established. EXPAR is very similar to SDA but utilizes the former target to produce large amounts of single-stranded DNA. The mechanism of EXPAR is based on the theory that long oligonucleotides can bind to the template and form stable duplexes, whereas short duplexes is unstable and can be dissociated easily from each other, as reported by previous studies to determine a certain length [169] (Fig. 10.4). The EXPAR depends on cleavage to create a nick site, and the remaining duplex is below the certain length of a stable duplex and the thereby extension to regenerate the amplified template. The amplified and displaced oligonucleotides are fully stable at 60 °C. Therefore, the oligonucleotide dissociates spontaneously from the template and leaves the space for following extension. This cycle generates large amounts of oligonucleotides with the same sequence. If the sequence is devised to produce oligonucleotides with the same sequence as the initial primer, it can in turn anneal to a new target template and extend and create a chain reaction to perform as an exponential style.

The Keck Graduate Institute, the founder of EXPAR, then evaluated the performance of randomly selected triggers using a computational tool to show that templates with similar thermodynamic characteristics related to trigger–template binding and template self-hybridization can perform quite differently in this reaction [170]. Researchers found it difficult to design optimal triggers and templates under the simple thermodynamic rules and therefore combined two different computational models, a standard PWM approach and a Naïve Bayes machine learning algorithm using positions motifs, into one computational tool.

EXPAR differs from other strand displacement methods in calculating the stability to achieve spontaneous release from the duplex and in evaluating the possibility of primer binding to the target and amplification, thus ensuring that amplification can occur at a low speed once the primer binds to the target in a chemical equilibrium status. The speed of the reaction and the flexibility to develop into multiple coupled reactions are all the advantages in the point-of-care detection. Due to these advantages, new emerging biosensing techniques seek to identify the target in an ultrasensitive way, updating through the efficient signal amplification caused by EXPAR, despite its imperfect design rules [171, 172]. EXPAR cannot amplify long sequence, which limited the practical application on nucleic acid detection, and is thereby restricted to the signal medium rather than the signal amplification area.

EXPAR application is limited in the food safety detection, partly due to its preference for short target sequences (Table. 10.1). However, EXPAR can be used as a signal transformation medium in the highly sensitive biosensing platforms. A few publications related to EXPAR identification are shown in the table, mainly involving virus, microRNA folate reporters, and telomerase detection.

10.2.2 Single-Oligonucleotide Duplication Mechanism

The SDA-based isothermal amplification often requires a high-capacity DNA polymerase with strand displacement activity, which creates complex and strict requirements for the system. Scientists have tried to imitate the inherent DNA duplication mechanism in the homothermal surroundings and then develop a new type of isothermal amplification technique. A single-oligonucleotide duplication mechanism is then applied to create a similar system to the one in the living organisms and facilitate DNA duplication in a biological way. The techniques detailed here are helicase-dependent amplification (HDA), nucleic acid sequence-based amplification (NASBA), recombinase polymerase amplification (RPA), and signal-mediated amplification of RNA technology (SMART).

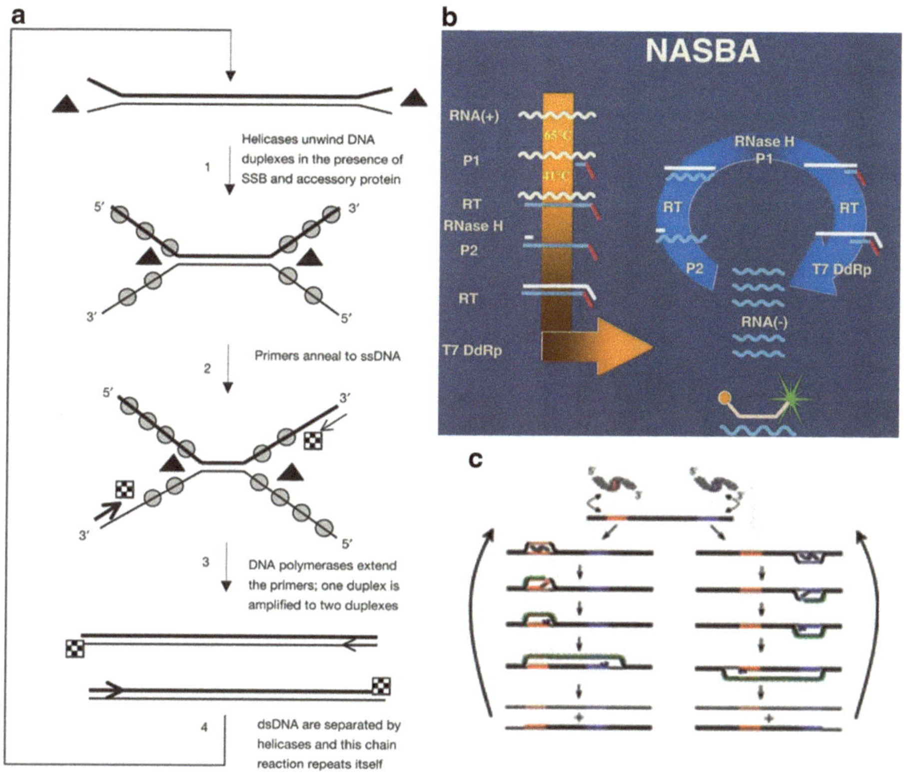

Fig. 10.5 (**a**) Mechanism of helicase-dependent amplification (HDA). Two complementary DNA strands are shown as *two lines*: the thick one is the top strand and the thin one is the bottom strand. *1* A helicase (*black triangle*) separates the two complementary DNA strands, which are bound by SSB (*gray circles*). *2* Primers (*lines* with *arrowheads*) hybridize to the target region on the ssDNA template. *3* A DNA polymerase (squares with mosaic patterns) extends the primers hybridized on the template DNA. *4* Amplified products enter the next round of amplification (Reproduced from Ref. [175] by permission of John Wiley & Sons Ltd). (**b**) Mechanism of nucleic acid sequence-based amplification (NASBA). The *orange arrow* represents the initiation phase and the blue arrow represents the cyclic phase. The activities of reverse transcriptase (RT), RNase H and T7 DNA-dependent RNA polymerase (T7 DdRp), and the primer-binding activities are indicated. The polarity of T7 DdRp product, amplified in the cyclic phase, is complementary to that of the target nucleic acid (Reprinted from Ref. [183], with kind permission from Springer Science+Business Media). (**c**) Mechanism of recombinase polymerase amplification (RPA). Recombinase–primer filaments scan template DNA for homologous sequences (*red*/*blue*). Following strand exchange, the displaced strand is bound by gp32 (*green*); primers are extended by Bsu polymerase (*blue*). The net result of two opposing primer binding/extension events is the generation of one complete copy of the amplicon in addition to the original template. Repetition of the process results in exponential DNA amplification (Reproduced from Ref. [186] with kind permission of PLOS)

10.2.2.1 Helicase-Dependent Amplification (HDA)

DNA is replicated by DNA helicase to separate duplex DNA and DNA polymerases with various accessory proteins to duplicate another new strand in vivo. HDA is how the DNA replication proceeds in vitro. *E. coli* UvrD helicase was attempted to unwind the DNA fragments. In the HDA system, strands of duplex DNA are separated by a DNA helicase and coated by single-stranded DNA-binding proteins (SSBs). Two specific primers hybridize to each site of the target DNA and extend to produce a dsDNA at the annealing temperature. The two newly synthesized dsDNA products are then used as templates by DNA helicases, initiating a new round of duplication. Thus, a polymerase chain reaction proceeds at a stable temperature, resulting in exponential amplification of the target sequence (Fig. 10.5).

HDA is actually an isothermal DNA amplification method. With the help of DNA helicase, dsDNA does not need to be separated in the melting process. In addition, various forms of templates (DNA, mRNA, and microRNA) can all in some way act as the template, which then serves as the input signal for amplification. The HDA has been used to detect target on crude samples and has the potential for use as a diagnostic tool [173, 174]. The limitations of speed (20 bp/s) and constancy (less than 100 bp per binding) may be the reason that the current UvrD HDA system is unable to amplify long templates [175]. The optimal temperature for UvrD helicase proves to be 60–65 °C, at which specificity is improved. Other applicable helicase can also be utilized in the HDA process.

HDA also has the potential and should be optimized for practical diagnosis, as shown by the efforts to detect SNPs in genotyping through real-time amplification [176]. HDA has been used on targets in complex biological matrices (such as crude bacterial lysates or blood) and can achieve one million-fold qualification and quantitation [177]. Adaptation of the HDA system to detection the pathogens *N. gonorrhoeae* and *S. aureus* was also performed (Table). The creator of HDA, BioHelix Corporation, has developed an isothermal helicase-dependent amplification platform for reverse transcription to detect Ebola virus RNA in a novel tube [178].

Otherwise, HDA has also been made into a device to combine the DNA extraction and template amplification [109, 179]. The most famous is by a scientific group from Boston University, which have devised a combination of integrated microfluidic chip and HDA [109]. The HDA was performed in microtubes to establish a combination of DNA extraction and amplification, to determine the optimal concentrations and conditions for the reactions, and to build up a smart passive fluidic control. Because the need for instrumentation to control the flow rate and temperature changes is reduced or eliminated, the device can easily use 10 CFU of *E. coli* [109–112].

10.2.2.2 Nucleic Acid Sequence-Based Amplification (NASBA)

Nucleic acid sequence-based amplification (NASBA) was first developed to detect an RNA target (HIV-1) [180–182]. This amplification is accomplished by two primers and the simultaneous enzymatic activity of AMV reverse transcriptase, T7 RNA polymerase and RNase H, and a RNA hydrolase that specifically cleaves RNA from a DNA/RNA duplex (Fig. 10.5). One primer used in NASBA contains a T7 promoter sequence (Pl). Taking RNA as a template, NASBA firstly extends the template using AMV reverse transcriptase to form an RNA/DNA duplex, which is then degraded by RNase H. Subsequently, the other primer (P2) can anneal to the leaving cDNA, and the complementary DNA strand is synthesized by AMV-RT, which has DNA-dependent DNA polymerase activity. The resulting double-stranded DNA includes a T7 RNA polymerase promoter sequence [183], facilitating RNA extension. The T7 RNA polymerase gives a 100–1,000-fold increase in a specific RNA [184]. The resulting RNA acts as the new template to initiate another round of amplification.

NASBA utilizes three specific enzymes: AMV reverse transcriptase, T7 RNA polymerase, and RNase H and was initially designed for RNA identification. It is similar to the duplication process in the viruses, as most viruses are genetically transformed via RNA. However, NASBA performs the artificial reverse transcription so that the cycling is DNA based. This feature may also be why NASBA is not stable in primer binding and reverse transcription efficiency. DNA can also serve as the initial template of NASBA, only if DNA is heated to denature dsDNA and to obtain single-stranded DNA intermediates. In the cycling amplification, the DNA template is amplified in the same way as RNA, except one condition that Pl and P2 are incorporated in the reverse order. NASBA can thus produce only approximately 100 bp length amplicons, inhibited by complex enzymatic catalysis and RNA interference in the reaction.

The duration of NASBA is 1–2 h at a reaction temperature of 37–42 °C, which is faster and more convenient than PCR. There have been commercialized applications using this technique, involving HIV-1 virus and hepatitis virus RNA [185], enterovirus RNA and GMOs, etc. (Table. 10.1).

10.2.2.3 Recombinase Polymerase Amplification (RPA)

RPA is based on the replication performed by enterobacteria phage T4 and requires enzymatic recombination to form a primer bind site. In RPA, the characteristic innovative application is recombinase. Recombinase is an enzyme that carries out site-specific genetic changes in the genome. In RPA, recombinase can scan the DNA template and facilitate strand exchange at the recognition site. In this way, Tm is not the restricted demand for primer designing, and primers even do not need to be complementary to the target [186]. The unwinding DNA is stabilized by single-stranded DNA-binding proteins (SSBs) to prevent the ejection of the primer by branch migration [187]. The ssDNA is then duplicated by the primer and polymerase by large fragment of *Bacillus subtilis* Pol I (Bsu, [188]). The key to RPA is the establishment of a dynamic reaction environment that balances the formation and disassembly of recombinase–primer filaments and the supply of recombination energy. Adenosine triphosphate (ATP) is added to the reaction, allowing the release of the chemical energy stored in the high-energy phosphoanhydride bonds and producing work in the form of mechanical energy. However, RPA is not stable owing to its high background noise. There has been developed to a new designed primer that contains a specific *Escherichia coli* endonuclease IV (Nfo) cleavage site as a solution. Only after it has been cleared by the endonuclease to reveal a 3′-OH end, DNA polymerase initiates to extend and replicate the template. The primer can also be labeled with a fluorophore and quencher so that the real-time cleavage reaction can be monitored [2, 186]. The technique can further be extended to real-time monitoring, facilitating sensitive and quantitative detection.

RPA is viewed as a progressive target amplification method and is expected to replace conventional PCR as a fast (15–45 min) and accurate technique, and several devices have been developed. However, RPA system is too complex to stabilize the recombinase catalytic activity and therefore amplify the whole genome as desired. A new method has conquered the problem by using a digital platform to simplify the reaction system, which may need further validation and feasible application. There are as yet only a few reports on RPA application in molecular diagnose, let alone the food safety detection. The targets are limited to medical viruses, GMOs screening elements, food pathogens, and biothreat agents (Table. 10.1).

10.2.2.4 Signal-Mediated Amplification of RNA Technology (SMART)

Another RNA technology, termed signal-mediated amplification of RNA technology (SMART), is based on a three-way junction (3WJ) structure and the transformation from long DNA to short RNA detection [135]. SMART consists of two single-stranded oligonucleotide probes, each of which includes a region that can hybridize to the target and a shorter region to bind each other. The longer probe also contains a T7 RNA promoter and the transcription template. This structure is called 3WJ structure. The Bst DNA polymerase extends the short probe to produce a T7 RNA polymerase promoter duplex. The resulting promoter allows T7 RNA polymerase to generate multiple copies of an RNA analyte for the following detection. Each RNA amplicon may itself be amplified by binding to a second template oligonucleotide and initiating RNA transcription, producing large amounts of RNA. This process is in fact a signal amplification method in which the target sequence is not itself amplified.

SMART has a sensitivity of as low as 50 pM analyte; however, the process takes several hours. The DNA genome can be analyzed through crude samples without sophisticated DNA extraction procedures. SMART has been applied to detect *E. coli* and marine cyanophage virus; however, SMART does not amplify the target amplification itself, which reduces its inherent specificity [189, 190]. This reduction may be inherent to the signal amplification technology.

10.3 Isothermal Hybridization Technologies

Nucleic acids act as a genetic information transmitter in the living organisms, forming a regular pattern of hydrogen bond-linked base pairing, which has inspired researchers to identify nucleic acids by similar means. Hybridization has attracted much scientific attention, coupling target nucleic acids and the corresponding pre-devised oligonucleotides to initiate biochemical detection [191].

A distinct achievement is the hybridization chain reaction (HCR), created by Robert M. Dirks and Niles A. Pierce from California Institute of Technology. "HCR came about not from a desire to compete with PCR but as a byproduct of engineering DNA hairpins that could be used as fuel packets to power DNA mechanical machines. By eliminating the machines from the design scenario, they found that the hairpins could be triggered to undergo a chain reaction. The realization that got us excited was the idea that this could be used as an amplifying signal transducer," says Pierce in the 26th issue of the Proceedings of the National Academy of Sciences [192].

We classify hybridization techniques here as isothermal nucleic acid detection technologies here for the following reasons:

- Hybridization usually occurs at isothermal temperature using multiple reagents; normally the temperature is maintained at room temperature.
- Hybridization usually involves signal accumulation. The signal detected consists of target-related oligonucleotides, similar to PCR amplification products.
- Many combinations of PCR and hybridization are being applied for the detection of potential dangerous factor. The design of such combinations can achieve specific and ultrasensitive target identification.

The hybridization techniques discussed in this chapter are all based on the principle of HCR, so the terms hybridization and HCR are used interchangeably.

10.3.1 Colorimetric Detection

Double-stranded and single-stranded oligonucleotides (dsDNA and ssDNA) have different electrostatic properties. ssDNA can be uncoiled randomly to expose its bases, whereas dsDNA always presents the negatively charged phosphate backbone because of its stable duplex structure. These properties can be utilized for colorimetric detection through gold nanoparticle aggregation. In the presence of dsDNA, gold nanoparticles are typically stabilized by adsorbed negative ions whose repulsion prevents the strong van der Waals attraction between gold particles from causing them to aggregate. These nanoparticle aggregates with distances greater than the average particle diameter appear red, but as the interparticle distances in these aggregates decrease to less than approximately the average particle diameter, the color becomes blue. It has been shown that ssDNA can stabilize the gold nanoparticles against aggregation at environments of high concentrations of salt, while dsDNA causes nanoparticles aggregation, causing the color to change from red to blue. As the color change is distinguishable with the naked eye, this detection can be performed without special instruments [193].

In this research, one strand of dsDNA should be the target, and the other one is designed oligonucleotide, followed by the HCR process to produce more dsDNA strands.

10.3.2 Fluorescence Detection

Two components of HCR, H1 and H2, are partially complementary and will bind each other to form a long double-stranded oligonucleotide, providing a possible modification site for fluorescence energy resonance transfer (FRET). H1 and H2 can be fluorophore labeled. The first hairpin opens up when it binds to the complementary target, leading to the binding reaction of hairpin (H2). This structure will trigger a chain reaction of H1 and H2 opening and hybridization. The distance between the fluorophore and quencher decreases to trigger FRET. An example is the enzyme-free amplification platform, which involves recognition of DNA and functional nucleic acids and triggers a hybridization chain reaction (Fig. 10.6). The

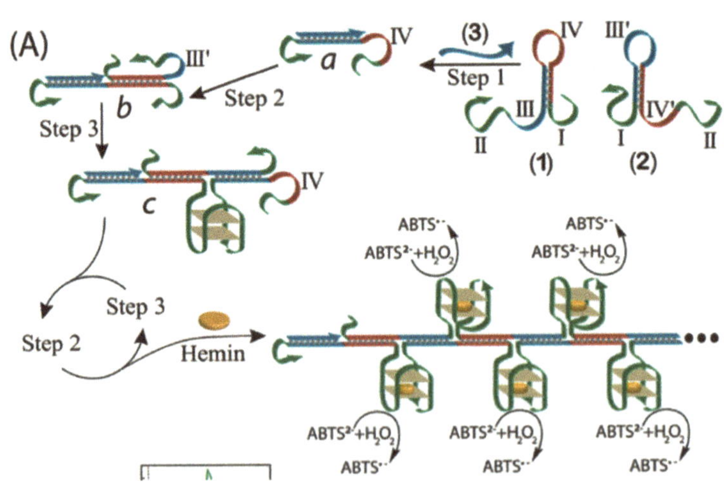

Fig. 10.6 Scheme for the amplified detection of a target DNA 3 using two functional hairpin structures (*1* and *2*) that upon recognition of the target DNA trigger the autonomous generation of nanowires consisting of the HRP-mimicking DNAzyme (Reprinted with permission from Ref. [194]. Copyright 2011 American Chemical Society)

functional nucleic acids form a hemin/G-quadruplex DNAzyme structure to make a color change [194].

Incorporating the fluorescent or electrochemical properties of advanced materials into the hybridization reaction offers the opportunity for ultrasensitive qualitative and quantitative biosensing detection, which will be detailed in other chapters (Chaps. 11 and 13).

10.4 Combinations of PCR and HCR

Theoretically, HCR cannot be stopped if enough H1 and H2 existed. However, the low speed of HCR encourages overnight incubation to achieve efficient detectable hybridization. On the other hand, detection by HCR alone has difficulty in specifically capturing the target. Combining high-efficiency enzymatic isothermal amplification and endless repeated hybridization is a welcome way to enhance the assay sensitivity significantly.

A Chinese group developed an RCA-based HCR system for the detection of site-specific methylation and MTase activity. The system fully took full advantage of branched RCA (BRCA) to make the amplified signal and HCR to amplify and recognize the signal for biosensing way [195]. The RCA products were then hybridized by designed hairpins (H1 and H2) for signal accumulation. Furthermore, the multiple HRP tags captured by the biotinylated hairpin and the formed dsDNA strands formed in the RCA products provide additional colorimetric detection.

Short oligonucleotides in the HCR process provide an ideal template for the majority of isothermal amplification because of their inherent limitations in amplifying long target sequences. Based on Liu's report on supersandwich damage in a biological ion channel, our group is now trying to destroy the long nicked dsDNA strand in HCR and amplify fallen hairpins through simple strand displacement amplification for the detection of mycotoxins.

10.5 Prospects

Isothermal nucleic acid techniques can accelerate the meaningful progress in point-of-care detection for food safety. From the time when chemical components of food samples were determined by chromatographic analysis in the professional laboratories to the time when nucleic acids in the food were identified via molecular biological methods, we have the history of the change from sophisticated to simple detection manifested by polymerase catalytic reactions. However, this stage is not the end of the story. Standardization, characterization, and benchmarking of the performance of isothermal systems are also necessary for continued development within research and commercial parts and will be important for the next step toward fast and ultrasensitive detection by biosensing platforms involved.

Acknowledgments This work is supported by the Ministry of Science and Technology of Beijing (XX2014B069). Many thanks to Chenguang Wang for his kindly help in the manuscript conception and preparation.

References

1. Weissensteiner T, Nolan T, Bustin SA, Griffin HG, Griffin A. PCR technology: current innovations. Boca Raton: CRC press; 2003.
2. Yan L, Zhou J, Zheng Y, Gamson AS, Roembke BT, Nakayama S, Sintim HO. Isothermal amplified detection of DNA and RNA. Mol BioSyst. 2014;10(5):970–1003.
3. Innis MA, Gelfand DH, Sninsky JJ, White TJ. PCR protocols: a guide to methods and applications. San Diego: Academic; 2012.
4. Kim J, Easley CJ. Isothermal DNA amplification in bioanalysis: strategies and applications. Bioanalysis. 2011;3(2):227–39.
5. Logan JMJ, Edwards KJ, Saunders NA. Real-time PCR: current technology and applications. Norfolk: Horizon Scientific Press; 2009.
6. Li J, Macdonald J. Advances in isothermal amplification: novel strategies inspired by biological processes. Biosens Bioelectron. 2015;64:196–211.
7. Deng H, Gao Z. Bioanalytical applications of isothermal nucleic acid amplification techniques. Anal Chim Acta. 2015;853:30–45.
8. Gill P, Ghaemi A. Nucleic acid isothermal amplification technologies—a review. Nucleosides Nucleotides Nucleic Acids. 2008;27(3):224–43.
9. Craw P, Balachandran W. Isothermal nucleic acid amplification technologies for point-of-care diagnostics: a critical review. Lab Chip. 2012;12(14):2469–86.
10. Chang C-C, Chen C-C, Wei S-C, Lu H-H, Liang Y-H, Lin C-W. Diagnostic devices for isothermal nucleic acid amplification. Sensors. 2012;12(6):8319–37.
11. Asiello PJ, Baeumner AJ. Miniaturized isothermal nucleic acid amplification, a review. Lab Chip. 2011;11(8):1420–30.
12. Saiki RK, Gelfand DH, Stoffel S, Scharf SJ, Higuchi R, Horn GT, Mullis KB, Erlich HA. Primer-directed enzymatic amplification of DNA with a thermostable DNA polymerase. Science. 1988;239(4839):487–91.
13. Saiki RK, Scharf S, Faloona F, Mullis KB, Horn GT, Erlich HA, Arnheim N. Enzymatic amplification of beta-globin genomic sequences and restriction site analysis for diagnosis of sickle cell anemia. Science. 1985;230(4732):1350–4.
14. Maruyama F, Kenzaka T, Yamaguchi N, Tani K, Nasu M. Detection of bacteria carrying the stx2 gene by in situ loop-mediated isothermal amplification. Appl Environ Microbiol. 2003;69(8):5023–8.
15. Hill J, Beriwal S, Chandra I, Paul VK, Kapil A, Singh T, Wadowsky RM, Singh V, Goyal A, Jahnukainen T. Loop-mediated isothermal amplification assay for rapid detection of common strains of Escherichia coli. J Clin Microbiol. 2008;46(8):2800–4.
16. Hara-Kudo Y, Yoshino M, Kojima T, Ikedo M. Loop-mediated isothermal amplification for the rapid detection of Salmonella. FEMS Microbiol Lett. 2005;253(1):155–61.
17. Imai M, Ninomiya A, Minekawa H, Notomi T, Ishizaki T, Van Tu P, Tien NTK, Tashiro M, Odagiri T. Rapid diagnosis of H5N1 avian influenza virus infection by newly developed influenza H5 hemagglutinin gene-specific loop-mediated isothermal amplification method. J Virol Methods. 2007;141(2):173–80.
18. Kono T, Savan R, Sakai M, Itami T. Detection of white spot syndrome virus in shrimp by loop-mediated isothermal amplification. J Virol Methods. 2004;115(1):59–65.

19. Zhao X, Li Y, Wang L, You L, Xu Z, Li L, He X, Liu Y, Wang J, Yang L. Development and application of a loop-mediated isothermal amplification method on rapid detection Escherichia coli O157 strains from food samples. Mol Biol Rep. 2010;37(5):2183–8.
20. Ohtsuka K, Yanagawa K, Takatori K, Hara-Kudo Y. Detection of Salmonella enterica in naturally contaminated liquid eggs by loop-mediated isothermal amplification, and characterization of Salmonella isolates. Appl Environ Microbiol. 2005;71(11):6730–5.
21. Wang L, Shi L, Alam MJ, Geng Y, Li L. Specific and rapid detection of foodborne Salmonella by loop-mediated isothermal amplification method. Food Res Int. 2008;41 (1):69–74.
22. Li X, Zhang S, Zhang H, Zhang L, Tao H, Yu J, Zheng W, Liu C, Lü D, Xiang R. A loop-mediated isothermal amplification method targets the phoP gene for the detection of Salmonella in food samples. Int J Food Microbiol. 2009;133(3):252–8.
23. Ueda S, Kuwabara Y. The rapid detection of Salmonella from food samples by loop-mediated isothermal amplification (LAMP). Biocontrol Sci. 2009;14(2):73–6.
24. Chen S, Wang F, Beaulieu JC, Stein RE, Ge B. Rapid detection of viable salmonellae in produce by coupling propidium monoazide with loop-mediated isothermal amplification. Appl Environ Microbiol. 2011;77(12):4008–16.
25. Yamazaki W, Taguchi M, Ishibashi M, Kitazato M, Nukina M, Misawa N, Inoue K. Development and evaluation of a loop-mediated isothermal amplification assay for rapid and simple detection of Campylobacter jejuni and Campylobacter coli. J Med Microbiol. 2008;57(4):444–51.
26. Tian CD, Zhang YZ, Ma XY, Zhang W, Wang J. Study on detection of vibrio parahaemolyticus in shellfish by use of loop-mediated isothermal amplification method. J Food Saf. 2011;31(3):371–8.
27. Nemoto J, Sugawara C, Akahane K, Hashimoto K, Kojima T, Ikedo M, Konuma H, Hara-Kudo Y. Rapid and specific detection of the thermostable direct hemolysin gene in Vibrio parahaemolyticus by loop-mediated isothermal amplification. J Food Prot. 2009;72 (4):748–54.
28. Yamazaki W, Taguchi M, Kawai T, Kawatsu K, Sakata J, Inoue K, Misawa N. Comparison of loop-mediated isothermal amplification assay and conventional culture methods for detection of Campylobacter jejuni and Campylobacter coli in naturally contaminated chicken meat samples. Appl Environ Microbiol. 2009;75(6):1597–603.
29. Yamazaki W, Kumeda Y, Misawa N, Nakaguchi Y, Nishibuchi M. Development of a loop-mediated isothermal amplification assay for sensitive and rapid detection of the tdh and trh genes of Vibrio parahaemolyticus and related Vibrio species. Appl Environ Microbiol. 2010;76(3):820–8.
30. Yamazaki W, Kumeda Y, Uemura R, Misawa N. Evaluation of a loop-mediated isothermal amplification assay for rapid and simple detection of Vibrio parahaemolyticus in naturally contaminated seafood samples. Food Microbiol. 2011;28(6):1238–41.
31. Yamazaki W, Ishibashi M, Kawahara R, Inoue K. Development of a loop-mediated isothermal amplification assay for sensitive and rapid detection of Vibrio parahaemolyticus. BMC Microbiol. 2008;8(1):163.
32. Zhao X, Wang L, Chu J, Li Y, Li Y, Xu Z, Li L, Shirtliff ME, He X, Liu Y. Rapid detection of Vibrio parahaemolyticus strains and virulent factors by loop-mediated isothermal amplification assays. Food Sci Biotechnol. 2010;19(5):1191–7.
33. Wang L, Zhong Q, Li Y. Ethidium monoazide-loop mediated isothermal amplification for rapid detection of Vibrio parahaemolyticus in viable but non-culturable state. Energy Procedia. 2012;17:1858–63.
34. Fang X, Liu Y, Kong J, Jiang X. Loop-mediated isothermal amplification integrated on microfluidic chips for point-of-care quantitative detection of pathogens. Anal Chem. 2010;82(7):3002–6.
35. Karanis P, Thekisoe O, Kiouptsi K, Ongerth J, Igarashi I, Inoue N. Development and preliminary evaluation of a loop-mediated isothermal amplification procedure for sensitive

detection of Cryptosporidium oocysts in fecal and water samples. Appl Environ Microbiol. 2007;73(17):5660–2.
36. Tang M-J, Zhou S, Zhang X-Y, Pu J-H, Ge Q-L, Tang X-J, Gao Y-S. Rapid and sensitive detection of Listeria monocytogenes by loop-mediated isothermal amplification. Curr Microbiol. 2011;63(6):511–6.
37. Shan X, Zhang Y, Zhang Z, Chen M, Su Y, Yuan Y, Alam MJ, Yan H, Shi L. Rapid detection of food-borne Listeria monocytogenes by real-time quantitative loop-mediated isothermal amplification. Food Sci Biotechnol. 2012;21(1):101–6.
38. Wan C, Yang Y, Xu H, Aguilar ZP, Liu C, Lai W, Xiong Y, Xu F, Wei H. Development of a propidium monoazide treatment combined with loop-mediated isothermal amplification (PMA-LAMP) assay for rapid detection of viable Listeria monocytogenes. Int J Food Sci Technol. 2012;47(11):2460–7.
39. Wang D, Huo G, Ren D, Li Y. Development and evaluation of a loop-mediated isothermal amplification (LAMP) method for detecting listeria monocytogenes in raw milk. J Food Saf. 2010;30(2):251–62.
40. Niessen L, Gräfenhan T, Vogel RF. ATP citrate lyase 1 (acl1) gene-based loop-mediated amplification assay for the detection of the Fusarium tricinctum species complex in pure cultures and in cereal samples. Int J Food Microbiol. 2012;158(3):171–85.
41. Denschlag C, Rieder J, Vogel RF, Niessen L. Real-time loop-mediated isothermal amplification (LAMP) assay for group specific detection of important trichothecene producing Fusarium species in wheat. Int J Food Microbiol. 2014;177:117–27.
42. Pavan Kumar P, Agarwal RK, Thomas P, Sailo B, Prasannavadhana A, Kumar A, Kataria JL, Singh DK. Rapid detection of Salmonella enterica subspecies enterica serovar Typhimurium by loop mediated isothermal amplification (LAMP) test from field chicken meat samples. Food Biotechnol. 2014;28(1):50–62.
43. Zhai S, Liu C, Zhang Q, Tao C, Liu B. Detection of two exogenous genes in transgenic cattle by loop-mediated isothermal amplification. Transgenic Res. 2012;21(6):1367–73.
44. Zhang G, Brown EW, González-Escalona N. Comparison of real-time PCR, reverse transcriptase real-time PCR, loop-mediated isothermal amplification, and the FDA conventional microbiological method for the detection of Salmonella spp. in produce. Appl Environ Microbiol. 2011;77(18):6495–501.
45. Zhu S-m, Wu J-j, Xu C, Qu J, Cheng W, Chen F-s. Rapid detection of salmonella spp. by loop-mediated isothermal amplification method. Mod Food Sci Technol. 2008;7:036.
46. Techathuvanan C, D'Souza DH. Reverse-transcriptase loop-mediated isothermal amplification as a rapid screening/monitoring tool for Salmonella Enterica detection in liquid whole eggs. J Food Sci. 2012;77(4):M200–5.
47. Techathuvanan C, Draughon FA, D'Souza DH. Loop-mediated isothermal amplification (LAMP) for the rapid and sensitive detection of Salmonella Typhimurium from pork. J Food Sci. 2010;75(3):M165–72.
48. Techathuvanan C, Draughon FA, D'Souza DH. Comparison of reverse transcriptase PCR, reverse transcriptase loop-mediated isothermal amplification, and culture-based assays for Salmonella detection from pork processing environments. J Food Prot. 2011;74(2):294–301.
49. Ravan H, Yazdanparast R. Development of a new loop-mediated isothermal amplification assay for prt (rfbS) gene to improve the identification of Salmonella serogroup D. World J Microbiol Biotechnol. 2012;28(5):2101–6.
50. Liu C, Zheng W, Zhang H, Hou Y, Liu Y. Sensitive and rapid detection of Enterobacter sakazakii in infant formula by loop-mediated isothermal amplification method. J Food Saf. 2009;29(1):83–94.
51. Liu X, Fang J, Zhang M, Wang X, Wang W, Gong Y, Xi X, Li M. Development of a loop-mediated isothermal amplification assay for detection of Cronobacter spp. (Enterobacter sakazakii). World J Microbiol Biotechnol. 2012;28(3):1013–20.

52. DeGuo W, LiJie Y, GuiCheng H, FuGui W, YongGang L, Daxi R. Development of a loop-mediated isothermal amplification (LAMP) method for detecting food-borne Shigella in raw milk. Milchwissenschaft. 2009;64(3):264–7.
53. Li C, Ying Q, Su X, Li T. Development and application of reverse transcription loop-mediated isothermal amplification for detecting live Shewanella putrefaciens in preserved fish sample. J Food Sci. 2012;77(4):M226–30.
54. Ohtsuki R, Kawamoto K, Kato Y, Shah M, Ezaki T, Makino SI. Rapid detection of Brucella spp. by the loop-mediated isothermal amplification method. J Appl Microbiol. 2008;104 (6):1815–23.
55. DeGuo W, GuiCheng H, JunLiang C, Sui Z, Daxi R, YongGang L. Development and evaluation of a loop-mediated isothermal amplification (LAMP) method for detecting Staphylococcus aureus in raw milk. Milchwissenschaft. 2009;64(4):368–71.
56. Suwanampai T, Pattaragulvanit K, Pattanamahakul P, Sutheinkul O, Okada K, Honda T, Thaniyavarn J. Evaluation of loop-mediated isothermal amplification method for detecting enterotoxin A gene of Staphylococcus aureus in pork. Southeast Asian J Trop Med Public Health. 2011;42(6):1489.
57. Nagarajappa S, Thakur MS, Manonmani HK. Detection of enterotoxigenic staphylococci by loop-mediated isothermal amplification method. J Food Saf. 2012;32(1):59–65.
58. Sakuma T, Kurosaki Y, Fujinami Y, Takizawa T, Yasuda J. Rapid and simple detection of Clostridium botulinum types A and B by loop-mediated isothermal amplification. J Appl Microbiol. 2009;106(4):1252–9.
59. Kaneko I, Miyamoto K, Mimura K, Yumine N, Utsunomiya H, Akimoto S, McClane BA. Detection of enterotoxigenic Clostridium perfringens in meat samples by using molecular methods. Appl Environ Microbiol. 2011;77(21):7526–32.
60. Chen S, Ge B. Development of a toxR-based loop-mediated isothermal amplification assay for detecting Vibrio parahaemolyticus. BMC Microbiol. 2010;10(1):41.
61. Hayashi N, Arai R, Tada S, Taguchi H, Ogawa Y. Detection and identification of Brettanomyces/Dekkera sp. yeasts with a loop-mediated isothermal amplification method. Food Microbiol. 2007;24(7):778–85.
62. Ahmed MU, Hasan Q, Hossain MM, Saito M, Tamiya E. Meat species identification based on the loop mediated isothermal amplification and electrochemical DNA sensor. Food Control. 2010;21(5):599–605.
63. Yang L, Fu S, Peng X, Li L, Song T, Li L. Identification of pork in meat products using real-time loop-mediated isothermal amplification. Biotechnol Biotechnol Equip. 2014;28 (5):882–8.
64. Li M, Wong Y-L, Jiang L-L, Wong K-L, Wong Y-T, Bik-San Lau C, Shaw P-C. Application of novel loop-mediated isothermal amplification (LAMP) for rapid authentication of the herbal tea ingredient Hedyotis diffusa Willd. Food Chem. 2013;141(3):2522–5.
65. Ran G, Ren L, Han X, Liu X, Li Z, Pang D, Ouyang H, Tang X. Development of a rapid method for the visible detection of pork DNA in Halal Products by loop-mediated isothermal amplification. Food Analytical Methods:1–6. 2015.
66. Zahradnik C, Martzy R, Mach RL, Krska R, Farnleitner AH, Brunner K. Loop-mediated isothermal amplification (LAMP) for the detection of horse meat in meat and processed meat products. Food Analytical Methods:1–6. 2014.
67. Luo J, Vogel RF, Niessen L. Development and application of a loop-mediated isothermal amplification assay for rapid identification of aflatoxigenic molds and their detection in food samples. Int J Food Microbiol. 2012;159(3):214–24.
68. Luo J, Taniwaki MH, Iamanaka BT, Vogel RF, Niessen L. Application of loop-mediated isothermal amplification assays for direct identification of pure cultures of Aspergillus flavus, A. nomius, and A. caelatus and for their rapid detection in shelled Brazil nuts. Int J Food Microbiol. 2014;172:5–12.
69. Storari M, Rohr R, Pertot I, Gessler C, Broggini GAL. Identification of ochratoxin A producing Aspergillus carbonarius and A. niger clade isolated from grapes using the loop-

mediated isothermal amplification (LAMP) reaction. J Appl Microbiol. 2013;114 (4):1193–200.
70. Yuan Y, Wei S, Liu G, Xie S, Chai Y, Yuan R. Ultrasensitive electrochemiluminescent aptasensor for ochratoxin A detection with the loop-mediated isothermal amplification. Anal Chim Acta. 2014;811:70–5.
71. Chen X, Wang X, Jin N, Zhou Y, Huang S, Miao Q, Zhu Q, Xu J. Endpoint visual detection of three genetically modified rice events by loop-mediated isothermal amplification. Int J Mol Sci. 2012;13(11):14421–33.
72. Li Q, Fang J, Liu X, Xi X, Li M, Gong Y, Zhang M. Loop-mediated isothermal amplification (LAMP) method for rapid detection of cry1Ab gene in transgenic rice (Oryza sativa L.). Eur Food Res Technol. 2013;236(4):589–98.
73. Di H, Shi L, Shen H, Yan H, Meng H, Li L, Alam MJ, Yamasaki S, Ye L. Rapid detection of genetically modified ingredients in soybean products by real-time loop-mediated isothermal amplification. J Food Nutr Res. 2014;2(7):363–8.
74. Zhang M, Liu Y, Chen L, Quan S, Jiang S, Zhang D, Yang L. One simple DNA extraction device and its combination with modified visual loop-mediated isothermal amplification for rapid on-field detection of genetically modified organisms. Anal Chem. 2012;85(1):75–82.
75. Zhang F, Wang L, Fan K, Wu J, Ying Y. The detection of T-Nos, a genetic element present in GMOs, by cross-priming isothermal amplification with real-time fluorescence. Anal Bioanal Chem. 2014;406(13):3069–78.
76. Guan X, Guo J, Shen P, Yang L, Zhang D. Visual and rapid detection of two genetically modified soybean events using loop-mediated isothermal amplification method. Food Anal Methods. 2010;3(4):313–20.
77. Lee D, La Mura M, Allnutt TR, Powell W. Detection of genetically modified organisms (GMOs) using isothermal amplification of target DNA sequences. BMC Biotechnol. 2009;9 (1):7.
78. Fukuta S, Mizukami Y, Ishida A, Ueda J, Hasegawa M, Hayashi I, Hashimoto M, Kanbe M. Real-time loop-mediated isothermal amplification for the CaMV-35S promoter as a screening method for genetically modified organisms. Eur Food Res Technol. 2004;218 (5):496–500.
79. Chen L, Guo J, Wang Q, Kai G, Yang L. Development of the visual loop-mediated isothermal amplification assays for seven genetically modified maize events and their application in practical samples analysis. J Agric Food Chem. 2011;59(11):5914–8.
80. Kiddle G, Hardinge P, Buttigieg N, Gandelman O, Pereira C, McElgunn CJ, Rizzoli M, Jackson R, Appleton N, Moore C. GMO detection using a bioluminescent real time reporter (BART) of loop mediated isothermal amplification (LAMP) suitable for field use. BMC Biotechnol. 2012;12(1):15.
81. Xu J, Zheng Q, Yu L, Liu R, Zhao X, Wang G, Wang Q, Cao J. Loop-mediated isothermal amplification (LAMP) method for detection of genetically modified maize T25. Food Sci Nutr. 2013;1(6):432–8.
82. Huang S, Xu Y, Yan X, Shang Y, Zhu P, Tian W, Xu W. Development and application of a quantitative loop-mediated isothermal amplification method for detecting genetically modified maize MON863. J Sci Food Agric. 2015;95(2):253–9.
83. Li F, Yan W, Long L, Qi X, Li C, Zhang S. Development and application of loop-mediated isothermal amplification assays for rapid visual detection of cry2Ab and cry3A genes in genetically-modified crops. Int J Mol Sci. 2014;15(9):15109–21.
84. Cheng Y, Zhang M, Hu K, Sun F, Tao R, Gao X, Luan F. Loop-mediated isothermal amplification for the event-specific detection of wheat B73-6-1. Food Anal Methods. 2014;7(2):500–5.
85. Tao C, Zhang Q, Zhai S, Liu B. Detection of HbsAg and hATIII genetically modified goats (Caprahircus) by loop-mediated isothermal amplification. Mol Biol Rep. 2013;40 (11):6177–82.

86. He J-L, Wu Z-S, Zhou H, Wang H-Q, Jiang J-H, Shen G-L, Yu R-Q. Fluorescence aptameric sensor for strand displacement amplification detection of cocaine. Anal Chem. 2010;82 (4):1358–64.
87. Walker GT, Nadeau JG, Spears PA, Schram JL, Nycz CM, Shank DD. Multiplex strand displacement amplification (SDA) and detection of DNA sequences from Mycobacterium tuberculosis and other mycobacteria. Nucleic Acids Res. 1994;22(13):2670–7.
88. Hellyer T, DesJardin L, Teixeira L, Perkins M, Cave M, Eisenach K. Detection of viable Mycobacterium tuberculosis by reverse transcriptase-strand displacement amplification of mRNA. J Clin Microbiol. 1999;37(3):518–23.
89. Ge B, Larkin C, Ahn S, Jolley M, Nasir M, Meng J, Hall R. Identification of Escherichia coli O157: H7 and other enterohemorrhagic serotypes by EHEC-hly A targeting, strand displacement amplification, and fluorescence polarization. Mol Cell Probes. 2002;16(2):85–92.
90. Kiesling T, Cox K, Davidson EA, Dretchen K, Grater G, Hibbard S, Lasken RS, Leshin J, Skowronski E, Danielsen M. Sequence specific detection of DNA using nicking endonuclease signal amplification (NESA). Nucleic Acids Res. 2007;35(18), e117.
91. Bi S, Cui Y, Li L. Dumbbell probe-mediated cascade isothermal amplification: a novel strategy for label-free detection of microRNAs and its application to real sample assay. Anal Chim Acta. 2013;760:69–74.
92. Yin B-C, Liu Y-Q, Ye B-C. One-step, multiplexed fluorescence detection of microRNAs based on duplex-specific nuclease signal amplification. J Am Chem Soc. 2012;134 (11):5064–7.
93. Hun X, Liu F, Mei Z, Ma L, Wang Z, Luo X. Signal amplified strategy based on target-induced strand release coupling cleavage of nicking endonuclease for the ultrasensitive detection of ochratoxin A. Biosens Bioelectron. 2013;39(1):145–51.
94. Murakami T, Sumaoka J, Komiyama M. Sensitive isothermal detection of nucleic-acid sequence by primer generation–rolling circle amplification. Nucleic Acids Res. 2009;37(3): e19.
95. Long Y, Zhou X, Xing D. Sensitive and isothermal electrochemiluminescence gene-sensing of Listeria monocytogenes with hyperbranching rolling circle amplification technology. Biosens Bioelectron. 2011;26(6):2897–904.
96. Sato K, Tachihara A, Renberg B, Mawatari K, Sato K, Tanaka Y, Jarvius J, Nilsson M, Kitamori T. Microbead-based rolling circle amplification in a microchip for sensitive DNA detection. Lab Chip. 2010;10(10):1262–6.
97. Haible D, Kober S, Jeske H. Rolling circle amplification revolutionizes diagnosis and genomics of geminiviruses. J Virol Methods. 2006;135(1):9–16.
98. Niel C, Diniz-Mendes L, Devalle S. Rolling-circle amplification of Torque teno virus (TTV) complete genomes from human and swine sera and identification of a novel swine TTV genogroup. J Gen Virol. 2005;86(5):1343–7.
99. Schubert J, Habekuß A, Kazmaier K, Jeske H. Surveying cereal-infecting geminiviruses in Germany—diagnostics and direct sequencing using rolling circle amplification. Virus Res. 2007;127(1):61–70.
100. Zhou X, Kong F, Sorrell TC, Wang H, Duan Y, Chen SC. Practical method for detection and identification of Candida, Aspergillus, and Scedosporium spp. by use of rolling-circle amplification. J Clin Microbiol. 2008;46(7):2423–7.
101. Smolina I, Lee C, Frank-Kamenetskii M. Detection of low-copy-number genomic DNA sequences in individual bacterial cells by using peptide nucleic acid-assisted rolling-circle amplification and fluorescence in situ hybridization. Appl Environ Microbiol. 2007;73 (7):2324–8.
102. Mahmoudian L, Kaji N, Tokeshi M, Nilsson M, Baba Y. Rolling circle amplification and circle-to-circle amplification of a specific gene integrated with electrophoretic analysis on a single chip. Anal Chem. 2008;80(7):2483–90.

103. Cheng Y, Zhang X, Li Z, Jiao X, Wang Y, Zhang Y. Highly sensitive determination of microRNA using target-primed and branched rolling-circle amplification. Angew Chem. 2009;121(18):3318–22.
104. Ma C, Wang W, Yang Q, Shi C, Cao L. Cocaine detection via rolling circle amplification of short DNA strand separated by magnetic beads. Biosens Bioelectron. 2011;26(7):3309–12.
105. Tong P, Zhao W-W, Zhang L, Xu J-J, Chen H-Y. Double-probe signal enhancing strategy for toxin aptasensing based on rolling circle amplification. Biosens Bioelectron. 2012;33 (1):146–51.
106. Tang S, Tong P, Li H, Tang J, Zhang L. Ultrasensitive electrochemical detection of Pb 2+ based on rolling circle amplification and quantum dots tagging. Biosens Bioelectron. 2013;42:608–11.
107. Li W, Yang Y, Chen J, Zhang Q, Wang Y, Wang F, Yu C. Detection of lead (II) ions with a DNAzyme and isothermal strand displacement signal amplification. Biosens Bioelectron. 2014;53:245–9.
108. Yu Y, Chen Z, Jian W, Sun D, Zhang B, Li X, Yao M. Ultrasensitive electrochemical detection of avian influenza A (H7N9) virus DNA based on isothermal exponential amplification coupled with hybridization chain reaction of DNAzyme nanowires. Biosens Bioelectron. 2015;64:566–71.
109. Mahalanabis M, Do J, ALMuayad H, Zhang JY, Klapperich CM. An integrated disposable device for DNA extraction and helicase dependent amplification. Biomed Microdevices. 2010;12(2):353–9.
110. Chow WHA, McCloskey C, Tong Y, Hu L, You Q, Kelly CP, Kong H, Tang Y-W, Tang W. Application of isothermal helicase-dependent amplification with a disposable detection device in a simple sensitive stool test for toxigenic Clostridium difficile. J Mol Diagn. 2008;10(5):452–8.
111. Andresen D, von Nickisch-Rosenegk M, Bier FF. Helicase dependent OnChip-amplification and its use in multiplex pathogen detection. Clin Chim Acta. 2009;403(1):244–8.
112. Gill P, Amini M, Ghaemi A, Shokouhizadeh L, Abdul-Tehrani H, Karami A, Gilak A. Detection of Helicobacter pylori by enzyme-linked immunosorbent assay of thermophilic helicase-dependent isothermal DNA amplification. Diagn Microbiol Infect Dis. 2007;59 (3):243–9.
113. Morisset D, Dobnik D, Hamels S, Žel J, Gruden K. NAIMA: target amplification strategy allowing quantitative on-chip detection of GMOs. Nucleic Acids Res. 2008;36(18):e118.
114. Uyttendaele M, Schukkink R, Van Gemen B, Debevere J. Development of NASBA®, a nucleic acid amplification system, for identification of Listeria monocytogenes and comparison to ELISA and a modified FDA method. Int J Food Microbiol. 1995;27(1):77–89.
115. Blais BW, Turner G, Sooknanan R, Malek LT. A nucleic acid sequence-based amplification system for detection of Listeria monocytogenes hlyA sequences. Appl Environ Microbiol. 1997;63(1):310–3.
116. Loeffler J, Hebart H, Cox P, Flues N, Schumacher U, Einsele H. Nucleic acid sequence-based amplification of aspergillus RNA in blood samples. J Clin Microbiol. 2001;39(4):1626–9.
117. Jean J, D'souza DH, Jaykus L-A. Multiplex nucleic acid sequence-based amplification for simultaneous detection of several enteric viruses in model ready-to-eat foods. Appl Environ Microbiol. 2004;70(11):6603–10.
118. Jean J, Blais B, Darveau A, Fliss IL. Detection of hepatitis A virus by the nucleic acid sequence-based amplification technique and comparison with reverse transcription-PCR. Appl Environ Microbiol. 2001;67(12):5593–600.
119. Lamhoujeb S, Fliss I, Ngazoa SE, Jean J. Evaluation of the persistence of infectious human noroviruses on food surfaces by using real-time nucleic acid sequence-based amplification. Appl Environ Microbiol. 2008;74(11):3349–55.
120. Rutjes SA, van den Berg HH, Lodder WJ, de Roda Husman AM. Real-time detection of noroviruses in surface water by use of a broadly reactive nucleic acid sequence-based amplification assay. Appl Environ Microbiol. 2006;72(8):5349–58.

121. Uyttendaele M, Schukkink R, Van Gemen B, Debevere J. Detection of Campylobacter jejuni added to foods by using a combined selective enrichment and nucleic acid sequence-based amplification (NASBA). Appl Environ Microbiol. 1995;61(4):1341–7.
122. Fykse EM, Skogan G, Davies W, Olsen JS, Blatny JM. Detection of Vibrio cholerae by real-time nucleic acid sequence-based amplification. Appl Environ Microbiol. 2007;73 (5):1457–66.
123. D'Souza D, Jaykus LA. Nucleic acid sequence based amplification for the rapid and sensitive detection of Salmonella enterica from foods. J Appl Microbiol. 2003;95(6):1343–50.
124. Cook N, Ellison J, Kurdziel A, Simpkins S, Hays J. A nucleic acid sequence-based amplification method to detect Salmonella enterica serotype enteritidis strain PT4 in liquid whole egg. J Food Prot. 2002;65(7):1177–8.
125. Zhao X, Dong T. Multifunctional sample preparation kit and on-chip quantitative nucleic acid sequence-based amplification tests for microbial detection. Anal Chem. 2012;84 (20):8541–8.
126. Mollasalehi H, Yazdanparast R. Development and evaluation of a novel nucleic acid sequence-based amplification method using one specific primer and one degenerate primer for simultaneous detection of Salmonella Enteritidis and Salmonella Typhimurium. Anal Chim Acta. 2013;770:169–74.
127. Zhao X, Dong T, Yang Z, Pires N, Høivik N. Compatible immuno-NASBA LOC device for quantitative detection of waterborne pathogens: design and validation. Lab Chip. 2012;12 (3):602–12.
128. Fykse EM, Nilsen T, Nielsen AD, Tryland I, Delacroix S, Blatny JM. Real-time PCR and NASBA for rapid and sensitive detection of Vibrio cholerae in ballast water. Mar Pollut Bull. 2012;64(2):200–6.
129. Moore C, Telles J-N, Corden S, Gao R-B, Vernet G, Van Aarle P, Shu Y-L. Development and validation of a commercial real-time NASBA assay for the rapid confirmation of influenza A H5N1 virus in clinical samples. J Virol Methods. 2010;170(1):173–6.
130. Xu C, Li L, Jin W, Wan Y. Recombinase Polymerase Amplification (RPA) of CaMV-35S promoter and nos terminator for rapid detection of genetically modified crops. Int J Mol Sci. 2014;15(10):18197–205.
131. Shen F, Davydova EK, Du W, Kreutz JE, Piepenburg O, Ismagilov RF. Digital isothermal quantification of nucleic acids via simultaneous chemical initiation of recombinase polymerase amplification reactions on SlipChip. Anal Chem. 2011;83(9):3533–40.
132. Kersting S, Rausch V, Bier FF, von Nickisch-Rosenegk M. Multiplex isothermal solid-phase recombinase polymerase amplification for the specific and fast DNA-based detection of three bacterial pathogens. Microchim Acta. 2014;181(13–14):1715–23.
133. Santiago-Felipe S, Tortajada-Genaro L, Puchades R, Maquieira A. Recombinase polymerase and enzyme-linked immunosorbent assay as a DNA amplification-detection strategy for food analysis. Anal Chim Acta. 2014;811:81–7.
134. del Río JS, Adly NY, Acero-Sánchez JL, Henry OY, O'Sullivan CK. Electrochemical detection of Francisella tularensis genomic DNA using solid-phase recombinase polymerase amplification. Biosens Bioelectron. 2014;54:674–8.
135. Wharam SD, Marsh P, Lloyd JS, Ray TD, Mock GA, Assenberg R, McPhee JE, Brown P, Weston A, Cardy DL. Specific detection of DNA and RNA targets using a novel isothermal nucleic acid amplification assay based on the formation of a three-way junction structure. Nucleic Acids Res. 2001;29(11):e54.
136. Walker GT, Little MC, Nadeau JG, Shank DD. Isothermal in vitro amplification of DNA by a restriction enzyme/DNA polymerase system. Proc Natl Acad Sci. 1992;89(1):392–6.
137. Walker GT, Fraiser MS, Schram JL, Little MC, Nadeau JG, Malinowski DP. Strand displacement amplification—an isothermal, in vitro DNA amplification technique. Nucleic Acids Res. 1992;20(7):1691–6.
138. Walker GT. Empirical aspects of strand displacement amplification. Genome Res. 1993;3 (1):1–6.

139. Spargo CA, Fraiser MS, Van Cleve M, Wright DJ, Nycz CM, Spears PA, Walker GT. Detection of M. tuberculosis DNA using thermophilic strand displacement amplification. Mol Cell Probes. 1996;10(4):247–56.
140. Hellyer TJ, Fletcher TW, Bates JH, Stead WW, Templeton GL, Cave MD, Eisenach KD. Strand displacement amplification and the polymerase chain reaction for monitoring response to treatment in patients with pulmonary tuberculosis. J Infect Dis. 1996;173 (4):934–41.
141. Cosentino LA, Landers DV, Hillier SL. Detection of Chlamydia trachomatis and Neisseria gonorrhoeae by strand displacement amplification and relevance of the amplification control for use with vaginal swab specimens. J Clin Microbiol. 2003;41(8):3592–6.
142. Paez JG, Lin M, Beroukhim R, Lee JC, Zhao X, Richter DJ, Gabriel S, Herman P, Sasaki H, Altshuler D. Genome coverage and sequence fidelity of φ29 polymerase-based multiple strand displacement whole genome amplification. Nucleic Acids Res. 2004;32(9):e71.
143. Zhang L, Cui X, Schmitt K, Hubert R, Navidi W, Arnheim N. Whole genome amplification from a single cell: implications for genetic analysis. Proc Natl Acad Sci. 1992;89 (13):5847–51.
144. Lasken RS. Genomic sequencing of uncultured microorganisms from single cells. Nat Rev Microbiol. 2012;10:631–40.
145. Notomi T, Okayama H, Masubuchi H, Yonekawa T, Watanabe K, Amino N, Hase T. Loop-mediated isothermal amplification of DNA. Nucleic Acids Res. 2000;28(12):e63.
146. Mori Y, Notomi T. Loop-mediated isothermal amplification (LAMP): a rapid, accurate, and cost-effective diagnostic method for infectious diseases. J Infect Chemother. 2009;15 (2):62–9.
147. Nagamine K, Hase T, Notomi T. Accelerated reaction by loop-mediated isothermal amplification using loop primers. Mol Cell Probes. 2002;16(3):223–9.
148. Mori Y, Nagamine K, Tomita N, Notomi T. Detection of loop-mediated isothermal amplification reaction by turbidity derived from magnesium pyrophosphate formation. Biochem Biophys Res Commun. 2001;289(1):150–4.
149. Poon LLM, Wong BWY, Ma EHT, Chan KH, Chow LMC, Abeyewickreme W, Tangpukdee N, Yuen KY, Guan Y, Looareesuwan S. Sensitive and inexpensive molecular test for falciparum malaria: detecting Plasmodium falciparum DNA directly from heat-treated blood by loop-mediated isothermal amplification. Clin Chem. 2006;52(2):303–6.
150. Fall J, Chakraborty G, Kono T, Maeda M, Suzuki Y, Itami T, Sakai M. Quantitative loop-mediated isothermal amplification method for the detection of Vibrio nigripulchritudo in shrimp. Fish Sci. 2011;77(1):129–34.
151. Mekata T, Sudhakaran R, Kono T, Supamattaya K, Linh NTH, Sakai M, Itami T. Real-time quantitative loop-mediated isothermal amplification as a simple method for detecting white spot syndrome virus. Lett Appl Microbiol. 2009;48(1):25–32.
152. Aoi Y, Hosogai M, Tsuneda S. Real-time quantitative LAMP (loop-mediated isothermal amplification of DNA) as a simple method for monitoring ammonia-oxidizing bacteria. J Biotechnol. 2006;125(4):484–91.
153. Toriniwa H, Komiya T. Rapid detection and quantification of Japanese encephalitis virus by real-time reverse transcription loop-mediated isothermal amplification. Microbiol Immunol. 2006;50(5):379–87.
154. Han F, Ge B. Quantitative detection of Vibrio vulnificus in raw oysters by real-time loop-mediated isothermal amplification. Int J Food Microbiol. 2010;142(1):60–6.
155. Heid CA, Stevens J, Livak KJ, Williams PM. Real time quantitative PCR. Genome Res. 1996;6(10):986–94.
156. Kong H, Higgins LS, Dalton MA, Kucera RB, Schildkraut I, Wilson GG. N. bstnbi nicking endonuclease and methods for using endonucleases in single-stranded displacement amplification. Washington, DC. U.S Patent. 2003.

157. Zyrina NV, Zheleznaya LA, Dvoretsky EV, Vasiliev VD, Chernov A, Matvienko NI. N. BspD6I DNA nickase strongly stimulates template-independent synthesis of non-palindromic repetitive DNA by Bst DNA polymerase. Biol Chem. 2007;388(4):367–72.
158. Liang X, Jensen K, Frank-Kamenetskii MD. Very efficient template/primer-independent DNA synthesis by thermophilic DNA polymerase in the presence of a thermophilic restriction endonuclease. Biochemistry. 2004;43(42):13459–66.
159. Connolly AR, Trau M. Isothermal detection of DNA by beacon-assisted detection amplification. Angew Chem. 2010;122(15):2780–3.
160. Connolly AR, Trau M. Rapid DNA detection by beacon-assisted detection amplification. Nat Protoc. 2011;6(6):772–8.
161. Xu W, Wang C, Zhu P, Guo T, Xu Y, Huang K, Luo Y. Real-time quantitative nicking endonuclease-mediated isothermal amplification with small molecular beacons. Analyst. 2016;141:2542–52.
162. Zhao W, Ali MM, Brook MA, Li Y. Rolling circle amplification: applications in nanotechnology and biodetection with functional nucleic acids. Angew Chem Int Ed. 2008;47 (34):6330–7.
163. Ali MM, Li F, Zhang Z, Zhang K, Kang D-K, Ankrum JA, Le XC, Zhao W. Rolling circle amplification: a versatile tool for chemical biology, materials science and medicine. Chem Soc Rev. 2014;43(10):3324–41.
164. Zhao Y, Chen F, Li Q, Wang L, Fan C. Isothermal amplification of nucleic acids. Chem Rev. 2015;115(22):12491–545.
165. Jarvius J, Melin J, Göransson J, Stenberg J, Fredriksson S, Gonzalez-Rey C, Bertilsson S, Nilsson M. Digital quantification using amplified single-molecule detection. Nat Methods. 2006;3(9):725–7.
166. Zhou L, Ou L-J, Chu X, Shen G-L, Yu R-Q. Aptamer-based rolling circle amplification: a platform for electrochemical detection of protein. Anal Chem. 2007;79(19):7492–500.
167. Liu H, Li L, Duan L, Wang X, Xie Y, Tong L, Wang Q, Tang B. High specific and ultrasensitive isothermal detection of microrna by padlock probe-based exponential rolling circle amplification. Anal Chem. 2013;85(16):7941–7.
168. Van Ness J, Van Ness LK, Galas DJ. Isothermal reactions for the amplification of oligonucleotides. Proc Natl Acad Sci. 2003;100(8):4504–9.
169. Li JJ, Geyer R, Tan W. Using molecular beacons as a sensitive fluorescence assay for enzymatic cleavage of single-stranded DNA. Nucleic Acids Res. 2000;28(11):e52.
170. Qian J, Ferguson TM, Shinde DN, Ramírez-Borrero AJ, Hintze A, Adami C, Niemz A. Sequence dependence of isothermal DNA amplification via EXPAR. Nucleic Acids Res. 2012;40(11):e87.
171. Nie J, Zhang D-W, Zhang F-T, Yuan F, Zhou Y-L, Zhang X-X. Reporter-triggered isothermal exponential amplification strategy in ultrasensitive homogeneous label-free electrochemical nucleic acid biosensing. Chem Commun. 2014;50(47):6211–3.
172. Nie J, Zhang D-W, Tie C, Zhou Y-L, Zhang X-X. G-quadruplex based two-stage isothermal exponential amplification reaction for label-free DNA colorimetric detection. Biosens Bioelectron. 2014;56:237–42.
173. Jeong Y-J, Park K, Kim D-E. Isothermal DNA amplification in vitro: the helicase-dependent amplification system. Cell Mol Life Sci. 2009;66(20):3325–36.
174. Motré A, Li Y, Kong H. Enhancing helicase-dependent amplification by fusing the helicase with the DNA polymerase. Gene. 2008;420(1):17–22.
175. Vincent M, Xu Y, Kong H. Helicase-dependent isothermal DNA amplification. EMBO Rep. 2004;5(8):795–800.
176. Li Y, Jortani SA, Ramey-Hartung B, Hudson E, Lemieux B, Kong H. Genotyping three SNPs affecting warfarin drug response by isothermal real-time HDA assays. Clin Chim Acta. 2011;412(1):79–85.

177. Kivlehan F, Mavré F, Talini L, Limoges B, Marchal D. Real-time electrochemical monitoring of isothermal helicase-dependent amplification of nucleic acids. Analyst. 2011;136 (18):3635–42.
178. Goldmeyer J, Kong H, Tang W. Development of a novel one-tube isothermal reverse transcription thermophilic helicase-dependent amplification platform for rapid RNA detection. J Mol Diagn. 2007;9(5):639–44.
179. Ramalingam N, San TC, Kai TJ, Mak MYM, Gong H-Q. Microfluidic devices harboring unsealed reactors for real-time isothermal helicase-dependent amplification. Microfluid Nanofluid. 2009;7(3):325–36.
180. Kievits T, van Gemen B, van Strijp D, Schukkink R, Dircks M, Adriaanse H, Malek L, Sooknanan R, Lens P. NASBA TM isothermal enzymatic in vitro nucleic acid amplification optimized for the diagnosis of HIV-1 infection. J Virol Methods. 1991;35(3):273–86.
181. Compton J. Nucleic acid sequence-based amplification. Nature. 1991;350(6313):91–2.
182. Guatelli JC, Whitfield KM, Kwoh DY, Barringer KJ, Richman DD, Gingeras TR. Isothermal, in vitro amplification of nucleic acids by a multienzyme reaction modeled after retroviral replication. Proc Natl Acad Sci. 1990;87(5):1874–8.
183. Deiman B, van Aarle P, Sillekens P. Characteristics and applications of nucleic acid sequence-based amplification (NASBA). Mol Biotechnol. 2002;20(2):163–79.
184. Milligan JF, Groebe DR, Witherell GW, Uhlenbeck OC. Oligoribonucleotide synthesis using T7 RNA polymerase and synthetic DNA templates. Nucleic Acids Res. 1987;15 (21):8783–98.
185. Yates S, Penning M, Goudsmit J, Frantzen I, Van De Weijer B, Van Strijp D, Van Gemen B. Quantitative detection of hepatitis B virus DNA by real-time nucleic acid sequence-based amplification with molecular beacon detection. J Clin Microbiol. 2001;39(10):3656–65.
186. Piepenburg O, Williams C, Stemple D, Armes N. DNA detection using recombination proteins. PLoS Biol. 2006;4(7):e204.
187. Harris LD, Griffith JD. Formation of D-loops by the UvsX protein of T4 bacteriophage: a comparison of the reaction catalyzed in the presence or absence of gene 32 protein. Biochemistry. 1988;27(18):6954–9.
188. Okazaki T, Kornberg A. Enzymatic synthesis of deoxyribonucleic acid XV. Purification and properties of a polymerase from Bacillus subtilis. J Biol Chem. 1964;239(1):259–68.
189. Hall M, Wharam S, Weston A, Cardy D, Wilson W. Use of signal-mediated amplification of RNA technology (SMART) to detect marine cyanophage DNA. Biotechniques. 2002;32 (3):604–11.
190. Wharam SD, Hall MJ, Wilson WH. Detection of virus mRNA within infected host cells using an isothermal nucleic acid amplification assay: marine cyanophage gene expression within Synechococcus sp. Virol J. 2007;4:52.
191. Ikbal J, Lim GS, Gao Z. The hybridization chain reaction in the development of ultrasensitive nucleic acid assays. TrAC Trends Anal Chem. 2015;64:86–99.
192. Evanko D. Hybridization chain reaction. Nat Methods. 2004;1(3):186–7.
193. Li H, Rothberg L. Colorimetric detection of DNA sequences based on electrostatic interactions with unmodified gold nanoparticles. Proc Natl Acad Sci U S A. 2004;101(39):14036–9.
194. Shimron S, Wang F, Orbach R, Willner I. Amplified detection of DNA through the enzyme-free autonomous assembly of hemin/G-quadruplex DNAzyme nanowires. Anal Chem. 2011;84(2):1042–8.
195. Bi S, Zhao T, Luo B, Zhu J-J. Hybridization chain reaction-based branched rolling circle amplification for chemiluminescence detection of DNA methylation. Chem Commun. 2013;49(61):6906–8.

Chapter 11
Aptamers and Their Application

Abstract Aptamers are oligonucleotides (single-strand RNA or DNA) with three-dimensional shapes that can bind to their target with conformational changes. Recently, aptamers have attracted increased attention in analysis because they have all the advantages of antibodies, as well as unique merits, such as thermal stability, small size, reversible binding, low cost, and easy signal transduction. In this chapter, several types of aptamer selection processes, which can be categorized into two groups (SELEX (systematic evolution of ligands by exponential enrichment) strategies and SELEX variant strategies), are explained in detail. Furthermore, we briefly review various applications of aptamers for the diagnosis of diseases, drug release, antimicrobial activity, bioanalysis, and serotyping.

Keywords Aptamer • SELEX • Diagnosis • Antimicrobial activity • Bioanalysis • Serotyping

11.1 Introduction

Aptamers are single-stranded nucleic acids (ssDNA or RNA) that can bind to various non-nucleic acid targets, such as small molecules, drugs, or proteins, with high affinity and specificity [1]. The affinities and selectivities of aptamers rival those of antibodies. Aptamers have been adapted to diagnostic and detection assay applications not only as alternatives to antibodies but also as unique recognition agents to recognize various targets. Aptamers are selected through an in vitro process, generating less batch-to-batch variation and lower costs than antibodies in vivo production. Even toxins and molecules that do not elicit a good immune response can generate high-affinity aptamers by an in vitro selection system [2]. The in vitro selection system termed the SELEX process (systematic evolution of ligands by exponential enrichment) was first developed by Ellington and Szostak in 1990 [3]. SELEX consists of screening large random oligonucleotide libraries (10^{15}different oligonucleotides) through several cycles of a binding/washing process and PCR amplification. In general, after typically 8–15 repeated selections using the obtained enriched pool as a starting library, the best aptamer sequences can be acquired. In addition to having high affinity and specificity, aptamers can be readily modified during chemical synthesis to incorporate signal reporters or

W. Xu, *Functional Nucleic Acids Detection in Food Safety*,
DOI 10.1007/978-981-10-1618-9_11

coupling moieties. Additionally, aptamer secondary structures can be induced to experience target-dependent conformational changes, opening up a wealth of possible signal transduction patterns. These merits allow aptamers to be applied in a wide range of assay methods, such as optical, electrochemical, and mass-based assays. In addition, aptamers are nucleic acids in nature; they are easily adapted to sequence (signal) amplification methods. In addition to assay applications, there are clinical applications of aptamers, including their use as therapeutic agents [4, 5], anti-infective agents [6], and drug delivery molecules [7] and for cancer detection [8–10]. There are also several aptamers that have been developed for clinical diagnostics and as infection inhibitor agents for microorganisms or protein targets.

In this chapter, methods for the production of aptamers and their potential applications are reviewed.

11.2 Aptamer Selection Strategies

11.2.1 General Strategies of SELEX Selection

The basic process is outlined as followed:

Construction of Random Oligonucleotide Libraries In a SELEX process, firstly, (Fig. 11.1) a randomized single-stranded library of oligonucleotides was chemically synthesized; apart from 20 to 80 nt random sequence in the central region of oligonucleotide, each oligonucleotide is flanked by 18–21 nt fixed sequence at its 5′ and 3′ end. This flanking sequence is the primer binding sites for PCR. Normally, the library contains approximately 10^{15} different sequences at the starting round; this sequence is then either transcribed for RNA selection or strand separated for DNA selection.

Separation of the Target Nucleic Acid Complex Aptamers generally adopt versatile 3D structures as a result of the diversity of random sequence. To isolate aptamers of a given target, the starting DNA library is incubated with the target. The rare nucleic acid, which binds to the specific target, is partitioned from unbound and weakly bound oligonucleotides by filtration through nitrocellulose (for protein target) or by affinity chromatography (for small molecule). This partitioning is a crucial aspect of SELEX to generate a high-affinity aptamer sequence.

Amplification The molecules that bind to the target are amplified by RT-PCR (for RNA selection) or PCR (for DNA selection) to generate a new lower-complexity pool. which will be in vitro transcribed and used for the next selection cycle. The enrichment of target sequence requires several repetitions of the successive steps, consisting of selection (binding, partition, and elution) and amplification. A negative selection step is necessary needed by passing the nucleic acid pool over a carrier without the target to eliminate the nonspecificity of aptamer. Then cloning

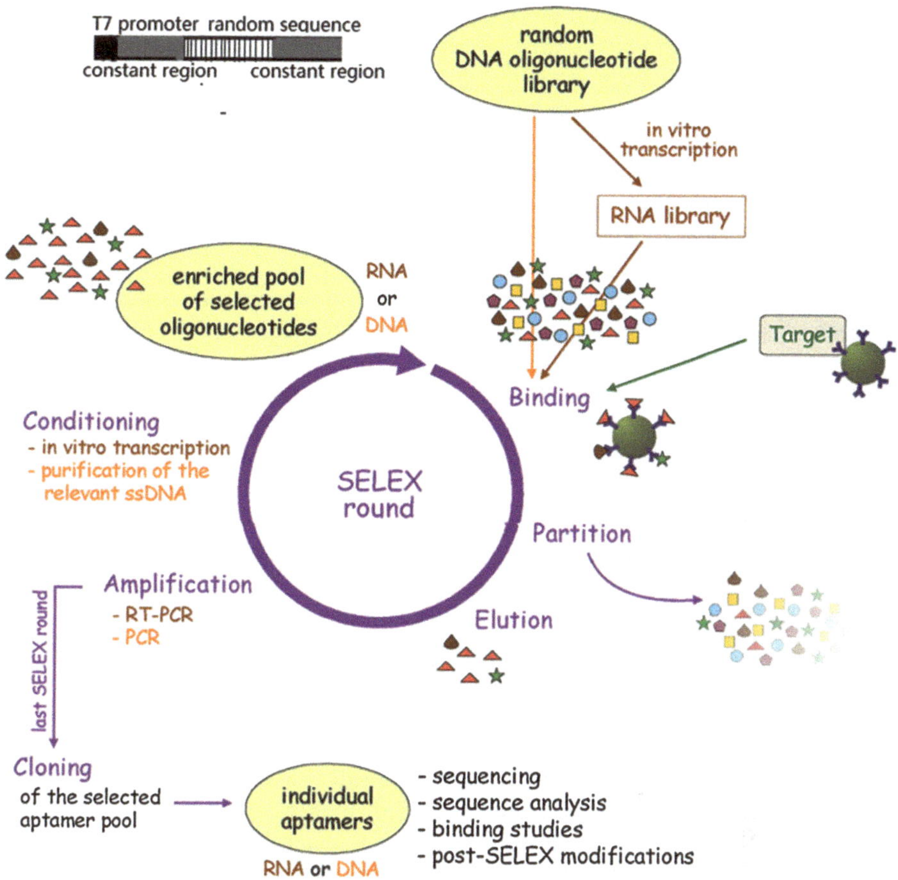

Fig. 11.1 Scheme of the systematic evolution of ligands by exponential enrichment (SELEX) (Reprinted from Stoltenburg et al. [11], Copyright 2007, with permission from Elsevier)

and sequencing of the selected clones will reveal the sequence and structure of the aptamers. After cloning and sequencing, the aptamer can be generated by chemical synthesis. Then its affinity and specificity for the target can be confirmed using flow cytometer or FRET (Förster resonance energy transfer).

SELEX has been applied into various targets. The first SELEX procedures generate aptamers against organic dyes and T4 DNA polymerase. Since then, a variety of biomolecules have been targeted. Most of these are summarized by Ellington et al. in the Aptamer Database. In addition to this database, several publications have also summarized the various target molecules [11, 12].

Large molecules, which can offer a multifunctional and large surface area for interaction with nucleic acids, are best suited to SELEX. Target molecules must also be easily modified to enable partitioning. Hence, proteins are good targets, as reflected by abundant literature reports. Many aptamers have been selected by

SELEX to bind small target molecules, such as Zn^{2+}, AMP, ATP, tyrosinamide, arginine, theophylline, ochratoxin A, and cocaine. However, these aptamers have not been successful. These aptamers generally have poor binding affinities, in contrast to the aptamers of larger targets. To enhance the affinity of these aptamers, many researchers have engineered some new techniques for selecting aptamers against small target molecules. Plummer et al. bound the target to a carrier protein, thus enabling the K_d of the selected aptamer to be less than 50 nM. Ethanolamine is the smallest molecular target so far to have been used for aptamer selection [13].

In addition to purified target molecules, SELEX can also be applied for complex pathogenic microorganisms, whole cells, and cellular targets, interesting areas of research. The complexity of the final aptamer pool generally depends on the abundance and number of potential target molecules and the affinity of the aptamers. The dominant-target aptamers in the final pool are cell surface molecules, often proteins. These selected aptamers are often used for monitoring structural alterations on the cell surface caused by environmental changes or diseases [10, 14].

The fact that aptamer selection techniques can be performed even in the absence of detailed target knowledge and characterization demonstrates a primary advantage in applying SELEX to complex target mixtures, such as cell protein mixtures. Some aptamers have also shown an inhibitory effect on their targets, including the inhibition of bacterial growth and whole-cell differentiation. However, the selection of aptamer ligands against a whole cell was "open-ended," in that no particular cell surface protein was strategically targeted; thus, SELEX against complex cellular targets has drawbacks. Ligand-mediated target purification using an aptamer is not always possible.

The number of targets requires careful consideration due selection favoring abundant targets. Rare targets, even those with high aptamer affinity, often fail the selection. The selection of aptamers against many purified targets before selection against more complex targets can help resolve such problems. Cerchia et al. developed SELEX techniques directed to overexpressed recombinant proteins on the mammalian cell surface, and the whole cell was selected as the target. The Kd values of these aptamers for their targets are 1–70 nM. Further refinement of such techniques will enhance the utility of SELEX against complex cellular targets. Cells for which the surface target molecules are known and can be overexpressed are more suitable for SELEX.

In addition, SELEX technology has been applied to nucleic acid structures. Tertiary RNA structures play an important role in several biological processes, including interfering with biological processes; aptamers against tertiary RNA could be used as therapeutic oligonucleotides [15, 16].

The special properties of SELEX indicate that these strategies can be applied to selection aptamers for virtually any target. However, there are some limits for diver's target to successfully select aptamers with high affinity and specificity. For single-target molecules, a sufficient amount of purified targets to minimize the nonspecific binding to increase the specificity should be present. Targets that have positively charged groups and hydrogen bond donors [17, 18] tend to facilitate aptamer selection. Targets with a largely hydrophobic character and those for negatively charged molecules (e.g., containing phosphate groups) may be more difficult to apply in SELEX due to these characters going against the basic principles of the intermolecular interactions in an aptamer-target complex.

As the improved requirements for specificity and specific functionalities, Kimoto et al. incorporate an unnatural nucleotide into the random sequences to increase the structural diversity and to enhance specificity and binding affinity of the resultant aptamer. They add up to three residues of the hydrophobic base to the library. The insertion of unnatural nucleotides allows the aptamer for interferon-γ (IFN-γ) with >100-fold improved binding affinities. This significant advancement in aptamer affinity shows an advantage in the development of new aptamer. The ability to perform PCR with the unnatural bases is critical to the success of this approach.

11.2.2 SELEX Variant Strategies

11.2.2.1 Non-SELEX Selection Strategies

Whereas SELEX for in vitro selection is robust for the identification of high-affinity aptamers against a variety of important macromolecule targets, continued improvements in selection technology are urgently needed for small molecules and the generation of aptamers with even higher affinities. Traditional SELEX involves multiple rounds of two steps: (i) the separation of aptamers from non-aptamers by an affinity method and (ii) the amplification of aptamers by polymerase chain reaction (PCR) (Fig. 11.2, left). Maxim Berezovski's research group first reported the non-SELEX selection of aptamers, a process that involves repetitive steps of separation with no amplification between them (Fig. 11.2, right). A highly efficient affinity method, capillary electrophoresis (CE)-based SELEX, was used for separation. This group found that three steps of CE-based separation in the non-SELEX approach were sufficient to improve the affinity of a DNA library to a target protein by more than four orders of magnitude. The advantage of CE over traditional selection methods is that the partition coefficient appears to be higher than that for filtration or affinity chromatography; single-stranded DNA (ssDNA) aptamers against HIV-1 RT isolated by CE-SELEX had dissociation constants (Kd) of 340 pM, better than those of previously selected ssDNA aptamers.

Remarkably, CE-based non-SELEX selection took only 1 h to complete, in contrast to the several days or several weeks required for a typical SELEX

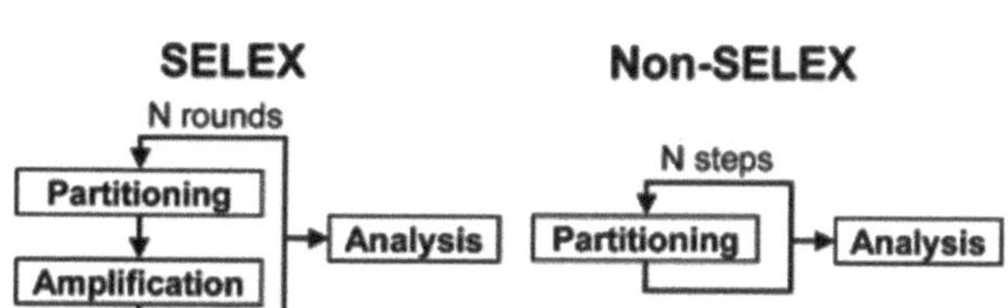

Fig. 11.2 The schematic of SELEX and non-SELEX (Reprinted with the permission from Berezovski et al. [30]. Copyright 2006 American Chemical Society)

procedure by conventional separation methods. In addition, CE-based non-SELEX permitted an accurate measurement of the number of aptamers in the library.

Despite significant efforts in aptamer development over the past 20 years, only a few aptamer-target pairs have been intensively used in proof-of-principle assays. By far, anti-thrombin aptamer is the most frequently used aptamer; the number of its related publications is >1000, followed by adenosine, cocaine's aptamer, all above account for more than one-third of the total aptamer research works. The limitation in target's species severely blocks the application of aptamers in assays and biosensors. Two reasons may account for this problem. First, the intrinsic property of natural DNA with limited chemical variation of interaction may lead to a difficulty in selecting aptamers against various targets. Incorporating various functional groups can increase the diversity of a randomized library to some extent, thereby improving the success rate of aptamer generation. Another reason is the intrinsic limitations of the conventional aptamer selection technique, known as SELEX, because this technique is time-consuming and labor-intensive. Therefore, a variation aptamer selection approach is urgently needed.

11.2.2.2 Toggle SELEX

The aptamers selected by traditional SELEX have high binding affinities with typical dissociation constants (Kd) ranging from picomolar to nanomolar. Furthermore, the produced aptamers also show a high specificity for binding the target. Species cross-reactivity is often desired during the preclinical evaluation of a molecule in animal models, as being too specific may be a problem. There is also a Toggle SELEX method to generate species cross-reactive aptamers (Generation of Species Cross-reactive Aptamers Using "Toggle" SELEX). Rebekah White's group described an in vitro selection strategy, called "toggling" SELEX, to generate aptamers that can bind both human and porcine thrombin. In their research, RNA aptamers that would be used both clinically in humans and testable in animal models were selected by "toggling" the protein target between human and porcine thrombin during alternating rounds of selection. After the "toggle" selection, a family of aptamers, which can bind human and porcine thrombin with high affinity simultaneously, can be yielded. If appropriate targets are available, the toggle strategy is a simple method to promote cross-reactivity. This strategy should facilitate the isolation of ligands with special cross-reactivity-requiring therapeutic and diagnostic applications (Fig. 11.3).

11.2.2.3 Photo-SELEX

Photo-SELEX is a derivative of SELEX in which substrates for either DNA or RNA polymerase include a 5-iodo or 5-bromo-substituted dUTP or UTP to satisfy the cross-linking of the aptamer to its targeted protein and normal selection. The substitute method involves incorporating a modified nucleotide that is activated

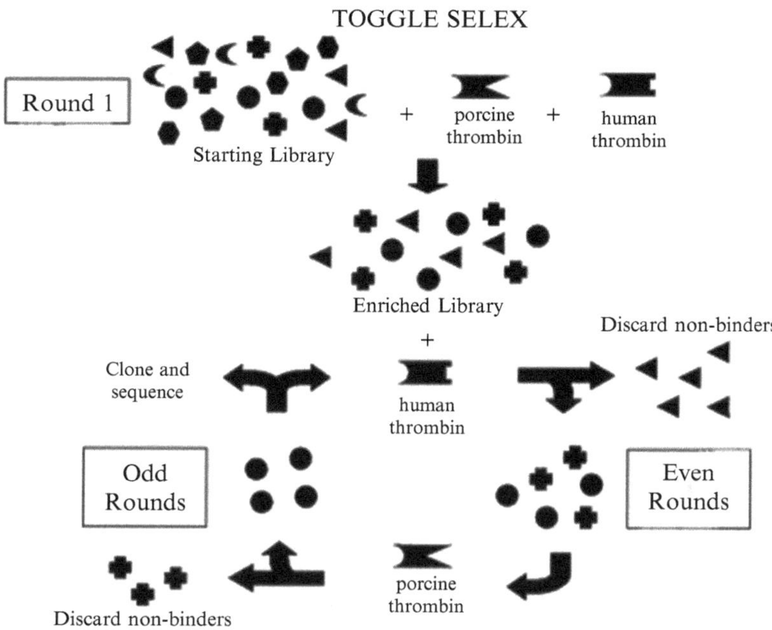

Fig. 11.3 Toggle SELEX. Aptamers which can bind both human and porcine thrombin were selected by "toggling" the protein target between human and porcine thrombin (Reprinted by permission from Macmillan Publishers Ltd: White et al. [31], copyright 2001)

by the absorption of light, in place of a native base, in either RNA- or ssDNA-randomized oligonucleotide libraries. The protocol of photo-SELEX requires the generation of aptamers that can be specifically photo-cross-linked to their cognate proteins.

The evolved aptamers of such interesting approaches may be able to simultaneously quantify a large number of analyte molecules. The aptamers that are selected through photo-SELEX have the ability to form a photo-induced covalent bond with the target molecule, thus showing greater sensitivity and specificity than those of the aptamers that are selected with conventional SELEX.

The importance of photo-cross-linking for diagnostics is that it makes independent the cross-linking ability parameter from the binding parameter. This independence is enough to generate multiplicative background reduction, leading to aptamers arrays on a solid surface to assemble an aptamer "chip," which can be incubated with blood or any other body fluid. Proteins will sort themselves in a manner analogous to affinity chromatography. Aptames evolving from photo-SELEX, which are specific for many different marker proteins, such as basic fibroblast growth factor (bFGF) and interleukin IL-6, could be fixed to a solid support and incubated with the patient serum. After washing, the chip is irradiated with UV light, causing covalent bond formation between the aptamer and the target, after which a fluorescent dye is added to conjugate the specific amino acids of the covalently bound conjugate to quantitatively detect the target molecules. This system has several advantages, such as minimal sample volume, very high specificity, and the ability to quantify multiple proteins. Moreover, this type of assay eliminates the need for secondary binding agents, such as detection antibodies, while still reaching the detection limits necessary for the detection of the target molecules at their physiological level.

11.2.2.4 Monoclonal Surface Display SELEX

A typical SELEX includes two major steps: enrichment from a large initial library and the screening of aptamer candidates from the enriched library. Over the past two decades, many innovative methods have been developed for the rapid and efficient enrichment of aptamers, including fluorescence-activated cell sorting SELEX [19, 20], capillary electrophoresis-SELEX [21, 22], and microfluidic-SELEX [23, 24]. However, the screening process remains unexplored. Normally, the enriched library is cloned by bacterial plasmids. To obtain aptamer candidates, growing bacteria are selected and sequenced. After the analysis of a large number of sequences, the high-repeat sequences with consensus are synthesized to evaluate their binding affinities individually. This sequencing-synthesis-screening protocol is inefficient, time-consuming, expensive, and labor-intensive. Krylov et al. reported an elegant non-SELEX method to address this problem. This non-SELEX method separates and screens the individual clones' affinity by non-equilibrium CE before sequencing. Zhu et al. developed a similar approach for aptamer screening by separating individual aptamer candidates from an enriched library and then transferring the separated aptamers into single agarose droplets for single-molecule PCR. This method screens the individual aptamers for their affinity to the target before sequencing and DNA synthesis. However, there are also some differences between the non-SELEX method and Zhu's group's method. The former approach evaluates a group of aptamer candidates, while the latter approach evaluates the individual sequences from an enriched library. Identifying high-affinity ligands hidden in a large group still requires a time-consuming screening process. Therefore, new SELEX methods are more suitable for the efficient enrichment and identification of aptamers with high affinity, effectively avoiding cloning, sequencing, chemical synthesis, and screening procedures to greatly speed up the selection process.

Recently, Zhu et al. designed a novel method called monoclonal surface display SELEX (MSD-SELEX) for the simple, rapid, and efficient enrichment and identification of aptamers from a library (Fig. 11.4). This library is composed of monoclonal DNA-displaying beads that are produced by single-molecule emulsion PCR. In MSD-SELEX, individual sequences with millions of repeats of the random ssDNA library are displayed on the microbead surface. The interaction of DNA with the target can be directly visualized under a microscope due to microbeads, which functionalize with large target particles, such as cells or molecules, and can convert the former interaction to an interaction between beads and targets. After

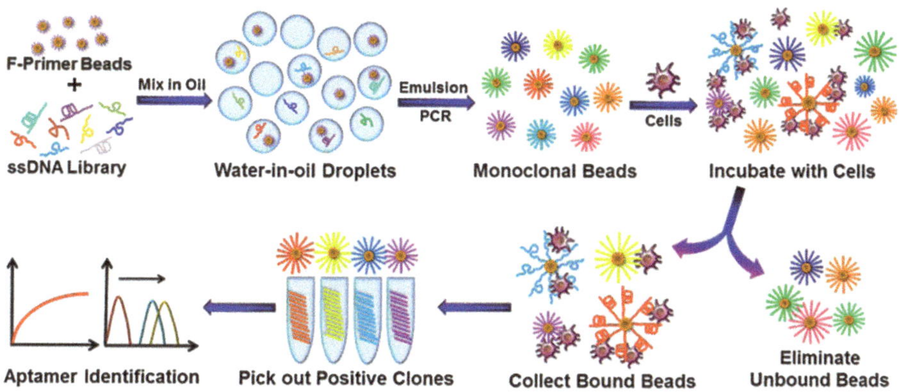

Fig. 11.4 Scheme of the MSD-SELEX technology for aptamer selection and identification (Reprinted with the permission from Zhu et al. [19]. Copyright 2014 American Chemical Society)

washing, nonbinding or weakly binding beads can be easily removed to enrich the library; sequences with high binding affinity are retained. In addition, based on the one-bead, one-sequence nature of the monoclonal beads, the affinity of the aptamer candidate sequences displayed on the surfaces of the enriched beads can be immediately confirmed without the need for sequencing and synthesis [19].

Compared to the conventional SELEX strategy, MSD-SELEX avoids the sequencing-chemical synthesis-screening work flow, thus offering a new approach for simple and rapid aptamer selection. This approach has successfully been applied for many different types of targets, including cancer biomarker protein and mycotoxin.

11.2.2.5 Complex-Target SELEX

When the target requires the presence of the cell membrane or a coreceptor to fold properly, it is difficult to program the SELEX experiment using a purified, soluble target. The use of SELEX has recently been extended from comparatively simple purified proteins to complex protein mixtures in situ on the surfaces of living cells, offering the potential to discover aptamers against intractable targets. In addition, because some cell type-specific diagnostic is sought, the most desirable target on the cell surface may not be known. Complex-target SELEX techniques can be performed in complex protein mixtures, even without detailed target information and characterization. Aptamers are advantageous in cellular detection because they can identify specific cell types or cell stages without having any information about the structural changes or membrane molecules related to that cell type or stage [25, 26]. In 1988, Morris et al. first reported a complex-target SELEX model for selecting cellular aptamers targeting human red blood cell membranes

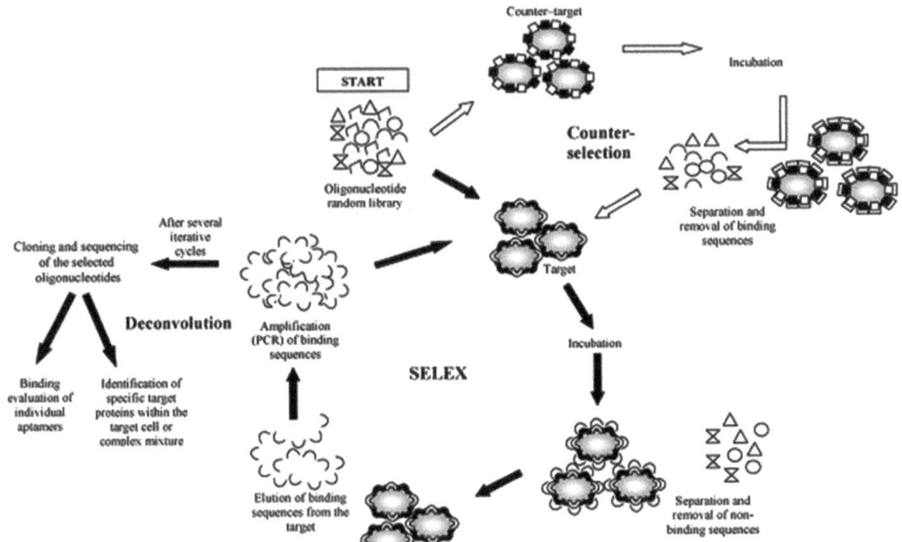

Fig. 11.5 Schematic of complex-target SELEX strategies. Positive selection procedure: the random library is first incubated with the target. Binding sequences are first separated. Counter-selection step: the random library is incubated closely related to the target to eliminate cross-reactivity. Deconvolution step: after cloning and sequencing, specific DNAs are analyzed to evaluate the affinity or to determine the specific binding sites (Reprinted from Fitter and James [28], Copyright 2009, with permission from Elsevier)

[27]. This report used a deconvolution-SELEX strategy to isolate and identify ligands that target to specific molecules in the complex mixtures. Deconvolution is a secondary selection step after the first enrichment of aptamer pool, permitting the partitioning of the pool of aptamers enriched from the SELEX process that evolved against multiple targets [28] (Fig. 11.5). This step involves (a) the evaluation of an aptamer's ability to specifically bind, capture, and identify the target or (b) the incubation of the last selected aptamer pool with the target to identify and isolate potential specific protein targets within the complex mixture using magnetic capture beads and SDS/PAGE analysis. Cellular aptamers can be used to isolate, concentrate, and identify particular cells from biological matrices [29].

11.2.2.6 Affinity Chromatography and Magnetic Bead-Based SELEX

Affinity chromatography is a method for separating and purifying the components of a biochemical mixture. This technique is first applied for the purification of recombinant proteins via a specific biological interaction (e.g., receptor/ligand and antigen/antibody interactions). Agarose-based beads with the target

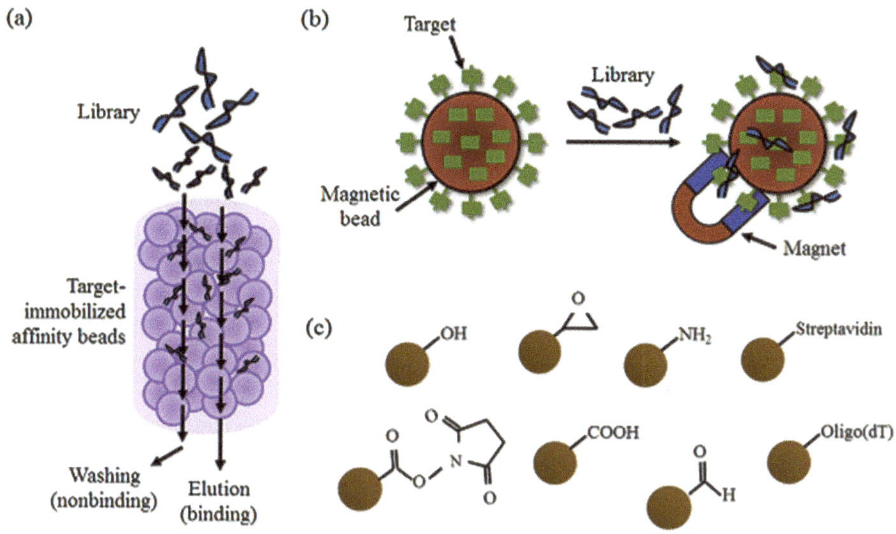

Fig. 11.6 (**a**) A schematic illustration of the selection step from a library using an affinity column; (**b**) the process of selection using magnetic beads; several types of functional group-activated beads; (**c**) such as tosyl-activated beads and epoxy-activated beads

molecules on the surface (components of the immobile phase) are packed onto a column for the binding and elution processes. In the binding and separation steps of SELEX, affinity chromatography can hold only the sequences with a high affinity for the target (Fig. 11.6). There are many ways to immobilize a target molecule; various tags, such as the His-tag and glutathione S-transferase (GST), are used when the target is a protein. The targets are covalently fixed onto the beads via a chemical reaction when the target is a small organic molecule [20]. Several aptamers have been successfully selected using this method by generating an affinity column containing target-immobilized beads, including small molecules as well as proteins [20–22]. However, if the target lacks the functional group or the affinity tag, it cannot couple to the beads, which is the disadvantage of this method.

Magnetic beads can also be used to immobilize the target via a chemical reaction or an interaction between an affinity tag and the substrate on the beads. It is very easy to separate the magnetic beads from the target-immobilized beads using a magnet (Fig. 11.6). Recently, this magnetic bead-based strategy has been applied in the SELEX technique. Particularly because the target-bound aptamer is simply and easily separated from the unbound one with a magnet, many attempts at selecting target-specific aptamers have been made.

A large number of variations of SELEX strategies were established by Tuerk and Gold. Some of these methods were developed to shorten the selected time and to increase the affinity or specificity of aptamers, while other methods were

Table 11.1 Examples of SELEX variants

Designation	Description	References
Non-SELEX	Use of capillary electrophoresis (CE) for separation: a process that involves repetitive steps of partitioning with no amplification between them; non-equilibrium capillary electrophoresis of equilibrium mixtures (NECEEM) is used for partitioning	[30]
Toggle SELEX	Switching ("toggling") between targets during alternating rounds of selection	[31]
Photo-SELEX	ssDNA aptamers capable of photo cross-linking the target molecule to enhance the affinity	[32]
MSD-SELEX	Belongs to non-SELEX, avoiding large-scale DNA sequencing, expensive and time-consuming DNA synthesis, and labor-intensive screening of large populations of candidates	[19]
Complex-target SELEX	To generate aptamers for complex targets, discriminates between relevant aptamers (binding to distinct target structures within the complex mixture), and irrelevant oligonucleotides	[33]

developed to optimize the procedure. Table 11.1 provides an overview of the SELEX variants.

11.3 Properties and Categories of Aptamers

Aptamers are single-stranded nucleic acids (RNA, DNA, and modified RNA or DNA) possessing unique characteristics for binding to their targets, unlike traditional nucleic acids. Aptamers are isolated from 10^{12} to 10^{15}combinatorial oligonucleotide (DNA or RNA) libraries that are chemically synthesized by a process known as in vitro selection. Over multiple rounds of selection, quite large populations ($>10^{13}$ different sequences) can be sieved, and the few "fittest" nucleic acid species can be isolated. The binding affinities of aptamers are highly target dependent and range from the picomolar scale (1×10^{-12} M) to the high nanomolar scale (1×10^{-7} M) for various protein targets. When small organics are targeted, the dissociation constants are higher (typically micromolar), as might be expected given the smaller number of interactions that form. In either case, interactions tend to be extremely specific, and aptamers can discriminate between even closely related compounds, such as (a) the anti-theophylline aptamer, which in turn can discriminate against caffeine based on the presence of a single methyl group, or (b) the anti-kinase aptamers, which can distinguish between closely related isozymes. In addition to these features, aptamers have shown extraordinary promise in analytical applications because they can be readily produced by chemical synthesis and can thereby be easily modified with reporter molecules, linkers, and other functional groups (8). Aptamers can also be joined to nucleic acid enzymes (e.g., ribozymes and deoxyribozymes) to create allosteric enzymes or so-called aptazymes.

11.4 Application of Aptamers

There are many applications for aptamers. The following sections address some aptamer applications, including assembly aptasensors for the detection and diagnosis of target molecules, as chiral selectors in separation techniques and as bacteriostatic agents.

11.4.1 Aptasensors

Because aptamers have been selected to bind to a wide variety targets, ranging from small molecules to macromolecules, such as proteins, different assay configurations have been designed to transduce bio-recognition events. Nevertheless, the majority of these designs fall into two categories of configuration: (1) single-site binding and (2) dual-site binding. The principle for assembling sensors depends on the inherently different recognition modes of each aptamer-target pair. For small molecular targets, these sensors are often buried within the binding pockets of aptamer structures (Fig. 11.7a), leaving little room for interaction with a second molecule. Due to this limitation, small-molecule targets are often assayed using the single-site binding configuration, which depends on aptamer structural transformation. In addition, macromolecular or cellular targets are structurally complicated, allowing for the interplay of various discriminatory contacts (stacking, shape complementarity, electrostatic interactions, and hydrogen bonding). As a result, macromolecular or cellular targets can be detected through both single-site binding (Fig. 11.7b) and dual-site binding (Fig. 11.7c). It is noteworthy that dual-site binding relies on the availability of a pair of aptamers that bind to different regions of the target and is commonly termed "sandwich assay." In sandwich mode, one aptamer is often

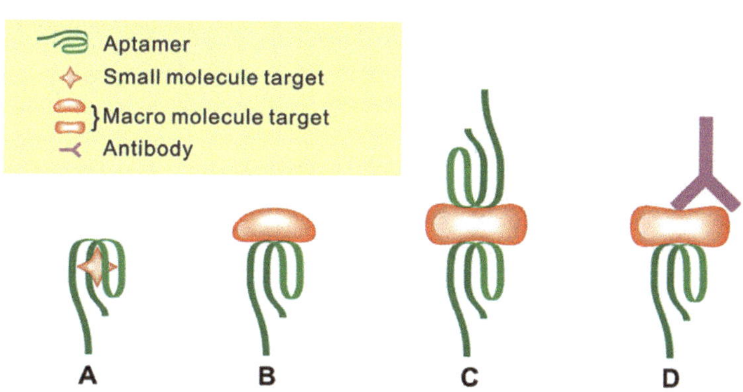

Fig. 11.7 Aptamer-based assay formats. (**a**) Small-molecule target buried within the binding pockets of aptamer structures; (**b**) single-site format; (**c**) dual-site (sandwich) binding format with two aptamers; and (**d**) "sandwich" binding format with an aptamer and an antibody

immobilized on the surface of solid supports (e.g., electrodes, glass chips, nanoparticles, or microparticles) to act as a capture probe, while the other aptamer is often conjugated with signaling moieties (e.g., fluorophores, enzymes, or nanoparticles (NPs)) to act as a reporter probe. If the pair of aptamers shares identical or overlapping binding sites on the target of interest, it is possible to use an antibody as the second "aptamer" (Fig. 11.7d).

11.4.1.1 Adaptation of Aptamers to Sequence Amplification

One of the most obvious analytical formats for aptamers is simply to detect them as one detects other nucleic acids: via any of a variety of amplification formats, such as polymerase chain reaction (PCR), rolling-circle amplification (RCA), nucleic acid sequence-based amplification, loop-mediated isothermal amplification, or the proximity ligation assay (PLA). Moreover, using detection reagents that can signal in real time, such as molecular beacons, one can also follow amplification cascades in real time. Additionally, the sensitivity of detection can be improved by looking at the evolution of signals, as with real-time PCR.

Affinity PCR

Two methods based on affinity PCR have been developed for protein detection. One of these methods relies on the proximate ligation of adjacent target-bound aptamers, while the other method does not. Ligation assays require that the target protein has two binding sites (epitopes) that are able to bind to two of the same aptamers or to two different aptamers. Each aptamer has an oligonucleotide extension. Upon the binding of the aptamers to the target, the aptamer extensions come into close proximity and are ligated by a connector ligand. The ligated sequences are then PCR-amplified and detected. The dual recognition of proteins results in excellent target selectivity, but only a few targets have the required pair of binding sites (Fig. 11.8).

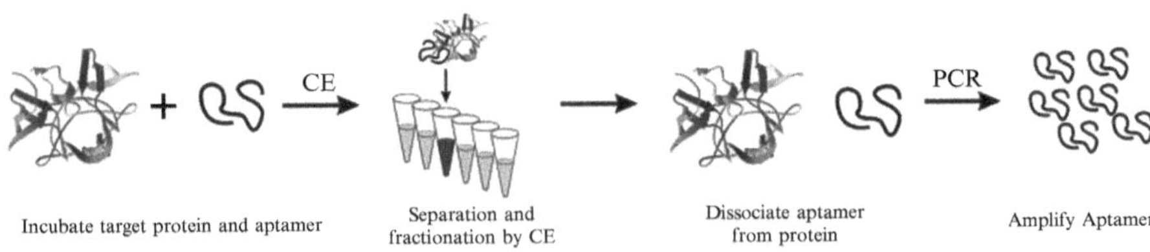

Fig. 11.8 Affinity aptamer PCR assay for proteins. The protein is first incubated with the aptamer to form an aptamer–protein complex. Separate the protein–aptamer complex from excess aptamers by capillary electrophoresis (CE). After collecting the aptamer–protein complex, the aptamer is dissociated from the protein and amplified by PCR. The amplified aptamers are indirect measures of the protein concentration (Reprinted from Hamula et al. [26], Copyright 2006, with permission from Elsevier)

Ligation assays using aptamers were first demonstrated for analysis of thrombin and a platelet-derived growth factor (PDGF-BB). Using ligation methods, researchers were able to detect thrombin down to picomolar concentrations and achieved linear dynamic ranges from picomolar to nanomolar concentrations. Interfering proteins had no effect on target protein detection.

Affinity PCR assays are fast and extremely sensitive due to the PCR amplification step. However, PCR is also a liability because contamination can be a problem. One limitation of ligation assays is that the target must have two distinct binding sites and appropriate aptamers to bind to the two sites. An advantage of this requirement is the increased specificity of the method.

Aptamers and Rolling-Circle Amplification

RCA assays are based on the ability of DNA or RNA polymerases to continuously replicate around a circular, single-stranded nucleic acid template under isothermal conditions. RCA is often employed as a signal amplification technique because the product contains hundreds to thousands of complements of the original template, and hybridization of probes to these sites can increase the original signal by several orders of magnitude with concomitantly lower detection limits.

The amplified nucleic acid can also serve as much more than a landing site for a hybridizing probe. Cheglakov et al. used a molecular beacon to initiate RCA on a circular template encoding a deoxyribozyme that mimicked the activity of HRP. When the substrate $ABTS^{2-}$ [2,2-azinobis(3-ethylbenzothiazoline)-6-sulfonic acid] was added in the presence of H_2O_2, a colorimetric change from clear to green was observed. The assay was also adapted to chemiluminescence by the addition of luminal and H_2O_2. As little as 1 pM of phage DNA was detected. The RCA products were ultimately detected by capturing alkaline phosphatase that led to the production of ascorbic acid, which in turn led to silver deposition at the same site.

Optical Sensors

Structural studies have shown that aptamers frequently undergo significant conformational changes upon binding to their cognate ligands. By introducing single reporters, such as organic fluorophores, into conformationally labile regions of aptamers, one can transduce ligand binding into a change in the chemical environment of the fluorophore and hence to a change in fluorescence characteristics, such as intensity and anisotropy.

Strategies that rely on two reporters are also practical in that they allow one to attempt optical signal transduction based on fluorescence resonance energy transfer (FRET). As more and varied fluorophores and aptamer modifications have become commercially available, more and varied design strategies have been developed (Fig. 11.9). The simplest format is to label aptamers with a quencher and either one

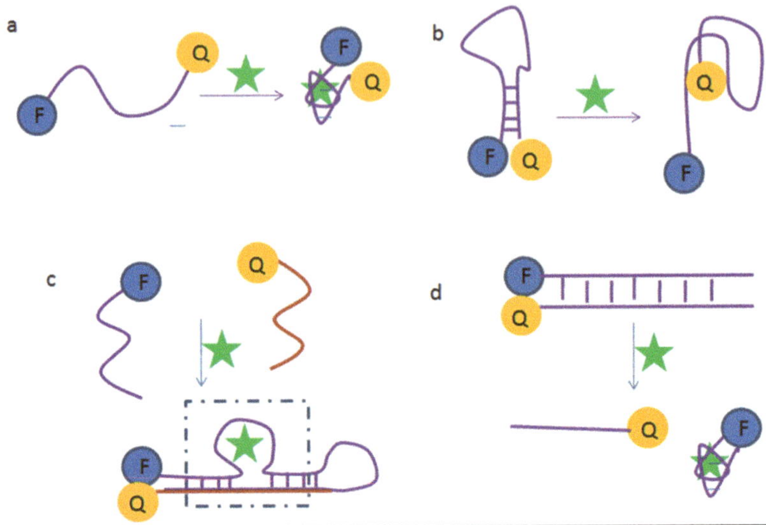

Fig. 11.9 Schemas for optical sensors that incorporate two reporters. (**a**) Aptamer beacon with quencher properties. (**b**) Fluorescence resonance energy transfer (FRET) aptamer beacon. (**c**) Assembly aptamer beacon. (**d**) Disassembly aptamer beacon. *F* and *Q* stand for fluorophore and quencher, respectively

or two fluorophores (Fig. 11.9a, b), enabling both quenching and "light-up" strategies. Aptamers can also be engineered to refold rather than to fold. In one experiment, a sequence was added to an anti-thrombin DNA aptamer that fixed it in a molecular beacon-like hairpin structure. Upon the addition of thrombin, the conformational equilibrium was shifted from the quenched hairpin to the thrombin-bound quadruplex, resulting in a dequenching of the fluorophore and the creation of an optical signal (Fig. 11.9b). Quaternary structural rearrangements that result in the assembly or disassembly of aptamers can also be exploited for signaling. Aptamers can be split into two pieces, and fluorophores can then be added to each piece (Fig. 11.9c). In the absence of the target ligand, the two oligomers largely exist independently in solution, but target binding brings the oligomers together and leads to ternary complex stabilization, ultimately resulting in a quenched optical signal. Disassembly strategies are also possible. Aptamer beacons have been generated in which a complementary, antisense DNA strand is used to intentionally denature the aptamer (Fig. 11.9d). Upon the addition of ligand, the native aptamer structure is stabilized, and the equilibrium is concomitantly shifted away from the denatured, antisense duplex and toward the native structure. The aptamer can be labeled with a fluorophore and the antisense oligonucleotide with a quencher, leading to a quenched fluorescent signal in the absence of a target, similar to the unimolecular aptamer beacons described above.

Quantum dots (QDs) and nanoparticles offer a number of advantages over standard fluorescent dyes for monitoring biological systems in real time, including greater photo-stability, larger effective Stokes shifts, longer fluorescent lifetimes, and sharper emission bands than those of traditional organic fluorophores. First, all QDs respond to the same excitation wavelength but emit at different wavelengths, which assists with multiplexing. Second, gold nanoparticles (GNPs or Au-NPs) experience an absorption property change depending on the aggregation state. In particular, the color changes associated with aggregation (red) and disaggregation (purple) are easily visible to the naked eye without the need for sophisticated instruments. In addition, GNPs are more biocompatible, easy to bioconjugate, and less toxic than QDs are. Analyte-dependent aptamer conformational changes have been coupled to GNP aggregation. An anti-ATP DNA aptamer was hybridized to a short complementary inhibitory oligonucleotide that in turn was attached to a GNP. The highly negatively charged phosphate groups in the nucleic acids repelled one another and stabilized the GNP to aggregation (Fig. 11.10a). As ATP molecules were bound, aptamers were displaced. The GNPs aggregated, leading to a color change from red to purple. With larger targets, avidity-mediated aggregation may be possible. Recently, Medley et al. coupled aptamers selected against cancer cell surfaces to GNPs (Fig. 11.10b). Assembly on the cell surface apparently places the aptamer-GNP conjugates in close enough proximity that a shift in optical properties can be detected. The use of GNPs in many assays is difficult because of the particles' strong tendency to aggregate over time. To address this limitation, disaggregation methods have been developed. An aptamer hybridizes to and bridges two oligonucleotide-coupled GNPs, and in the presence of the appropriate analyte, it undergoes a conformational change and is released (Fig. 11.10c). This

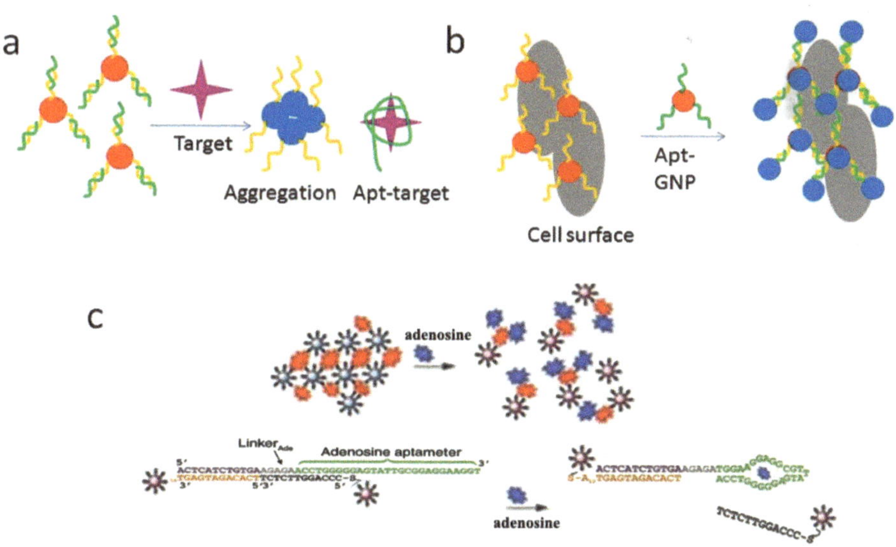

Fig. 11.10 Schemas for optical sensors that incorporate gold nanoparticles (GNPs). (**a**) Aptamer release and GNP aggregation. (**b**) Avidity-mediated aggregation. (**c**) Aptamer release and disaggregation

disaggregation method has been demonstrated for the detection of ATP, cocaine, Pb^{2+}, K^{+}, and their combinations.

11.4.1.2 Electrochemical Aptasensors

There is an increasing number of analytical applications for aptamers in electrochemical analysis fields because electrochemical techniques have high sensitivities and because the instrumentation is simple, miniaturized, and low cost. Recently, various electrochemical detection methods have been used with aptamers, including electrochemical impedance spectroscopy (EIS), potentiometry with ion-selective electrodes (ISEs), electro generated chemiluminescence (ECL), cyclic voltammetry, and amperometry. Detection in the femtomolar range has been achieved with the aptamer capture of a protein target followed by EIS and chemical amplification (10 fM), as well as with protein-induced oligonucleotide displacement coupled with ECL (1 fM). Many of the optical-sensing strategies are relevant to electrochemical sensing. The difference is that electroactive reporters, such as methylene blue (MB), ferrocene, ferrocene-bearing polymers, ruthenium complexes, and Fe $(CN)_6{}^{4-}/^{3-}$, are used for signal transduction in electrochemical aptamer-based sensors. A major difference among optical methods is that electrochemical sensors are limited to heterogeneous assays because the aptamers, complementary oligonucleotides, or target molecules need to be attached to electrodes. In other words, analytical methods may be more important than molecular engineering.

11.4.2 Aptamers for Diagnostics and Therapeutics

Since the SELEX protocol was codified, aptamers have been selected against an extremely wide variety of targets, ranging from small inorganic molecules to whole organisms. Numerous aptamers have been selected against therapeutic targets, such as IgE, IFN-g, alpha-thrombin, and PTPase, and have shown great efficacy in tissue culture experiments and animal models. An anti-thrombin aptamer was used in place of heparin for anticoagulation during heart bypass surgery in canines, and aptamers against inflammation factor human neutrophil elastase significantly reduced lung inflammation in rats and had better specificity for their target than for an anti-elastase IgG control. Because aptamers are thought to be able to penetrate tumors rapidly and to be cleared quickly from the blood due to their small size (8–15 kDa), they have potential for use as anticancer therapeutics. In addition, aptamers are non-immunogenic and nontoxic, thereby allowing for their application as drugs.

Many therapeutics aptamers are targets for extracellular or cell surface components, facilitating their development and delivery because aptamers can potentially be injected and will not require cellular internalization. Several aptamers appear to interact with viral coats and might be promising antiviral agents. The HIV-1 protein gp120 interacts with the host cell receptor CD4 and coreceptor CCR5 during viral entry. The CCR5 binding site on gp120 is difficult for antibodies to reach due to steric hindrance. James's group at the University of Oxford selected 20-fluoropyrimidine-substituted RNA aptamer that was specific for the conserved coreceptor region of the surface envelope glycoprotein (gp120) of the R5 strain of HIV-1.

Cellular and viral proteins can be involved in pathogenesis, and aptamers have been selected against the insoluble, amyloid conformer of the prion protein (PrPSc, also known as scrapie) that is present in transmissible spongiform mencephalopathies, such as bovine spongiform encephalopathy and Creutzfeldt–Jakob disease. Although it is difficult to raise antibodies against the scrapie conformer, the selected aptamers can readily distinguish the two forms. Cell-based studies have shown that minimized aptamers can prevent the accumulation of the insoluble forms of PrP.

Cell surface signal transduction proteins, such as receptor tyrosine kinases (RTKs), play a key role in regulating cell growth and proliferation in several different types of cancers, making them attractive targets for ameliorating tumor progression. Aptamers can also be directed against intracellular targets, and RNA and DNA aptamers specific for HIV-1 reverse transcriptase (RT) have been selected and shown to inhibit viral replication. In conjunction with the observation that aptamers might be able to access epitopes or tissues that antibodies cannot, it would seem that aptamers are neither too large nor too small, but rather just the right size for therapy.

To apply aptamers for diagnostics in animal models of disease and in humans, they need to be modified. First, sugar modifications of nucleoside triphosphates are necessary to render the resulting aptamers resistant to nucleases found in serum. Changing the 2′OH groups of ribose to 2′ for 2′NH groups yields aptamers that can live long in blood. The relatively low molecular weight of aptamers (8000–12,000) results in rapid clearance in the blood. Aptamers can be kept in circulation from hours to days by conjugating them to a higher-molecular-weight carrier. Proteins can include the Fc fragment of IgG for opsonization and the first component of complement (C1q) to trigger complement-mediated lysis of antibiotic-resistant Gram-negative bacteria, cancer cells, and possibly some parasites during vulnerable stages. In addition, the 3′protein adduct may be a biotoxin or enzyme or may simply be human serum albumin (HSA) or a drug known to bind HSA, thereby retarding the kidney and other organ clearance and inhibiting serum exonucleases. In addition, the PCR amplification of aptamers could dramatically decrease the current $2000/g cost of parallel chemical oligonucleotide synthesis, thereby enabling the mass production of aptamer-3′-protein or drug conjugates to better compete against expensive humanized monoclonal antibodies. It should be mentioned that the modification should have no harmful side effects. After being modified, conjugated aptamers are injected into animals, inhibiting physiological functions known to be associated with their target proteins.

11.4.3 Antimicrobial Aptamers for the Detection and Inhibition of Microbial Pathogen Growth

According to the WHO, microbial infections are one of the major causes of death worldwide. Assessments of alternative sources of antimicrobial agents are essential in the ongoing battle against microbial pathogens because conventional drug development methods are at a bottleneck. These challenges include the following: limited chemical libraries, the barrier that Gram-negative organisms have to the intracellular accumulation of drug candidates, and the presence of nonselective efflux pumps on and the charge and size of drug candidate molecules. The discovery of alternative sources of antimicrobial agents is essential in the ongoing battle against microbial pathogens. Legislative and scientific challenges considerably hinder the discovery and use of new antimicrobial drugs, and new approaches are in urgent demand. The technological advancements in aptamer selection and recent examples of antimicrobial properties of some aptamers suggest that this new source of antimicrobials can help to overcome some of the challenges in traditional antibiotic discovery. Because the major action mechanism is the strong affinity of an aptamer toward a specific target, antimicrobial aptamers are not affected by the aforementioned challenges. Therefore, aptamers can be used as antimicrobial agents to neutralize pathogens either passively by binding to surface antigens or actively by targeting intracellular key biochemical processes. However, the major attractive feature of antimicrobial aptamers is the combinatorial nature of aptamer selection, in which antimicrobial aptamers can be obtained without any prior knowledge regarding the target pathogen and for any type of bacteria (Gram-negative or Gram-positive). Thus, affinity oligonucleotides (aptamers) provide unique opportunities for the development of nanotechnological solutions for medical applications. Aptamers specifically recognizing microbial cells and viruses have also shown great potential in a range of analytical and therapeutic applications.

Olga [34] selected some species-specific bacteriostatic DNA aptamers for *Salmonella* using a cell-SELEX technique (Fig. 11.11).

11.4.4 Riboswitches as Drug Targets

Riboswitches are regulatory sequences of mRNAs that can bind to target molecules and regulate gene activity. Because aptamers are defined as target-binding nucleic acid sequences, the target-binding domains of riboswitches can be considered natural aptamer sequences. The majority of clinically used antibiotics target ribosomes, among others, such as enzymes related to folic acid biosynthesis or DNA topoisomerases. In addition, studies on ribosome–antibiotic complexes have revealed that antibiotics target rRNAs in the 30S or 50S ribosomal subunits

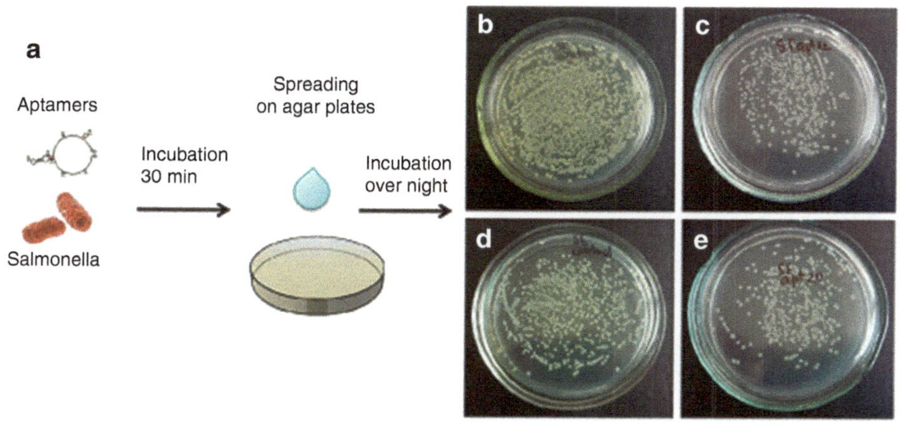

Fig. 11.11 Effect of aptamer pools on colony formation on Petri dishes with LB agar. (**a**) Bacterial culturability test after the incubation of salmonella with aptamers. Aptamer pools, clones, or an ssDNA library (1 μM) in DPBS was incubated with S. *Typhimurium* or *S. enteritidis* is for 30 min. The bacteria were diluted, plated on Petri dishes with LB agar, and incubated for 18 h at 37 °C in a dry incubator (repeated in triplicate). The effect was estimated by counting the number of colonies. (**b**) Intact *S. Typhimurium*; (**c**) *S. Typhimurium* plated after a 30 min incubation with 1 μM synthetic aptamer ST-12; (**d**) Intact *S. enteritidis*; (**e**) *S. enteritidis* plated after a 30 min incubation with 1 μM synthetic aptamer SE-20. To obtain the same number of bacteria at the beginning of incubation, (**b–e**), were initially diluted from the same aliquot of *S. enteritidis* or *S. Typhimurium*, respectively (Reprinted with the permission from Kolovskaya et al. [34]. Copyright 2013 American Chemical Society)

rather than ribosomal proteins. In fact, the catalytic core of the ribosome is a ribozyme, made entirely of RNA; therefore, RNA is a validated drug target. Riboswitches are sequences in the untranslated regions of mRNAs that bind to specific metabolites; this binding leads to changes in the activity of the same gene. Each riboswitch consists of an aptamer domain (binding to a metabolite), switching sequence and expression platform. Thus, riboswitches contain aptamer sequences binding specifically to a metabolite. These sequences are natural aptamers compared to artificially selected aptamers as described in the previous section. Riboswitches are common metabolism control elements in pathogens with 20 different identified classes. Because riboswitches are involved in fundamental metabolism in most pathogenic microorganisms, they are valuable targets to develop novel antimicrobial agents. Therefore, interfering agents with the aptamer regions of riboswitches might be a new strategy for inhibiting the infection of some pathogens. Several studies have investigated the potential of ligand analogs for riboswitch classes as antimicrobial drugs. The thiamin pyrophosphate analog pyrithiamine pyrophosphate can inhibit bacterial growth as an analog binding to the thiamin pyrophosphate riboswitch suppresses the expression of thiamin biosynthesis (Fig. 11.12).

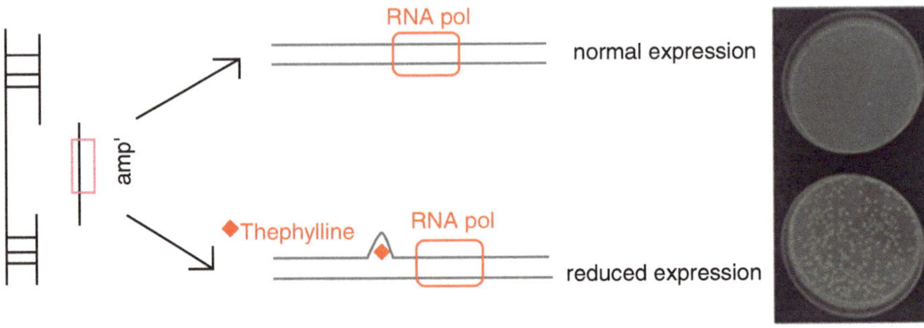

Fig. 11.12 A sequence was transformed into the genome of a microorganism inside the b-lactamase gene and interferes with antibiotic resistance in the presence of small molecular. Amp' above means ampicillin resistance gene

11.4.5 Aptamers Interfering with Biochemical Processes

Most drugs act by inhibiting specific biochemical reactions inside the cells. Therefore, the first challenge is to transfer drug molecules across the cell membrane; only then can drug molecules find their target molecule inside pathogenic bacteria or human host cells. Aptamers can be selected as antagonists to a receptor protein to inhibit receptor action. In this scenario, the binding should inhibit the microorganism's infection ability. One good example is antibiotic-resistant bacteria, which are a major challenge in the pharmaceutical industry. The metallo-b-lactamase enzyme is one of the most difficult targets for researchers to inactivate. This enzyme catalyzes the hydrolysis of the b-lactam ring of lactam antibiotics, which inactivates the antibiotics; therefore, bacteria with this enzyme are resistant to lactam-type antibiotics, such as ampicillin. These enzymes are extremely heterogeneous; more than 700 different types of metallo-b-lactamases have been identified. As a strategy to maintain the activity of lactam-type antibiotics, an aptamer was selected to bind to metallo-b-lactamases and inhibit antibiotic degradation. This type of strategy relies on the inhibitory action of aptamers, which is similar to that of many other drug molecules that are actually enzyme inhibitors. Aptamers can be selected to bind tightly to specific enzymes or its substrate to prevent normal enzyme functioning, thus acting as an enzyme inhibitor. However, the real challenge is to transport the aptamers into the cell, where metabolic pathways occur. Although aptamers are small molecules compared to antibodies (~5 kDa vs 150 kDa), an active transport mechanism is required for aptamers to pass through biological membranes. Specific aptamers are selected for targeting marker receptors on specific cell types and for subsequent internalization into intracellular compartments. Therefore, future applications can be envisioned as chimeric aptamer sequences with domains acting for enzyme inhibition and domains acting for cell targeting and penetrating functions.

11.4.6 Detection of Biological Terrorist Threat Agents [Aptamer-Based Assays for Diagnostics and Environmental and Food Analysis]

The detection of biological terrorist threat agents, such as bacterial cells, spores, viruses, and toxins, is a significant military and civilian challenge. Traditional analytical techniques for these targets are mainly based on immunological methods, such as conventional ELISA, immunomagnetic electrochemiluminescence assays, and time-resolved fluorescence assays. However, in this particular field, aptamers can have a great advantage because all these methods are very dependent on the ability to produce specific antisera for these toxic materials in animals. Aptamers, due to their synthetic nature, are independent of animals and can be selected for these toxic molecules. Aptamers that are specific for these particular targets, such as anthrax spores, cholera toxin, staphylococcal enterotoxin B, ricin, and abrin toxin, have been selected in recent years and used to develop different detection systems.

11.4.7 Alternative Serotyping Methods with Aptamers

There is a debate surrounding the ATP bioluminescence method and similar generic methods regarding their reliability to monitor cleanliness because ATP is an indicator of organic matter rather than actual microbiological contamination. Considering the species-specific capabilities of immunological or DNA markers, monitoring methods can be developed to identify individual varieties of microorganisms. Depending on the application purposes, multiplex detection may be necessary to differentiate between serovars to provide information regarding microbial contamination. The serotype is the distinct cell surface antigens of bacteria, which are either somatic (O-type) or flagellar (H-type). The epidemiologic classification of pathogens depends on the classification of a microorganism species to subgroups based on serotypes because bacterial serotypes are often associated with a specific disease. Although standard culturing methods are reliable for identifying a pathogen, conventional methods suffer from a number of drawbacks, including being labor-intensive, high cost, and time-consuming. More importantly, culturing methods require additional biochemical tests for determining specific virulence factors. Therefore, serological O-antigen typing by antibody-based methods has been the most common method in clinical diagnostic applications for pathogens. Another serotyping method is the use of a DNA sequence signature for each serotype. PCR-based diagnostic methods are reliable methods and eliminate problems associated with culturing. Quantitative PCR or microarray methods are fast (2–3 h to obtain results) but are not instantaneous. Aptamers can be selected to bind specifically to the surface proteins of bacterial cells to obtain serotype-specific aptamer sequences. Subsequently, serotype-specific aptamers can be used to construct diagnostic assays for a particular serovar of pathogens.

E. coli, a common Gram-negative bacterium, is found in the intestinal tract, and most of its serotypes are harmless. The *E. coli* O157:H7 serotype is a common enterohemorrhagic microorganism and one of the most dangerous foodborne pathogens. A cell-based subtractive SELEX procedure was used to identify a nuclease-resistant RNA aptamer sequence that binds specifically to the cell surface of the O157:H7 serotype. Streptococcus pyo genes are a Gram-positive bacterium, also known as group A streptococcus (GAS), and are responsible for many clinically serious diseases, such as rheumatic fever and acute glomerulonephritis. Hamula et al. reported DNA aptamer sequences that selectively recognize the ten most common serotypes of GAS.

An RNA aptamer for *Salmonella enteritidis* has been proposed as an alternative serotyping method. *Salmonella* species are important pathogens causing a variety of foodborne illnesses worldwide. Salmonella enterica has six subspecies, each with more than 2500 serotypes. There are wide variations in *Salmonella serovars* regarding disease manifestation, host specificity, and virulence. Therefore, aptamer technology can provide specific affinity agents for diagnosis, and more importantly, new serovars can be easily incorporated into detection formats.

11.5 Future Perspectives

In this review, the advantages of aptamers and their applications have been presented through descriptions of several excellent experiments. An aptamer is a convincing substitute for an antibody because it is very stable under hot or hazardous conditions. Because aptamers are mostly oligonucleotides, they can easily be synthesized in quantity with high purity and can be modified with various molecules using simple chemical reactions. Therefore, aptamers have shown great strength in a wide range of applications. Moreover, aptamers are non-immunogenic and nontoxic materials that can be exploited in medical applications, such as for the diagnosis and treatment of diseases. It is also noteworthy that the types of target molecules of aptamers are unlimited. Diverse SELEX methods have been designed as targets to select aptamers conveniently and quickly. Affinity chromatography and magnetic bead-based SELEX are powerful techniques for relatively large molecules, such as proteins, because of their high affinity to the target and easy separation from dispensable molecules. Moreover, capillary electrophoresis-based SELEX is suitable for small molecules and is a rapid method. Aptamers against biomarker-related diseases can be used to diagnose these diseases. In the ALISA and RDT methods, aptamers have been used as capture-probe molecules based on their affinity and selectivity to the targets. As a result, sensors have shown excellent performance in the diagnosis of diseases. For the detection of small molecules, aptamers will be utilized pivotally because antibody-based detection has considerable limitations, while the applications of aptamers are relatively broad. Numerous sensor techniques incorporating aptamers have been described, such as electrochemical, colorimetric, optical, and mass-sensitive methods. These aptasensors have been employed in various fields, including the medical and industrial fields. In addition, the use of aptamers as drugs, therapeutics, bio-imaging materials, and analytical reagents has been investigated, and as a result, the drug Macugen was generated. In conclusion, it is certain that aptamers are strong and versatile alternatives to antibodies. Of course, aptamers still have issues that need to be resolved for the purposes of biological applications. For example, if they are used as agents for in vivo applications, unmodified aptamers are highly susceptible to degradation by several nucleases, and their small size makes them liable to renal filtration, but it can be expected that the potential of aptamers for bio-applications will increase, based on their merits. Thus, further investigations into aptamers and aptasensor systems need to be carried out to quickly overcome existing limits.

Acknowledgments This work is supported by the Ministry of Science and Technology of Beijing (XX2014B069). Many thanks to Longjiao Zhu, for her kindly help in manuscript conception and preparation.

References

1. Jayasena SD. Aptamers: an emerging class of molecules that rival antibodies in diagnostics. Clin Chem. 1999;45(9):1628–50.
2. O'Sullivan CK. Aptasensors–the future of biosensing? Anal Bioanal Chem. 2002;372(1):44–8.
3. Ellington AD, Szostak JW. In vitro selection of RNA molecules that bind specific ligands. Nature. 1990;346(6287):818–22.
4. Lee JF, Stovall GM, Ellington AD. Aptamer therapeutics advance. Curr Opin Chem Biol. 2006;10(3):282–9.
5. Ng EW, Shima DT, Calias P, Cunningham ET, Guyer DR, Adamis AP. Pegaptanib, a targeted anti-VEGF aptamer for ocular vascular disease. Nat Rev Drug Discov. 2006;5(2):123–32.
6. Ulrich H, Magdesian MH, Alves MJM, Colli W. In vitro selection of RNA aptamers that bind to cell adhesion receptors of Trypanosoma cruzi and inhibit cell invasion. J Biol Chem. 2002;277(23):20756–62.
7. Bagalkot V, Zhang L, Levy-Nissenbaum E, Jon S, Kantoff PW, Langer R, Farokhzad OC. Quantum dot-aptamer conjugates for synchronous cancer imaging, therapy, and sensing of drug delivery based on bi-fluorescence resonance energy transfer. Nano Lett. 2007;7(10):3065–70.
8. Tang Z, Shangguan D, Wang K, Shi H, Sefah K, Mallikratchy P, Chen HW, Li Y, Tan W. Selection of aptamers for molecular recognition and characterization of cancer cells. Anal Chem. 2007;79(13):4900–7.
9. Shangguan D, Li Y, Tang Z, Cao ZC, Chen HW, Mallikaratchy P, Sefah K, Yang CJ, Tan W. Aptamers evolved from live cells as effective molecular probes for cancer study. Proc Natl Acad Sci. 2006;103(32):11838–43.
10. Blank M, Weinschenk T, Priemer M, Schluesener H. Systematic evolution of a DNA aptamer binding to rat brain tumor microvessels selective targeting of endothelial regulatory protein pigpen. J Biol Chem. 2001;276(19):16464–8.
11. Stoltenburg R, Reinemann C, Strehlitz B. SELEX—a (r) evolutionary method to generate high-affinity nucleic acid ligands. Biomol Eng. 2007;24(4):381–403.
12. Wilson DS, Szostak JW. In vitro selection of functional nucleic acids. Annu Rev Biochem. 1999;68(1):611–47.

13. Mann D, Reinemann C, Stoltenburg R, Strehlitz B. In vitro selection of DNA aptamers binding ethanolamine. Biochem Biophys Res Commun. 2005;338(4):1928–34.
14. Wang C, Zhang M, Yang G, Zhang D, Ding H, Wang H, Fan M, Shen B, Shao N. Single-stranded DNA aptamers that bind differentiated but not parental cells: subtractive systematic evolution of ligands by exponential enrichment. J Biotechnol. 2003;102(1):15–22.
15. Ducongé F, Toulmé J-J. In vitro selection identifies key determinants for loop–loop interactions: RNA aptamers selective for the TAR RNA element of HIV-1. RNA. 1999;5(12):1605–14.
16. Toulmé J-J, Darfeuille F, Kolb G, Chabas S, Staedel C. Modulating viral gene expression by aptamers to RNA structures. Biol Cell. 2003;95(3):229–38.
17. Wilson C, Nix J, Szostak J. Functional requirements for specific ligand recognition by a biotin-binding RNA pseudoknot. Biochemistry. 1998;37(41):14410–9.
18. Rimmele M. Nucleic acid aptamers as tools and drugs: recent developments. ChemBioChem. 2003;4(10):963–71.
19. Zhu Z, Song Y, Li C, Zou Y, Zhu L, An Y, Yang CJ. Monoclonal surface display SELEX for simple, rapid, efficient, and cost-effective aptamer enrichment and identification. Anal Chem. 2014;86(12):5881–8.
20. Mayer G, Ahmed M-SL, Dolf A, Endl E, Knolle PA, Famulok M. Fluorescence-activated cell sorting for aptamer SELEX with cell mixtures. Nat Protoc. 2010;5(12):1993–2004.
21. Mendonsa SD, Bowser MT. In vitro evolution of functional DNA using capillary electrophoresis. J Am Chem Soc. 2004;126(1):20–1.
22. Berezovski M, Drabovich A, Krylova SM, Musheev M, Okhonin V, Petrov A, Krylov SN. Nonequilibrium capillary electrophoresis of equilibrium mixtures: a universal tool for development of aptamers. J Am Chem Soc. 2005;127(9):3165–71.
23. Qian J, Lou X, Zhang Y, Xiao Y, Soh HT. Generation of highly specific aptamers via micromagnetic selection. Anal Chem. 2009;81(13):5490–5.
24. Cho M, Xiao Y, Nie J, Stewart R, Csordas AT, Oh SS, Thomson JA, Soh HT. Quantitative selection of DNA aptamers through microfluidic selection and high-throughput sequencing. Proc Natl Acad Sci. 2010;107(35):15373–8.
25. Daniels DA, Chen H, Hicke BJ, Swiderek KM, Gold L. A tenascin-C aptamer identified by tumor cell SELEX: systematic evolution of ligands by exponential enrichment. Proc Natl Acad Sci. 2003;100(26):15416–21.
26. Hamula CL, Guthrie JW, Zhang H, Li X-F, Le XC. Selection and analytical applications of aptamers. TrAC Trends Anal Chem. 2006;25(7):681–91.
27. Morris KN, Jensen KB, Julin CM, Weil M, Gold L. High affinity ligands from in vitro selection: complex targets. Proc Natl Acad Sci. 1998;95(6):2902–7.
28. Fitter S, James R. Deconvolution of a complex target using DNA aptamers. J Biol Chem. 2005;280(40):34193–201.
29. Torres-Chavolla E, Alocilja EC. Aptasensors for detection of microbial and viral pathogens. Biosens Bioelectron. 2009;24(11):3175–82.
30. Berezovski M, Musheev M, Drabovich A, Krylov SN. Non-SELEX selection of aptamers. J Am Chem Soc. 2006;128(5):1410–1.
31. White R, Rusconi C, Scardino E, Wolberg A, Lawson J, Hoffman M, Sullenger B. Generation of species cross-reactive aptamers using "toggle" SELEX. Mol Ther. 2001;4(6):567–74.
32. Johnson L, Gershon PD. RNA binding characteristics and overall topology of the vaccinia poly (A) polymerase-processivity factor-primer complex. Nucleic Acids Res. 1999;27(13):2708–21.
33. Shamah SM, Healy JM, Cload ST. Complex target SELEX. Acc Chem Res. 2008;41(1):130–8.
34. Kolovskaya OS, Savitskaya AG, Zamay TN, Reshetneva IT, Zamay GS, Erkaev EN, Wang X, Wehbe M, Salmina AB, Perianova OV. Development of bacteriostatic DNA aptamers for salmonella. J Med Chem. 2013;56(4):1564–72.

Chapter 12
Lateral Flow Nucleic Acid Biosensors

Abstract Lateral flow nucleic acid biosensors (LFNABs) provide a variety of advantages, including being affordable, sensitive, specific, user-friendly, rapid, equipment-free, and deliverable to end users; thus, LFNABs have been very popular in biomedical, food, agriculture, environmental sciences, and so forth. This chapter discusses the latest progress on LFNABs, which primarily unite with a series of variants, such as constructions, signal systems, recognition systems, and amplification systems. Furthermore, we imagine LFNABs as an excellent candidate for available, fast, simple, and inexpensive detection with commendable applications.

Keywords Lateral • Biosensors • Nucleic acid • Amplification • Recognition • Signal

12.1 Introduction

A large number of analytical methods that can be applied to detect hazardous chemicals or pathogenic organisms correlated with human health, food safety, and the environment suffer from being time-consuming and too complex to use [1]. We need to use more average techniques that comparatively unskilled staff can learn easily and quickly. Lateral flow biosensors are an excellent candidate because they are the earliest and simplest designs to use as an analytical device platform.

The first lateral flow biosensor designed for the determination of glucose in urine was invented in 1957 by Singer and Plotz [1]. The first merchant lateral flow immunochromatographic biosensor was applied for human chorionic gonadotropin (HGC) determination in the early 1980s, which can be regarded as one of the most commonly used point-of-care biosensors [2, 3]. The speed of observing the results, detected immediately by the naked eye, and the usefulness of a membrane strip as the immunosorbent supplied an analytical platform that permits one-step, low-cost, and fast analysis. Since then, the fundamental principle of lateral flow immunochromatographic biosensors has been further refined and commonly applied in various investigatory domains, such as the environment, veterinary diagnostics, and food supervisory control [4]. Some previous investigators have put forward a number of retrospect papers depicting lateral flow immunoassays,

W. Xu, *Functional Nucleic Acids Detection in Food Safety*,
DOI 10.1007/978-981-10-1618-9_12

their creation, forms, SWOT (strengths, weaknesses, opportunities, and threats) analysis, and adhibition [2, 4–7].

Currently, the progress of lateral flow biosensors has been expanded toward nucleic acid-based reactions, that is to say, lateral flow nucleic acid biosensors (LFNABs). According to the reaction mechanism, LFNABs primarily can be classified into nucleic acid detection and functional nucleic acid-based reactions for the detection of other appointed materials detection, uniting the unique advantages of nucleic acids and chromatography. As a matter of fact, LFNABs hold excellent potential for the transmission of fast detection on the basis of being *ASSURED*, which is **a**ffordable, paper is inexpensive and abundant; **s**ensitive and **s**pecific, they can be based on immunoreactions, nucleic acid hybridizations, or functional nucleic acid-based reactions; **u**ser-friendly, the best candidate for point-of-care biosensors; **r**apid, the response is developed within a few minutes; **e**quipment-free, they are mainly read with the naked eye, or if quantitative detection is required, the equipment is small and cheap; and **d**eliverable to end users, they are quite stable to a wide range of temperatures and times [2, 3].

This chapter is intended to provide an overview of recent advances on LFNABs, which basically combines a number of variants, such as constructions, amplification systems, recognition systems, and signal systems.

12.2 Constructions

Lateral flow nucleic acid biosensors are typically composed of five different parts: sample pad, conjugation pad, nitrocellulose membrane, absorbent pad, and backing pad (Fig. 12.1); a brief description of each is given as follows.

12.2.1 Sample Pad

The sample pad is made of porous materials, such as cellulose fiber or glass fiber [2]. It is used to preprocess the sample before its conveyance and then to deliver the sample. The sample pad can also be disseminated with, e.g., buffer salts,

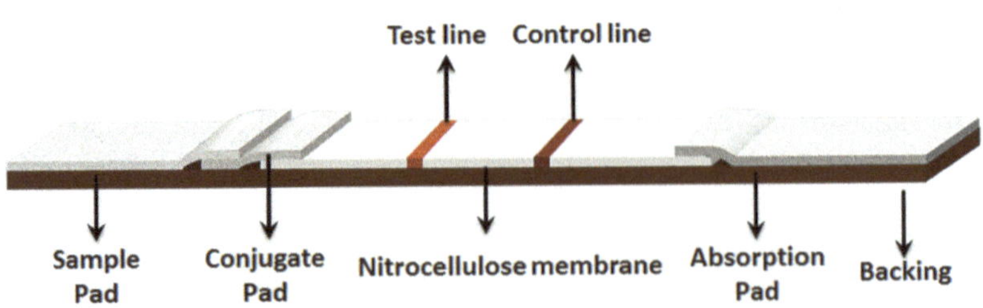

Fig. 12.1 Schematic illustration of the constructions of lateral flow nucleic acid biosensors

proteins, detergents, and viscosity enhancers to impact the rate of flow of the sample. The objective of pretreatment is to guarantee the best analytical status in the course of all the flux, possibly including segregation of a constituent part of the sample, regulation of pH, removal of interferences, etc. [6]. The delivery is requested in a kindred manner and continuously smooth. The pores in the sample pad can be symmetrically, homogeneously distributed, or asymmetrically distributed, providing an initial filter to remove coarse material, e.g., whole cells.

12.2.2 Conjugate Pad

The conjugate pad is the place where labeled bio-recognition molecules are distributed. It consists of cellulose, glass fiber, polyesters, cross-linked silica, and some other materials, which should promptly liberate a labeled conjugate to contact the floating liquid sample. The lack of preprocessing of a labeled conjugate can disadvantageously affect the susceptibility of the test. The labeled conjugate of the span of lateral flow biosensors should be sustainably stable [3]. Any alteration in the distribution of drying or liberation of the conjugate can change the test results dramatically. Specific reciprocity will be originated here and will continue during the chromatographic course.

12.2.3 Nitrocellulose Membrane

A nitrocellulose membrane is highly probative in detecting the susceptibility of LFNABs because it offers a place for both detection and reaction during the test, as developed by Whatman, Pierce, and Millipore. Test and control lines are dispensed in this segment of the membrane. Thus, a perfect membrane should offer support and proper binding to seize molecules (e.g., aptamers, antibodies, nucleic acid probes, etc.). Nonspecific adsorption over test and control lines may dramatically influence the consequence of the assay; therefore, a good membrane has the feature of less nonspecific adsorption in the range of test and control lines. There are some other membrane polymeric materials, including polyethylene (Porex), polyethersulfone (Pall Biosciences or Nalgene Nunc), fused silica (Fusion5, Whatman), and nylon (Millipore, Pall Biosciences), but these are not commonly used [8]. All together, these membranes are easy to operate, low-cost, and supply high-affinity binding for nucleic acids and proteins.

12.2.4 Adsorbent Pad

The adsorbent pad supplies traveling power on account of capillary effect at the end of the strip and wicks the fluid through the membrane to assure a sustaining flow. It is also applied to maintain the flow rate of the sample liquid over the membrane and prevent the back flow of the sample. Absorbent capacity has a significant role in holding liquid to keep a clear background for the consequences of the assay. When we apply an absorbent pad in the test, we can add an amount of sample to result in increased susceptibility. Moreover, the most common material is a cellulose filter.

12.2.5 Backing Pad

The backing pad serves as a certain mechanical support appliance, and it makes the operation of the strip test easier. Materials for the backing card are not strictly required because they play an unimportant role in LFNABs, except for offering a platform for proper fitting of all of the strip components.

All of these strip components are fixed or installed over a backing card after distributing the agents at their proper pads. Each two neighboring components overlap with each other slightly with the purpose of coordinating the fluid flow [3]. Then, after the incision and laminate, a practicable LFNAB is generated. The special step needed is the adhibition of sample to acquire results.

12.3 Amplification Systems

For the sensitivity and specificity of LFNABs to be improved, a pre-amplification step is commonly adopted by using classical PCR or isothermal amplification. The LFNAB is then used as a detection method for the amplified product, which may increase the sensitivity over six orders of magnitude compared to direct detection of nucleic acids by the LFNAB [8].

12.3.1 Classical PCR Amplification

Classical PCR amplification has had an enormous impact on both basic and diagnostic aspects of molecular biology because it can produce large amounts of a specific DNA fragment from small amounts of a complex template [9]. Combinations of various classical PCR and LFNABs have been reported by many research groups, which can be listed based on the reaction principle (Table 12.1).

Table 12.1 Classical PCR amplification methods relevant to LFNABs

Assay	Primers	Reaction duration (min)	Target	Amplification product	References
PCR	Two: forward primer and reverse primer	90–100	ds-DNA	ds-amplicon	[10]
RT-PCR	Two: forward primer and reverse primer	150	ss-RNA	ds-amplicon	[11]
Colony PCR	Two: forward primer and reverse primer	80–90	ds-DNA	ds-amplicon	[13]
Triprimer PCR	Three: a forward and two reverse primers	80–90	ds-DNA	ds-amplicon	[14]
Multi-PCR	Multiple: forward primers and reverse primers	180	ds-DNA	ds-amplicon	[15]
Multi-RT-PCR	Multiple: forward primers and reverse primers	150	ss-RNA	ds-amplicon	[12]
Multi-nested-PCR	Twice: forward primers and reverse primers	80–90	ds-DNA	ds-amplicon	[16]

12.3.1.1 PCR

PCR-based LFNAB is the most classic combination. Aveyard et al. used this detection method to screen for the existence of sole stranded target DNA [10]. They have demonstrated how unpurified PCR products can be tested with the installation of lateral flow that does not depend in any way on the forthcoming antibody hapten science technique. It was feasible to detect amplification products that were equivalent to 1000 template molecules (0.002 amol) by the naked eye.

12.3.1.2 RT-PCR and Multi-RT-PCR

Yong Tae Kim et al. constructed a reverse transcription polymerase chain reaction (RT-PCR) reactor and LFNAB for colorimetric detection of gene expression of influenza A virus subtype H1N1 in 2012 [11]. RT-PCR was executed to amplify the target H1 gene (102 bp). The following step elutes the resultant amplicons to the LFNAB. If a violet color is observed in the test line, the H1N1 target virus can be confirmed. The research method was raised on a portable genetic analyzer within 2.5 h and the on-site colorimetric pathogen was identified with a test sensitivity of 14.1 pg of RNA templates. Then, this team further covered a multiplex RT-PCR-based LFNAB for testing influenza A virus and another for hemagglutinin subtyping to determine the viral strain among H3N2, H5N1, and H1N1 in 2014 [12]. A multiplex single-step RT-PCR reaction was executed by employing a Qiagen OneStep RT-PCR kit. Influenza A, H5N1, H3N2, and H1N1 could be distinguished by a limit of analysis of 102 transcripts of RNA templates in 10 min.

12.3.1.3 Colony PCR

Fei Chen et al. exploited a susceptive, featured, and fast LFNAB integrated with colony PCR for the analysis of *Cronobacter* [13]. One hundred positive powdered infant formula samples spiked with 108 cfu mL^{-1} of *Cronobacter* and 20 negative samples with no bacteria were detected by the colony-exploited PCR-based strip, and the susceptibility and specificity of the detection were both as high as 100 %. This technology showed promise for microbial analysis as it relates to clinical diagnosis or food safety.

12.3.1.4 Triprimer PCR

The *JAK2V617F* mutation was incorporated as a new clonal marker in the 2008 revision of the WHO diagnostic criteria. Jessica K et al. introduced a lateral flow nucleic acid dipstick that comprises a triprimer PCR for direct detection of the *JAK2V617F* allele within minutes [14]. Target DNA was amplified by triprimer PCR in the presence of a forward and two reverse primers (outer and inner reverse primers). The analytic reagents serve as a dry situation on the dipstick; therefore, there is not any need for incubation steps or multiplicate pipetting.

12.3.1.5 Multi-PCR

A combination of multiplex PCR and LFNAB was also described. Shin Nihonyanagi et al. developed a new multiplex PCR lateral flow method that allows specific amplification of *femA* to detect *S. aureus*, *mecA* to detect SCCmec, and *kdpC* to detect SCCmec type II; moreover, PCR products can be evaluated visually within approximately 3 h [15]. The multi-PCR lateral flow for methicillin-resistant *Staphylococcus aureus* (MRSA) detection had high sensitivity, specificity, and speed, and its clinical application as a method for early diagnosis of MRSA infections appears to be feasible.

12.3.1.6 Multi-nested-PCR

Po-Chi Soo et al. exploited and detected a multiplex-nested PCR LFNAB for simultaneously and directly testing *Mycobacterium tuberculosis* complex (MTBC) and *Mycobacterium tuberculosis* (MTB) immediately at the same time from 1500 clinical sputum specimens [16]. Primers applied in PCR were devised from *IS6110* or *Rv3618* belonging to RD9 in the *M. tuberculosis* genome. The multiplex-nested PCR LFNAB test consequences in comparison with biochemical identification results and traditional culture as well as patients' clinical assessments demonstrate excellent potential in clinical applications.

12.3.2 *Isothermal Amplification*

Considering the nucleic acid detection in particular, major efforts will probably focus on the use of isothermal amplification techniques instead of classical PCR. In fact, the isothermal technique uses a single reaction temperature without losing sensitivity or sensibility, which can also be cheaper and easier to use in countries with fewer resources [2].

Many isothermal amplification methods coupled with LFNABs have been reported, which can be grouped based on the reaction principle (Table 12.2 and Fig. 12.1), including NASBA, helicase-dependent amplification (HDA), RPA, LAMP, RT-LAMP, cross-priming amplification (CPA), RCA, ISDPR, SDA, NEMA, ICA, and other isothermal amplification methods.

12.3.2.1 NASBA

Natalya V and coworkers, studying the rapid serotype-specific detection of Dengue virus, reported a multi-analyte LFNAB according to liposome signal amplification [17]. The multi-analyte LFNAB uses generic and serotype-specific DNA probes, which mingled with dengue RNA amplified by the isothermal nucleic acid result-based amplification (NASBA) reaction. After RNA amplification, testing of dengue virus-specific serotypes can be finished using a single analysis within 25 min. The NASBA-based biosensor had a test limit of 50,000 molecules for serotype1, 50 RNA molecules for serotype 2, and 500 RNA molecules for serotypes 3 and 4.

12.3.2.2 HDA

Et et al. analyzed *Staphylococcus aureus* and its methicillin resistance immediately from Gram-positive cocci in cluster-containing blood culture medium via LFNAB upon helicase-dependent amplification (HDA) [18]. The sample was attenuated and calcified before amplification of mecA and the nuc genes with isothermal HAD, which is a characteristic isothermal DNA amplification technique that has been exploited by employing a DNA helicase enzyme to unfasten double-stranded DNA and RNA-DNA crossbreeds. Amplicons were analyzed using a casual analysis installation. The detected susceptibility of the experiments was 50 CFU per reaction, and the clinical susceptibility and specificity were both 100 % for *S. aureus* determination and 100 % and 98 % for methicillin resistance detection, respectively.

12.3.2.3 RPA

Olaf Piepenburg et al. discussed an integration of recombinase polymerase amplification (RPA) and LFNAB for the pathogen methicillin-resistant *Staphylococcus*

Table 12.2 Isothermal PCR amplification methods relevant to LFNABs

Assay	Principle	Reaction temperature (°C)	Reaction duration (min)	Target	Amplification product	References
NASBA	RNA transcription	41	105	ss-RNA	ss- amplicon	[17]
HDA	Enzymatic duplex melting/primer annealing	65	75–90	ds-DNA	ds-amplicon	[18]
RPA	Enzymatic duplex melting/primer annealing	30–42	20	ds-DNA	ds-amplicon	[19]
LAMP	DNApolymerase-mediated strand displacement	60–65	60–90	ds-DNA	ds-amplicon	[20]
RT-LAMP	DNA-polymerase-mediated strand displacement	60–65	40	ds-DNA	ds-amplicon	[21]
CPA	DNA-polymerase-mediated strand displacement	65	60	ds-DNA	ds-amplicon	[22]
ISDPR	DNA-polymerase-mediated strand displacement	42	75	ss-DNA	ds-amplicon	[23]
NASDP	Polymerase extension/strand displacement	37	75	ss-DNA	ss- amplicon	[24]
CPT	Cyclic probe cleavage	55	30–90	ss-DNA	non-amplicon	[25]
RIDA	Cyclic probe cleavage	55	25	ss-DNA	non-amplicon	[26]

aureus analysis [19]. In RPA, scientists bind the adverse oligonucleotide primers to template DNA and their extension by a DNA polymerase to achieve the isothermal amplification of specific DNA fragments. The technique is demonstrated to be susceptive to fewer than ten copies of genomic DNA. Furthermore, the outcome can be determined in an easy sandwich assay, creating an instrument-free DNA analytical system. This special integration of properties is meaningful progress in the exploitation of portable and generally accostable nucleic acid-based determinations.

12.3.2.4 LAMP

A combined loop-mediated isothermal amplification lateral flow dipstick (LAMP-LFD) format was evaluated in the detection of human infective trypanosome DNA from clinical samples by Zablon Kithinji Njiru [20]. The LAMP-LFD showed analytical sensitivity equivalent to 0.01 tryps/mL, levels that were identical to those using gel electrophoresis and SYBR® Green I dye. LAMP-LFD showed superior specificity to SYBR® Green I when supernatant prepared from boiled human biological samples was used as template. The exploited LAMP-LFD format is fast and easy and has hidden value in the diagnosis of sleeping sickness in the future.

12.3.2.5 RT-LAMP

An integration of multiplex reverse transcription loop-mediated isothermal amplification (RT-LAMP) was exploited with an LFNAB for subtyping influenza A virus by Jae [21]. The HA (H1, H3, and H5 gene) and conserved M gene were amplified through the most ideal primer design, reaction time, temperature, concentration of enzymes (Bst DNA polymerase and AMV reverse transcriptase), and dNTP. The exploited detection could be finished within 40 min, and the pathotyping and subtyping of influenza A virus were executed even with only ten transcripts of viral RNA template. The subtyping of a clinical specimen exhibited extreme practicability of our methodology for actual sample detection with high simplicity, velocity, and susceptibility.

12.3.2.6 CPA

Fang and coauthors evaluated the cross-priming amplification (CPA) technique for tuberculosis diagnosis from sputum specimens [22]. The consequence of each CPA test was ascertained by inspecting the visible bands on the LFNAB kit, including the presence (positive result) or absence (negative result) of bands. The susceptibility of CPA from smear- and liquid culture-positive samples was 96.9 % and that from smear-negative and liquid culture-positive specimens was 87.5 %. The specificity of CPA in culture-negative specimens was 98.8 %.

12.3.2.7 ISDPR

He developed an LFNAB to visually detect the mutation in the R156H gene of keratin. The authors first amplified the gene with an isothermal strand displacement polymerase reaction (ISDPR) using primers modified with digoxin and biotin [23]. Then, they could detect the products using a sandwich assay based on AuNPs functionalized with an antibody specific for digoxin and an antibody specific for the biotin in the test line. The LoD was as low as 1 fM. This approach provided a simple, sensitive, and low-cost tool for the detection of gene mutations.

12.3.2.8 NASDP

NEase-assisted strand displacement polymerization (NASDP) is based on a hairpin probe-dependent biologic circuit where the accessional polymerization elicited by NEase supplies high susceptibility and rapid detecting speed. An ecumenical biosensor was developed by Liu, which can be used for multiplex DNA detection upon NASDP with a determination limit of 100 aM in 75 min [24]. The achievement of tag/anti-tag DNA and gold nanoparticle reporters allows for an ecumenical terrace for multiplex genotyping without instrumentation.

12.3.2.9 CPT

Cyclic probe technology (CPT) is an isothermal reaction that is not prone to cross contamination because the target is not amplified. The target can hybridize circularly with the intact probe molecule while cut probe fragments accumulate. RNase H cleavage of the RNA portion of the probe-target duplex is the key aspect of CPT. The CPT assay with an LFNAB was developed for the detection of the *mecA* gene from methicillin-resistant *Staphylococcus aureus* (MRSA) cultures by Whalley [25]. A screen of 324 *S. aureus* clinical isolates by the CPT-LFNAB assay showed 99.4 % sensitivity and 100 % specificity compared to the results of PCR for the detection of the *mecA* gene.

12.3.2.10 RIDA

The rapid isothermal nucleic acid detection assay (RIDA), another type of cyclic probe technology, is a "probe amplification" assay that uses the single-strand nicking activity of restriction nicking endonucleases to repeatedly cleave synthetic probes hybridizing to the same target sequences. N.BstNBI, an enzyme identified from *Bacillus stearothermophilus* that can recognize the sequence of 5′-GAGTC-3′ and cleaves 4 bp downstream of the 3′ end, is the key aspect of RIDA as developed by Wenjuan Gao et al. [26]. The RIDA-LFNAB combined assay can specifically detect nucleic acid targets without sophisticated equipment, which is a flexible simple assay that could be applied for point-of-care detection.

12.3.3 Direct Detection Without Amplification

With the development of trace or ultrasensitive LFNABs, direct detection of target without amplification has become popular in recent years.

Mao et al. hybridized the sample solution of target DNA with DNA-AuNP probes by using a sandwich DNA hybridization reaction at the conjugate pad [27]. The shaped product kept on moving through the test line, where the second hybridization was observed between the immobilized DNA probes and the target DNA, displaying a red band. At the control line, a second red line was intuitional due to the hybridization between additional DNA-labeled AuNPs and the other DNA probes. This nucleic acid LFA could test the target DNA in the scope of 1–100 nM in less than 15 min without extra mixing, washing steps, or incubation.

Cheng-Che Liu et al. utilized 16S rDNA/rRNA probe-conjugated gold nanoparticles to develop a rapid and sensitive sandwich lateral flow immunoassay that can detect *Salmonella* in food or clinical samples [28]. Without any device for target amplification, the nucleic acids purified from 10^4*Salmonella* bacteria can be detected using this assay with silver enhancement after nucleic acid purification in 40 min. In developing countries, this method has a potential to detect *Salmonella* for food safety and screening in the field.

Xuefei Gao et al. reported a DNA-gold nanoparticle (DNA-GNP)-based lateral flow nucleic acid biosensor for visual detection of microRNA (miRNA)-215 in aqueous solutions and biological samples without any amplification [29]. - Sandwich-type hybridization reactions among biotin-modified DNA probes, miRNA-215, and GNP-labeled DNA probe were executed in the lateral flow facility. The amassment of GNPs in the detection region of the biosensor enables the optical analysis of miRNA-215. After systematic optimization, the biosensor was able to test a minimum concentration of 60 pM miRNA-215, which shows excellent promise for clinical adhibition and biomedical diagnosis in some vicious diseases.

12.4 Recognition Systems

LFNABs perform endpoint detection sandwich assays using passive fluidics [30]. The formats of LFNABs are classified into three types according to the recognition principle, including antibody-, hybridization-, and functional nucleic acid-dependent formats.

12.4.1 Antibody-Dependent Format

The antibody-dependent format employs an antibody capture line and a labeled amplicon or oligonucleotide probe of complementary sequence to the amplicon. The target is an amplified double-stranded nucleic acid sequence (ds-amplicon) specific to the analyte using primers with two different tags (Fig. 12.2). The recognition of the specific tagged ds-amplicon is performed by a combination of tag-antibody interactions or a biotin–streptavidin (or avidin) system. Typical combinations of tag-antibody interactions relevant to LFNABs are summarized in Table 12.3.

In a typical layout developed for the visual detection of a mutation in the R156H gene of keratin, the gene was amplified with an isothermal strand displacement polymerase reaction (ISDPR) using primers modified with digoxin and biotin (Fig. 12.3a). Then, they could detect the products using a sandwich assay based on AuNPs functionalized with an antibody specific for the digoxin and an antibody specific for the biotin in the test line [23]. In another case (Fig. 12.3b), the detection of an amplified double-stranded nucleic acid sequence specific to a target organism can be achieved by using a primer with two different tags (e.g., biotin and fluorescence isothiocyanate). Recognition of the analyte is achieved by binding to a tag-specific antibody (anti-fluorescein antibody) sprayed previously on a

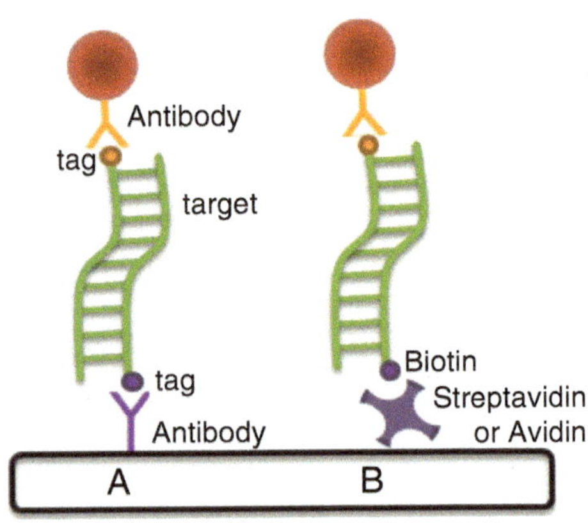

Fig. 12.2 Schematic representation of two detection principles in the antibody-dependent format of LFNABs, including (**a**) a tag-antibody interaction and (**b**) a biotin–streptavidin or biotin–avidin system

Table 12.3 Typical combination of tag-antibody interactions relevant to LFNABs

Tag	Antibody	References
Bio	Anti-biotin antibody (streptavidin/avidin)	[11, 16, 31–33]
Dig	Anti-digoxigenin antibody	[14, 16, 33–35]
FITC	Anti-fluorescein isothiocyanate antibody	[16, 23, 34, 36, 37]
TxR	Anti-Texas red antibody	[11, 12, 21, 34]
DNP	Anti-2,4-dinitrophenol antibody	[34]
Cy5	Anti-Cy5 antibody	[34]

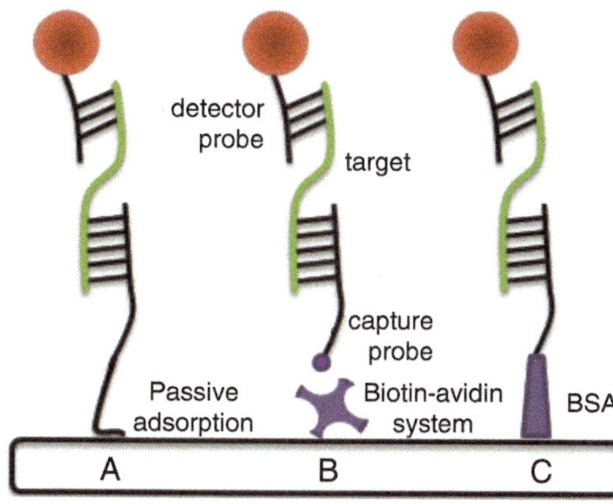

Fig. 12.3 Schematic illustrations for the antibody-dependent format of LFNABs

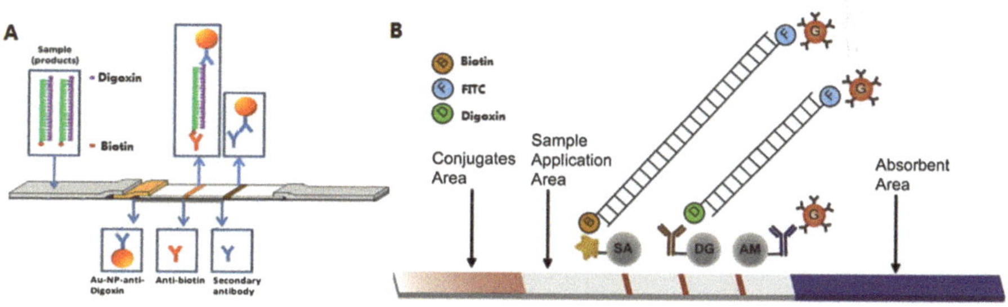

Fig. 12.4 Schematic representation of three detection principles in the hybridization-dependent format of LFNABs: (**a**) through passive adsorption, (**b**) via a biotin–avidin system (Reprinted from Refs. [23, 37], Copyright 2012 and 2011, with permission from Elsevier)

nitrocellulose membrane, and gold nanoparticle-labeled avidin is used as reported, enabling the visualization [37].

12.4.2 Hybridization-Dependent Format

Specific nucleic acid hybridization of amplicons with immobilized complementary probes is another option. The hybridization-dependent format requires two types of oligonucleotide probes, i.e., a detector probe and a capture probe, which are both complementary with a target single-stranded amplicon (ss-amplicon). Several convenient approaches are possible in the immobilization of the oligonucleotide capture probe to the nitrocellulose membrane of such a format: (1) through passive adsorption (Fig. 12.4a) and (2) via a biotin–avidin system (Fig. 12.4b) [4].

Cheng-Che Liu proposed a NANAB to detect *Salmonella* using gold nanoparticles conjugated with a DNA probe that is complementary to the 16S ribosomal RNA and DNA of *Salmonella* (Fig. 12.5a) [28]. The detection limit is

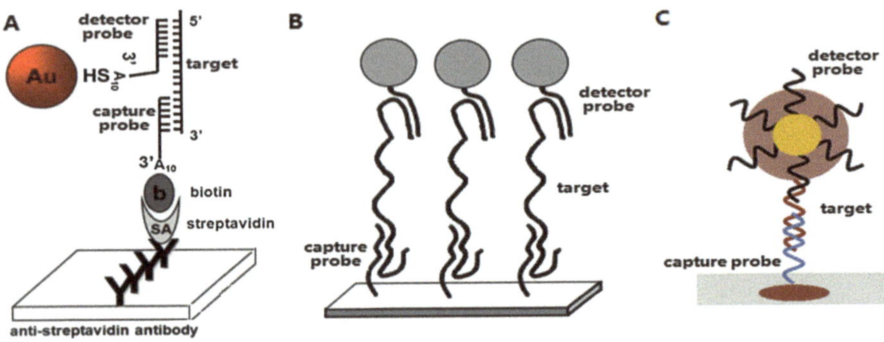

Fig. 12.5 Schematic illustrations for the antibody-dependent format of LFNABs (**a** Reprinted from Ref. [28], Copyright 2013, with permission from Elsevier **b** Reprinted from Ref. [17], with kind permission from Springer Science+Business Media, **c** Reprinted from Ref. [38] with permission from PLoS ONE)

5 fmol for the synthetic single-stranded DNA. For the *Salmonella* cultured samples, the nucleic acids from 10^7 bacteria were rapidly detected in 30 min. After silver enhancement, the detection limit was as low as 10^4 cells, which is lower than 10^5 bacteria cells, the human infective dose of foodborne *Salmonella.* In another case (Fig. 12.5b), Natalya V. Zaytseva developed a multi-analyte biosensor based on nucleic acid hybridization and liposome signal amplification for the rapid serotype-specific detection of dengue virus [17]. The generic probe (reporter probe) is coupled to dye-entrapping liposomes and can hybridize to all four dengue serotypes, whereas the serotype-specific probes (capture probes) are immobilized through biotin–streptavidin interactions on the surface of a polyethersulfone membrane strip in separate locations. A mixture of amplified dengue virus RNA sequences and liposomes is applied to the membrane and allowed to migrate along the test strip. After the liposome-target sequence complexes hybridize to the specific probes immobilized in the capture zones of the membrane strip, the dengue serotype present in the sample can be determined. In the third example (Fig. 12.5c), Brittany A. Rohrman et al. present a lateral flow assay that employs gold nanoparticle probes and gold enhancement solution to detect amplified HIV RNA quantitatively [38]. Preliminary results show that, when coupled with nucleic acid sequence-based amplification (NASBA), this assay can detect concentrations of HIV RNA that match the clinically relevant range of viral loads found in HIV patients. The lateral flow test is inexpensive, simple, and rapid to perform and requires few resources.

12.4.3 Combination of Antibody- and Hybridization-Dependent Formats

The antibody-dependent format fits for double-stranded nucleic acid sequence (ds-amplicon) as targets, whereas the hybridization-dependent format is suitable

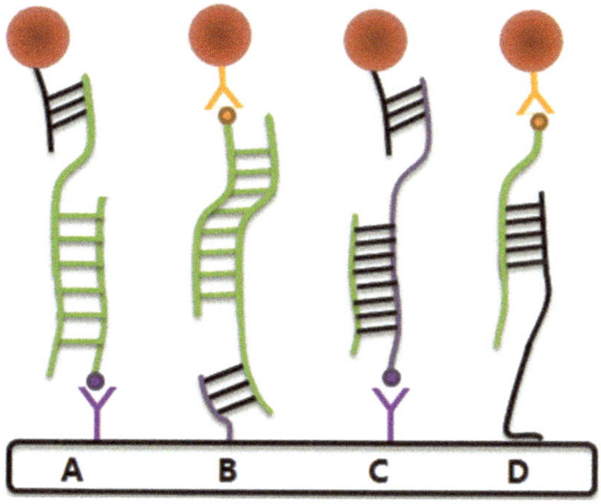

Fig. 12.6 Schematic representation of four detection principles in the combination of antibody- and hybridization-dependent formats of LFNABs: (**a**, **b**) ds-amplicon and (**c**, **d**) ss-amplicon

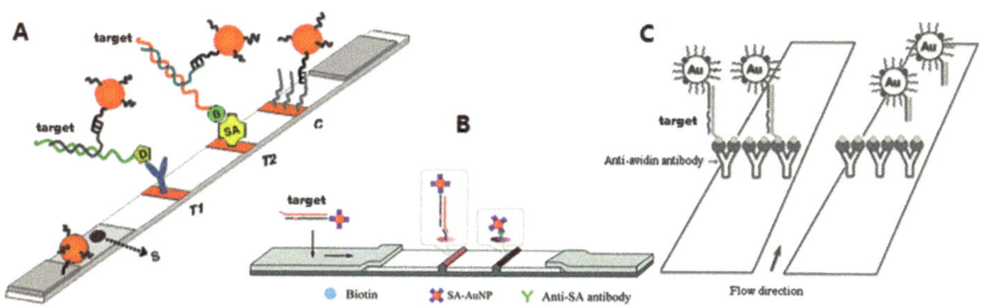

Fig. 12.7 Schematic illustrations for the combination of antibody- and hybridization-dependent formats of LFNABs (**a** Reprinted from Ref. [14], with kind permission from Springer Science+Business Media **b** Reprinted from Ref. [24] by permission of The Royal Society of Chemistry **c** Reprinted from Ref. [39], Copyright 2012, with permission from Elsevier)

for single-stranded nucleic acid sequence (ss-amplicon) as targets, which shows that they both have a certain degree of restriction. Therefore, some researchers combine antibody and hybridization together to build a novel type of format, which can detect both ds-amplicons (Fig. 12.6a, b) and ss-amplicons (Fig. 12.6c, d).

One typical case for ds-amplicon detection is that Jessica K. Konstantou et al. develop a dry-reagent strip biosensor for *JAK2V617F* mutation analysis by hybridization and an antibody [14]. Digoxin- or biotin-modified PCR products are hybridized with a poly(dA)-tailed oligo and applied on the strip, which is then immersed in the appropriate buffer (Fig. 12.7a). As the buffer migrates upward, it rehydrates the nanoparticles that are linked to the target DNA through poly(dA)/(dT) hybridization. Capture of the hybrids by immobilized anti-digoxin antibody and streptavidin in the test zone of the strip generates a characteristic red band. A second red band is formed, by hybridization, in the control zone of the strip to indicate proper test performance. Another case is presented by Jie Liu et al. [24]. Streptavidin–biotin–anti-tag DNA and anti-streptavidin antibodies are dispensed on the nitrocellulose membrane to form a test and a control zone, respectively (Fig. 12.7b). The amplification products are first mixed with streptavidin-modified gold nanoparticles (SA–AuNPs), forming the complexes of dsDNA–biotin–SA–AuNPs, and then the mixture is applied on the sample pad. The resulting complexes are captured in the test zone by the reaction between tag DNA and the complexes. Consequently, the accumulation of AuNPs in the test zone is visualized as a characteristic red band. The excess of the sample continues to migrate and is captured in the control zone. As for ss-amplicon detection, Shao-Yi Hou et al. built a microRNA detection method using lateral flow nucleic acid strips by hybridization and an antibody [39]. They tested microRNA, and the detection probe perfectly matches with the capture probe (Fig. 12.7c) instead of the traditional sandwich methods in which the tested oligonucleotide matches with the detection and capture probes. Mung bean nuclease, a single-strand-specific nuclease, catalyzes the degradation of the capture probe if there is no tested miRNA in the sample. The proposed method is simple, convenient, fast, and suitable for point-of-care.

12.4.4 Functional Nucleic Acid-Dependent Format

Functional nucleic acids, mainly including aptamers, DNAzymes, and aptazymes, are nucleic acids that have functions not limited to nucleic acid hybridization. The functional nucleic acid-dependent formats are flexible and interesting.

12.4.4.1 Aptamers

Aptamers, similar to antibodies, are essentially nucleic acid molecules that can fold into complex three-dimensional shapes, forming binding pockets and clefts for the specific recognition and tight binding of any given molecular target, from metal ions and small chemicals to large proteins and higher-order protein complexes, whole cells, viruses, or parasites [40]. Moreover, aptamers are easy to produce and compatible with AuNPs. Thus, as an alternative to antibodies, aptamers were also adopted to label AuNPs for preparing the conjugation pad of the LFNAB. The aptamer opportunities in LFNABs were also stated in a review by Ailiang Chen et al. in 2015 [41]. Therefore, we present several typical cases as follows.

As shown in Fig. 12.8a, Hui Xu et al. developed a dry-reagent lateral flow biosensor for visual detection of thrombin according to the conventional sandwich format [42]. Two aptamers were bound to thrombin at two different sites. When the detection aptamer was dipped into a solution containing thrombin, thrombin would migrate by capillary action and bind to gold NP–aptamer conjugates. The complexes continued to migrate along the strip and were captured by the secondary aptamer via an interaction between the secondary aptamer and thrombin. A characteristic red band could be observed because of the accumulation of gold NPs in the test zone. In other words, the more target thrombin that was in the solution, the stronger the intensity was on the test line. Ochratoxin A (OTA) is a type of small

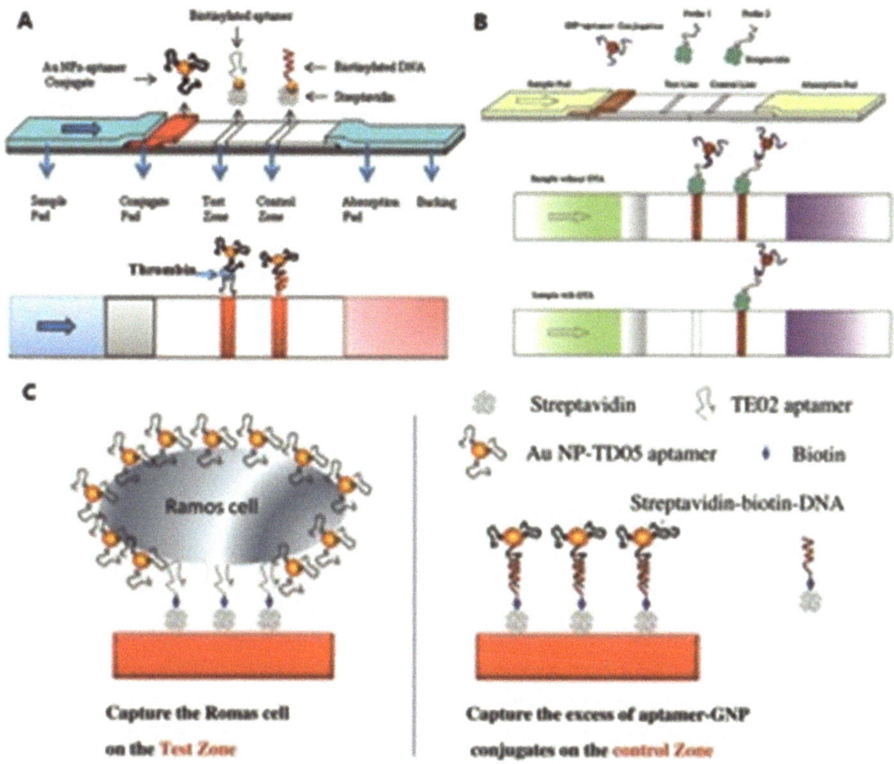

Fig. 12.8 Schematic illustrations for the aptamer-dependent format of LFNABs (**a**, **c**) Reprinted with the permission from Refs. [42, 44]. Copyright 2009 American Chemical Society [43] (**b**) Reprinted from Ref. [44], Copyright 2011, with permission from Elsevier)

molecular mycotoxin with only one aptamer; a competitive (or inhibition) format based on aptamer-linked gold nanoparticles for visual detection of ochratoxin A was developed by Chuanlai Xua et al. [43]. In this system, the thiol-modified OTA aptamer with a polyA tail was conjugated on the colloidal gold surface. The biotin-modified complementary strand, named DNA probe 1, was immobilized through streptavidin binding onto the test zone of the NC membrane. The biotin-modified polyT, named DNA probe 2, was immobilized through streptavidin binding onto the control zone of the NC membrane. The aptamer-based strip assay was implemented by the competitive reaction between DNA probe 1 (test line) and the target OTA to combine with aptamers. Once found in the detection solution, OTA would combine with the aptamer-GNP probe, decreasing the amount of aptamer-GNP that could hybridize with DNA probe 1 on the test line and cause the red color intensity to become weaker. In other words, the more OTA that was in the solution, the weaker the intensity was on the test line.

In addition to macromolecules and micromolecules, aptamer-based LFNAB can also be used to detect whole cells. Guodong Liu et al. developed an aptamer-AuNP strip biosensor for the detection of circulating cancer cells, reaching a detection limit of 4000 Ramos cells with the naked eye and 800 Ramos cells with a portable strip reader within 15 min [44]. As shown in Fig. 12.8c, a thiolated aptamer is immobilized on the GNPs, and a biotinylated aptamer is immobilized in the test zone of the NC membrane through streptavidin. Interacting with the aptamer probes of the AuNP–aptamer conjugates, Ramos cells form AuNP–aptamer cell complexes and continue to move along the test strip. Then, a large number of AuNPs accumulate in the test zone and produce a distinctive red band, which can be used for qualitative (visual) evaluation or quantitative cell detection with a portable strip reader. Furthermore, the cells were also detected in human blood samples.

12.4.4.2 DNAzymes and Aptazymes

Most DNAzymes possess enzyme activity in the presence of specific mental ions, and this mechanism can be utilized for sensing given metal ions by measuring the enzymatic activity [45]. A combination of DNAzymes and aptamers results in aptazymes. These have been employed for colorimetric sensing of metal ions and biochemical molecules in LFNABs [40]. Generally speaking, metal ions mediate DNAzyme and aptazyme reactions to produce a single-stranded nucleic acid for LFNAB detection in the hybridization-dependent format (Sect. 12.4.2). How can the production of a single-stranded nucleic acid be induced from metal ions? Several good examples are presented as answers.

As shown in Fig. 12.9a, Zhiyuan Fang et al. developed a lateral flow nucleic acid biosensor based on copper-dependent DNA-cleaving DNAzyme and gold nanoparticles for the visual detection of copper ions (Cu^{2+}) in an aqueous solution with a detection limit of 10 nM [46]. In the presence of Cu^{2+}, the substrate is irreversibly cleaved at the cleavage site (the guanine is in red). The cleaved piece (in green and purple) at the 30 end of the substrate is released due to decreased affinity to the enzyme (Fig. 12.9a(a)). The cleaved product is then applied to the LFNAB in which three DNA probes are immobilized (Fig. 12.9a(b)). Similarly, easy-to-use dipstick tests for lead have been developed by immobilizing nanoparticle–DNAzyme conjugates on lateral flow devices and their application for detecting lead in paints is demonstrated [47]. In the presence of Pb^{2+}, the 17E enzyme catalyzes the cleavage of the substrate, 17S, at the single ribo-linkage (shown in red) (Fig. 12.9b(a)). A modified 8–17 construct conjugated to AuNPs (called Enz-SubAuNP) is used for the dipstick tests (Fig. 12.9b(b)), and an assembled lateral flow device is shown in Fig. 12.9b(c). In the absence of Pb^{2+}, AuNP-uncleaved substrate is captured at the control zone via a streptavidin–biotin interaction, producing a single red line (Fig. 12.9b(d)). Substrate is cleaved in the presence of Pb^{2+} and the AuNP-cleaved product migrates beyond the control zone to be captured at the test zone by hybridization to complementary DNA. Two red lines are produced.

In addition to metal ion-dependent DNA-cleaving DNAzyme, the binding of metal ions by pyrimidine base pairs in DNA duplexes is another important task. Thymine–thymine (T–T) and cytosine–cytosine (C–C) base pairs selectively

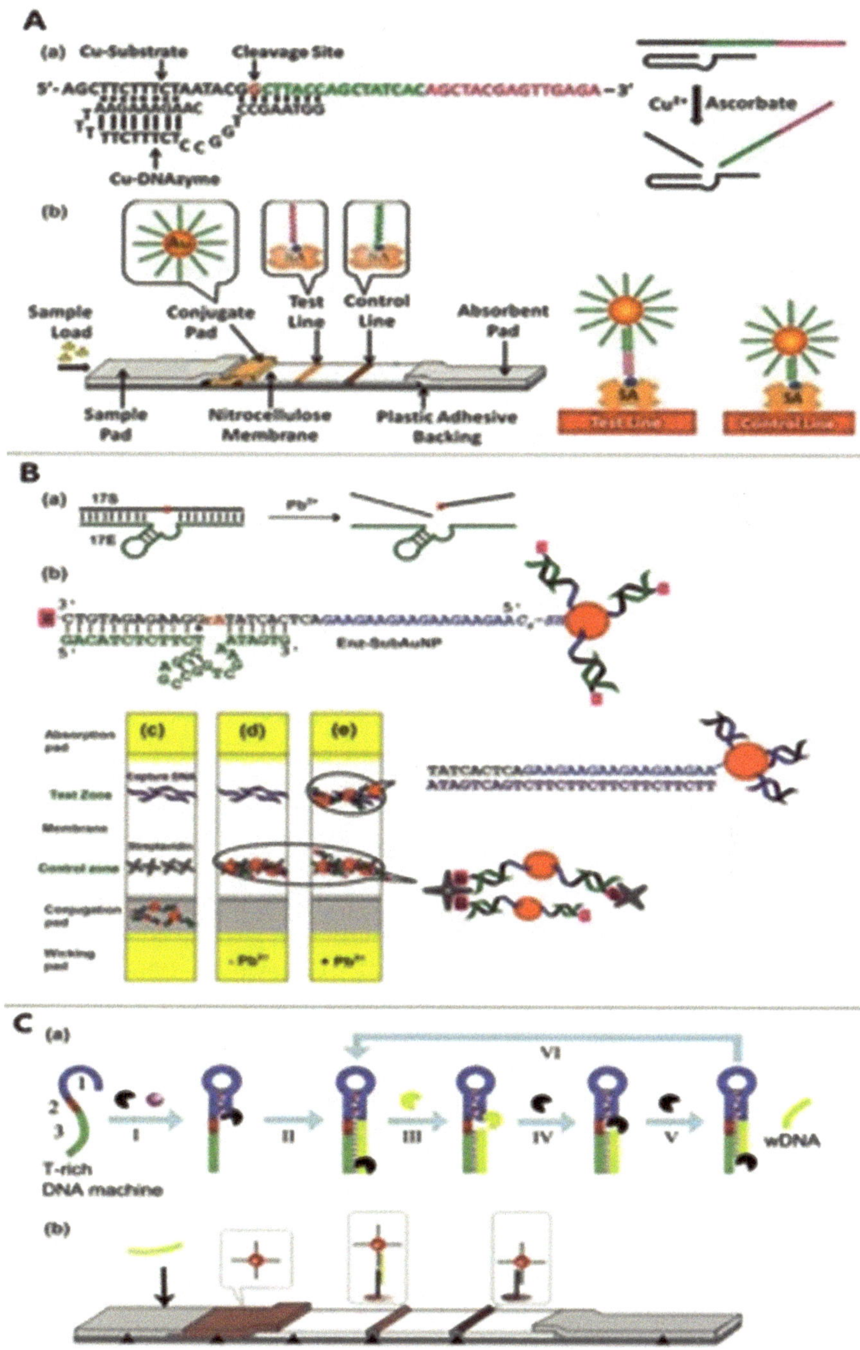

Fig. 12.9 Schematic illustrations for the DNAzyme-dependent format of LFNABs (Reprinted from Refs. [47–49] by permission from The Royal Society of Chemistry)

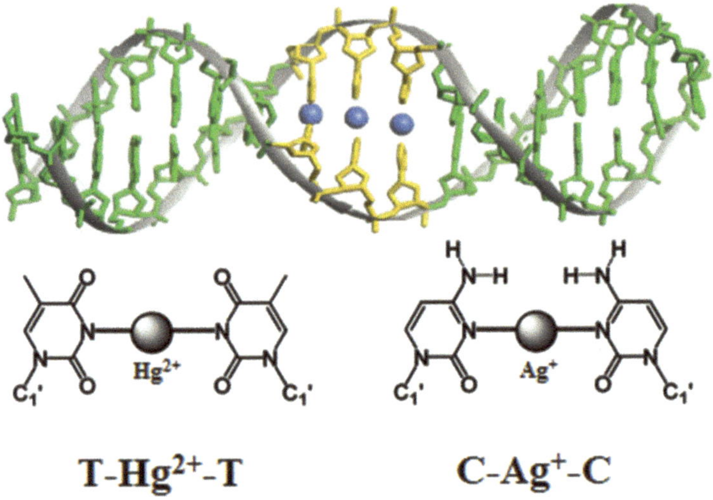

Fig. 12.10 Binding of metal ions by pyrimidine base pairs in DNA duplexes (Reprinted by permission from Macmillan Publishers Ltd: Ref. [50], copyright 2010)

capture Hg(II) and Ag(I) ions, respectively, and the metallo-base pairs T-Hg(II)-T and C-Ag(I)-C are formed in DNA duplexes (Fig. 12.10) [48].

Jie Liu et al. published a good case on this point as shown in Fig. 12.9c [49]. An autonomous thymine-rich DNA machine as an amplification unit was developed for the sensitive detection of mercury ions with high specificity. Combined with a lateral flow biosensor, the amplified signal of Hg^{2+} can be read out by the naked eye. The DNA machine is ignited by Hg^{2+} to form a self-primed template. Subsequently, KF initiates the replication at the 3′ end of the structure. Upon replication, Nt.BbvCI recognizes its binding site and nicks the replicated strand. After nicking, KF moves to the 3′ end of the nicked DNA and primes a new round of replication. During this replication, the complementary strand of domain 3 is displaced as the machine's waste product (wDNA). Once the DNA machine is ignited by Hg^{2+}, the repeated operation of scission and replication continues to produce numerous wDNAs, which are subsequently applied on an LFB for visual detection. The intensity increased with the concentration of Hg^{2+}, and no red line was observed on the test zone in the absence of Hg^{2+}. The integration of the LFB with the DNA machine enabled instrument-free readout with a detection limit of 5 nM, which is much lower than the maximum level (30 nM) of Hg^{2+} in drinking water permitted by the WHO.

12.5 Signal Systems

The list of materials used as a label in LFNAB analytical signal systems is vast, which includes gold nanoparticles, colored latex beads, carbon nanoparticles, selenium nanoparticles, textile dyes, liposomes, quantum dots, up-converting phosphors, organic fluorophores, magnetic particles, and others. Any material that is used as a label should be detectable at very low concentrations and should retain its properties upon conjugation with bio-recognition molecules. This conjugation is also expected to not change features of the bio-recognition probes. Ease in conjugation with biomolecules and stability over a longer period of time are desirable features for a good signal label. Recently, several reviews focused on signal systems of lateral flow biosensors have been published completely and in great detail [2, 3, 7, 50, 51]; therefore, it is not necessary to repeat them specifically in this chapter. Table 12.4 presents the advantages and limitations of different signaling systems.

Most published reports on LFNABs employed gold nanoparticles (AuNPs) as label [2]. AuNPs have many properties that make them excellent labels: good stability, easy functionalization, easy manipulation, biocompatibility, a strong red color, a characteristic surface Plasmon resonance, and remarkable enhancement of signal in nanoscale geometries that can be used in optical detection of ions/small molecules, proteins, nucleic acids, and even cancer cells. Optical properties of gold nanoparticles are dependent on size and shape. The particle sizes can be tuned through use of suitable chemical additives. The size of the gold particles varies between 2 and 150 nm, but generally, 15–40 nm particles were used [7]. The optical properties of gold nanoparticles enhance the sensitivity of analysis in LFNABs. GNPs still suffer from some drawbacks: the sensitivity is insufficient for low abundance targets, and the signal amplification process can be costly [51]. The choice of an appropriate reporter is a critical parameter in LFNAB development. Typical operation for the preparation of AuNPs, antibody-AuNP conjugates, and DNA-AuNP conjugates are listed in Table 12.5.

12.6 My Typical Research

12.6.1 Accurate and Easy-to-Use Assessment of Contiguous DNA Methylation Sites Based on Proportion Competitive Quantitative PCR and Lateral Flow Nucleic Acid Biosensor

Many types of diagnostic technologies have been reported for DNA methylation, but they require a standard curve for quantification or only show moderate accuracy. Moreover, most technologies have difficulty providing information on the level of methylation at specific contiguous multi-sites, not to mention easy-to-use detection to eliminate labor-intensive procedures. We have addressed these limitations and report here a cascade strategy that combines proportion competitive quantitative PCR (PCQ-PCR) and lateral flow nucleic acid biosensor (LFNAB), resulting in accurate and easy-to-use assessment. The P16 gene with specific multi-methylated sites, a well-studied tumor suppressor gene, was used as the target DNA sequence model [52]. First, PCQ-PCR provided amplification products with an

Table 12.4 Summary of the advantages and disadvantages of different signaling systems of LFNABs

Signal type	Details	Advantages	Limitations
Visual signal	Colloidal gold nanoparticles	Widely used, colored, shape- and size-dependent optical properties, ease of preparation and functionalization, good biocompatibility, cost-effective, visually interpretable, and easy-to-read results	Qualitative or semi-quantitative, low signal intensity, poor sensitivity, false-positives or false-negatives, background interference, and strip reader required for quantitation
	Colloidal carbon	Colored, high signal-to-noise ratio (black to white background), the cheapest labels, very stable, ease of preparation and functionalization, nontoxic, no need for activation, and high sensitivity	Qualitative or semi-quantitative, background interference, and reader required for quantitation
	Colloidal selenium	Rust-colored, low-cost, easy-to-prepare and operate	Qualitative or semi-quantitative, difficult to discriminate the result by the naked eye
Fluorescent signal	Quantum dots	Robust photo-stability, wide absorption spectra and narrow emission bands, size-dependent Stokes shifts, strong stability against photobleaching and chemical degradation, high fluorescent quantum yield, large molar extinction coefficients, and high sensitivity	Bad chemical and colloidal instability in biological environments, fluorescence quenching, toxic, high auto-fluorescence background, and fluorescence reader required for quantitation
	Up-converting phosphor nanoparticles	Sharp emission spectra (<10 nm), large Stokes shifts (>150 nm), long fluorescence lifetimes, excellent up-conversion fluorescence, ease of functionalization, high chemical stability and fluorescent quantum yield, tunable emission colors, high photo-stability, low toxicity,	Fluorescence reader required for quantitation

(continued)

Table 12.4 (continued)

Signal type	Details	Advantages	Limitations
		no background fluorescent interference, and high sensitivity	
	Dye-doped NPs	Less aggregation, little dye leakage, less photobleaching and photo-degradation and higher optical signal compared with free dye molecule, ease of surface functionalization, and high sensitivity	Difficulty of preparation, fluorescence dye quenching, and fluorescence reader required for quantitation
Magnetic signal	Magnetic particles	Ease of surface functionalization, ease of manipulation, magnetic properties, entire magnetic signal detectable from MNPs, no degradation of signals over a long time period, low background noise	Magnetic reader or giant magnetoresistive sensor required
Electrochemical signal	Redox reactions: electron transfer between molecules or particles Non-redox reactions: changes of electrical properties, such as impedance, resistance, conductance, and potential	High sensitivity and selectivity, low-cost, and portability	Need electrochemical workstation for quantitation

Reproduced from Huang et al. [51] with permission from Elsevier

accurate proportion of multi-methylated sites following the principle of proportionality, and double-labeled duplex DNA was synthesized. Then, a LFNAB strategy was further employed for amplified signal detection via immune affinity recognition, and the exact level of site-specific methylation could be determined by the relative intensity of the test line and internal reference line. This combination resulted in all recoveries being greater than 94 %, which are pretty satisfactory recoveries in DNA methylation assessment. Moreover, the developed cascades show significantly high usability as a simple, sensitive, and low-cost tool. Therefore, as a universal platform for sensing systems for the detection of contiguous multi-sites of DNA methylation without external standards and expensive instrumentation, this PCQ-PCR-LFNAB cascade method shows great promise for the point-of-care diagnosis of cancer risk and therapeutics.

Table 12.5 Typical operation for the preparation of AuNPs, antibody-AuNP conjugates, and DNA-AuNP conjugates

Preparation	Typical operation	References
AuNPs	All glassware used in this preparation was thoroughly cleaned in aqua regia (three parts HCl and one part HNO_3), rinsed in doubly distilled water, and oven dried prior to use. Briefly, in a 500 mL, round-bottom flask, 4 mL of 1 % sodium citrate was added to 96 mL of a rapidly stirred and boiling aqueous solution of 0.01 % $HAuCl_4$. The solution turned deep blue within 20 s, and the final color changed to wine-red 60 s later. After turning wine-red, the solution was boiled for 6 additional minutes and then cooled to room temperature. The resulting AuNP solution was stored in dark bottles at 4 °C	[23]
Antibody-AuNPs conjugates	Briefly, conjugation reactions were conducted by adding 5 μL of 10 mM anti-biotin antibody and 20 μL of 10 % BSA solution to 1 mL of AuNP solution (pH 9.0 with borate buffer solution) followed by incubation at room temperature with periodic gentle mixing for 1 h. Then, the solution was centrifuged at 13,000 × g for 15 min. Two phases can be obtained: a clear to pink supernatant of unbound antibodies and a dark red, loosely packed sediment of the anti-biotin AuNP conjugates. The resulting anti-biotin AuNP conjugates were stored at 4 °C before further use	[16]
DNA-AuNPs conjugates	First, dATP was added to 1 mL of concentrated AuNP solution (final concentration of dATP is 7.05 μM). The mixture was incubated at room temperature for 20 min. Then, 15 μL of 1 % SDS was slowly added to the mixture and incubated on a shaker for 10 min. Then, 50 μL of 2 M NaCl was dropped into the mixture at a rate of 2 μL/2 min. Then, 0.25 OD of thiolated DNA (probe 1) was added, and the mixture was incubated for 3 h in a water bath at 60 °C. The use of dATP was to protect AuNPs from salt-induced aggregation by mononucleotide (A) adsorption in the preparation of thiol-DNA/GNP conjugates. After the incubation, the mixture was centrifuged at 12,000 rpm for 15 min, and the supernatant was discarded. The mixture was then washed with 1 mL of PBS 3 times; the as-obtained ruby sediments were resuspended in 1 mL of eluent buffer (20 nM of $Na_3PO_4 \cdot 12H_2O$ containing 5 % BSA, 0.25 % Tween 20 and 10 % sucrose)	[29]

12.6.2 *Point-of-Care and Visual Detection of* P. aeruginosa *and Its Toxin Genes by Multiple LAMP and Lateral Flow Nucleic Acid Biosensor*

This study describes a simple and sensitive approach for visual and point-of-care detection of *Pseudomonas aeruginosa* (*P. aeruginosa*) and its toxin genes based on multiple loop-mediated isothermal amplification (mLAMP) and lateral flow nucleic acid biosensor (LFNAB) [53]. Differentiation of the internal standard gene *ecfX* and toxin genes (*ExoS* and *ExoU*) in *P. aeruginosa* was determined using FITC-, hex-, and digoxin-modified primers in the mLAMP process. In the presence of biotin- and

FITC- (hex-, digoxin-) modified primers and Bst DNA polymerase large fragments, the mLAMP produced numerous biotin- and FITC- (hex-, digoxin-) attached duplex DNA products. The products were detected by LFNAB through dual immunoreactions (anti-biotin antibodies on the gold nanoparticle (AuNP) and biotin on the duplex, anti-FITC (hex, digoxin) antibodies on the LFNAB test line and FITC (hex, digoxin) on the duplex). The accumulation of AuNPs produced a characteristic red band, enabling visual detection of *P. aeruginosa* and its toxin genes without instrumentation. After systematic optimization of LFNAB preparation and detecting conditions, the current approach was capable of detecting concentrations as low as 20 CFU/mL *P. aeruginosa* or its toxin genes within 50 min without complicated instrument, which is more sensitive than PCR. Therefore, this approach provides a simple, pollution-free, sensitive, and low-cost point-of-care test for the detection of *P. aeruginosa* and its toxin genes.

12.7 Conclusions and Prospects

In the last few years, LFNABs have been extensively utilized to generate POC diagnostics owing to their scale-up in production and low cost. LFNABs have the advantages of fast detection, no need for expert personnel, a simple test procedure, a low cost of operation, and a low sample volume requirement. With reciprocity between biomedical engineering and materials science, LFNABs are becoming more susceptive, simpler, and multipurpose. It is confirmed that theoretical simulations and optimization can be beneficial to the improvement of LFNABs.

Nevertheless, this technique still requires further enhancement. For instance, a decreased number of applications have been found in multiresidue analysis. As for future research, we can focus on the improvement of novel constructions with more sample capture on the test with the purpose of testing several analytes in complicated specimens. When binding paper diagnostics with mobile phone-based detection by the naked eye, telemedicine is extremely significant in enhancing medical care services in resource-limited environments. On-site detection often requires quantitative measurements of proteins, nucleic acids, and other biomarkers. However, visual colorimetric measurements by the naked eye are not sufficient for quantitative purposes due to the variation in visual perception of color among end users and under different lighting conditions. To achieve quantitative analysis in LFNABs, cameras or scanners are used to record the color intensity due to its relationship with the amount of analyses, which is the direction future research is going. Moreover, PNA has a high connection for its supplementary DNA or RNA molecules, primarily owing to the lack of electrostatic repulsion between the backbone of the natural nucleic acid and that of uncharged PNA [52]. PNA applications with LFNABs will be a primary breakthrough to enhance executive stability. Novel remarkable signal materials with high detection sensitivity are also highly desired in the future researches, such as nanozyme, which was reported in 2015. Colloidal gold was replaced by the Fe3O4 magnetic nanoparticles (MNPs) in their study to detect the glycoprotein of *Ebola virus* (EBOV) as low as 1 ng/mL, which is 100-fold more sensitive than the standard strip method [53]. Quality control by definition advances the performance properties (sensitivity, positive predictive value, negative predictive value, cross-reactivity, specificity, etc.) of a specific test. With internationalization of molecules, new generations of detective equipment have to be fast, precise, cost-effective, and easy to operate. The lateral flow test strip technique could offer a hopeful method for achieving this goal. Opportunities are also primarily connected with off-laboratory adhibitions. A well-known successful case is the pregnancy test. Currently, many companies make home testing available; these are nonprescription tests that can be obtained through the Internet or at a drugstore; nevertheless, there is still plenty of potential left for more advances. Fast detection results are extremely important in many situations, such as testing for drugs of abuse for law enforcement or for the existence of pathogenic organisms in food and testing in the emergency sickroom at a hospital or in the consulting room of an ordinary physician. Above all, LFNABs are an extraordinarily perfect candidate for fast, cost-effective, simple, and accurate analysis with excellent foreground.

Acknowledgments This work is supported by the Ministry of Science and Technology of Beijing (XX2014B069). Many thanks to Nan Cheng, for her kindly help in manuscript conception and preparation.

References

1. Free AH, Adams EC, Kercher ML, Free HM, Cook MH. Simple specific test for urine glucose. Clin Chem. 1957;3(3):163–8.
2. Parolo C, Merkoçi A. Paper-based nanobiosensors for diagnostics. Chem Soc Rev. 2013;42 (2):450–7.
3. Hu J, Wang S, Wang L, Li F, Pingguan-Murphy B, Lu TJ, Xu F. Advances in paper-based point-of-care diagnostics. Biosens Bioelectron. 2014;54:585–97.
4. Ngom B, Guo Y, Wang X, Bi D. Development and application of lateral flow test strip technology for detection of infectious agents and chemical contaminants: a review. Anal Bioanal Chem. 2010;397(3):1113–35.
5. Posthuma-Trumpie GA, Korf J, van Amerongen A. Lateral flow (immuno) assay: its strengths, weaknesses, opportunities and threats. A literature survey. Anal Bioanal Chem. 2009;393 (2):569–82.
6. Karakus C, Salih BA. Comparison of the lateral flow immunoassays (LFIA) for the diagnosis of Helicobacter pylori infection. J Immunol Methods. 2013;396(1):8–14.
7. Sajid M, Kawde A-N, Daud M. Designs, formats and applications of lateral flow assay: a literature review. J Saudi Chem Soc. 2015;19(6):689–705.
8. Sun J, Xianyu Y, Jiang X. Point-of-care biochemical assays using gold nanoparticle-implemented microfluidics. Chem Soc Rev. 2014;43(17):6239–53.
9. Erlich HA. Polymerase chain reaction. J Clin Immunol. 1989;9(6):437–47.
10. Aveyard J, Mehrabi M, Cossins A, Braven H, Wilson R. One step visual detection of PCR products with gold nanoparticles and a nucleic acid lateral flow (NALF) device. Chem Commun. 2007;41:4251–3.
11. Kim YT, Chen Y, Choi JY, Kim W-J, Dae H-M, Jung J, Seo TS. Integrated microdevice of reverse transcription-polymerase chain reaction with colorimetric immunochromatographic

detection for rapid gene expression analysis of influenza A H1N1 virus. Biosens Bioelectron. 2012;33(1):88–94.
12. Kim YT, Jung JH, Choi YK, Seo TS. A packaged paper fluidic-based microdevice for detecting gene expression of influenza A virus. Biosens Bioelectron. 2014;61:485–90.
13. Chen F, Ming X, Chen X, Gan M, Wang B, Xu F, Wei H. Immunochromatographic strip for rapid detection of Cronobacter in powdered infant formula in combination with silica-coated magnetic nanoparticles separation and 16S rRNA probe. Biosens Bioelectron. 2014;61:306–13.
14. Konstantou JK, Iliadi AC, Ioannou PC, Christopoulos TK, Anagnostopoulos NI, Kanavakis E, Traeger-Synodinos J. Visual screening for JAK2V617F mutation by a disposable dipstick. Anal Bioanal Chem. 2010;397(5):1911–6.
15. Nihonyanagi S, Kanoh Y, Okada K, Uozumi T, Kazuyama Y, Yamaguchi T, Nakazaki N, Sakurai K, Hirata Y, Munekata S. Clinical usefulness of multiplex PCR lateral flow in MRSA detection: a novel, rapid genetic testing method. Inflammation. 2012;35(3):927–34.
16. Soo P-C, Horng Y-T, Hsueh P-R, Shen B-J, Wang J-Y, Tu H-H, Wei J-R, Hsieh S-C, Huang C-C, Lai H-C. Direct and simultaneous identification of Mycobacterium tuberculosis complex (MTBC) and Mycobacterium tuberculosis (MTB) by rapid multiplex nested PCR-ICT assay. J Microbiol Methods. 2006;66(3):440–8.
17. Zaytseva NV, Montagna RA, Lee EM, Baeumner AJ. Multi-analyte single-membrane biosensor for the serotype-specific detection of Dengue virus. Anal Bioanal Chem. 2004;380 (1):46–53.
18. Goldmeyer J, Li H, McCormac M, Cook S, Stratton C, Lemieux B, Kong H, Tang W, Tang Y-W. Identification of Staphylococcus aureus and determination of methicillin resistance directly from positive blood cultures by isothermal amplification and a disposable detection device. J Clin Microbiol. 2008;46(4):1534–6.
19. Piepenburg O, Williams CH, Stemple DL, Armes NA. DNA detection using recombination proteins. PLoS Biol. 2006;4(7):e204.
20. Njiru ZK. Rapid and sensitive detection of human African trypanosomiasis by loop-mediated isothermal amplification combined with a lateral-flow dipstick. Diagn Microbiol Infect Dis. 2011;69(2):205–9.
21. Jung JH, Oh SJ, Kim YT, Kim SY, Kim W-J, Jung J, Seo TS. Combination of multiplex reverse-transcription loop-mediated isothermal amplification with an immunochromatographic strip for subtyping influenza A virus. Anal Chim Acta. 2015;853:541–7.
22. Fang R, Li X, Hu L, You Q, Li J, Wu J, Xu P, Zhong H, Luo Y, Mei J. Cross-priming amplification for rapid detection of Mycobacterium tuberculosis in sputum specimens. J Clin Microbiol. 2009;47(3):845–7.
23. He Y, Zeng K, Zhang S, Gurung AS, Baloda M, Zhang X, Liu G. Visual detection of gene mutations based on isothermal strand-displacement polymerase reaction and lateral flow strip. Biosens Bioelectron. 2012;31(1):310–5.
24. Liu J, Chen L, Lie P, Dun B, Zeng L. A universal biosensor for multiplex DNA detection based on hairpin probe assisted cascade signal amplification. Chem Commun. 2013;49(45):5165–7.
25. Fong WK, Modrusan Z, Mcnevin JP, Marostenmaki J, Zin B, Bekkaoui F. Rapid solid-phase immunoassay for detection of methicillin-resistant Staphylococcus aureus using cycling probe technology. J Clin Microbiol. 2000;38(7):2525–9.
26. Gao W, Li X, Zeng L, Peng T. Rapid isothermal detection assay: a probe amplification method for the detection of nucleic acids. Diagn Microbiol Infect Dis. 2008;60(2):133–41.
27. Mao X, Ma Y, Zhang A, Zhang L, Zeng L, Liu G. Disposable nucleic acid biosensors based on gold nanoparticle probes and lateral flow strip. Anal Chem. 2009;81(4):1660–8.
28. Liu C-C, Yeung C-Y, Chen P-H, Yeh M-K, Hou S-Y. Salmonella detection using 16S ribosomal DNA/RNA probe-gold nanoparticles and lateral flow immunoassay. Food Chem. 2013;141(3):2526–32.

29. Gao X, Xu H, Baloda M, Gurung AS, Xu L-P, Wang T, Zhang X, Liu G. Visual detection of microRNA with lateral flow nucleic acid biosensor. Biosens Bioelectron. 2014;54:578–84.
30. Yager P, Domingo GJ, Gerdes J. Point-of-care diagnostics for global health. Annu Rev Biomed Eng. 2008;10:107–44.
31. Glynou K, Ioannou PC, Christopoulos TK, Syriopoulou V. Oligonucleotide-functionalized gold nanoparticles as probes in a dry-reagent strip biosensor for DNA analysis by hybridization. Anal Chem. 2003;75(16):4155–60.
32. Sapountzi EA, Tragoulias SS, Kalogianni DP, Ioannou PC, Christopoulos TK. Lateral flow devices for nucleic acid analysis exploiting quantum dots as reporters. Anal Chim Acta. 2015;864:48–54.
33. Corstjens P, Zuiderwijk M, Brink A, Li S, Feindt H, Niedbala RS, Tanke H. Use of up-converting phosphor reporters in lateral-flow assays to detect specific nucleic acid sequences: a rapid, sensitive DNA test to identify human papillomavirus type 16 infection. Clin Chem. 2001;47(10):1885–93.
34. Noguera P, Posthuma-Trumpie G, Van Tuil M, Van der Wal F, De Boer A, Moers A, Van Amerongen A. Carbon nanoparticles in lateral flow methods to detect genes encoding virulence factors of Shiga toxin-producing Escherichia coli. Anal Bioanal Chem. 2011;399 (2):831–8.
35. Mens PF, van Amerongen A, Sawa P, Kager PA, Schallig HD. Molecular diagnosis of malaria in the field: development of a novel 1-step nucleic acid lateral flow immunoassay for the detection of all 4 human Plasmodium spp. and its evaluation in Mbita, Kenya. Diagn Microbiol Infect Dis. 2008;61(4):421–7.
36. Blažková M, Koets M, Rauch P, van Amerongen A. Development of a nucleic acid lateral flow immunoassay for simultaneous detection of Listeria spp. and Listeria monocytogenes in food. Eur Food Res Technol. 2009;229(6):867–74.
37. Chua A, Yean CY, Ravichandran M, Lim B, Lalitha P. A rapid DNA biosensor for the molecular diagnosis of infectious disease. Biosens Bioelectron. 2011;26(9):3825–31.
38. Rohrman BA, Leautaud V, Molyneux E, Richards-Kortum RR. A lateral flow assay for quantitative detection of amplified HIV-1 RNA. PLoS One. 2012;7(9):e45611.
39. Hou S-Y, Hsiao Y-L, Lin M-S, Yen C-C, Chang C-S. MicroRNA detection using lateral flow nucleic acid strips with gold nanoparticles. Talanta. 2012;99:375–9.
40. Famulok M, Hartig JS, Mayer G. Functional aptamers and aptazymes in biotechnology, diagnostics, and therapy. Chem Rev. 2007;107(9):3715–43.
41. Chen A, Yang S. Replacing antibodies with aptamers in lateral flow immunoassay. Biosens Bioelectron. 2015;71:230–42.
42. Xu H, Mao X, Zeng Q, Wang S, Kawde A-N, Liu G. Aptamer-functionalized gold nanoparticles as probes in a dry-reagent strip biosensor for protein analysis. Anal Chem. 2008;81(2):669–75.
43. Wang L, Ma W, Chen W, Liu L, Ma W, Zhu Y, Xu L, Kuang H, Xu C. An aptamer-based chromatographic strip assay for sensitive toxin semi-quantitative detection. Biosens Bioelectron. 2011;26(6):3059–62.
44. Liu G, Mao X, Phillips JA, Xu H, Tan W, Zeng L. Aptamer– nanoparticle strip biosensor for sensitive detection of cancer cells. Anal Chem. 2009;81(24):10013–8.
45. Lu Y, Liu J. Functional DNA nanotechnology: emerging applications of DNAzymes and aptamers. Curr Opin Biotechnol. 2006;17(6):580–8.
46. Fang Z, Huang J, Lie P, Xiao Z, Ouyang C, Wu Q, Wu Y, Liu G, Zeng L. Lateral flow nucleic acid biosensor for Cu2+ detection in aqueous solution with high sensitivity and selectivity. Chem Commun. 2010;46(47):9043–5.
47. Mazumdar D, Liu J, Lu G, Zhou J, Lu Y. Easy-to-use dipstick tests for detection of lead in paints using non-cross-linked gold nanoparticle–DNAzyme conjugates. Chem Commun. 2010;46(9):1416–8.
48. Johannsen S, Megger N, Böhme D, Sigel RK, Müller J. Solution structure of a DNA double helix with consecutive metal-mediated base pairs. Nat Chem. 2010;2(3):229–34.

49. Liu J, Chen L, Chen J, Ge C, Fang Z, Wang L, Xing X, Zeng L. An autonomous T-rich DNA machine based lateral flow biosensor for amplified visual detection of mercury ions. Anal Methods. 2014;6(7):2024–7.
50. Quesada-González D, Merkoçi A. Nanoparticle-based lateral flow biosensors. Biosens Bioelectron. 2015;73:47–63.
51. Huang X, Aguilar ZP, Xu H, Lai W, Xiong Y. Membrane-based lateral flow immunochromatographic strip with nanoparticles as reporters for detection: a review. Biosens Bioelectron. 2016;75:166–80.
52. Briones C, Moreno M. Applications of peptide nucleic acids (PNAs) and locked nucleic acids (LNAs) in biosensor development. Anal Bioanal Chem. 2012;402(10):3071–89.
53. Duan D, Fan K, Zhang D, Tan S, Liang M, Liu Y, Zhang J, Zhang P, Liu W, Qiu X. Nanozyme-strip for rapid local diagnosis of Ebola. Biosens Bioelectron. 2015;74:134–41.

Chapter 13
Nucleic Acid Biosensors for Food Safety

Abstract Biosensors that are based on nucleic acids as recognition elements or as target analytes are called nucleic acid-based biosensors (NABs). NABs mainly use deoxyribonucleic acid (DNA), ribonucleic acid (RNA), aptamers (both DNA and RNA), and peptide nucleic acid (PNA) as oligonucleotide probes. Different methodologies can be used to construct a variety of NABs, including electrochemical, optical, and mass-sensitive biosensors. Over the last few decades, nucleic acid-based analysis has gained much interest for various detection and monitoring food safety applications. Nucleic acid-based biosensors (NABs) are preferred over other sensor types due to their high sensitivity and selectivity. Moreover, among nucleic acids, aptamers represent new, promising recognition elements for biosensor development.

Keywords Functional nucleic acid • Biosensor • Optical • Electrochemistry • Amplification • Food safety

13.1 Current State of Nucleic Acid-Based Biosensors (NABs) in Food Safety

Sensors are devices that respond to physical or chemical stimuli and produce detectable signals. Currently, they are a critical extension of human perception of the world in many aspects; this is largely because humans are much more sensitive to the physical environment (e.g., light, pressure, temperature, or humidity) than to the chemical or biological environment. The development of highly sensitive and selective sensors to recognize important analytes has long been a research focus in many areas, including environmental monitoring, medical diagnostics, food safety, and industrial quality control.

A sensor includes both signal transduction and target recognition components. Signal transduction elements are responsible for converting molecular recognition events into physically detectable signals, such as changes in fluorescence, color, electrochemistry, or magnetic resonance images. Target recognition elements can be any chemical or biological entities, such as small organic molecules, peptides, proteins, nucleic acids, carbohydrates, or even whole cells. Different structural DNA motifs are shown in Fig. 13.1. Ideally, these elements should have high

W. Xu, *Functional Nucleic Acids Detection in Food Safety*,
DOI 10.1007/978-981-10-1618-9_13

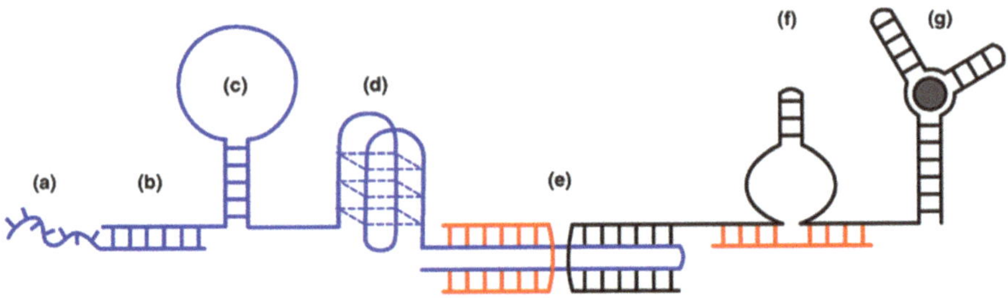

Fig. 13.1 Different structural DNA motifs: (*a*) single-stranded sticky end, (*b*) duplex, (*c*) hairpin, (*d*) quadruplex, (*e*) crossover, (*f*) DNAzyme, and (*g*) aptamer–substrate complex

specificity (low interference), high affinity (low detection limit), wide dynamic range, long shelf life, fast response time, and good generality for detecting a broad range of analytes with the same recognition element class. The antibodies of protein-based binding molecules have long been used for target recognition because they meet most of the above criteria.

Biosensors are analytical tools that have been applied in the fields of environmental monitoring, disease monitoring, food safety monitoring, and medical diagnostics. Over the last few decades, nucleic acid-based determinations have acquired much focus on different monitoring and analyzing food safety inhibition. On account of their selectivity and high sensitivity, nucleic acid-based biosensors (NABs) precede other sensor types. The application of nucleic acids (NAs) as biocomponents in biosensors offers a multitude of advantages over classical affinity sensing methods mainly based on antibodies; such possible advantages include easy regeneration of immobilized FNA function, homogeneous preparation, and utilization of different detection methods enabled by easy labeling. In the following sections, different biosensors that have been realized using FNAs as biorecognition elements are discussed, as well as different transducers that have been used for biosensor development. Novel synthetic probes (e.g., PNAs, aptamers) have been utilized to make point-of-care NABs. Because of low costing and flexible fabrication, they are evolving as fast and credible tools which can be applied to monitor food safety.

NABs mainly use deoxyribonucleic acid (DNA), ribonucleic acid (RNA), aptamers (both DNA and RNA), and peptide nucleic acid (PNA) as oligonucleotide probes. According to Chargaff's rules of base pairing (i.e., for DNA, A to T and G to C), the basal principle behind NABs relies on sequence complementarity, apart from the case of aptamers. The principle of aptamer-based determination resembles more to that of receptor–ligand interactions or antigen–antibody.

Some DNA-based biosensors that can analyze the pathogenic microorganisms, infectious agents such as genetic polymorphisms, viruses, and point mutations (SNPs) have entered the bazaar currently [1]. The ability to amplify desired target DNA from a host pathogen using PCR and consequently augment the signal generated by the biosensor is a major advantage of DNA-based probes. DNA forms complex biological recognition layers more simple than other sensing

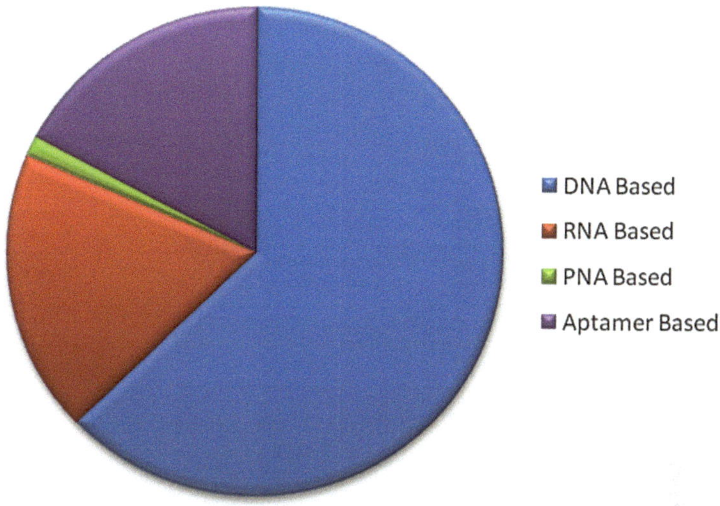

Fig. 13.2 Proportion of publications related to different NABs for clinical applications over the last 5 years (Reproduced from Bora et al. 2013 [3] with permission)

molecules, such as antibodies or enzymes, and these layers are highly stable and reusable after by briefly thermally melting the DNA duplexes [2]. DNA precedes other nucleic acid-based probes, which are used as a sensing tool (Fig. 13.2).

More progress is expected in the evolution of highly accurate and susceptive NABs (nucleic acid-based biosensors) that can simultaneously determinate multiple analytes, thereby decreasing assay time and cost.

Hybridization-based biosensors depend on nucleic acid duplex construction. Hybridization generally occurs between a known DNA sequence, i.e., target DNA, and an unknown counterpart, i.e., probe DNA, while RNA–RNA and DNA–RNA hybridizations can also happen [3]. The basic NABs' pattern is described in Fig. 13.3.

Several chemical approach, PCR included, are effective for the compound of DNA probes. However, for RNA, a probe can be gained by the reverse transcription (RT) of a previously isolated or figured messenger RNA (mRNA) or by forecasting its nucleotide sequence applying the amino acid sequence of the protein which is expressed by that DNA. Although several strategies has been affirmed for this kind of probe, its application is restricted by codon degeneracy. Conventional nucleic acid hybridization approaches include gel electrophoresis as well as Southern and Northern blotting, which are generally labor-intensive and lengthy in comparison with hybridization procedure that takes place immediately on the surface of a physical biosensor.

NABs are physicochemical transducer [4] which is constructed by immobilizing nucleic acids (i.e., DNA, RNA, PNA, oligonucleotides) to a carrier via adsorption, covalent bonding, or ionic interaction. A variety of immobilization methods are usable, such as magnetic particles, adsorption to inert carriers or chemical cross-linking to macroscopic beads, microencapsulation in nanospheres, and physical

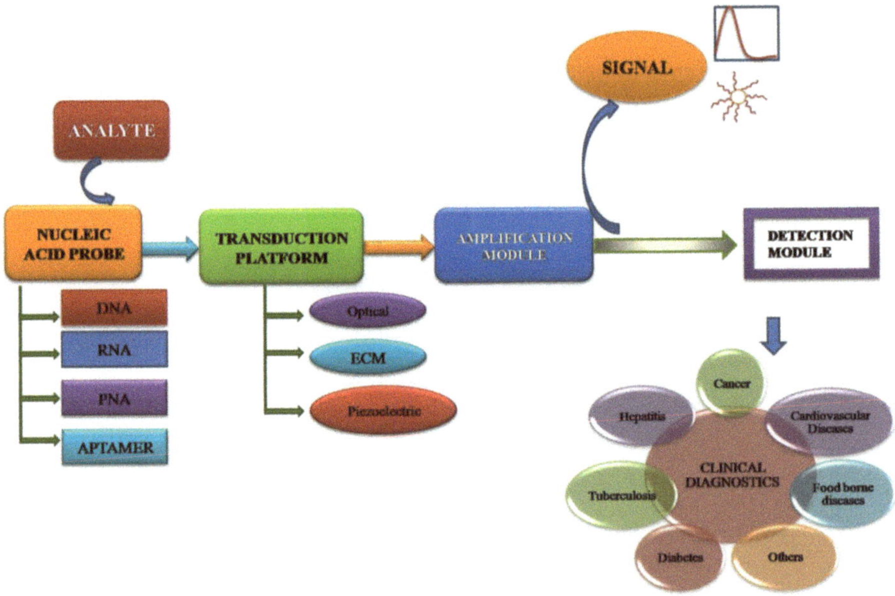

Fig. 13.3 Schematic representation of NABs (Reproduced from Bora et al. 2013 [3] with permission)

entrapment in gel-like matrix. The immobilization tenor also promotes the orientation and prepares accessibility of the probes to the target elements [5].

The transducer is a key biosensor component which transforms biochemical responses into output signals that are very easy to measure. NABs can be assembled with multifarious types of transducers to generate systems upon optics, electrochemistry, piezoelectricity, and microfluidics.

13.2 Categories of Nucleic Acid-Based Biosensors Based on Signal Transduction

13.2.1 Nucleic Acid and Biosensor Signal Transduction

13.2.1.1 Nucleic Acids (NAs)

Information encoded in the DNA base sequence provides a description of the structure and functional properties of this biopolymer. The structural information of this biopolymer includes the formation of duplexes, G-quadruplexes, supramolecular crossover tiles, base–metal ion complexes, and i-motifs [6]. The functional information encoded in DNA is reflected by specific binding (aptamers) or catalytic properties (DNAzymes). Recently, advances in DNA-based biosensors and amplified biosensing achieved by tailoring DNA structures have been reported.

A DNAzyme with Peroxidase Activity for Colorimetry

DNAzymes are oligonucleotides, and similar to protein enzymes, some DNAzymes can oxidize or reduce substrates by recruiting hemin as a cofactor with peroxidase activity [7]. Sen and his coworkers performed an aptamer selection against N-methyl mesoporphyrin IX (NMM) and separated many guanine-rich aptamers [8]. Then, it was found that these aptamers could be used as DNAzymes to catalyze porphyrin metalation [9]. Further characterization revealed that the G-rich sequence/hemin complex exhibited peroxidase activity [10]. The peroxidase activity of the DNAzyme was 250-fold greater than that of free hemin [7]. As a result, common peroxidase substrates could be used to observe the activity of the DNAzyme. For instance, in the presence of H_2O_2, this enzyme can catalyze the conversion of luminol or 2,2′-azino-bis(3-ethylbenzothiazoline-6-sulfonic acid) (ABTS), resulting in the emission of chemiluminescence or green-colored radical ions ($ABTS^{\bullet +}$, $\varepsilon = 3.6 \times 10^4\ M^{-1}\ cm^{-1}$) (Fig. 13.4) [7, 11].

Metal-Specific Nucleic Acid Enzymes (NAEs) for Cleavage

There are also other species of nucleic acid enzymes (NAEs), which are isolated through in vitro selection [12–14]. Most of these NAEs, including uranium, Pb^{2+}, Cu^{2+}, and Zn^{2+} NAEs [15, 16], have demonstrated a high specificity for transition metal ions. These NAEs can have DNA or RNA catalytic motifs and are termed DNA-cleaving enzymes (DNAzymes) or RNA-cleaving enzymes (RNAzymes), respectively. These NAEs commonly have two strands. In contrast with DNAzymes, RNAzymes may have special action sites. As shown in Fig. 13.5, one strand is the catalyst strand (green), and the other is the substrate strand (black);

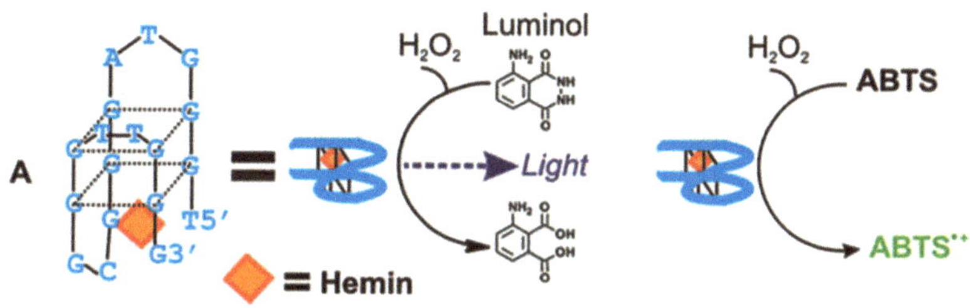

Fig. 13.4 Sensing with a peroxidase DNAzyme for signal transformation. (**a**) Proposed secondary structure of the 18-nucleotide peroxidase DNAzyme (*left*); the DNAzyme can catalyze the conversion of luminol to generate chemiluminescence (*middle*) and the conversion of ABTS to generate color (*right*). DNA detection is based on inhibited DNAzyme activity (Reprinted with the permission from Ref. [7]. Copyright 2009 American Chemical Society)

Fig. 13.5 The general reaction scheme of a substrate-cleaving NAE (Reprinted with the permission from Ref. [7]. Copyright 2009 American Chemical Society)

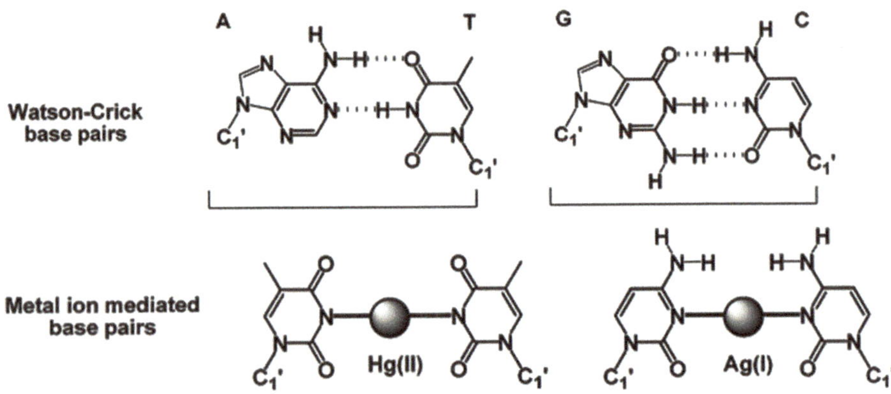

Fig. 13.6 Metal ion-mediated base pair (Reproduced from Ref. [17] by permission of The Royal Society of Chemistry)

a special RNAzyme cleavage site is shown in red. At low metal ion (e.g., M^{n+}) concentrations (i.e., levels ranging from nanomolar to low micromolar), the catalyst would cleave the substrate into two strands [7, 17].

Pyrimidine Base Pairs in DNA Duplexes for Binding Metal Ions

It has been reported that pyrimidine cytosine–cytosine (C-C) and thymine–thymine (T-T) base pairs selectively capture Ag (I) and Hg (II) ions, respectively; the metallo-base pairs, C-Ag (I)-C and T-Hg (II)-T, form DNA duplexes. In these metallo-base pairs, the metal–base bonds replace the hydrogen bonds of the Watson–Crick (W-C)-type base pairs found in natural DNA. The formation of a DNA triplex containing a C-Ag (I)-G-C triad as well as RNA duplexes containing U-Hg (II)-U and C-Ag (I)-C triads has also been reported. Metal ion binding motifs can be synthesized from commercially available DNA fragments and can be readily used in many laboratory procedures. The methods used for the detection of metal ion–duplex interactions and structures, as well as the thermodynamic properties of the metallo-base pairs, are summarized. The oligonucleotides included in this manuscript are commercially available and can be used in many biosensors (Fig. 13.6) [17].

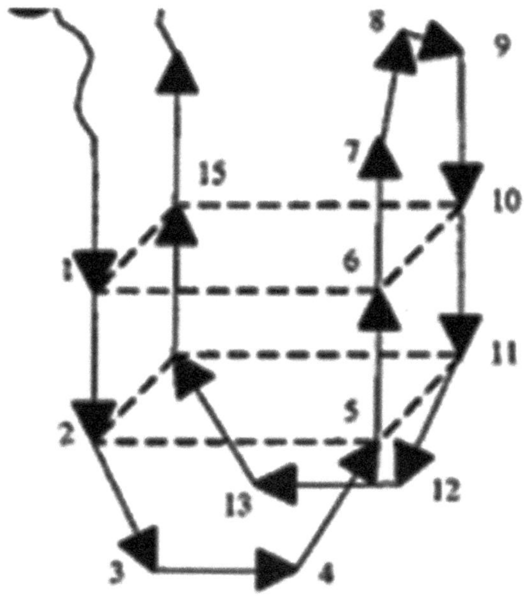

Fig. 13.7 Schematic of the intramolecular G-quadruplex structure formed by TBA (Reprinted from Ref. [18], Copyright 2002, with permission from Elsevier)

Aptamers

Aptamers are artificial, short, single-stranded DNA or RNA ligands. Unlike traditional nucleic acids, aptamers exhibit unique target binding characteristics. They are selected from a combinatorial random nucleic acid library of 10^{13}–10^{16} sequences using an in vitro selection process known as systematic evolution of ligands by exponential enrichment (SELEX), which involves iterative binding, separation, and amplification. Many high-affinity and high-specificity aptamers have been selected for their interactions with multiple types of target molecules, including small organics, peptides, and proteins, as well as supramolecular complexes, such as viruses or cells. Due to the inherent affinity of aptamers, they are capable of detecting various protein targets at concentrations ranging from the picomolar scale (1×10^{-12} M) to the high-nanomolar scale (1×10^{-7} M). Most aptamers form a G-quadruplex structure upon binding with a target (Fig. 13.7).

Triplex DNA Structures

In 1957, Felsenfeld et al. first observed the formation of a three-stranded, or triple-helix, nucleic acid structure [19]. Triplex DNA structures have demonstrated stable and specific binding of polyuridine/polyadenosine duplexes to single-stranded polyuridine oligonucleotides. When a DNA or RNA oligonucleotide binds to a homopurine region of DNA, specifically in the major DNA groove, Hoogsteen or reverse Hoogsteen hydrogen bonds are created with bases in the purine-rich strand

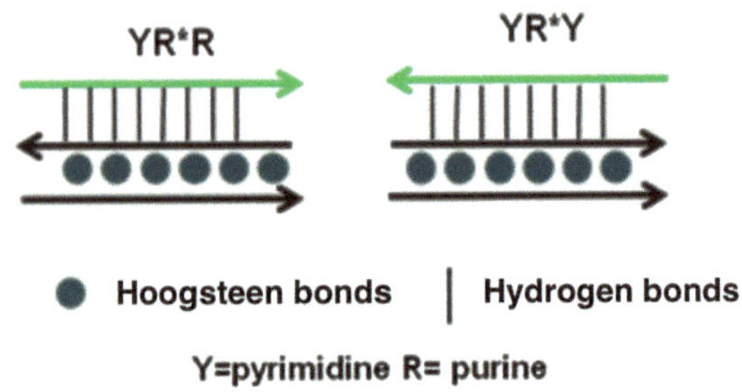

Fig. 13.8 Scheme of different types of triplex DNA structures

(Fig. 13.8), forming a triplex structure. According to its base composition and binding orientation relative to its DNA target site, a TFO can be categorized as having either a pyrimidine or purine motif. Triplex DNA has been introduced into and is currently being used in several analysis fields.

13.2.1.2 Signal Transduction Elements

Fluorescence Based on Fluorescence Resonance Energy Transfer (FRET)

FRET is a physical process of energy transfer from an excited donor chromophore (DC) to an acceptor chromophore (AC) through intermolecular, long-range dipole–dipole coupling. The necessary distance for effective transfer ranges from 10 to 100 A. Effective FRET requires an adequate overlap of the fluorescence spectrum of the DC and the absorbance spectrum of the AC [20]. Ultimately, the quantum yield of the DC (Φ0) and the absorption coefficient of the AC (A) can be very high (e.g., $\Phi\,0 \geq 0.1$ and $A \geq 1{,}000$). Moreover, to achieve the dipole–dipole interaction, the transition dipoles of the DC and AC must have favorable orientations relative to each other, or one (or both) must have a certain degree or extent of fast rotational freedom; this latter principle generally applies to chromophores in conjunction with biomolecules in solution. This type of energy transfer is non-radiative, and no photons are transferred from the DC to the AC [21].

Gold Nanoparticles

Gold nanoparticles with localized surface plasmon resonance (LSPR) are special nanomaterials with very high extinction coefficients (e.g., $2.7 \times 10^8\ \mathrm{M}^{-1}\ \mathrm{cm}^{-1}$) that can be 1000 times higher than those of conventional organic chromophores [22]. Because of the distance-dependent surface plasmon properties and combination DNA programmable nature of GNPs, they have been widely used in the design of biosensor interfaces and signal transduction events [23]. Functional DNA that can change conformation or perform catalytic reactions in the presence of specific targets, such as inorganic/organic molecules or biomolecules or even bacteria, cancer cells, or viruses, has also been applied for GNP distance control. The excellent optical properties of GNPs and the programmable nature of DNA are great advantages in the construction of optical biosensors, which offer a convenient and attractive analysis method as the results can be easily monitored by the naked eye without the use of advanced instruments [24]. The high surface area-to-volume ratio, catalytic properties, conductivity, and biocompatibility of GNPs have also led to great advancements in the design of piezoelectric and electrochemical biosensors; signals arising from nanomolar concentrations of nanoparticles can be measured, allowing sensitive detection of trace analytes [24].

13.2.2 Optical Nucleic Acid-Based Biosensors (ONABs)

13.2.2.1 Fluorescence NABs

Fluorescent detection is widely employed because aptamers can easily be labeled with fluorescent dyes; thus, the availability of many different fluorophores and quenchers and the inherent capability for real-time detection have increased. To convert aptamers into fluorescent signaling probes, two strategies have been developed, one of which is often referred to as dual labeling of signaling aptamers (Fig. 13.9). This strategy is based on fluorescence resonance energy transfer (FRET) and optical signal transduction relying on more than one reporter. In addition, dual-labeled signaling aptamers can be quenched. More design strategies are shown in Fig. 13.9.

These signaling aptamers are modified versions of traditional molecular beacons, which are called aptamer-based molecular beacons (aptabeacons). One format consists of an aptamer sequence that is designed into a hairpin structure and end-labeled with a fluorophore and a quencher. Target binding damages the stem of

Fig. 13.9 Signal-on fluorescence NABs. (**a**) An aptamer sequence in a molecular beacon-like hairpin structure is end-labeled with a fluorophore (*F*) and a quencher (*Q*). Target binding disrupts the stem, separating the *F* from the *Q* and leading to an increase in fluorescence. (**b**) A fluorophore-labeled aptamer with a quencher-labeled complementary DNA sequence in a duplex structure is separated after a binding event, leading to fluorescent signals. (**c**) The fluorescence emission of an aptamer labeled with one pyrene monomer at either end is triggered upon target binding (Reprinted from Ref. [28], Copyright 2008, with permission from Elsevier)

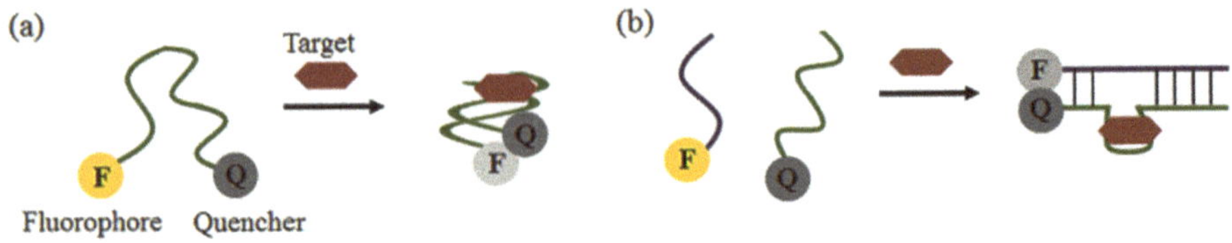

Fig. 13.10 Schemes for signal-off fluorescence NABs that incorporate two fluorophores. (**a**) Aptamer beacons or aptabeacons. An aptamer (*blue line*) is designed into a hairpin structure that is labeled with a fluorophore (*F*) and a quencher (*Q*), one at each end. When the target is added, the binding event leads to a fluorescence signal through Watson–Crick base pairs. (**b**) Structure-switching signaling aptamers. A fluorophore-labeled aptamer forms a duplex structure with an antisense DNA strand labeled with a quencher. Target binding changes the structure from the DNA duplex to the aptamer–target complex, which is accompanied by a decrease in fluorescence signal (Reproduced from Song et al. [25] with permission)

the hairpin, separating the fluorophore from the quencher and leading to a fluorescence signal (Fig. 13.10a). Another adopted format is that of an aptamer with a fluorophore and quencher arranged head-to-head, forming a modified duplex structure with a complementary DNA sequence; target binding dissociates the complementary strand from the aptamer and leads to an increase in fluorescence (Fig. 13.10b). The above fluorescence-based assay methods have enabled only qualitative or semiquantitative bioassays. However, quantum dots (QDs) have been employed in place of organic fluorescent dyes to achieve improved assay performance.

Fluorescence NABs can also be produced by introducing organic fluorophores into conformationally labile regions of aptamers, whereby binding events can be transduced into chemical environment changes for the fluorophores, which can generate changes in fluorescence characteristics, including intensity and anisotropy. For example, signaling aptamers have been created by labeling anti-adenosine DNA aptamers with a fluorophore adjacent to the adenosine binding site. The apparent Kd based on the dose-responsive change in fluorescence intensity was much higher than that of the parental aptamers (i.e., fluorophore-free aptamers). Loss of affinity can be considered a general disadvantage of aptamers that rely upon conformational transduction for signaling.

Aptamers can be labeled in numerous ways, and the nature of the label as well as the site of introduction tends to influence signal transduction. For example, a boron–dipyrromethene fluorophore can be linked to three different aptamers (e.g., anti-AMP, anti-tyrosinamide, and anti-argininamide) through the ribose 2′-hydroxyl group, which converts them into signaling aptamers. The boron–dipyrromethene fluorophore is very sensitive to the local environment, and a change in the local environment results in a significant change in its fluorescence quantum yield. For example, an anti-argininamide aptamer has been shown to generate a 3.7-fold increase in fluorescence via this conversion.

Apart from internally or externally modifying aptamers with fluorescent dyes, signal transduction can also be achieved by replacing nonfluorescent nucleotides with fluorescent nucleotide analogs. The quantum yield of nucleotide analogs is highly dependent upon the local environment, especially the extent of base stacking. Thus, the modified nucleotides are most suitable for the study of nucleic acid structure. However, it may be difficult to identify the aptamer site to label. Some selection experiments have described directly inserting labels into the original random sequence pool to select the binding sequences.

13.2.2.2 Colorimetric NAB

AuNP-Based Colorimetric NABs

The intersection between nanomaterial science and biology may provide a wide array of platforms with which to promote basic biomaterial research as well as practical bionanotechnology applications. The interaction between gold nanoparticles (AuNPs) and DNA is a good example of such a platform. Because AuNPs have distance-dependent surface plasmon properties and large extinction coefficients, strong color changes will be generated, and the optical properties may rival or even exceed those of most organic dyes. While difficult, it is very important to control the distance between AuNPs to produce color changes precisely. DNA allows for such precise control via its programmable nature.

With the development of nanomaterials (e.g., AuNPs), colorimetric NABs based on wavelength shifts have been designed for simple visual target detection. Two types of AuNP-based colorimetric NABs have been introduced.

One type uses unmodified AuNPs to differentiate unfolded and folded DNA. This assay format provides a new way to monitor aptamer-based biorecognition processes colorimetrically. In solution, unbound aptamers form random coil-like structures, which can absorb onto the surface of the AuNPs and protect them from salt-induced aggregation. However, bound aptamers form 3D structures and can no longer absorb onto the surface of the AuNPs to prevent them from aggregating. As the wavelengths of aggregated and dissociated AuNPs in solution are different, the shift from unfolded aptamers to folded aptamers will produce color change signals visible to the naked eye. This assay format has been applied successfully in the detection of various targets, including potassium ions [26] (Fig. 13.11) and proteins.

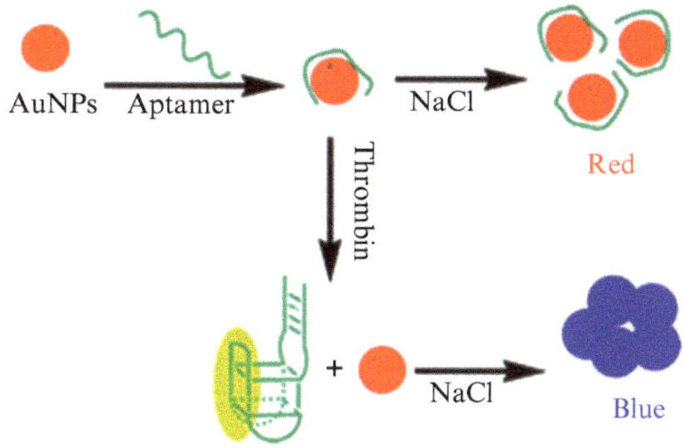

Fig. 13.11 Unmodified gold nanoparticle biosensors (Reproduced from Ref. [26] by permission of The Royal Society of Chemistry)

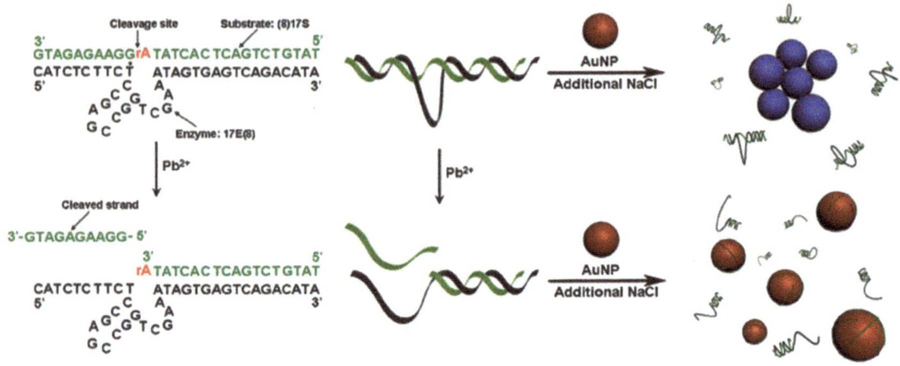

Fig. 13.12 DNA aptamer can tightly bind to AuNPs and stabilize them against salt-induced aggregation. Upon binding to its target, the aptamer forms a 3D structure and leads to AuNP aggregation. (B) *Left*: Secondary structure of the DNAzyme complex, which consists of an enzyme strand (17E(8)) and a substrate strand ((8)17S). After lead-induced cleavage, 10-mer ssDNA strands are released and can absorb onto the AuNP surface. *Right*: Schematic of the label-free colorimetric sensor. The lead-treated and untreated complexes and NaCl are mixed with the AuNPs. The AuNPs aggregate in the absence of lead but remain dispersed in the presence of lead (Reproduced from Ref. [27] by permission of John Wiley & Sons Ltd)

This assay strategy has also been applied successfully in the detection of lead ions (Fig. 13.12) with the use of DNAzymes. DNAzymes are also known as catalytic DNA or DNA molecules with high metal ion specificities; thus, DNAzymes can be used for developing metal ion sensors. In 2004, Rothberg et al. [27] found that ssDNA and double-stranded DNA (dsDNA) exhibited different adsorption properties onto AuNPs. Because ssDNA is flexible and can partially coil in solution, it can tightly bind to AuNPs and enhance the electrostatic repulsion between AuNPs with adsorbed ssDNA, which confers high salt tolerance to the AuNPs. In contrast, as dsDNA is stiffer and negatively charged, there is strong repulsion force between the negatively charged AuNPs and dsDNA that largely prevents binding. This interaction confers low salt tolerance and cannot prevent salt-induced AuNP aggregation. Based on this phenomenon, the Wang group developed an optical biosensor using a DNAzyme for lead ions [27].

The other types of AuNP-based colorimetric NABs use aptamer-modified AuNP probes. In this strategy, the ssDNA complementary to the aptamers is first covalently attached onto the surface of the AuNPs, functionalizing them. Then, the aptamers act as linkers to assemble the AuNPs. In the presence of target, the aptamers capture the target and change conformation, leading to dissociation of the AuNPs (Fig. 13.13a, b). As a result, blue-colored AuNP aggregates disperse into red-colored individual AuNPs. This signal-on assay format can be applied to analyze various molecules by replacing the aptamer sequences and has been successfully used to detect cocaine and adenosine in serum. Additionally, another study has reported the self-assembly of AuNPs with aptamers hybridized with short complementary ssDNA; this strategy provided a guaranteed repulsion force

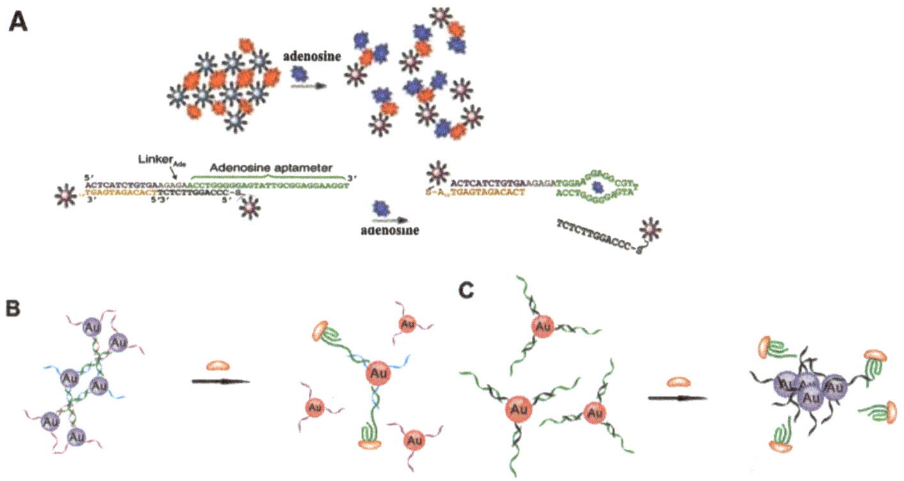

Fig. 13.13 (**a**) A signal-on colorimetric AuNP-based NAB [29]. Adenosine-induced disassembly of nanoparticle aggregates for the colorimetric detection of adenosine. Nanoparticles 1 and 2 are functionalized with two different DNA molecules through thiol–gold chemistry. The two types of nanoparticles are linked by LinkerAde to form aggregates. In the presence of adenosine, the aggregates disassemble to yield dispersed red nanoparticles. (**b**) A signal-on colorimetric AuNP-based NAB. In the presence of the target molecules, purple-colored, DNA-modified AuNP aggregates are dispersed into red-colored individual AuNPs (Reprinted from Ref. [28], Copyright 2008, with permission from Elsevier). (**c**) A signal-off colorimetric AuNP-based NAB. Upon binding the target, AuNPs functionalized with short, complementary ssDNA and aptamers become unstable against salt-induced aggregation, leading to a red-to-purple color change (Reprinted from Ref. [28], Copyright 2008, with permission from Elsevier)

between the hybrid-functionalized AuNPs, which were well dispersed in solution and red. In the presence of the target, the aptamer binds the target and undergoes a conformational change that results in the aptamers dissociating from the AuNPs. This unhybridization renders the AuNPs unstable at the same salt concentration, and they aggregate immediately. As a result, a rapid red-to-purple color change is generated (Fig. 13.13c) [28].

The methods described above use intact aptamer-modified AuNPs for detecting targets with multiple binding sites, such as proteins and whole cells through a basic "sandwich" binding format. In fact aptamers cannot form the necessary sandwich interaction with small molecules that have just one binding site. Split aptamers are capable of this task, as they are comprised of two DNA strands that only bind to one another in the presence of a specific small-molecule target.

The Sheng group [30] has developed double-functionalized AuNPs with split aptamers for the detection of adenosine triphosphate using the sandwich format. Of all the strategies for target-directed assembly or disassembly of AuNPs, the traditional design of DNA–AuNP conjugates yields AuNPs at the two ends of the folded structures, similar to Assembly Type 1, shown in Fig. 13.14a. Clearly, Assembly Type 2 (Fig. 13.14a) may produce more color change because the distance between

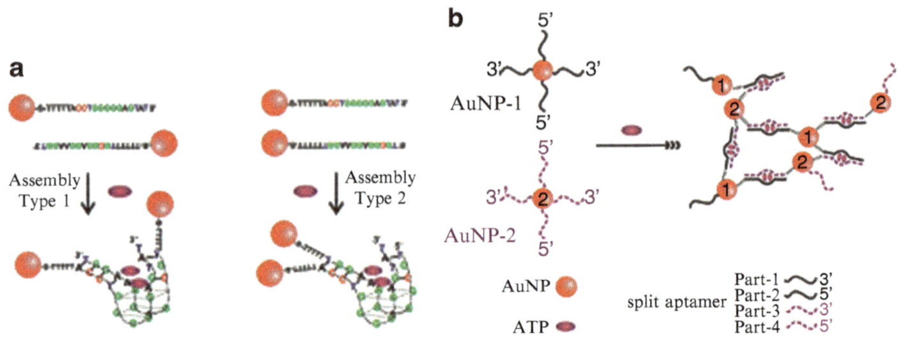

Fig. 13.14 (**a**) Two assembly types of AuNPs functionalized with different split ATP aptamers. (**b**) Sensing scheme of the double-functionalized AuNP-split ATP aptamer system (Reprinted from Ref. [30], Copyright 2013, with permission from Elsevier)

the AuNPs is much smaller than that in Type 2 after the assembly binds the target. However, Sheng et al. found that the assembly rate of Type 2 was not sufficient for sensing applications. Hence, they designed a new functionalization strategy by combining Assembly Types 1 and 2, yielding a double-functionalized AuNP–DNA system for the detection of adenosine triphosphate (ATP). As shown in Fig. 13.14b, each split aptamer fragment binds to the same AuNPs at the 5′ end and 3′end. The results showed that this double-functionalized AuNP–DNA system can produce more color change than the traditional system.

Peroxidase-Like DNAzyme-Based Colorimetric NABs

DNAzymes are single-stranded DNA molecules that exhibit catalytic activity. The peroxidase-like DNAzymes have recently attracted great interest and have great potential in the production of signal in bioanalytical chemistry. To ensure catalytic activity, oligonucleotides need to bind a hemin molecule to adopt a G-quadruplex structure. This principle facilitates a redox reaction between the hydrogen peroxide and reduction substrate, which results in a green–blue oxidized product.

DNAzymes with peroxidase-like activity have been successfully applied into bioanalytical assay and include the detection of specific DNA sequences via colorimetric methods. DNAzymes have also been applied in telomerase or methyltransferase activity assays, which are potential targets of anticancer therapy. Other applications include the determination of metal cations, such as Ag^{+}, Hg^{2+}, Pb^{2+}, or Cu^{2+}, through metal ion-based mismatch and RCA-based amplified detection of small molecules, such as adenosine or cocaine. In the last decade, DNAzymes have been widely used in many areas of science, including chemistry and biology.

The Yang group [31] has designed hairpin-based bioassays that include G-rich oligonucleotides, the aptamer for OTA, and a blocking tail. The stem regions include a part of the aptamer for OTA and a part of the DNAzyme. The blocking

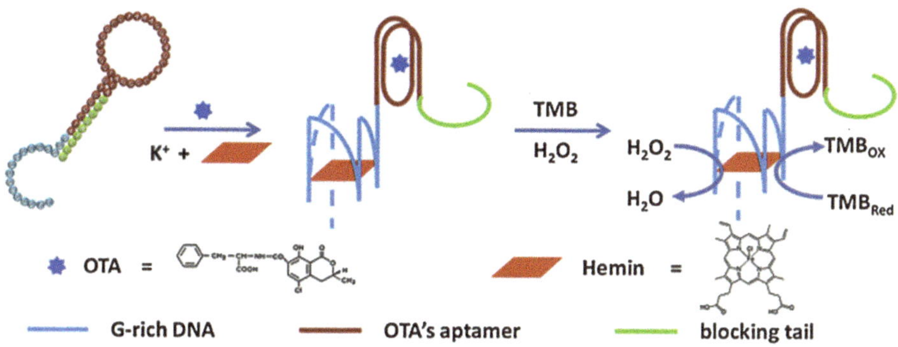

Fig. 13.15 The principle of the bioassay for the detection of OTA (Reprinted from Ref. [31], Copyright 2012, with permission from Elsevier)

tail inactivates the HRP-mimicking DNAzyme. In the presence of target OTA, the OTA–aptamer complex is formed and opens the hairpin, resulting in the self-assembly and activation of the HRP-mimicking DNAzyme.

A DNAzyme-based Ag^+ sensor was developed by the Zhou group [32]. This method creates Ag^+-mediated C-C mismatches by forming C-Ag^+-C base pairs. Then, per the design, only one unlabeled oligonucleotide strand is required to complete the assay. When the sample does not contain the target, the oligonucleotide strand forms an intramolecular duplex through hydrogen bonding. The G-rich DNAzyme sequence in the oligonucleotide becomes partially caged in this duplex structure and cannot fold into the G-quadruplex structure, even in the presence of hemin. Upon the addition of Ag^+, another intramolecular duplex tends to form via C-Ag^+-C base pairs, leading to a conformational change and release of the G-rich sequence. As a result, the released G-rich sequence can fold into a G-quadruplex, bind the hemin, and form a catalytically active DNAzyme. With this method, optical density at 450 nm is significantly increased in the H_2O_2–ABTS reaction system. The detection limit of this method can be as low as 6.3 nM with a high specificity (Fig. 13.15).

13.2.2.3 Paper-Based NABs

Numerous other aptasensors have been exploited in combination with various types of analytical matrices, such as paper.

Recently, lateral flow biosensors, also called dry-reagent strip biosensors, using chromatography combined with conventional immunoassay techniques, have gained increasing attention for analysis and clinical diagnosis applications. Such lateral flow biosensors have several advantages, including user-friendly formats, very short assay times (generally several minutes), less interference due to chromatographic separation, long-term stability over a wide range of climates, relatively low costs, visual signals, and no requirements for skilled technicians. This

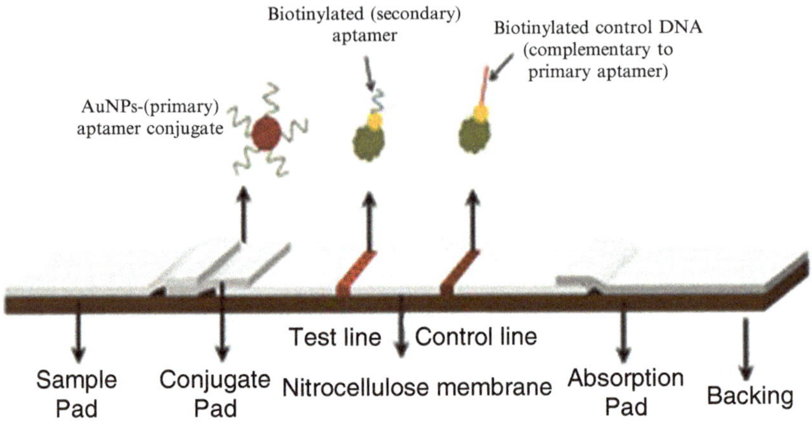

Fig. 13.16 A schematic illustration of an AuNP-based strip assay

ideal platform is suitable for on-site testing by untrained analysts. Liu's group has developed a dry-reagent strip biosensor that can visually detect protein within minutes for thrombin analysis. This method is based on aptamers and functionalized AuNPs, and the sensitivity and specificity of this strip biosensor are superior to those of sensors based on antibody interactions. Aided by the lateral flow, the target is captured by the AuNP–aptamer conjugate during the migration of the sample solution onto each pad. Then, via the sandwich format, the target–aptamer–AuNP complexes are captured by the secondary aptamers that are immobilized in the test zone. Finally, a red band can be observed in the test zone due to the accumulation of AuNPs (Fig. 13.16).

13.2.2.4 Chemiluminescence NABs

Willner et al. [33] reported a chemiluminescence approach for DNA detection (Fig. 13.17). A probe A that is partly complementary to the target DNA is immobilized onto the Au surface. Then, a reporter DNA strand B is added. This strand consists of two domains: one is responsible for catalytic activity (G-quadruplex) and the other is hybridized with the target DNA. As a result, the DNAzyme is localized on the Au surface via hybridization of the reporter strand (DNAzyme label) with the target DNA, and upon adding luminol and H_2O_2, a chemiluminescent signal can be generated.

This research group also presented a catalytic molecular beacon as a hybridization probe for DNA detection (Fig. 13.18) [34]. The hairpin structure contains two parts, fragments A (black) and B (gray), which can assemble into intact G-quadruplexes. As the hairpin structure of the DNAzyme sequences becomes partly caged through hybridization, the G-quadruplex structures cannot be formed. However, when a target DNA is present, it can hybridize with the hairpin loop, open the hairpin, and release fragment B, which generates a DNAzyme in the presence of

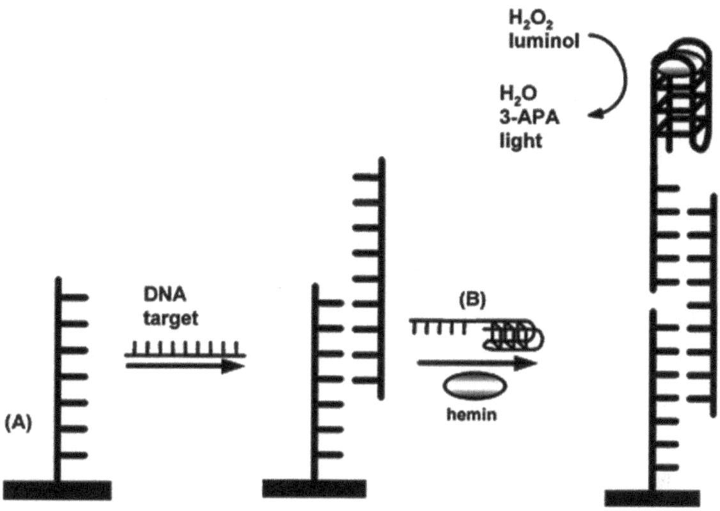

Fig. 13.17 A chemiluminescence approach for DNA detection in a sandwich assay on an Au surface (Reprinted with the permission from Ref. [33]. Copyright 2004 American Chemical Society)

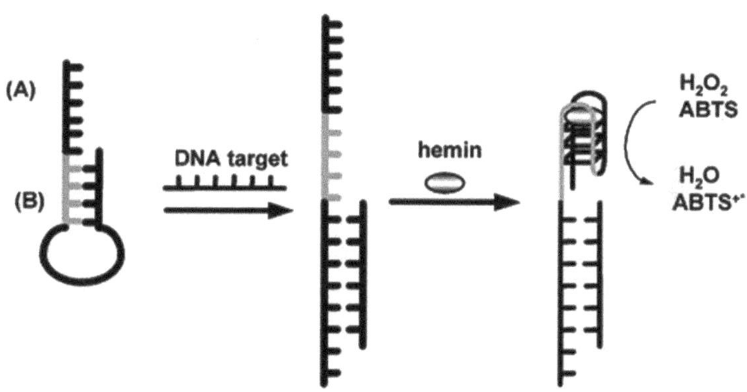

Fig. 13.18 A colorimetric DNA assay based on opening a DNA hairpin and generating a DNAzyme. The formation of this peroxidase-like DNAzyme-based colorimetric NAB is generally followed by a conformational change (Reprinted with the permission from Ref. [34]. Copyright 2004 American Chemical Society)

hemin. Upon adding ABTS and H_2O_2, the system then produces a colorimetric readout.

13.2.2.5 Mass-Sensitive NABs

A mass-sensitive biosensor is defined as any device that measures a mass associated property via a sensitive surface with assembled DNA capture probes. Aptamer-based, mass-sensitive biosensors are label-free bioassays and include evanescent wave-based sensors and acoustic wave-based sensors. For example, surface plasmon resonance (SPR) sensors are typical evanescent wave-based sensors. In addition, quartz crystal microbalances (QCMs), surface acoustic wave (SAW) devices, and micromechanical cantilever-based sensors are all types of acoustic wave-based sensors. These are referred to as "mass-sensitive" strategies because of the gravity or thickness measurements obtained from the thin films.

Acoustic Nucleic Acid-Based Biosensors (ANABs)

A general acoustic sensor is based on the detection of mass changes which is measured by the surface acoustic wave. This wave results from the binding of an analyte to an MRE immobilized on the surface of quartz (Fig. 13.19) [35].

A multifunctional Love-wave sensor using aptamer recognition to detect serine protease thrombin and Rev peptide has also been reported. The reported sensitivity of this method was calculated by dividing the measured phase shift by the determined mass loading. This method exhibited high sensitivity for both thrombin and Rev peptide; the detection limit was 72 and 77 pg/cm^2 for thrombin and Rev peptide, respectively. Interestingly, this aptamer-based, Love-wave sensor could potentially be used in an array format because the device has demonstrated a synergistic effect. Generally, the BIA core™ 3000 system is used with Love-wave-based sensors.

Another Love-wave biosensor array, which is a special type of surface acoustic wave (SAW) sensor, has been designed by immobilizing aptamers on the surface of a Love-wave sensor chip. The sensor array consists of five single-sensor elements and allows label-free and quantitative detection of five targets by binding fluorescently labeled molecules and correlating the mass of the bound molecules to fluorescence intensity. The reported detection limit was 75 pg/cm^2, and the specificity was high. The sensor can easily be regenerated by a simple washing process. The versatile applicability of the sensor is achieved through the immobilization of single-stranded DNA (ssDNA) for the detection of the corresponding complementary strand.

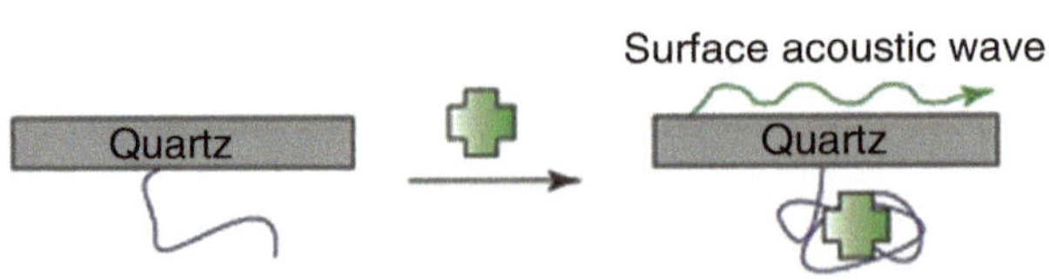

Fig. 13.19 Surface acoustic sensors. An aptamer is immobilized onto the surface of quartz, a piezoelectric material. Target binding generates a surface acoustic wave, represented by a curving green line with an arrowhead (Reprinted from Ref. [35], Copyright 2006, with permission from Elsevier)

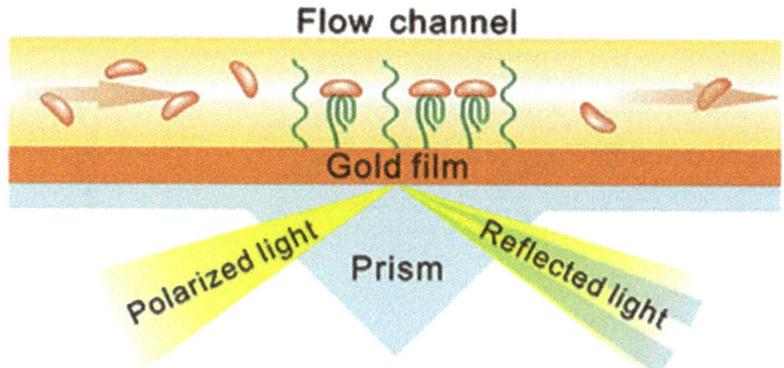

Fig. 13.20 SPR-based aptasensor (Reprinted from Ref. [28], Copyright 2008, with permission from Elsevier)

Surface Plasmon Resonance (SPR)

SPR sensors are devices that reflect mass changes via the associated changes in refractive index. SPR can determine the dissociation constants of aptamers from their targets, and this technology is often used in conjunction with SELEX. The SPR method is also applied in aptamer-based biosensors. In this type of biosensor, the aptamer is often immobilized to form a selective surface. In the presence of target, resonance angle changes occur at the surface. After the instrument measures these changes as a signal, the assay is complete (Fig. 13.20).

As the signal is proportional to the number of bound molecules, this method allows label-free detection in a single-binding site configuration. Recently, an anti-HIV-1 Tat protein aptamer was immobilized on SPR chips using the avidin–biotin bridge. The assay showed high specificity for HIV-1 Tat protein, which was identified using a control protein. An enzymatically amplified SPR imaging (SPRI) assay was also developed to detect thrombin and vascular endothelial growth factor (VEGF). In contrast with the former method, a horseradish peroxidase (HRP)-conjugated antibody was introduced instead of an aptamer, which produces a precipitate on the Au surface and can result in a significantly amplified SPRI response.

Cantilever-Based Biosensors

Gerber developed a label-free biosensor capable of detecting targets via the bending of microfabricated cantilevers coated with probe.

The sensor cantilever is often functionalized by a thiol-modified aptamer specific for a target. Interferometry can be used to determine the differential bending between the sensor cantilever and the reference cantilever. The differential cantilever bending could range from 3 to 32 nm. This method has been successfully applied in Taq DNA polymerase analysis, and these piezoelectric signal biosensors can be regenerated with 7 M urea (Fig. 13.21).

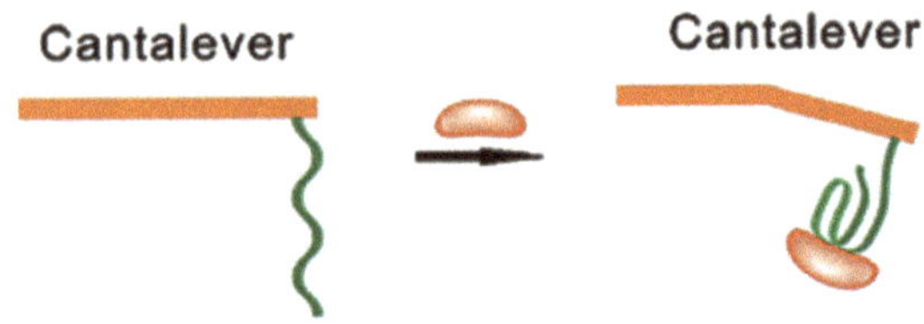

Fig. 13.21 The principle of cantilever-based biosensors (Reprinted from Ref. [28], Copyright 2008, with permission from Elsevier)

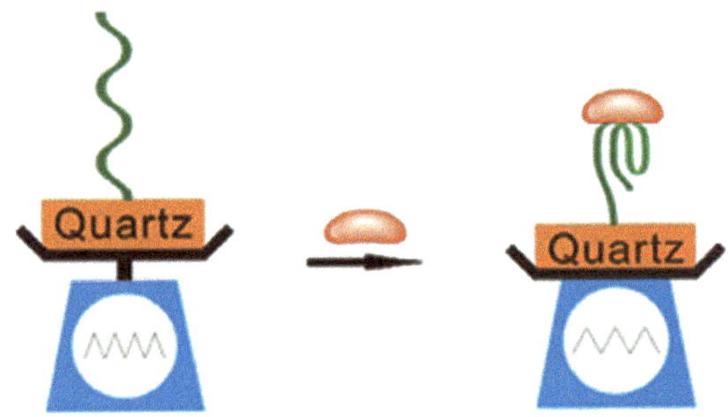

Fig. 13.22 QCM-based aptasensor (Reprinted from Ref. [28], Copyright 2008, with permission from Elsevier)

Quartz Crystal Microbalance NABs

The use of microgravimetric analysis for the detection of aptamer–target interactions on QCMs has been reported. As the mass changes of crystal will result in the frequency of a quartz crystal, thus, when the target bind to the aptamer-modified crystals, the mass on the transducer was increased, leading to a decrease in the resonance frequency of the crystal (Fig. 13.22). This system is a magnetoelastic signal biosensor.

Two similar methods using aptamer-functionalized Au/quartz crystal for the microgravimetric sensing of thrombin and HIV-1 Tat protein have also been reported. The LODs for thrombin and HIV-1 Tat protein were 1 nM and 0.25 ppm, respectively.

The parameters of these QCM-based aptasensors can also be evaluated and optimized to improve analytical performance. In addition to QCM applications, piezoelectric quartz crystals are also used to fabricate SAW aptasensors. Detailed information is presented in the "Acoustic nucleic acid-based biosensors" section.

13.2.2.6 Electrochemical NABs

The flexible, single-stranded chains of nucleic acid aptamers fold into three-dimensional (3D) shapes upon binding their targets. When immobilizing aptamer on a conductive support, this property enables redox-active moieties to be tethered to the aptamers. Through probing the electron transfer features of the redox moieties can identify the formation of aptamer–target rigidified 3D complexes. Various electrochemical aptasensors have been designed based on this principle.

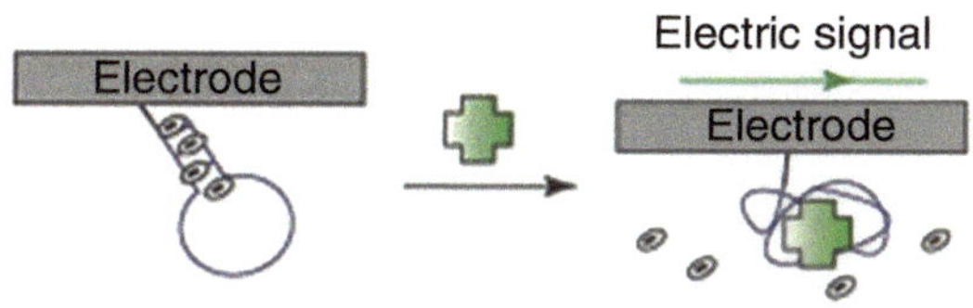

Fig. 13.23 An electrochemical sensor. An aptamer is designed into a hairpin structure into which a small-molecule electrochemical marker can intercalate. Target binding causes the release of these markers, resulting in an electrochemical signal (Reprinted from Ref. [35], Copyright 2006, with permission from Elsevier)

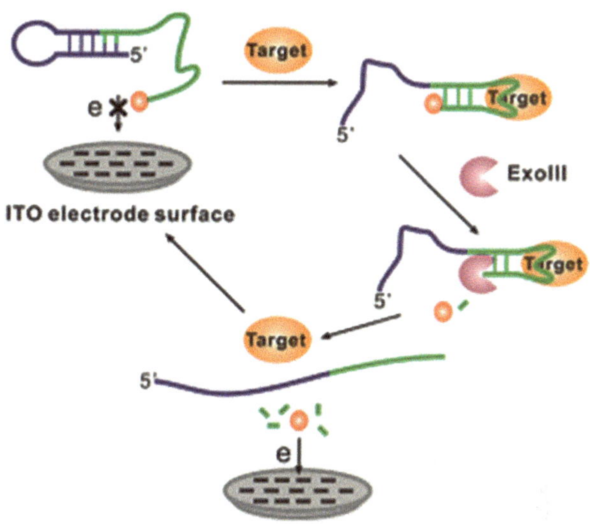

Fig. 13.24 Exonuclease-assisted amplification assays. Using hairpin aptamer probes to enable exonuclease-assisted signal amplification (Reprinted with the permission from Ref. [36]. Copyright 2014 American Chemical Society)

Kim et al. has reported an electrochemical sensor that can detect any aptamer–protein interaction (Fig. 13.23) [35]. Their electrode is modified with an antithrombin aptamer which configured into a hairpin structure which can bind the methylene blue (MB) (electrochemical marker) via intercalation. Upon adding thrombin, the target disrupts the hairpin stem and releases the MB, resulting in a decreased electrochemical signal.

Exo III-assisted assays have also been developed using structure-switching aptamers. Liu et al. designed a hairpin anti-ATP aptamer with a 3′ overhang to protect unbound aptamers from exo III digestion (Fig. 13.24) [36]. Upon target binding, the aptamer folded into its secondary structure, pairing the 3′ end. Exo III then hydrolyzed the 3′ end of the aptamer, destroying the complex and freeing ATP for further binding. The cyclic cleavage of the aptamer resulted in amplified electrochemical detection of ATP. Xu et al. designed a similar assay for ATP detection that used fluorescence detection. An alternative design also used aptamers with a hairpin; however, target binding disrupted the hairpin structure of the aptamer but did not enable pairing of its 3′ end. Instead of enzymatic cleavage of

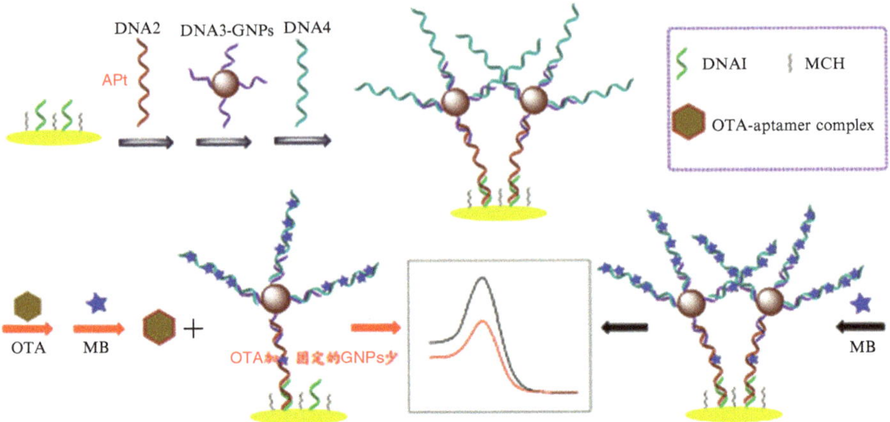

Fig. 13.25 Schematic representation of the electrochemical aptasensor designed for OTA detection (Reprinted from Ref. [37], Copyright 2014, with permission from Elsevier)

the 3′ end of the aptamer, the structural change exposed a complementary sequence to a signaling sequence. This initiated cyclic cleavage of the signaling sequence by exo III. This strategy amplified the detection of lysozyme and thrombin.

The Yang group [37] described an ultrasensitive electrochemical aptasensor for OTA based on a two-level cascaded signal amplification strategy with methylene blue (MB) as a redox indicator. In this method, capture DNA, aptamers, and reporter DNA-functionalized gold nanoparticles (GNPs) were immobilized on the electrode accordingly, where GNPs were used as the first-level signal enhancer. To obtain a more sensitive response, more guanine (G)-rich DNA molecules were bound to the surface of the GNPs, thereby providing abundant anchoring sites for MB and enabling the second-level signal amplification. By employing this novel strategy, a ~ 8.5 (±0.3)-fold amplification in signal intensity was obtained. Afterward, OTA was added to force a partial release of GNPs/G-rich DNA from the sensing interface, which decreased the electrochemical response. An effective sensing range from 2.5 pM to 2.5 nM and an extremely low detection limit of 0.75 (±0.12) pM were achieved. This amplification strategy has the potential to be the main technology for aptamer-based electrochemical biosensor applications in a variety of fields (Fig. 13.25).

It has been reported that nucleic acids can be used to control ion permeability through the nanopores to regulate current. Cytosine-rich nucleic acid strands or guanine-rich DNA [6] was attached to the nanopores by chemical etch. The G/C-rich DNA was sensitive to pH or K^+, respectively. Stimulated organization of G-quadruplexes was used to control the opening of the pores, and closing of the pores was controlled by pH/K^+-induced DNA self -assemble. Under basic conditions, the immobilized DNA existed in a randomly coiled state, which allowed the free diffusion of ions. The ion current could be switched off by changing the environment. The low pH or concentrated K^+ caused the assembly of DNA into a

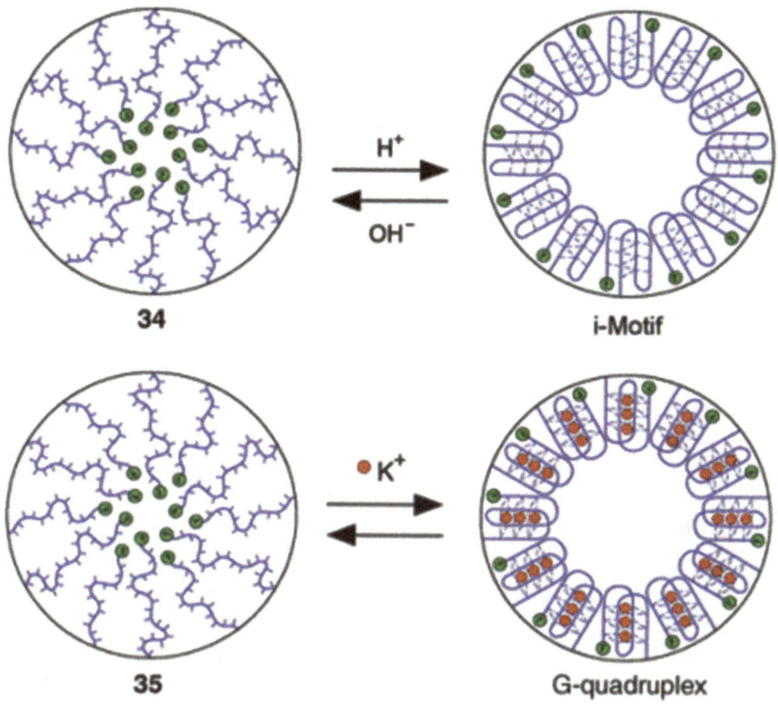

Fig. 13.26 Functional nucleic acids immobilized inside of etched nanopores: i-motif DNA 34 allowing pH-induced changes of ion permeability (*top*). Potassium-dependent opening and closing of G-quadruplex-modified nanopores transforming into K^+-unresponsive duplexes after the addition of complementary DNA (*bottom*) (Reprinted from Ref. [6], Copyright 2010, with permission from Elsevier)

rigid i-motif quadruplex structure, which efficiently blocked the diffusion of ions (Fig. 13.26).

13.3 Categories of Nucleic Acid-Based Biosensors Based on Amplification

13.3.1 Amplification-Based Optical NABs

Nucleic acid aptamers are obtained by in vitro selection methods; thus, they are capable of recognizing a wide range of target analytes, and they have been widely used in optical transduction approaches. Recently, an amplification approach has demonstrated that DNA can also play the transduction role. There are many types of nucleic acid amplification technologies, and these types can be divided into two groups: thermocycling and isothermal amplification methods. The main difference between the groups is that the former requires a thermocycling machine, which provides cyclical temperature changes during the amplification process. Although thermocycling amplification strategies can exponentially amplify a target sequence, there are also several drawbacks, such as susceptibility to contamination, tendency of false amplification, high cost, and easy sequence mismatches with typical error rates ranging from 1.3×10^{-6} to 5.5×10^{-5}, depending on the polymerase used.

Thus, some isothermal amplification techniques are emerging. We will introduce some enzyme-dependent isothermal amplification methods for nucleic acid biosensing, such as loop-mediated isothermal amplification (LAMP), rolling-circle amplification (RCA), and strand displacement amplification (SDA). Several other isothermal nucleic acid amplification techniques exist, such as the nicking enzyme amplification (NEA) reaction and the nucleic acid sequence-based amplification (NASBA), but these techniques are not covered here because they have not been reported in the context of biosensor assembly. One development of molecule technology is a nucleic acid amplification process that is triggered by a hybridization chain reaction; this method was designed by Robert et al. in 2004 and was termed HCR. This technology has been widely applied in small-molecule and nucleic acid biosensing since its development. In this section, we will review some nucleic acid biosensors based on amplification.

13.3.1.1 Enzyme-Dependent Isothermal Amplification

Rolling-Circle Amplification (RCA)

In 1998, Paul et al. developed an isothermal rolling-circle amplification strategy for mutation detection and single-molecule counting. This strategy is driven by DNA polymerase and can replicate circularized oligonucleotide probes with either linear or geometric amplification under isothermal conditions [38]. Once present, a hybridized circular primer can link to the linear oligonucleotide and continuously progress around the circular DNA probe to generate a complex pattern of DNA strand displacement that acts as a rolling circle. After 90 min, the target DNA can be amplified to 10^9 or more copies (Fig. 13.27). RCA is beneficial for subsequent manipulation and detection because the amplicons of RCA are single-stranded oligonucleotides that have repetitive sequences. For biosensor assembling, 17-nt DNAzyme complementary sequences are the best circularized oligonucleotide probes and can generate hundreds of thousands of DNAzyme units to transduce or amplify a signal.

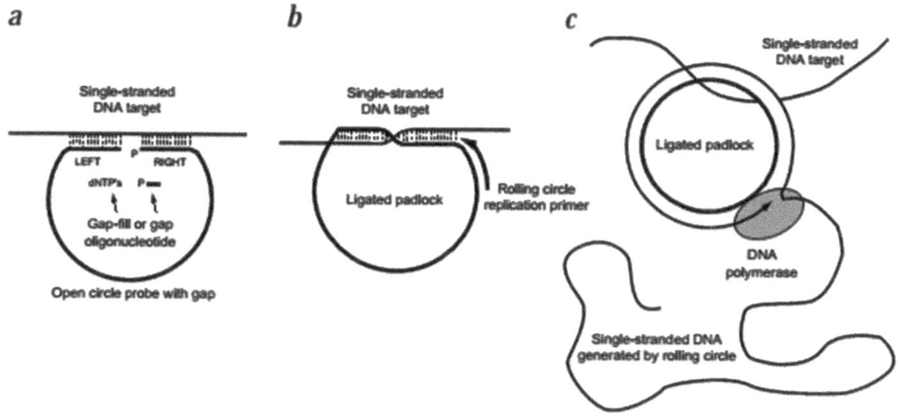

Fig. 13.27 The principle of the rolling-circle reaction (RCA). (**a**) Circularizable probe with a small gap. (**b**) Ligated (padlock) probe and binding of complementary primer for rolling-circle amplification. (**c**) Rolling-circle amplification of a padlock probe, catalyzed by a strand displacing DNA polymerase. (Reprinted by permission from Macmillan Publishers Ltd: Ref. [38], copyright 1998)

Loop-Mediated Isothermal Amplification (LAMP)

LAMP was first introduced as a novel isothermal nucleic acid amplification technique by Notomi [39]. This powerful and straightforward technique is an attractive method for detecting nucleic acids because it rapidly recognizes a specific DNA fragment with high selectivity and sensitivity using a set of four primers [40] and Bst DNA polymerase. The amplification occurs under isothermal conditions; thus, a simple water bath is sufficient for this method [41]. Furthermore, LAMP amplifies a few copies of the target to 10^9 copies in less than 1 h, even when large amounts of nontarget DNA are present [42]. LAMP has an advantage over methods that require pathogen-free samples; in fact, this method is convenient and effective for pathogen detection (Fig. 13.28). Most LAMP methods for monitoring products require gels or other apparatuses, such as tailor-made turbidimeters or high-resolution fluorescence detectors. Thus, it is not portable or affordable enough to be used for the simple and swift diagnosis of possible pathogens. In recent years, it has been reported that LAMP can be combined with aptamers for electrochemical biosensor or microfluidic applications.

Strand Displacement Amplification (SDA)

The third isothermal amplification technology introduced in this section is strand displacement amplification (SDA). SDA is an in vitro nucleic acid amplification technique. The process has two components. One component depends on the ability of a Hinc II restriction endonuclease to nick the unmodified strand of its recognition site in the target DNA. The other component is based on the action of an

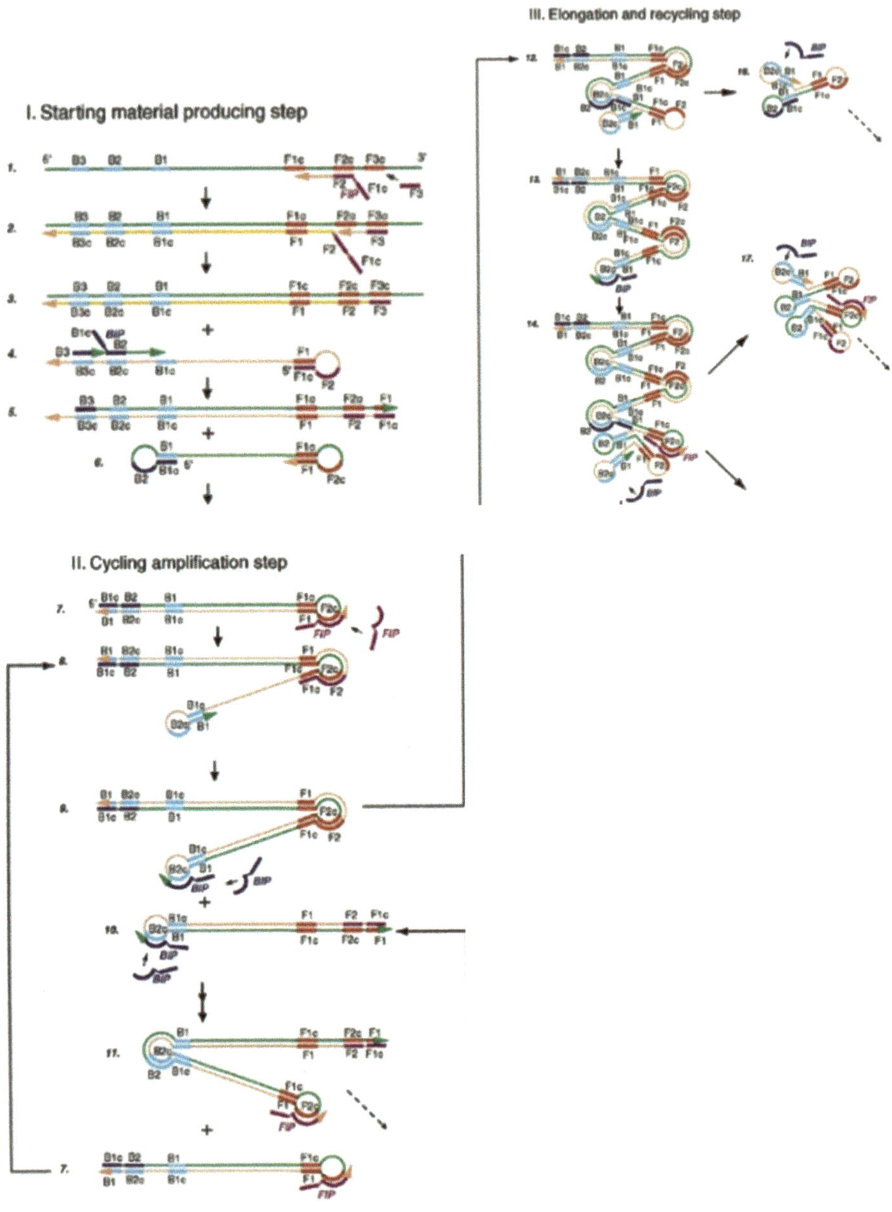

Fig. 13.28 Principle of loop-mediated isothermal amplification (Reproduced from Ref. [39] with permission of Oxford)

exonuclease-deficient DNA polymerase in extending the 3′ end at the nick and displacing the downstream DNA strand. The displaced strand serves as a template for an antisense reaction and vice versa, resulting in exponential amplification of the target DNA [43]. In recent years, SDA has often been combined with HRP-like

DNAzymes for target-induced, signal-amplifying colorimetric detection of nucleic acids [42].

Hybridization Chain Reaction (HCR)

To satisfy the need for the ultrasensitive, simple, low-cost, highly selective detection of nucleic acids, hybridization chain reaction (HCR) was developed. Unlike PCR and other enzyme-dependent amplification processes, which are target amplification techniques, HCR is a unique probe amplification technique. HCR amplifies target DNA indirectly through specially designed oligonucleotide hairpin probes. Using a pair of complementary DNA hairpins is a crucial factor for HCR, as these drive dsDNA strand propagation and the hybridization chain reaction events. A typical HCR includes three elements, a single-stranded target that acts as the initiator and two hairpins that act as energy sources. In the absence of the initiator, the DNA hairpins remain stable while storing potential energy. However, upon addition of the initiator, hairpin 1(H1) opens up due to interactions with the complementary regions of the initiator. The opening of the first hairpin triggers a cascade reaction (Fig. 13.29) [44]. Free energy gained from base pairing provides the driving force for this reaction. Due to its enzyme-free nature, isothermal and simple protocol, and excellent amplification power, this method has been extensively applied in nucleic acid biosensing [20].

13.3.1.2 Enzyme-Dependent Amplification for Optical NABs

PCR is the most widely used method to amplify and quantify nucleic acids. Because PCR amplification is exponential, it enables the detection of as few as ten nucleic

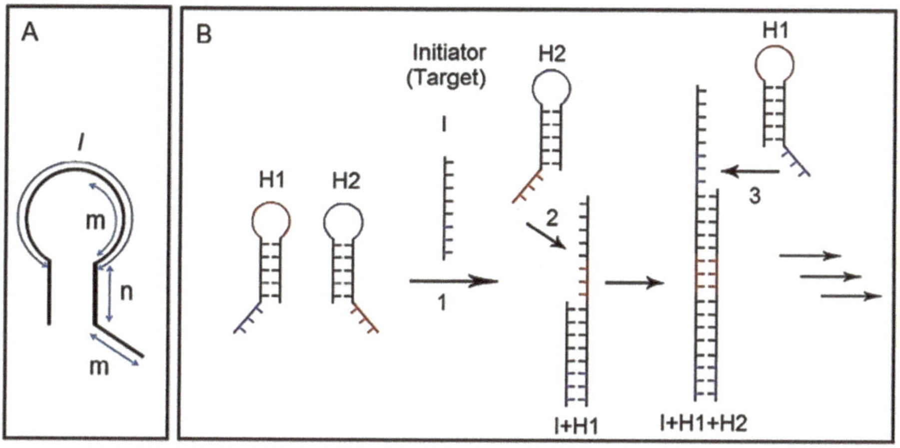

Fig. 13.29 The principle of hybridization chain amplification (Reprinted from Ref. [20], Copyright 2015, with permission from Elsevier)

acid molecules. As nucleic acid affinity ligands, aptamers possess a number of unique features that enable the use of PCR for the detection of non-nucleic acid biomolecules. For instance, aptamers can serve as both target binding molecules and PCR templates in immuno-PCR assays and can be easily extended or modified when required. Some of the unique features of aptamers that have been explored to develop various aptamer-based amplification assays include their ability to switch structures, strong negative charge, and stability against enzymatic degradation when bound to targets.

Aptamer-Based Immuno-PCR Assays

Immuno-PCR combines the exponential amplification power of PCR with the versatile and specific immunorecognition of ELISA to achieve highly sensitive antigen detection. Conventional immuno-PCR requires antibodies conjugated with specific DNA sequences to be amplified by PCR to generate a detection signal. The conjugation is usually cumbersome and laborious, which negatively impacts the applicability of immuno-PCR. This can be largely circumvented by using aptamers as the affinity ligands because aptamers can be easily extended and modified for direct amplification by PCR. Tok and coworkers demonstrated the use of DNA aptamers as both target recognition and signal generation molecules by developing an immuno-PCR assay for thrombin detection (Fig. 13.30a). Antibody-coated magnetic beads were used to capture thrombin from sample solutions. An aptamer probe consisting of a 15-nt antithrombin aptamer, a poly-A15 linker, and a short, 20-nt primer was then introduced to bind to the captured thrombin. After unbound aptamers were washed away, the bound aptamers were dissociated by being heated and then subjected to real-time PCR detection. A limit of detection (LOD) in the low pM range was obtained for the detection of thrombin in 10 % serum. A similar format was applied to the detection of other proteins, such as human IgG and PDGF-BB.

Structure-switching aptamers have been used to develop immuno-PCR methods for target molecules that can only be bound by a single aptamer molecule. When hybridized to complementary sequences, these aptamers exist as a duplex. Upon binding to their target molecules, the aptamers switch structures into specific secondary arrangements and dissociate from their complementary sequences. Xu and coworkers used structure-switching aptamers to demonstrate the sensitive detection of ochratoxin A (Fig. 13.30b). A biotinylated anti-ochratoxin A aptamer was immobilized on the surface of a streptavidin-coated PCR tube. The complementary sequence was then introduced to hybridize with the aptamer. Upon target addition, the aptamers bound to the target and switched structures. This disrupted the duplex structure, which released the complementary sequences from the tube surface. After the released complementary sequences were washed away, the remaining bound complementary sequences were amplified by real-time PCR. Ochratoxin A was then quantified by measuring the increase of cycle numbers of the PCR reaction due to the decreased amounts of DNA templates. This method was

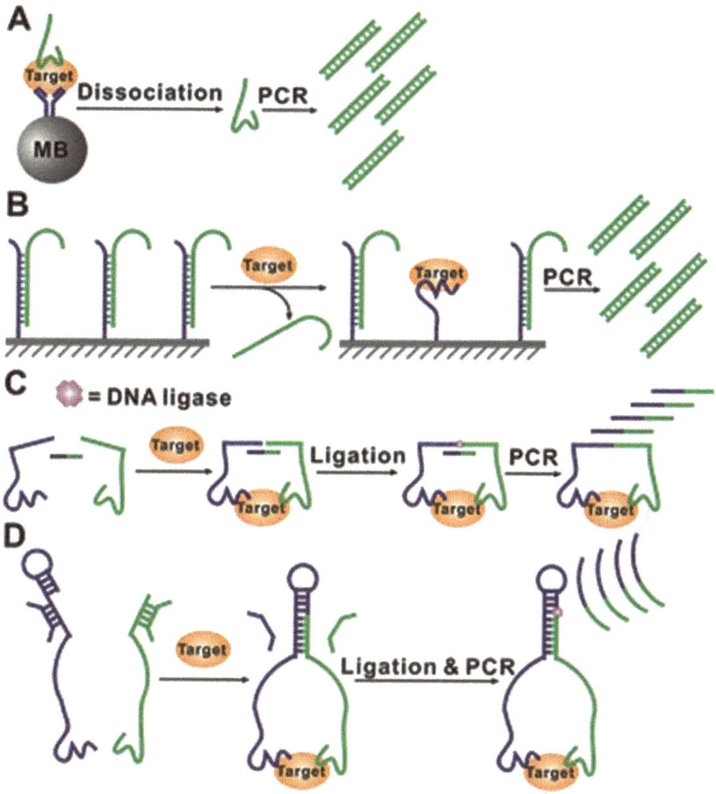

Fig. 13.30 Amplification assays using PCR. (**a**) Immuno-PCR assay using aptamers as both target recognition and signal generation molecules. (**b**) Structure-switching aptamers facilitate the detection of non-nucleic acid targets by PCR. (**c**) Proximity ligation assay enables the homogeneous detection of proteins and other biomolecules by PCR. (**d**) Application of binding-induced DNA assembly for the detection of proteins by PCR (Reprinted with the permission from Ref. [36]. Copyright 2014 American Chemical Society)

able to detect as low as 1 fg/mL ochratoxin A. An assay using a similar structure-switching aptamer was developed to detect aflatoxin B1 with an LOD of 25 fg/mL.

The proximity ligation assay (PLA) generates a new amplifiable DNA sequence from the binding of a target molecule by two aptamer probes without separation or washing. PLA was originally developed by Landegren and coworkers to detect PDGF-BB by using two aptamer probes (Fig. 13.30c). The two probes were prepared by extending anti-PDGF-BB aptamers with additional sequences at either the 5′ or 3′ end. When the two probes bound the same PDGF-BB molecule, the free ends of the probes were brought into close proximity. This allows an added connector DNA strand to juxtapose the two ends, which does not occur in the absence of the target molecule. The two ends were enzymatically ligated together, forming a new DNA sequence. This new sequence was then measured by real-time

PCR for the quantitative detection of PDGF-BB. The detection of as few as 2400 PDGF-BB molecules was achieved. The use of PLA to detect thrombin has also been demonstrated. Using proximity probes prepared by a single aptamer, PLA has also been used to detect adjacent cell surface antigens.

Motivated by PLA, we introduced the concept of binding-induced DNA assembly (BINDA). Distinct from DNA self-assembly that is dependent on the spontaneous hybridization of complementary sequences, BINDA is DNA assembly that only occurs when a specific target triggers a binding event. BINDA has been demonstrated to be an excellent strategy for developing versatile homogeneous assays. Using aptamers as affinity ligands, we used BINDA to develop a homogeneous assay for the ultrasensitive detection of PDGF-BB (Fig. 13.30d). Two probes were constructed by tagging two biotinylated aptamers with DNA F or R motifs using streptavidin as a connecting molecule. The two motifs each contained a short complementary sequence, such that hybridization is unstable in the absence of target molecules. To further minimize self-hybridization between the two complementary sequences, we designed a pair of blocking oligonucleotides to compete with the two complementary sequences for hybridization. Upon the addition of PDGF-BB molecules, the binding of the two probes to a single target molecule placed the two DNA motifs in close proximity, which dramatically increased their local concentrations and led to the hybridization of the complementary regions of the DNA motifs. Because the F motif consisted of a DNA hairpin structure at the 3′ end, the assembly of the DNA motifs placed the 3′ end of the hairpin sequence adjacent to the 5′ end of the complementary sequence of the R motif, allowing the two motifs to be enzymatically ligated together. This formed a new DNA sequence that was then detected by real-time PCR, enabling the ultrasensitive detection of PDGF-BB. A detection limit of 1 fM was achieved for the detection of PDGF-BB.

RCA-Based Optical NABs

RCA is an isothermal nucleic acid amplification process that synthesizes a long, single-stranded nucleic acid product from the continual and unidirectional replication of a circular, single-stranded nucleic acid template. Similar to immuno-PCR, the combination of RCA with ELISA is a common approach to develop RCA-based assays for the amplified detection of proteins and other targets. These assays usually use immobilized antibodies or aptamers to capture target molecules from sample solutions. Aptamer probes containing a primer region for annealing to the circular RCA template are then introduced to bind to the captured target molecules. Unbound aptamer probes are washed away, and RCA is performed. The RCA products are detected using various signal generation techniques. Yu and coworkers first described this approach using electrochemical detection to quantify RCA products. The assay detected as low as 10 fM PDGF-BB with a dynamic range of four orders of magnitude.

Yang et al. designed a structure-switching aptamer that can circularize upon target binding, similar to Fig. 13.31 showed [45]. A 13-nt sequence was added to

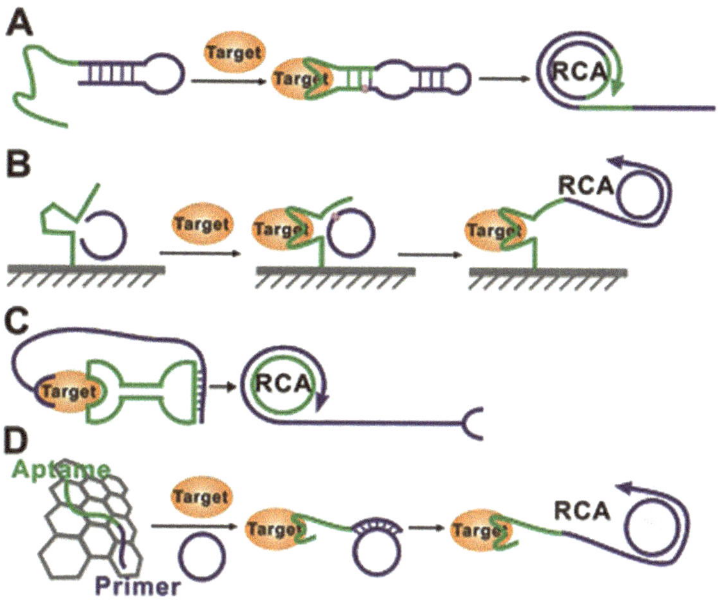

Fig. 13.31 Assays incorporating rolling-circle amplification (RCA). (**a**) RCA assay using a structure-switching aptamer that circularizes upon target binding. (**b**) Allosteric DNAzyme transducing target binding into the generation of a circular RCA template. (**c**) Proximity extension assay using a circularized aptamer probe. (**d**) RCA-mediated assay based on target binding-induced desorption of aptamers from graphene oxide (GO). Specially designed aptamers can also directly trigger RCA (Reprinted with the permission from Ref. [36]. Copyright 2014 American Chemical Society)

the 3′ end of an anti-PDGF-BB aptamer via a linker. This 13-nt sequence was complementary to the aptamer and disrupted its secondary structure by forming a hairpin. The aptamer was further modified by adding a 6-nt sequence to its 5′ end and altering the linker sequence, such that target binding resulted in a closed ligation junction. Therefore, enzymatic ligation generated circular DNA that served as an RCA template. This structure-switching aptamer was used for the homogeneous detection of PDGF-BB, and a detection limit of 0.4 nM was achieved. The same research group also designed an allosteric DNAzyme that transduced target binding into the generation of a circular template for RCA [46] (Fig. 13.31b). The allosteric DNAzyme consisted of an anti-ATP DNA aptamer linked to a DNAzyme in such a way that the DNAzyme is only activated when the aptamer binds to ATP. The selected DNAzyme catalyzes DNA ligation; thus, upon ATP binding, the allosteric DNAzyme enabled the ligation and circularization of a DNA substrate, which formed a padlock probe. The allosteric DNAzyme was used for the amplified detection of ATP by RCA. Up to a 1,500-fold increase in signal was obtained after 10 min of RCA. Aptamers are single-stranded nucleic acids and can thus be constructed into circular structures to directly serve as RCA templates. Di Giusto et al. described a proximity extension assay enabling the homogeneous detection of various proteins [47] (Fig. 13.31c). They selected thrombin to demonstrate the proof of concept because there are two known aptamers that bind to different epitopes of thrombin. They constructed the first aptamer into a circular structure and added a second aptamer to the 3′ end of a short RCA primer via a linker. In the absence of the target molecule, the annealing of the primer to the circular aptamer was unstable. When the two aptamers bound to the same thrombin molecule, the primer was brought into close proximity with the circular aptamer, which allowed the primer to anneal to the circular aptamer and initiate RCA. A detection limit of 30 pM was achieved for thrombin. Aptamers can absorb onto graphene oxide (GO) and desorb upon target binding. Liu and coworkers developed an RCA-mediated homogeneous assay for the amplified detection of proteins and small molecules using GO [48] (Fig. 13.31d). Aptamers containing RCA primers that adsorbed to a GO surface disabled the occurrence of RCA. Upon target binding, the aptamers were released from the GO, thereby exposing the primers for RCA template annealing and RCA initiation. The detection of RCA products represented the amplified detection of the target molecules. This assay was used to detect thrombin and ATP.

The rapid detection of trace biomarkers in a biologically complex mixture remains a major challenge. Tang's group [49] reported a novel protein detection aptasensor assay that integrates two signal amplification processes, namely, rolling-circle amplification (RCA) and a DNAzyme colorimetric reaction (Fig. 13.32).

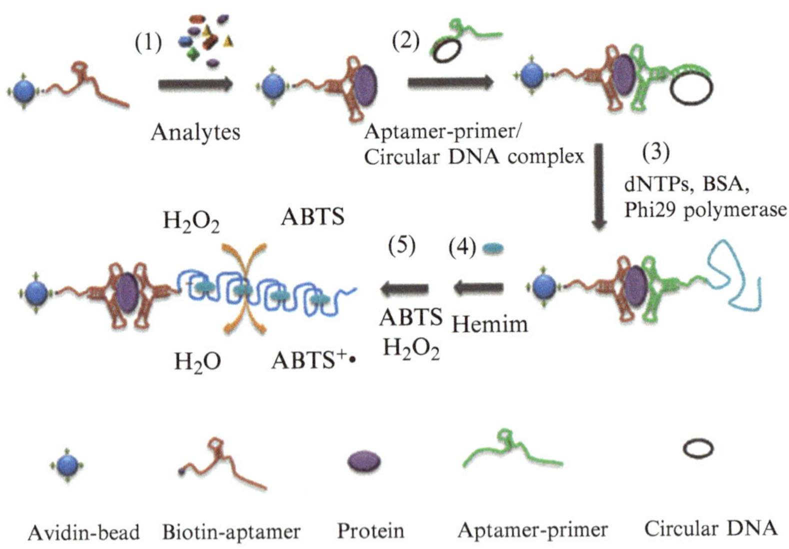

Fig. 13.32 Schematic of the colorimetric assay using sandwiched aptasensors and dual signal amplification. *1* Aptamer beads capture the target in sample solution. *2* The sandwich format. *3* Rolling-circle amplification. *4* Addition of hemin. *5* DNAzyme catalyzes the oxidation of ABTS, producing a *blue-green* colorimetric signal (Reprinted with the permission from Ref. [49]. Copyright 2012 American Chemical Society)

First, microbeads (MBs) are functionalized with primary aptamer to capture the target in a sandwich assay. Then, secondary aptamer, connected with RCA primers/circular DNA templates, is added to initiate the RCA reaction. The RCA reaction can transfer the single target binding events into a long DNA molecule containing multiple DNAzyme units. When hemin is added, the units function with peroxidase-like catalytic activity. In the presence of H_2O_2, DNAzyme catalyzes the oxidation of ABTS (2,2′-azino-bis(3-ethylbenzothiazoline-6-sulfonic acid)), which generates a green colorimetric signal. This new aptasensor platform permits label-free, ultrasensitive, visual, and rapid target detection. Using platelet-derived growth factor (PDGF) B-chain as a model system, the detection limit was as low as 0.2 pg/mL in a serum-containing medium, which is similar to the detection limit of traditional assays, such as ELISA. We expect this colorimetric biomarker detection assay to be very useful for point-of-care applications.

Strand Displacement Dual-Amplified NABs

Highly sensitive and specific DNA detection methods are necessary for clinical diagnosis and biomedical research. Wu's group [50] developed a label-free, isothermal colorimetric method for the ultrasensitive detection of DNA based on a target-triggered quadratic amplification strategy. With the presence of three designed hairpin structures and exonuclease III (exo III), the target DNA can trigger two independent reaction cycles, hairpin assembly and exo III cleavage reactions, which are designed to initiate target DNA recycling amplification and reporter DNA amplification, respectively. The sensitivity of this method dramatically increases through the two signal amplification steps, with a detection limit of 81 fM. It can also discriminate mismatched DNA from perfectly matched target DNA (Fig. 13.33).

13.3.1.3 Microcapillary-Based LAMP NABs

Zhang's group [51] improved upon the advantages of cLAMP, such as rapid detection and lack of contamination and bulky equipment, to provide microcapillary-based LAMP with new features, including sample pretreatment, stable storage of preloaded reagents, and easy operation. This truly integrated sample-to-answer microcapillary-based LAMP system (icLAMP) is used for the detection of the CYP2C19 gene from untreated, freshly drawn blood samples simply by using a positive displacement pipette tip and a handheld UV flashlight. As icLAMP only requires 200 nL of whole blood, the blood from a finger prick can be used and introduced into the capillary. This icLAMP method is also applicable for the collection, amplification, and detection of nucleic acids from bacteria.

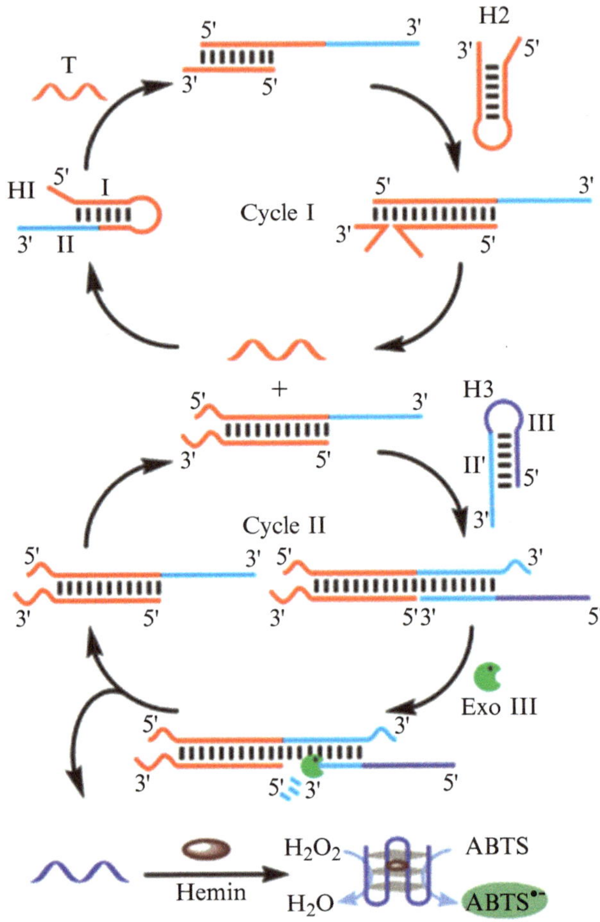

Fig. 13.33 Schematic of a new quadratic amplification strategy (Reprinted with the permission from Ref. [50]. Copyright 2012 American Chemical Society)

13.3.1.4 Enzyme-Free Amplification-Dependent Optical NABs

Simcha Shimron [52] described an enzyme-free amplified detection platform using a horseradish peroxidase (HRP)-mimicking DNAzyme as a label to functional two hairpin structures (Fig. 13.34). Firstly, the DNAzyme was split into two parts, one was in three-fourths and the other one was in one-fourth of the intact DNAzyme. In the absence of DNA target, the split DNAzyme was caged in hairpin not to stimulate the catalytic action. In the presence of the target DNA, one of the hairpins is opened, and this triggers the cascade cross-opening of the two hairpins via the hybridization-mediated strand displacement. As a result, nanowires consisting of numerous HRP-mimicking DNAzyme units were formed. The DNAzyme units in the resulting DNA nanowires act as catalytic labels for the colorimetric or chemiluminescent readout. This analysis method was "protein enzyme-free" and showed

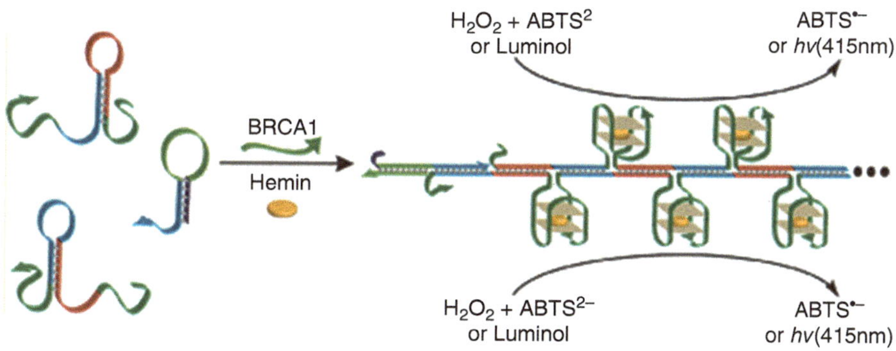

Fig. 13.34 Scheme for the amplified detection of target DNA using two functional hairpin structures (1 and 2), which upon recognition of the target DNA trigger the autonomous generation of nanowires consisting of the HRP-mimicking DNAzyme. The DNAzyme units act as readout labels for the colorimetric or chemiluminescent detection of the target DNA (Reprinted with the permission from Ref. [52]. Copyright 2012 American Chemical Society)

excellent thermostability. The analytical platform showed a detection limit down to 1×10^{-13} M.

13.3.2 Amplification-Dependent Electrochemical NABs

13.3.2.1 Enzyme Amplification-Dependent NABs

LAMP-Based NABs

Yuan's group [53] presented an efficient, accurate, rapid, simple, and ultrasensitive detection system for the small-molecule ochratoxin A (OTA) via the integration of the loop-mediated isothermal amplification (LAMP) technique and subsequent direct readout of LAMP amplicons with a signal-on electrochemiluminescent (ECL) system (Fig. 13.35). First, the dsDNA composed of OTA aptamer and its capture DNA are immobilized on an electrode. After target recognition, the OTA aptamer bonds with the target OTA and subsequently leaves the electrode, which decreases the number of immobilized OTA aptamers on the electrode. Then, the remaining OTA aptamers on the electrode serve as inner primers to initiate the LAMP reaction. Interestingly, the LAMP amplification is detected by monitoring the intercalation of DNA-binding $Ru(phen)_3^{2+}$ ECL indicators into newly formed amplicons with a set of integrated electrodes. The binding of the ECL indictor $Ru(phen)_3^{2+}$ to amplicons causes a reduction in ECL intensity due to the slow diffusion of $Ru(phen)_3^{2+}$–amplicon complexes to the electrode surface. Therefore, the presence of more OTA is expected to lead to the release of more OTA aptamers, which, in turn, leads to fewer OTA aptamers remaining on the electrode for the

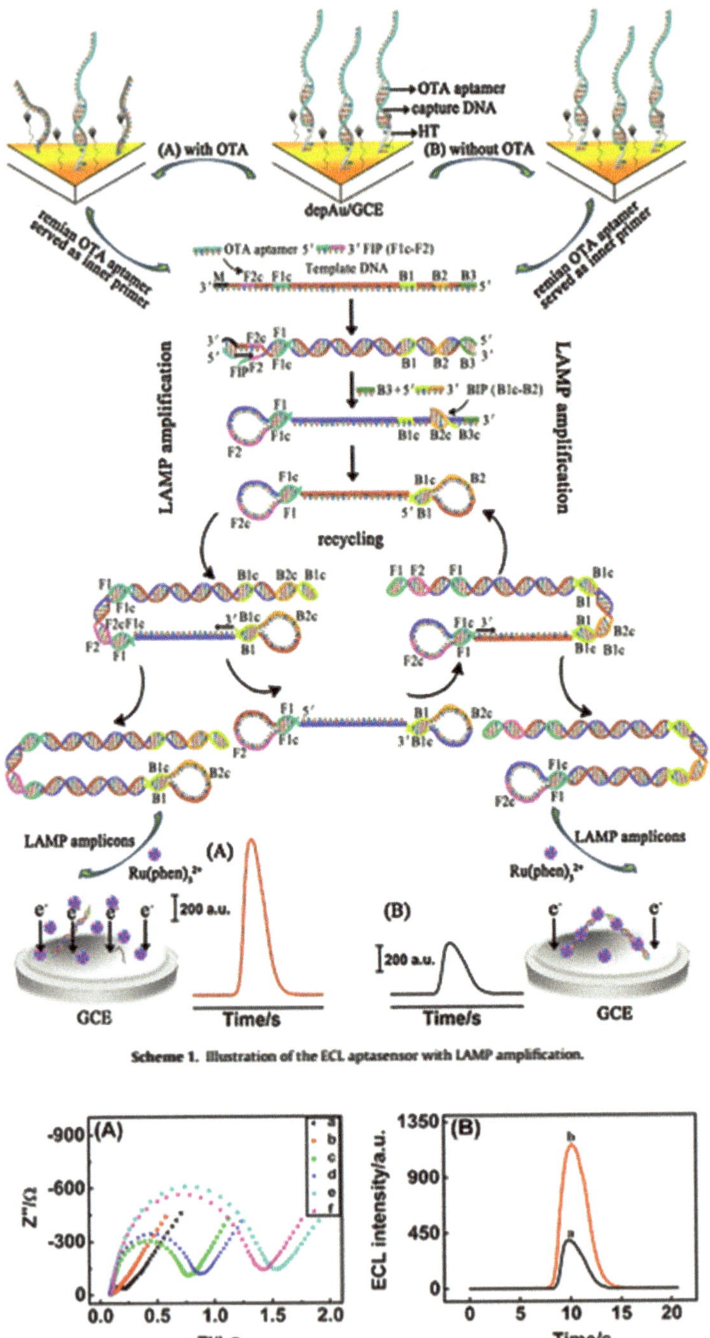

Fig. 13.35 The schematic of LAMP-mediated biosensors (Reprinted from Ref. [53], Copyright 2014, with permission from Elsevier)

production of LAMP amplicons. As a result, less $Ru(phen)_3^{2+}$ interlaces into the formed amplicons, and with a fixed $Ru(phen)_3^{2+}$ amount, an obviously increased ECL signal input can be observed. A detection limit as low as 10 fM for OTA has been achieved. This aptasensor also has good reproducibility and stability.

RCA Electrochemiluminescence NABs

Dong group developed a sensitive and selective DNA biosensing method with the detect limitation down to sub-femptomolar level [54]. This method combines horseradish peroxidase-mimicking DNAzyme amplification with chemiluminescent (CL) signaling. The DNAzyme synthesis was via combining a target DNA-mediated ligase reaction with rolling-circle amplification (RCA) (Fig. 13.36). As a result thousands of repeated horseradish peroxidase-mimicking DNAzyme units were produced. The CL signal of the luminol–H_2O_2 system was significantly enhanced by the RCA. This method can successfully apply in the genotyping or in allele discrimination. This novel chemiluminescence strategy

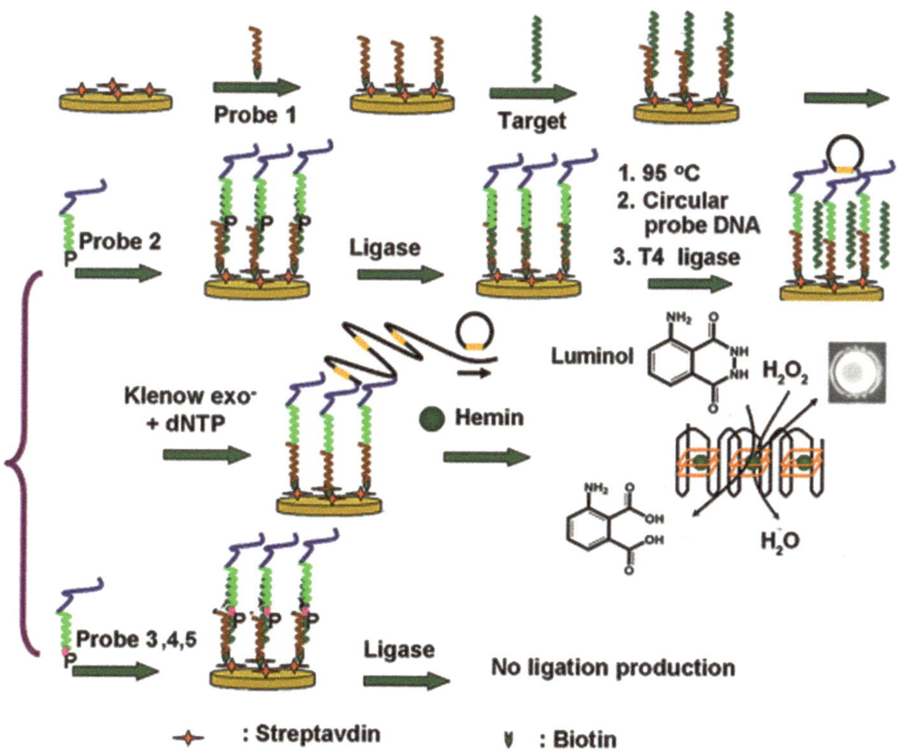

Fig. 13.36 Schematic of LAMP-mediated electronic biosensors (Reprinted from Ref. [54], Copyright 2013, with permission from Elsevier)

based on the ligation-mediated RCA synthesis of DNAzyme can discriminated single-base mismatches and efficiently promoted signal amplification for ultrasensitive target detection. The detection limit of this method was 0.26 fM and linear range was from 1×10^{-15} to 1×10^{-11} M with a detection limit of. This approach provided a robust, highly sensitive platform for DNA analysis.

13.3.2.2 Enzyme-Free Amplification-Dependent NABs

The Zhang group [55] successfully fabricated an electrochemical aptasensor with a novel signal amplification method for the sensitive detection of thrombin. The signal amplification method relies on the hybridization chain reaction (HCR) and an enzyme catalytic system (Fig. 13.37). Double-stranded DNA (dsDNA) polymers, which are produced by HCR, can form duplex structures that are effective for carrying large amounts of methylene blue (MB), an electron mediator, to produce a differential pulse voltammetry (DPV) signal. In addition, hemin can also be anchored by intercalating split DNAzyme into two induced single-stranded DNA (ssDNA) hairpins to form hundreds of thousands of hemin/G-quadruplex DNAzyme units. Upon the addition of NADH into the electrolytic cell, the hemin/G-quadruplexes function as NADH oxidase and HRP-mimicking DNAzyme in a pseudobienzyme amplifying system, which can catalyze the formation of H_2O_2 and result in significant signal enhancements. The binding reaction can be detected by a decrease in the integrated charge of MB, which adsorbs onto the dsDNA polymers. In the presence of target, the dsDNA polymers are disassembled, and the adsorbed MB and hemin/G-quadruplex structures are removed from the electrode surface, leading to a dramatic decrease in redox current. This method can quantitatively measure thrombin via DPV signals with a linear calibration range of

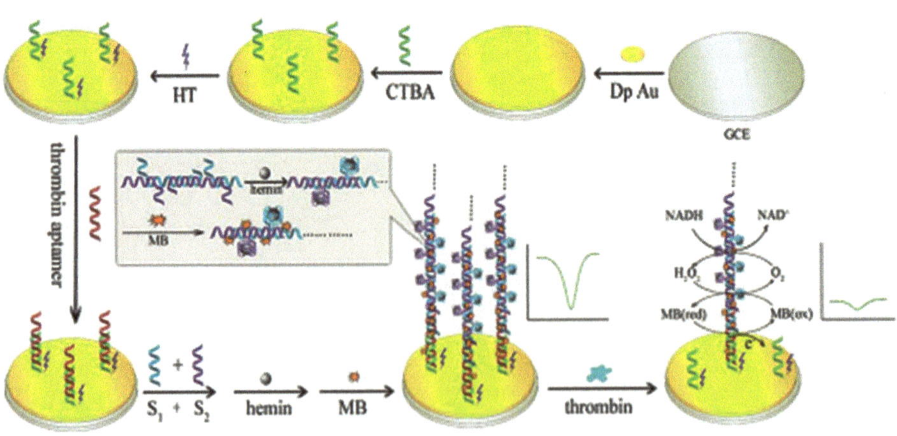

Fig. 13.37 Schematic illustration of the electrochemical aptasensor based on HCR with hemin/G-quadruplex DNAzyme amplification (Reproduced from Ref. [55] by permission of The Royal Society of Chemistry)

0.01–50 nM and a detection limit of 2 pM. This electronic aptasensor has shown good specificity, as well as acceptable reproducibility and stability, indicating a potential for use in various protein analysis applications.

13.4 NABs for Food Safety

13.4.1 NABs Against Biologic Contaminations

Toxins Mycotoxins are the main toxins found in our food. The damage to crops and agro-products caused by mycotoxins directly and indirectly induces serious economic losses on a global scale. Many mycotoxins can cause serious harm to human health, or even death, through mycotoxin-contaminated food; in particular, aflatoxins and ochratoxins are the most toxic and common contaminants. Ochratoxin A is a mycotoxin that can be detected successfully in food with high specificity using a specific DNA aptamer [56]. The advantages of this aptamer include its high accuracy and stability, as well as its easy modification with chemical products. It has also been integrated into several biosensor detection systems, including electrochemical, electrochemiluminescent, colorimetric, and fluorescent platforms. Bacterial endotoxins are also major contaminants causing septic shock in humans and animals. *Staphylococcus aureus* enterotoxin B (SEB), encoded by the SEB gene, is a crucial toxin causing food poisoning and an important biological warfare agent. Liu Nan created a new type of nucleic acid biosensor for the fast, sensitive, and specific detection of the SEB gene. In this sensor, bilayer lipid membranes (BLMs) served as sensitive molecular recognition elements that were fixed onto oligonucleotide probes specific to the SEB gene. The biosensor was fabricated by utilizing patch-clamp pipette technology, which is characterized in sensitive changes of ionic current through physiological cellular membranes. Seeds of the leguminous herb lupin have been widely used as a low-cost protein source, but the allergic reactions to lupin seeds are disturbing. A DNA aptamer made by Nadal et al. can be used in a colorimetric sensing system to detect lupin successfully. Endocrine-disrupting compounds (EDCs) can cause severe health hazards by disrupting normal endocrine functions in humans and aquatic organisms; one example is 17β-estradiol, which greatly affects the male reproductive system [57]. EDCs can also be detected in environmental water samples using a highly selective and rapid fluorescence-based DNA aptasensor.

Viruses The development of modern clinical diagnosis methods is moving toward new combinations of sensor technologies with molecular diagnostic techniques. Yao C et al. constructed a peptide nucleic acid (PNA) piezoelectric biosensor for the real-time monitoring of hepatitis B virus (HBV) genomic DNA hybridization [58]. The PNA probe can combine with target DNA sequences more effectively and specifically than can a DNA probe. The PNA probe was designed and immobilized onto the surface of the biosensor as a substitute for the conventional DNA probe, thereby enabling the direct detection of HBV genomic DNA without previous amplification by PCR. This label-free method is rapid and highly sensitive and can be a useful supplement to commercial assay methods in clinical chemistry [59]. In addition, several aptasensors have been developed to detect viral proteins. HIV-1 Tat protein was successfully detected by immobilizing an RNA aptamer on a piezoelectric quartz crystal; this method showed great sensitivity, specificity, and reproducibility. The observed detection limit of 0.25 ppm rivaled that of the available immunosensor with immobilized anti-Tat antibodies. The quartz crystal microbalance (QCM)-based aptasensor and the corresponding surface plasmon resonance (SPR)-based aptasensor were both developed via biotin–avidin linking onto the gold surface of the transducers (i.e., quartz crystals or chips) for immobilization. These two different platforms also have similar reproducibility, sensitivity, and specificity. A novel leaky surface acoustic wave (LSAW) bis-peptide nucleic acid (bis-PNA) biosensor with double two-port resonators has been constructed successfully for the quantitative detection of human papilloma virus (HPV). The bis-PNA probe can directly detect HPV genomic DNA without polymerase chain reaction (PCR) amplification, and it can bind to the target DNA sequences more effectively and specifically than can a DNA probe. The detection limit was 1.21 pg/L, and the clinical specificity was 97.22 % that of real-time PCR. The bis-PNA probe was able to distinguish sequences that differ in only one base. Both the intra- and interassay coefficients of variance (CVs) were <10 %, and the biosensor could be regenerated ten times without an appreciable loss of activity. Therefore, this technical LSAW biosensor platform can be applied for direct HPV detection in clinical samples [60].

Bacteria Using nucleic acid-based biosensors (NABs) to detect bacteria is a relatively new area. According to the literature, unmodified gold nanoparticles can be used in an aptamer-based, colorimetric biosensor to detect these bacteria effectively. Cholera is another food-borne disease, which is caused by Vibrio cholera, and a DNA biosensor has been developed for the detection of PCR amplicons of *Vibrio cholera*. Salmonellosis is a globally widespread disease caused by *Salmonella* sp. Recently, it has been reported that DNA biosensors can be used for the detection of an invA gene of *Salmonella* via an SPR detection method [61].

13.4.2 NABs for Chemical Contaminants

Heavy Metal Ions Great concern about heavy metal ions exists not only in the scientific community but also among the general public. Heavy metal ions causing soil and water pollution have become severe issues worldwide due to their high toxicity and persistent bioaccumulation [62]. Lead ions (Pb^{2+}), mercury ions (Hg^{2+}), and silver ions (Ag^{+}) are three widespread metal pollutants that may cause adverse effects on human health even at low concentrations. Recently, functional nucleic acid-based sensors have shown great potential for the detection of heavy metal ions. Functional nucleic acids are single-stranded oligonucleotides obtained from the systematic evolution of ligands by an exponential enrichment procedure

(SELEX); DNAzymes and aptamers are two major types of functional nucleic acids. DNAzymes have a catalytic capability similar to that of protein enzymes. And because their synthesis is low cost and their hydrolysis is more stable than that of protein enzymes, DNAzymes are widely applied for the detection of DNAs, proteins, and metal ions. In addition, most DNAzymes can be denatured and renatured many times without losing substrate binding ability or activity. For instance, a Pb^{2+}-specific 8–17 DNAzyme is able to specifically catalyze the cleavage of substrate RNA with Pb^{2+} as the cofactor. Aptamers show very high affinity and specificity toward proteins, cancer cells, and small molecules. In addition, aptamers are more easily obtained than antibodies and can be easily chemically modified with various chemical tags, such as fluorescent labels, electrochemical indicators, and nanoparticles. Notably, two simple and specific mismatched base pair aptamers, i.e., thymine–Hg^{2+}–thymine (T-Hg^{2+}-T) and cytosine–Ag^{+}–cytosine (C-Ag^{+}-C) [63], have been used for Hg^{2+} and Ag^{+} assays. Three typical heavy metal environmental pollutants are Pb^{2+}, Hg^{2+}, and Ag^{+}, and the interactions between them and with functional nucleic acids are being extensively investigated.

Antibiotics Antibiotics are often administered to farm animals along with their feed for prophylactic and therapeutic purposes. However, a large percentage of these antibiotics remain unmetabolized and either accumulates in tissues or is excreted into the environment. The presence of these types of antibiotics in the environment might support antibiotic resistance; in addition, these antibiotics may be transmitted to humans in several ways, e.g., via the food chain. Burke's group was the first to report the use of RNA aptamers against chloramphenicol [64]. Recently, Mehta et al. developed DNA aptamers to detect chloramphenicol with remarkable higher selectivity and affinity [65]. Tetracyclines are components of a group of broad-spectrum antibiotics; by interfering with the binding of aminoacyl-tRNA to ribosomes, this group is able to inhibit prokaryotic translation of A sites [66]. Tetracyclines are often used as a veterinary drug to promote animal growth. Thus, antibiotic residues can be detected in honey, milk, meat, and eggs. In addition, tetracyclines have also been reported to be hepatotoxic to pregnant women. Recently, researchers have developed aptamer-based systems for detecting tetracyclines. Berens et al. first screened an RNA aptamer (cb28) against tetracycline in a manner similar to its interaction with the small ribosomal subunit [67]. Recently, Kim et al. developed an ssDNA electrochemical aptasensor with a higher specificity for tetracyclines over other derivatives [68]. Aminoglycosides are a class of antibiotics that binds prokaryotic ribosomes and can eventually cause cell death by producing nonsense peptides due to frameshift mutations. Because of toxic side effects to the kidney and ear, its clinical use has been restricted. Tobramycin, kanamycin, and neomycin are antibiotic examples of aminoglycosides. Initially, Wang and Rando developed RNA aptamers against tobramycin. Recently, Song et al. have screened an ssDNA aptamer against kanamycin. The aptamers conjugated with gold nanoparticles have showed dissociation constants in the nanomolar range. Due to this observation, it is feasible to detect kanamycin using colorimetric methods [69].

13.4.3 NABs Against Genetically Modified Foods (GMFs)

Upon the recognition of specific nucleic acid fragments, we can detect genetically modified food components. Promoters and terminators regulate the expression of fresh genes introduced into GMOs. The 35S promoter (derived from cauliflower mosaic virus) and the nos terminator (derived from *Agrobacterium tumefaciens*) sequences are most widely used for this purpose. The recognition of one of these regulatory sequences in the DNA sample under examination serves as evidence demonstrating the existence of GMOs. Along with the progression of modern molecular techniques, screening for genetically modified food compounds has also developed. In this area, capillary electrophoresis chips, microtiter well-based hybridization assays, microarrays, and real-time or competitive PCR have provided very hopeful developments. In addition, biosensors are expected to be a leading approach available in the near future for the low-cost, rapid, and simple execution of routine GMO screening tests [70]. We can apply many types of biosensors, including surface plasmon resonance sensors, quartz crystal microbalance piezo-electric sensors, and electrochemical sensors, to detect DNA fragments that are specific to genetically modified food compounds. Carpini et al. [71] reported the use of an *enzyme-based* electrochemical sensor to detect the 35S promoter sequence (amplified by PCR) via oligonucleotide-modified screen-printed gold electrodes. With the application of a similar biosensor, Lucarelli et al. [72] also performed the enzyme-based electrochemical detection of PCR products; the sensor was used to analyze soy and maize powder samples containing 1 % and 5 % of genetically modified product. Moreover, Meric et al. designed another electrochemical sensor to detect the nos terminator sequence [73]. This biosensor functioned through immobilizing oligonucleotide probes on a screen-printed carbon electrode surface; methylene blue was used as a hybridization indicator. Although biosensing applications in GMO screening have been extensively researched, further investigation remains in high demand.

13.4.4 NABs Against Biomarkers

Generally, it is beneficial in the early diagnosis of any disease to detect biomarkers present in blood, urine, and other bodily fluids. Various working principles of aptasensor assays can be utilized for biological molecule detection. Pu et al. developed a high sensitivity and specificity smart polymeric transducer and aptamer/intercalating dye system that allow the label-free detection of thrombin [74]. A method based on the combination of ex situ polarization modulation FTIR measurements of the RNA monolayer and in situ surface plasmon resonance imaging (SPRI) measurements of hybridized DNA adsorption onto the surface has been developed to detect protein factor IXa.

13.5 Commercialization and Prospects of NABs

In food safety applications, it is important to achieve the timely monitoring and regulation of various parameters. Therefore, having reliable analytical devices that can perform rapid and accurate analyses is a pressing need. Biosensors using the specificity of a biological molecule in conjunction with a physicochemical transducer to convert a signal (usually very weak) into a comprehensible electrical signal have been developed [75]. There are relatively new and exciting areas in analytics for the application of nucleic acids as a recognition layer in biosensor project [76].

The development of biosensor technology in the context of food safety has made great progress recently. Because of their high sensitivity and selectivity, NABs are deemed to a powerful and inexpensive food safety monitoring [77].

Some commercialized commodities for pathogen detection of food safety monitoring have been generated through advancements in NABs [78]. The advancement in the nucleic acid-based probe technology has led to several commercialized products for pathogen detection mostly for clinically relevant samples (Table 13.1).

At present, the vast majority of commercialized products based on NABs have applications in clinical diagnostics and laboratory analysis. For food safety monitoring, NABs are in an early stage of development, but in near future, the advanced level of food safety monitoring will be essentially dependent on the state-of-the-art biosensors.

Table 13.1 A list of commercially available nucleic acid-based biosensors for pathogen detection with their mode of detection and the sample source

Organism	Company	Product name	Detection method	Sample source
Candida sp.	Becton Dickinson, Inc.	BD Affirm™ APIII	DNA hybridization	Vaginal swab
Chlamydia trachomatis	Qiagen	HC2 CT-ID	Chemiluminescence	Endocervical swab
	Gen-probe	APTIMA® CT	TMA[a]/16S RNA	Urine/urethral swab
	Gen-probe	PACE2 CT	HPA[b]	Endocervical swab
	Becton Dickinson, Inc.	BD ProbeTec™ CT	SDA[c]	Endocervical swab
	Roche	COBAS AMPLICOR CT	PCR	Endocervical/ urethral swab
Escherichia coli O157:H7	Qualicon, Inc.	BAX system	Real-time PCR	Water

(continued)

Table 13.1 (continued)

Organism	Company	Product name	Detection method	Sample source
Gardnerella	Becton Dickinson, Inc.	BD Affirm™ APIII	DNA hybridization	Vaginal swab
Mycobacterium avium	Accuprobe®	Gen-probe	TMA/RNA	Culture
Mycobacterium gordonae	Accuprobe®	Gen-probe	TMA/RNA	Culture
Mycobacterium intracellulare	Accuprobe®	Gen-probe	TMA/RNA	Culture
Mycobacterium kansasii	Accuprobe®	Gen-probe	TMA/RNA	Culture
Mycobacterium tuberculosis	Accuprobe® MTD	Gen-probe	TMA	Sputum
	BD ProbeTec™ ET	Becton Dickinson, Inc.	SDA	Respiratory and non-respiratory
	COBAS AMPLICOR MTB	Roche	PCR	Respiratory and non-respiratory
Neisseria gonorrhoeae	Qiagen	HC2 GC-ID	Chemiluminescence	Endocervical swab
	Gen-probe	APTIMA® GC	TMA/16S RNA	Urine/urethral swab
	Gen-probe	PACE2 GC	HPA	Endocervical swab
	Becton Dickinson, Inc.	BD ProbeTec™ GC	SDA	Endocervical swab
	Roche	COBAS AMPLICOR NG	PCR	Endocervical/urethral swab
Streptococci Group A	Gen-probe	GASDirect®	HPA	Pharyngeal swab
Streptococci Group B	Infectio Diagnostic Inc.	IDI-Strep B	Real-time PCR	Vaginal swab
Trichomonas vaginalis	Gen-probe	APTIMA®	TMA/16S RNA	Urine/vaginal swab
	Becton Dickinson, Inc.	BD Affirm™ APIII	DNA hybridization	Vaginal swab

[a]Transcription-mediated amplification (TMA)
[b]Hybridization probe assay (HBA)
[c]Strand displacement amplification (SDA)

Acknowledgments This work is supported by the Ministry of Science and Technology of Beijing (XX2014B069). Many thanks to Longjiao Zhu and Yuting Chen for their kind help in manuscript conception and preparation.

References

1. Dell'Atti D, Tombelli S, Minunni M, Mascini M. Detection of clinically relevant point mutations by a novel piezoelectric biosensor. Biosens Bioelectron. 2006;21(10):1876–9.
2. Mikkelsen SR. Electrochemical biosensors for DNA sequence detection. Electroanalysis. 1996;8(1):15–9.
3. Zhai J, Cui H, Yang R. DNA based biosensors. Biotechnol Adv. 1997;15(1):43–58.
4. Wang J. Survey and summary from DNA biosensors to gene chips. Nucleic Acids Res. 2000;28(16):3011–6.
5. Herne TM, Tarlov MJ. Characterization of DNA probes immobilized on gold surfaces. J Am Chem Soc. 1997;119(38):8916–20.
6. Teller C, Willner I. Functional nucleic acid nanostructures and DNA machines. Curr Opin Biotechnol. 2010;21(4):376–91.
7. Liu J, Cao Z, Lu Y. Functional nucleic acid sensors. Chem Rev. 2009;109(5):1948–98.
8. Li Y, Breaker RR. Deoxyribozymes: new players in the ancient game of biocatalysis. Curr Opin Struct Biol. 1999;9(3):315–23.
9. Li Y, Sen D. A catalytic DNA for porphyrin metallation. Nat Struct Mol Biol. 1996;3 (9):743–7.
10. Travascio P, Li Y, Sen D. DNA-enhanced peroxidase activity of a DNA aptamer-hemin complex. Chem Biol. 1998;5(9):505–17.
11. Wang M, Han Y, Nie Z, Lei C, Huang Y, Guo M, Yao S. Development of a novel antioxidant assay technique based on G-quadruplex DNAzyme. Biosens Bioelectron. 2010;26(2):523–9.
12. Wilson DS, Szostak JW. In vitro selection of functional nucleic acids. Annu Rev Biochem. 1999;68(1):611–47.
13. Breaker RR. In vitro selection of catalytic polynucleotides. Chem Rev. 1997;97(2):371–90.
14. Breaker RR. DNA aptamers and DNA enzymes. Curr Opin Chem Biol. 1997;1(1):26–31.
15. Liu J, Lu Y. A colorimetric lead biosensor using DNAzyme-directed assembly of gold nanoparticles. J Am Chem Soc. 2003;125(22):6642–3.
16. Liu J, Lu Y. A DNAzyme catalytic beacon sensor for paramagnetic Cu2+ ions in aqueous solution with high sensitivity and selectivity. J Am Chem Soc. 2007;129(32):9838–9.
17. Dichtl B, Pan T, DiRenzo AB, Uhlenbeck OC. Replacement of RNA hairpins by in vitro selected tetranucleotides. Nucleic Acids Res. 1993;21(3):531–5.
18. Nagatoishi S, Isono N, Tsumoto K, Sugimoto N. Loop residues of thrombin-binding DNA aptamer impact G-quadruplex stability and thrombin binding. Biochimie. 2011;93(8):1231–8.
19. Frank-Kamenetskii MD, Mirkin SM. Triplex DNA structures. Annu Rev Biochem. 1995;64 (1):65–95.
20. Ikbal J, Lim GS, Gao Z. The hybridization chain reaction in the development of ultrasensitive nucleic acid assays. TrAC Trends Anal Chem. 2015;64:86–99.
21. Clegg RM. Fluorescence resonance energy transfer. Curr Opin Biotechnol. 1995;6(1):103–10.
22. Jin R, Wu G, Li Z, Mirkin CA, Schatz GC. What controls the melting properties of DNA-linked gold nanoparticle assemblies? J Am Chem Soc. 2003;125(6):1643–54.
23. Li Y, Schluesener HJ, Xu S. Gold nanoparticle-based biosensors. Gold Bull. 2010;43 (1):29–41.
24. Zhou Y, Wang S, Zhang K, Jiang X. Visual detection of copper (II) by azide-and alkyne-functionalized gold nanoparticles using click chemistry. Angew Chem. 2008;120(39):7564–6.
25. Song K-M, Lee S, Ban C. Aptamers and their biological applications. Sensors. 2012;12 (1):612–31.
26. Wei H, Li B, Li J, Wang E, Dong S. Simple and sensitive aptamer-based colorimetric sensing of protein using unmodified gold nanoparticle probes. Chem Commun. 2007;36:3735–7.
27. Wang Z, Lee JH, Lu Y. Label-free colorimetric detection of lead ions with a nanomolar detection limit and tunable dynamic range by using gold nanoparticles and DNAzyme. Adv Mater. 2008;20(17):3263–7.

28. Song S, Wang L, Li J, Fan C, Zhao J. Aptamer-based biosensors. TrAC Trends Anal Chem. 2008;27(2):108–17.
29. Liu J, Lu Y. Preparation of aptamer-linked gold nanoparticle purple aggregates for colorimetric sensing of analytes. Nat Protocol Electron Ed. 2006;1(1):246.
30. Cheng S, Zheng B, Wang M, Lam MH-W, Ge X. Double-functionalized gold nanoparticles with split aptamer for the detection of adenosine triphosphate. Talanta. 2013;115:506–11.
31. Yang C, Lates V, Prieto-Simón B, Marty J-L, Yang X. Aptamer-DNAzyme hairpins for biosensing of Ochratoxin A. Biosens Bioelectron. 2012;32(1):208–12.
32. Zhou X-H, Kong D-M, Shen H-X. G-quadruplex–hemin DNAzyme-amplified colorimetric detection of Ag+ ion. Anal Chim Acta. 2010;678(1):124–7.
33. Pavlov V, Xiao Y, Gill R, Dishon A, Kotler M, Willner I. Amplified chemiluminescence surface detection of DNA and telomerase activity using catalytic nucleic acid labels. Anal Chem. 2004;76(7):2152–6.
34. Xiao Y, Pavlov V, Niazov T, Dishon A, Kotler M, Willner I. Catalytic beacons for the detection of DNA and telomerase activity. J Am Chem Soc. 2004;126(24):7430–1.
35. Navani NK, Li Y. Nucleic acid aptamers and enzymes as sensors. Curr Opin Chem Biol. 2006;10(3):272–81.
36. Li F, Zhang H, Wang Z, Newbigging AM, Reid MS, Li X-F, Le XC. Aptamers facilitating amplified detection of biomolecules. Anal Chem. 2014;87(1):274–92.
37. Yang X, Qian J, Jiang L, Yan Y, Wang K, Liu Q, Wang K. Ultrasensitive electrochemical aptasensor for ochratoxin A based on two-level cascaded signal amplification strategy. Bioelectrochemistry. 2014;96:7–13.
38. Lizardi PM, Huang X, Zhu Z, Bray-Ward P, Thomas DC, Ward DC. Mutation detection and single-molecule counting using isothermal rolling-circle amplification. Nat Genet. 1998;19 (3):225–32.
39. Notomi T, Okayama H, Masubuchi H, Yonekawa T, Watanabe K, Amino N, Hase T. Loop-mediated isothermal amplification of DNA. Nucleic Acids Res. 2000;28(12):e63.
40. Li B, Chen X, Ellington AD. Adapting enzyme-free DNA circuits to the detection of loop-mediated isothermal amplification reactions. Anal Chem. 2012;84(19):8371–7.
41. Parida M, Sannarangaiah S, Dash PK, Rao P, Morita K. Loop mediated isothermal amplification (LAMP): a new generation of innovative gene amplification technique; perspectives in clinical diagnosis of infectious diseases. Rev Med Virol. 2008;18(6):407–21.
42. Yan L, Zhou J, Zheng Y, Gamson AS, Roembke BT, Nakayama S, Sintim HO. Isothermal amplified detection of DNA and RNA. Mol BioSyst. 2014;10(5):970–1003.
43. Gill P, Ghaemi A. Nucleic acid isothermal amplification technologies—a review. Nucleosides Nucleotides Nucleic Acids. 2008;27(3):224–43.
44. Jeong Y-J, Park K, Kim D-E. Isothermal DNA amplification in vitro: the helicase-dependent amplification system. Cell Mol Life Sci. 2009;66(20):3325–36.
45. Yang L, Fung CW, Cho EJ, Ellington AD. Real-time rolling circle amplification for protein detection. Anal Chem. 2007;79(9):3320–9.
46. Cho EJ, Yang L, Levy M, Ellington AD. Using a deoxyribozyme ligase and rolling circle amplification to detect a non-nucleic acid analyte, ATP. J Am Chem Soc. 2005;127(7):2022–3.
47. Di Giusto DA, Wlassoff WA, Gooding JJ, Messerle BA, King GC. Proximity extension of circular DNA aptamers with real-time protein detection. Nucleic Acids Res. 2005;33(6):e64.
48. Liu M, Song J, Shuang S, Dong C, Brennan JD, Li Y. A graphene-based biosensing platform based on the release of DNA probes and rolling circle amplification. ACS Nano. 2014;8 (6):5564–73.
49. Tang L, Liu Y, Ali MM, Kang DK, Zhao W, Li J. Colorimetric and ultrasensitive bioassay based on a dual-amplification system using aptamer and DNAzyme. Anal Chem. 2012;84 (11):4711–7.
50. Wu H, Liu Y, Wang H, Wu J, Zhu F, Zou P. Label-free and ultrasensitive colorimetric detection of DNA based on target-triggered quadratic amplification strategy. Biosens Bioelectron. 2015;66:277–82.

51. Zhang L, Zhang Y, Wang C, Feng Q, Fan F, Zhang G, Kang X, Qin X, Sun J, Li Y. Integrated microcapillary for sample-to-answer nucleic acid pretreatment, amplification, and detection. Anal Chem. 2014;86(20):10461–6.
52. Shimron S, Wang F, Orbach R, Willner I. Amplified detection of DNA through the enzyme-free autonomous assembly of hemin/G-quadruplex DNAzyme nanowires. Anal Chem. 2011;84(2):1042–8.
53. Yuan Y, Wei S, Liu G, Xie S, Chai Y, Yuan R. Ultrasensitive electrochemiluminescent aptasensor for ochratoxin A detection with the loop-mediated isothermal amplification. Anal Chim Acta. 2014;811:70–5.
54. Dong H, Wang C, Xiong Y, Lu H, Ju H, Zhang X. Highly sensitive and selective chemiluminescent imaging for DNA detection by ligation-mediated rolling circle amplified synthesis of DNAzyme. Biosens Bioelectron. 2013;41:348–53.
55. Zhang J, Chai Y, Yuan R, Yuan Y, Bai L, Xie S, Jiang L. A novel electrochemical aptasensor for thrombin detection based on the hybridization chain reaction with hemin/G-quadruplex DNAzyme-signal amplification. Analyst. 2013;138(16):4558–64.
56. Cruz-Aguado JA, Penner G. Determination of ochratoxin A with a DNA aptamer. J Agric Food Chem. 2008;56(22):10456–61.
57. Chang H-S, Choo K-H, Lee B, Choi S-J. The methods of identification, analysis, and removal of endocrine disrupting compounds (EDCs) in water. J Hazard Mater. 2009;172(1):1–12.
58. Yao C, Zhu T, Tang J, Wu R, Chen Q, Chen M, Zhang B, Huang J, Fu W. Hybridization assay of hepatitis B virus by QCM peptide nucleic acid biosensor. Biosens Bioelectron. 2008;23(6):879–85.
59. Tang Z, Parekh P, Turner P, Moyer RW, Tan W. Generating aptamers for recognition of virus-infected cells. Clin Chem. 2009;55(4):813–22.
60. Lee S, Kim YS, Jo M, Jin M, Lee D-K, Kim S. Chip-based detection of hepatitis C virus using RNA aptamers that specifically bind to HCV core antigen. Biochem Biophys Res Commun. 2007;358(1):47–52.
61. Zhang D, Yan Y, Li Q, Yu T, Cheng W, Wang L, Ju H, Ding S. Label-free and high-sensitive detection of Salmonella using a surface plasmon resonance DNA-based biosensor. J Biotechnol. 2012;160(3):123–8.
62. Que EL, Domaille DW, Chang CJ. Metals in neurobiology: probing their chemistry and biology with molecular imaging. Chem Rev. 2008;108(5):1517–49.
63. Ono A, Cao S, Togashi H, Tashiro M, Fujimoto T, Machinami T, Oda S, Miyake Y, Okamoto I, Tanaka Y. Specific interactions between silver (I) ions and cytosine–cytosine pairs in DNA duplexes. Chem Commun. 2008;39:4825–7.
64. Burke DH, Gold L. RNA aptamers to the adenosine moiety of S-adenosyl methionine: structural inferences from variations on a theme and the reproducibility of SELEX. Nucleic Acids Res. 1997;25(10):2020–4.
65. Mehta J, Van Dorst B, Rouah-Martin E, Herrebout W, Scippo M-L, Blust R, Robbens J. In vitro selection and characterization of DNA aptamers recognizing chloramphenicol. J Biotechnol. 2011;155(4):361–9.
66. Spahn C, Prescott C. Throwing a spanner in the works: antibiotics and the translation apparatus. J Mol Med. 1996;74(8):423–39.
67. Berens C, Thain A, Schroeder R. A tetracycline-binding RNA aptamer. Bioorg Med Chem. 2001;9(10):2549–56.
68. Kim Y-J, Kim YS, Niazi JH, Gu MB. Electrochemical aptasensor for tetracycline detection. Bioprocess Biosyst Eng. 2010;33(1):31–7.
69. Song K-M, Cho M, Jo H, Min K, Jeon SH, Kim T, Han MS, Ku JK, Ban C. Gold nanoparticle-based colorimetric detection of kanamycin using a DNA aptamer. Anal Biochem. 2011;415(2):175–81.
70. Elenis DS, Kalogianni DP, Glynou K, Ioannou PC, Christopoulos TK. Advances in molecular techniques for the detection and quantification of genetically modified organisms. Anal Bioanal Chem. 2008;392(3):347–54.

71. Carpini G, Lucarelli F, Marrazza G, Mascini M. Oligonucleotide-modified screen-printed gold electrodes for enzyme-amplified sensing of nucleic acids. Biosens Bioelectron. 2004;20 (2):167–75.
72. Lucarelli F, Marrazza G, Mascini M. Enzyme-based impedimetric detection of PCR products using oligonucleotide-modified screen-printed gold electrodes. Biosens Bioelectron. 2005;20 (10):2001–9.
73. Meric B, Kerman K, Marrazza G, Palchetti I, Mascini M, Ozsoz M. Disposable genosensor, a new tool for the detection of NOS-terminator, a genetic element present in GMOs. Food Control. 2004;15(8):621–6.
74. Pu F, Huang Z, Hu D, Ren J, Wang S, Qu X. Sensitive, selective and label-free protein detection using a smart polymeric transducer and aptamer/ligand system. Chem Commun. 2009;47:7357–9.
75. Turner A, Karube I, Wilson GS. Biosensors: fundamentals and applications. New York: Oxford University Press; 1987.
76. Paleček E, Fojta M, Tomschik M, Wang J. Electrochemical biosensors for DNA hybridization and DNA damage. Biosens Bioelectron. 1998;13(6):621–8.
77. Teles F, Fonseca L. Trends in DNA biosensors. Talanta. 2008;77(2):606–23.
78. Singh A, Poshtiban S, Evoy S. Recent advances in bacteriophage based biosensors for food-borne pathogen detection. Sensors. 2013;13(2):1763–86.

Chapter 14
PCR Methods for Detecting GM Crops and Food in Agriculture and the Food Chain: A Review

Abstract Frequent trade circulation and controversy over edible and environmental safety make genetically modified organisms (GMOs) a social concern. Nations have proposed labeling requirements to satisfy the consumers' rights and guarantee national commercial interests. Besides basic situation in global GM crops cultivation and production, the detecting techniques are frequently used for detecting genetically modified organisms (GMOs) from raw products. The techniques mainly include polymerase chain reaction (PCR), quantitative PCR (qPCR), digital PCR (dPCR), enzyme-linked immunosorbent assay (ELISA), biosensors, and many others and are systematically described and discussed in this chapter. The chapter mainly discusses the development of detecting techniques, as well as the endogenous genes, standard substances, and restraining factors. The problems and prospects of detection techniques of GMOs are also noted.

Keywords GM food • PCR • Flanking sequence • Biosensors • High-throughput detection

14.1 Introduction

A main aim for crop breeding technologies is to increase crop production and resist natural diseases. Crossing different varieties and species to select dominant species has become a preferred solution. However, intra-variety combination and selection involves mixing thousands of genes, with some unpredicted results. Radiation and chemical mutants are likely to carry a baggage of uncharacterized genetic changes with unknown effects. Genetically modification technology is a fast-developing breeding biotechnology that produces directional varieties. It is an innovative effort that involves inserting specific genes into the host genome artificially and, more importantly, specifically [1, 2]. In 1977, Eugene Nester, Milton Gordon, and Mary-Dell Chilton found that genes from plasmids carried by *Agrobacterium tumefaciens* could be inserted into the host plant DNA [3, 4]. The source of genetically modified organisms can be traced back to Paul Berg. This Nobel Prize winner in chemistry in 1980 constructed a DNA molecule by cutting viral and bacterial DNA with restriction enzymes and then recombining them [5]. The ensuring scientific developments and high demands for food security have given rise to genetically modified

W. Xu, *Functional Nucleic Acids Detection in Food Safety*,
DOI 10.1007/978-981-10-1618-9_14

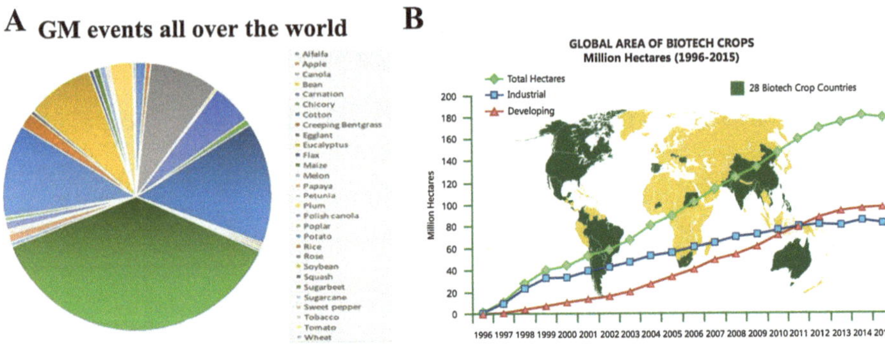

Fig. 14.1 GM crop situation in 2014. (**a**) GM crop categories commercialized all over the world, based on the ISAAA report. The top three are maize (*green*), soybean (*blue*), and canola (*yellow*). (**b**) Global area of biotech crops from 1996 to 2014 (Reproduced from James 2015)

organisms (GMOs). Now these GMOs led to the great progress, improving numerous species mainly with high yield, stress resistance, and superior agronomic performance as well as enhanced quality and nutrition [6–8].

Applicable outcomes of crop species are mainly maize, soybean, canola, cotton, sugar beet, alfalfa, and some other fruits and flowers [9]. The tendency of biotech crops to be adopted as a cultivation and import focus is a result of the national need for food, feed, clothing, and other diverse purposes, rather than solely related to scientific progress. Thus, the approval of GM events in various crops has varied from year to year since wide-scale commercialization began in 1996 [10–13]. Approval data can be obtained from the International Service for the Acquisition of Agri-biotech Applications (ISAAA) (http://www.isaaa.org). It features detailed information on GM events, including crops, traits, developer, years of approval for cultivation for domestic and nondomestic use, food and feed approvals for direct use or processing, lists of country authorizations and relevant dossiers sourced from credible government regulatory websites, and biosafety clearinghouses. Additionally, it links to applicable detection technologies worldwide (http://gmdd.shgmo.org and http://cera-gmc.org/GMCropDatabase) (Fig. 14.1).

A labeling system for genetically modified organism (GMO) regulation was further developed against the background of increasing imports and exports of GMOs as well as due to public concern about GMO for the safety issues. The regulation is mainly performed by governments or authorities (e.g., the European Union) and includes detecting GMO content, analyzing the nutritional and environmental changes, and determining the need for labeling. Sensitive and accurate detection techniques have therefore been developed to basically meet these requirements [14]. Over the past decade, several biotech crops have been developed, including insect pest-resistant transgenic cotton, maize, canola, and soybean, delayed-ripening tomato, antivirus peppers, and antivirus papaya. Rice (*Oryza sativa L.*) is the main grain in Southeast Asia, and two genetically modified species have been developed and certified for commercialization in China [15, 16]. The genetically modified potato made by Simplot Inc., which aims to eliminate acrylamide, a major cause of human cancer, has recently appeared on the market [17].

This review is to provide details on newly emerging detection methods for GMO identification. Ranging from PCR to biosensors, nucleic acid-based GMO detection techniques have developed into a diverse array of methods that combined newly emerging advances in sensitivity.

14.2 Qualitative PCR Detection Techniques

Polymerase chain reaction (PCR) is the common way to identify the GMOs in the grains or processed food products. PCR detection can be classified into four strategies: screening PCR, gene-specific PCR, construct-specific PCR, and event-specific PCR [18], in Fig. 14.2.

Screening PCR method is used to detect the screening elements of GMOs. Common screening elements include the promoters, the terminators, and some marker genes. 35S promoter of the cauliflower mosaic virus 35s promoter (CaMV 35S) and the terminator of *Agrobacterium tumefaciens* nopaline synthase (T-NOS) are the most common elements [19]. Screening PCR targets necessary elements used in transgenic technologies, however, is only suitable for the first round of sample screening because the 35S and ubiquitin promoter occur inherently in cauliflower mosaic virus and the maize. Therefore, experimental results using screening PCR can be compromised because of its tendency to cause false-positive results. Nevertheless, detection efficiency in the official work will be improved a lot by elements screening for a large amount of samples.

Gene-specific PCR detects the gene that expresses the main property of resistance, mainly derived from nature with slightly modification [20]. The main traits of the detected target can be obtained through gene-specific PCR. Therefore, gene-specific PCR is more specific than screening PCR. However, some studies have classified gene-specific PCR as screening PCR, stating that both of them are

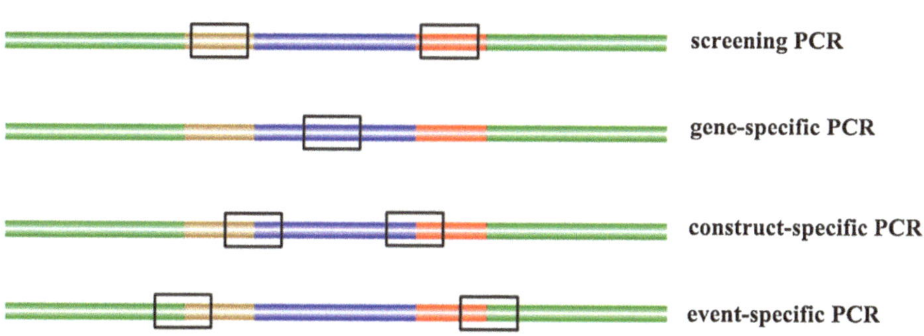

Fig. 14.2 Four levels of specificity of PCR analytical modules targeting DNA sequences (Reprinted from Ref. [18], Copyright 2012, with permission from Elsevier with slightly modified)

focused on a single element to provide clues to the identity of present commercialized GMOs [18].

Construct-specific PCR targets a complete transformation vector cassette which is more rigorous than gene-specific PCR. Using this method, the positive signals appear only in GMO material and the corresponding transformant [21]. Construct-specific modules provide indisputable evidence of the presence of GMOs and to some extent the corresponding GMO event. However, different GM crops may contain the same transformant but different insertion site (GM canola RF1 and RF2). Therefore, more specific methods need to be developed for events determination [22].

Event-specific PCR can detect the boundary between insertion vector and the host genome and, therefore, are able to identify the concrete GMO event [23]. Flanking sequence is the term to indicate the connecting region between inserted DNA fragment and the nearby host genome and is the first thing for molecular characterization of GMO. Thus, the flanking sequence is favorable for specific GMO detection [24]. Before approving a GMO for commercial use, governments or authorities always request the developer to provide the detailed insertion information (including flanking sequence), as well as a detection method specific for validation and further legal regulation. Therefore, event-specific PCR is suitable for the identification of authorized GMOs.

Additionally, nested and semi-nested PCR can be helpful in trace quantity detection of GMOs. Nested PCR involves the design of two pairs of primers to detect target genes. The first round of PCR is identical to the normal PCR, with a pair of primers to bind to the template and extend, and the amplicon is used as the template in the second round of PCR [25]. Semi-nested PCR is basically the same as nested PCR, except the utilization of only three primers. These two methods can amplify the positive signal by several orders of magnitude, improving the overall specificity and accuracy and reducing false-positive results. Huang and Luo have successfully detected GM soybeans and derived processed food products by nested and semi-nested PCR, showing that damaged or degraded DNA can be detected by both nested PCR and semi-nested PCR [26].

Multiplex PCR is often applied in GMOs screening in a high-throughput way. Multiplex PCR is the combination of several normal PCR, using more than one pair of primers targeting different elements or specific genes. Each pair of primers functions as the singleplex PCR does, and different lengths of amplicons are amplified in a single tube. Multiplex PCR provides a high throughput to screen possible GMOs with similar specificity and sensitivity. Multiple PCR can also provide a wide coverage of commercialized GMOs to screen positive sample in only one reaction [27–29]. Huang and Pan simultaneously detected MON810 and NK603 by multiplex PCR with a detection limit lower than 0.5 % [30]. Feng et al. [31] and Zhao et al. [32] built a multiplex PCR system to detect GMO components from soybeans, corn, and canola and applied it to practical samples (the derived products in the market). The most recent studies report more targets involved in the multiplex PCR method for screening or identification purposes [33, 34]. All research indicates that multiplex PCR is a simple, low-cost, and high-throughput technology for detecting GMOs.

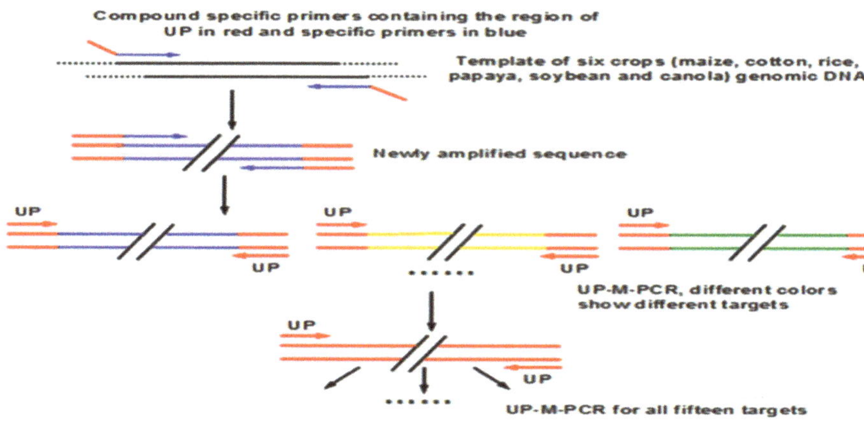

Fig. 14.3 The mechanism of UP-M-PCR. Each compound-specific primer contained a universal sequence at the 59-end (*red*) and the specific primer at the 39-end (*blue*). The amplified fragments with the primer pairs of different targets are individually marked in different colors. The amplified fragments only by the universal primer are marked in *red* (Reproduced from Xu et al. [36], with permission of PLOS)

Multiplex PCR still faces many challenges. Each pair of primers in the combination may not function at the same speed, resulting in different amplification efficiencies in the reaction. On the other hand, self-inhibition, complex manipulation, and amplification disparity are also evident in the multiplex system. Many modifications have been introduced to solve these problems. Xu and his coworkers have developed a universal primer multiplex PCR (UP-M-PCR) (Fig. 14.3). Specific primers in the reaction are less than in the normal multiplex reaction to prevent self-inhibition. The universal primer aims to amplify the target with the same efficiency, achieving high specificity and sensitivity [35, 36]. Shang and her coworkers developed a novel single universal primer multiplex ligation-dependent probe amplification (SUP-MLPA) technique to allow a target detection limit as low as 0.1 ng [37]. Guo et al. developed a quadruplex degenerate PCR to balance the annealing temperature for multiplex PCR [38].

Otherwise, multiplex PCR are now prone to screen for GMOs in the sample. Due to the increasing number of commercial GMO species each year, limited event-specific detection methods are not developing as quickly as new species are introduced. Screening methods are based on the first strategies of qualitative PCR for GMO detection, aiming at several common elements or marker genes. The screening results may not be accurate or sensitive, but they provide an alternative way for researchers to control the overall GMO pollution condition and to perform further species validation. Such successful applications are mostly focused on the promoters, terminators, and marker genes [35, 39–43].

14.3 Quantitative PCR

Quantitative PCR (qPCR) is a method for detecting actual GMO content. PCR was first quantified by competitive PCR, allowing the quantification of target DNA through the competition with known amounts of DNA. However, competitive PCR involves end point detection for the target quantity and cannot monitor the process of PCR.

The alternative real-time PCR technique solved the problem, which was first reported by Higuchi [44]. The first fully automated real-time PCR instrument in the world was produced by ABI company (now have been merged by Thermo Scientific Inc.) in 1996. This technology is dependent on the fluorescence accumulation throughput the whole process by adding a specific fluorophore. The fluorescence is shown as an amplification curve to show the amplification status on the process of PCR and helps calculate the quantity of the target. Compared with conventional PCR, real-time PCR can achieve sensitive, highly automated, and effective detection for GMOs identification (Fig. 14.4).

Fluorescent dye is the most common detection method for real-time quantitative PCR. SYBR Green I is most frequently used. SYBR Green I can bind to single-stranded DNA with lower performance and preferentially bind to the minor groove of double-stranded sequence to emit green light ($\lambda_{max} = 520$ nm) [45]. A detection probe with a labeled fluorophore is also added into the PCR reaction to replace fluorescent dye. This principle is based on fluorescence resonance energy transfer

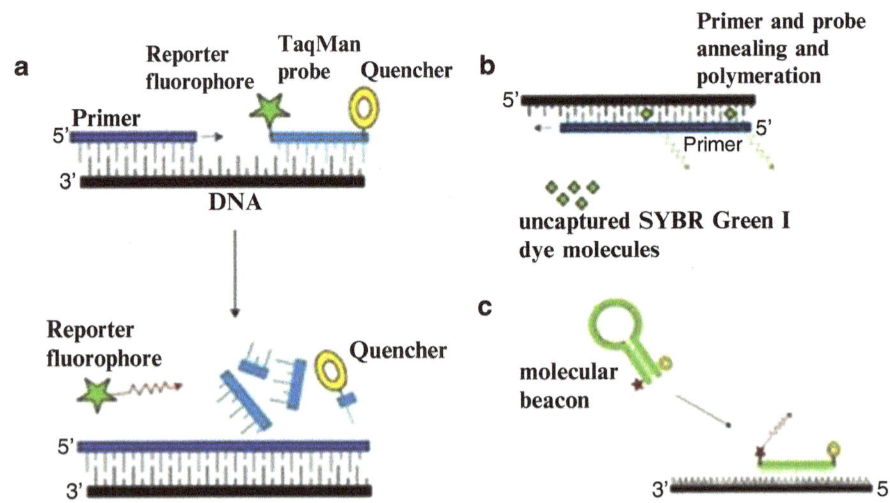

Fig. 14.4 Three types of qPCR detector. (**a**) *TaqMan* probe. The probe is labeled as a reporter on the 5′ end and a quencher on the 3′ end. When the primer extends, the *TaqMan* probe is cleavage to emit fluorescence. (**b**) SYBR Green I emit fluorescence when it binds to the double-stranded targets. (**c**) The molecular beacon makes the conformational change to emit the fluorescence on the presence of target

(FRET) phenomenon: when the excitation spectrum of a fluorophore overlaps with the excitation spectrum of another fluorophore (quencher), fluorophore is quenched by another fluorophore (quencher) and emits low fluorescence but emits high-density fluorescence when the excitation spectrum of fluorophore is apart from that of quencher [46]. Real-time PCR can capture the fluorescence on the process of amplification. Specific probe will be unclosed in the PCR reaction and being captured by the PCR instrument, thereby improving the specificity of quantification [47–51]. Common probes used in qPCR included the TaqMan probe, hybridization probe, and molecular beacon.

TaqMan probe binds at specific sites between the two primers using the principle of Watson–Crick base pairing. The reporter fluorophore (R), such as FAM or VIC, is labeled at the 5′ end of the probe, and the quencher group (Q), such as TAMRA or BHQ-1, is labeled at the 3′ end of the probe. The probe emits low fluorescence in the absence of target because of short distance between fluorophore and quencher [52]. When the template is added, the probe can bind to the corresponding target sequence and is cut off by polymerase with exonuclease activity. The distance between fluorophore and quencher increases, and high fluorescence is emitted because the reporter energy cannot be absorbed. Therefore, the fluorescent signal exhibits an exponential increase together with the amplicon. More probes are cleaved when more templates are added in the reaction, emitting higher fluorescence. The fluorescence is proportional to the initiate quantity of target, and the relationship helps to calculate the target copy number.

The threshold and Ct vale are important concepts in the data analysis. Baseline is the background noise of the reaction, normally identified as the mean fluorescence between 3 and 15 cycles. The threshold is an arbitrary fluorescence to demonstrate the beginning of the exponential amplification or the initial stage that can be quantified. The Ct is the PCR cycle at which each amplification plot reaches this threshold [53].

Real-time quantitative PCR needs to build a standard curve, which is built by detection of different concentrations of template [53]. Ct value is acquired from the amplification curve and being calculated into linear regression curve with the logarithm of initial concentration of target. A linear curve with correlation coefficient (R^2) >0.95 should be the optimal standard curve.

Digital PCR (dPCR) is a new PCR platform for target quantification. In dPCR reaction, the original PCR mixture is partitioned into a series of small reactions, each of which may contain several copies of target. The number of templates in the reaction totally follows a Poisson distribution, which means that most of the partitioned samples contain zero copies of the template, while others contain one or more copies [54]. PCR process is identical to the normal qPCR. By calculating the number of positive chambers and calculating them with Poisson distribution, the absolute copy number of the target templates in the volume can be determined. dPCR achieves the absolute quantification without reference materials and the standard curve. The dPCR method has been applied in GMO quantification because it can achieve actual absolute quantification [39, 55–59].

14.4 Event-Specific Detection of Genetically Modified Organisms

It has been acknowledged before that event-specific PCR is the most specific method for GMO event identification, not only for ordinary GMOs but also for stacked GMOs. According to the definition of the Organisation for Economic Cooperation and Development (OECD), a stacked GMO could be a retransformation of an existing transgenic line or hybridization between two or more GM transgenic lines [60]. Many GM crops have been successfully identified through event-specific detection, such as GTS 40-3-2 soybean [61–66], MON810 [30, 49, 50, 67], Bt11 [61, 68], MON863 [69–71], GA21 [24, 72], NK603 [73, 74], T25 corn [68, 75], and GT73 canola [24, 76, 77], etc. The core of event-specific detection method is to obtain the flanking sequences in GM crops. There have been several methods to evacuate flanking sequence in GMOs, such as inverse PCR (I-PCR), thermal asymmetric interlaced PCR (TAIL-PCR), ligation-mediated PCR (LM-PCR), etc.

I-PCR is a genome walking method that developed by Ochman [78]. During the experiment, DNA was firstly digested into pieces randomly using restriction endonuclease. Then, sequence was self-looped and ligated to form a circular target. Next, primers bind to the template as usual and extend the unknown parts of the genome. The amplicons contain the unknown sequence outside the two primers, which can be sequenced later to determine the unknown sequence. Flanking sequences of several species and corresponding event-specific method have been developed [79, 80]. I-PCR is efficient for unknown genome walking and is easy to design. However, it also requires a circular DNA template, which may reduce amplification efficiency.

TAIL-PCR is a modified method based on the thermal asymmetric PCR [81]. TAIL-PCR is accomplished by a series of nested sequence-specific primers and a short arbitrary primer. In TAIL-PCR, the nested sequence-specific primers are longer than the arbitrary primers, which have higher annealing temperature. The amplification of specific and nonspecific products can be controlled by varying the annealing temperature during the reaction. The long primers can anneal with a specific sequence at both high and low temperatures; however, the short arbitrary primers can only anneal with an unknown sequence at the low annealing temperature. Target genes are effectively amplified by alternating high and low annealing temperatures. The TAIL-PCR method can reduce the background of specific products of the non-flanking region and, at the same time, can produce more than two nested fragments. Compared with other methods, TAIL-PCR method is simple, specific, efficient, rapid, and sensitive. Hernandez et al. [50] and Yang et al. [69] obtained the flanking sequences of MON810 and MON863 corn, respectively, by TAIL-PCR.

LM-PCR was developed by Pfeifer et al. [82]. This technology is the combination of restriction enzymes digestion, synthetic joints ligation, linear PCR with one 5′ biotin-labeled specific primer targeting a known sequence, and exponential PCR with specific primers for the known sequence and adapter primers. In LM-PCR, one primer is designed based on the known sequence, while the other is the common adapter. Holck used this method to obtain the flanking sequence of GM corn MON810 [49]. However, LM-PCR has certain limitations: (1) LM-PCR requires enzymatic treatment to make a digested and ligated DNA, (2) 5′ site of the DNA template must be phosphorylated, and (3) linear PCR has low amplification efficiency to extend the overall reaction time. Therefore, scientists have tried to modify LM-PCR, and some of which have been reported. An A-T linker adapter PCR added an A at the 3′ ends of restriction fragments and modified the adapter to overcome the inherent limitations of the original ligation-mediated PCR method, such as low specificity and a lack of restriction enzyme choice [83]. A loop-linker PCR modification promotes the elongation of the 3′ end of the loop-linker adapter to generate a stem-loop structure in the ligation products in case of self-ligation of adapter [84]. Moreover, the suppressive effect of the stem-loop structure decreases nonspecific amplification and increases the success rate of the approach. Then, a processed flanking identification method called randomly broken fragment PCR with 5′ end-directed adaptor (RBF-PCR) circumvents the need for a restriction enzyme [85]. Details of these modifications are described in Chap. 6. Using these technologies, the flanking sequences of many species have been developed successfully.

14.5 Species-Specific Detection Issues to Be Settled

14.5.1 Endogenous Gene Detection

Many genes have been found as endogenous gene to be applied in GMO detection to prevent the false-positive amplification and provide a quality control in the amplification system, as well as to assist in GMO quantitation. Thus, endogenous genes of main cultivated crops all over the world have been developed, including the soybean lectin [86] and β-actin gene [87], corn invertase [40], ZssIIb and Zein genes [88, 89], canola HMG I/Y [90] and BnACCg8 gene [91], rice sps gene [92], cotton sad1 gene [93], papaya papain gene [94], and peach Lhcb2 gene [95].

Endogenous genes functioned as indicators in the species characterization and quality control in the GMO detection, which means problems in the PCR system if the endogenous genes detection got negative results. Furthermore, if endogenous genes are detected and transgenic elements are not, one can conclude that templates are GMO negative. Endogenous genes also function as reference for GMO quantification. An endogenous gene is expressed stably in the host genome, so its copy number is identical in all host species. Quantification of endogenous gene and target gene gives a GMO portion calculation. Copy number variation (CNV) and the primer specificity compromise may affect the quantification [96].

Currently, endogenous genes in many new species are being studied and some have got ideal results. On the other hand, many species still need development of their endogenous gene. Endogenous genes of some species are located in mitochondrial DNA (mtDNA) for its stable expression, making it difficult to determine accurate copy numbers. Ideally, every new GM event should have its own (at least one) suitable endogenous gene. The endogenous gene should be amplified on every wild type, and the corresponding copy numbers should be determined.

14.5.2 Production of Reference Materials

Reference materials are substances of which one or more properties are sufficiently well established to be used for the calibration of an apparatus, for the assessment of a measurement method, or for the verification of one or more measurements in other materials, based on the guidelines of the International Organization of Standardization (ISO 31: 2015). Certified reference materials (CRMs) are reference materials approved by authorities, which are used for quality control in the GMO detection. Main CRM authorities are the American Oil Chemists' Society (AOCS) and Institute for Reference Materials and Measurements (IRMM). Not all the GM events have CRMs, some of which are under interlaboratory validation. Currently, the CRMs for GMO detection include a list of 46 events from AOCS and 29 events from IRMM. CRMs are now offered as a concrete content of mixed powder and leaf tissue DNA. Based on Xu's perspective, certain factors, including matrix effects and GM exact percentage, can compromise the performance of CRM quantification, including the following [97]:

- CRMs are considered by weight as its unit. Most CRMs are listed as weight percentage in the package; however, the value of reported GM % depends on the calibrant used. In Europe, results calibrated with a known mass fraction lead to be expressed in mass fraction; likewise the calibration with a calibrant of a known copy number ratio leads to be expressed in copy number ratios.
- The number of CRMs is limited, and their preparation is time consuming.
- CRM materials must be mixed well and CRMs out of shelf life may not be convincing in DNA quantification.
- CRMs are prepared at lower concentrations (up to 10 %), and larger amounts of GMO content in the tested sample may result in difference in quantification.
- It is difficult to determine sensitivity from mixed samples using pure CRMs. On the real-time quantification of corn Bt11, the sensitivity of Bt11 changes when adding soybean DNA genome. Similar conclusion has also been drawn by Berdal and Holst-Jensen [61]. A combined standard material is needed [98].

Other types of standard materials are emerging to overcome those disadvantages, such as PCR amplification products and linear or circular plasmids [96, 99, 100]. PCR products are too inaccurate and unstable, while plasmids as references can circumvent those problems. The producing procedure is much similar to gene cloning in the field of molecular biology and mainly consists of enzyme digestion and ligation. Theoretically several elements or genes can be inserted into the plasmids. Huang and Pan have successfully made the quantitative detection of corn MON810 and corn NK603 based on the plasmid reference [30]. Kuribara et al. [101] have also constructed standard molecules as reference material for the quantitative detection of GM corn and soybean.

14.5.3 Techniques for Eliminating PCR Inhibitors

PCR inhibitors are substances that complicate ordinary PCR amplification and thereby identification of amplicons [102]. PCR can be inhibited from many sources, such as the bloods, fabrics, fats, or even glove powder in the laboratory. The inhibitors generally interfere with one or more steps in the reaction. Inhibitors have the ability of escaping from the DNA extraction through binding to the DNA, or they can protect cell from being lysed [103]. Many inhibitors can be influenced in the activity of DNA polymerase and blocking DNA from being recognized by primers [104, 105]. Therefore, PCR regents must be purified thoroughly to ensure that sufficient quantity of DNA can be amplified. There are methods to remove these contaminants, including dialysis and CsCl gradient density centrifugation for the purification of DNA [106]. Both methods are simple, but final DNA template concentration is relatively low. Other methods aim at one specific contaminant, such as molecular sieve separation, ion-exchange chromatography, and anion binding resin [107]. They are effective but also expensive and tend to increase the risk of cross contamination, consequently leading to false-positive results [108].

Researchers are targeting on finding effective solutions to these problems [109–111] or to improve the tolerance of PCR detection [112]. Other methods, such as nested PCR, are also suitable for overcoming PCR inhibitors.

14.6 Biosensing Techniques for Genetically Modified Organisms

New fundamental approaches offer a wide platform for qualitative GMO detection. The third-generation PCR detection method (digital PCR) based on limiting dilution and statistical analysis has been applied in qualitative GMO screening and event-specific detection [39]. Han et al. [113] inserted an event-specific reaction system into a liquid bead array system to construct a ten-plex sensitive (0.1 % LOD) detection system for genetically modified maize (Fig. 14.5a). Shao et al. [114] developed a multiplex PCR amplification and hybridization microarray (MACRO) system as a readout signal on a chip for specific and convenient GMO monitoring. Isothermal amplification provides the potential to achieve rapid point-of-care

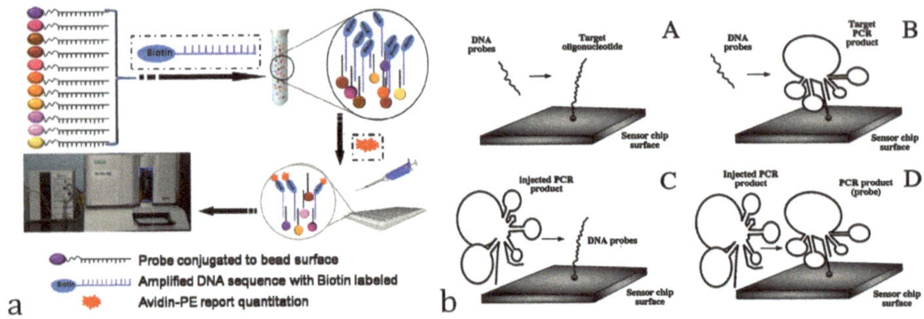

Fig. 14.5 (**a**) Outline of the suspension array method used for the simultaneous and rapid detection of ten GM maize (Reprinted from Ref. [113], Copyright 2013, with permission from Elsevier). (**b**) Experimental strategies and formats for GMO detection using SPR-based BIA and sensor chips carrying lectin or Roundup Ready oligonucleotides or PCR products (Reprinted with the permission from Ref. [120]. Copyright 2002 American Chemical Society)

diagnosis, and some reports offer some solutions with high specificity and simplicity [115–117]. Other chip-based detection methods provide an alternative choice for accurate GMO detection [29, 118].

Recently, biosensors have attracted many researchers for the use of their optical, electrochemical, piezoelectric transducers in GMO detection [119]. Surface plasmon resonance (SPR) is a powerful optical technique that occurs at the interface of a thin gold-coated prism in contact with the analyte solution, which can detect various molecular interactions [119]. Feriotto et al. [120, 121] have successfully accomplished quantitative analysis of GM soybean and corn based on SPR system (Fig. 14.5b). Mariotti et al. [122] firstly reported their application of an SPR-based biosensor for the screening analyses of GMOs. To quantitatively trace the genetically modified products, a chemiluminometric immunosensor array for the detection of recombinant marker proteins expressed in GMOs, i.e., 5-enolpyruvylshikimate-3-phosphate synthase (EPSPS), neomycin phosphotransferase II (NPT II), and phosphinothricin acetyltransferase (PAT), was also constructed. Electrochemical DNA sensors take advantage of the interactions between the solid electrode surface, recognition probe, and DNA and offer sensitivity, selectivity, and the reliable detection of target sequences upon hybridization [123].

14.7 Other Techniques

GMO detection techniques are undergoing increasing development, with different substances (RNA, protein, and metabolic substances) as the target. Samples can be analyzed using one screening method for multiple elements, resulting in a high-throughput possible identification [21, 124–126]. RNA can also be analyzed through RNA-seq or real-time PCR because the insertion gene has similar gene expression to the endogenous gene in the host. It is also recommended to apply high-performance liquid chromatography (HPLC) and mass spectrometry (MS) to analyze the chemical components and identify GMO products if the chemical components differ between the GMOs and the host one. In addition, some badly DNA-damaged products, such as plant oil, can be analyzed by HPLC for fatty acids and triglycerides identification if DNA cannot be detected easily [127]. Results showed much difference between wild canola and GM canola. However, this method must be performed after the chemical differences of a specific GMO have been fully analyzed. This idea is then transformed into omics analysis for safety assessment [128]. However, this instrumental method has difficulty in accurate detection and discrimination from mixed samples. Furthermore, infrared spectrum analysis can also be used to detect changes in fiber structures and therefore to identify positive GMOs [126, 129].

14.8 Prospect

The review is focused on the basic theories and recent developments in GMO detection, especially through nucleic acids identification. The inherent goals of transgenic detection techniques are high accuracy, high throughput, simplicity, and low cost, which are still being pursued. Scientists are mining multiple methods from other fields to achieve these goals. Upcoming forecasts may include the development of electrophoresis techniques into biosensing diagnosis, lab analysis into point-of-care analysis, and standard curve-based quantification into absolute quantification through digital PCR.

Studying the endogenous genes will also be a focus of GMO detection field. For example, 5S DNA has been developed as a specific internal gene [130].

Developers are eager to introduce both GMOs from the plant sources and GM animals into the market [131]. The US government has recently approved the first GM animals in the world. Therefore, detection methods for GM animals have to be considered. Normal methods used in GM plants can be used universally in the GM detection of animals, but new technologies for GM animals make it difficult to easily identify GM components in animal foods [132].

In view of the great breakthroughs in the magic gene editing technologies (ZFNs, TALENs, and CRISPR/Cas9 technologies), we can expect the production of new types of genetically modified organisms in the future, and the corresponding detection techniques will become a new and challenging topic [133–135].

Acknowledgments This work is supported by the National Science and Technology Major Project (2016ZX08012-001). Many thanks to Chenguang Wang, for his kindly help in manuscript conception and preparation.

References

1. Aldemita RR, Reaño IME, Solis RO, Hautea RA. Trends in global approvals of biotech crops (1992–2014). GM Crops Food. 2015;6(3):150–66.
2. Tingle N. World food: an encyclopedia of history, culture, and social influence from hunter-gatherers to the age of globalization. Ref User Serv Q. 2013;52(3):270.
3. Chilton M-D, Drummond MH, Merlo DJ, Sciaky D, Montoya AL, Gordon MP, Nester EW. Stable incorporation of plasmid DNA into higher plant cells: the molecular basis of crown gall tumorigenesis. Cell. 1977;11(2):263–71.
4. Drummond MH, Gordon MP, Nester EW, Chilton M-D. Foreign DNA of bacterial plasmid origin is transcribed in crown gall tumours. Nature. 1977;269:535–6.
5. Mulligan RC, Berg P. Expression of a bacterial gene in mammalian cells. Science. 1980;209 (4463):1422–7.
6. Bakshi S, Dewan D. Status of transgenic cereal crops: a review. Clon Transgen. 2013;3 (119):2.
7. Brookes G, Barfoot P. Environmental impacts of genetically modified (gm) crop use 1996–2013: impacts on pesticide use and carbon emissions. GM Crops Food. 2015;6:103–33.
8. Varshney RK, Tuberosa R. Translational genomics for crop breeding: abiotic stress tolerance, yield, and quality, an introduction. In: Translational genomics for crop breeding: volume 2-improvement for abiotic stress, quality and yield improvement. Ames: Wiley Blackwell; 2013. p. 1.
9. James C. Global status of commercialized biotech/GM crops: 2010. Ithaca: International Service for the Acquisition of Agri-biotech Applications (ISAAA); 2010.
10. Carpenter JE. Peer-reviewed surveys indicate positive impact of commercialized GM crops. Nat Biotechnol. 2010;28(4):319–21.
11. Grohmann L, Reiting R, Mäde D, Uhlig S, Simon K, Frost K, Randhawa GJ, Zur K. Collaborative trial validation of cry1Ab/Ac and Pubi-cry TaqMan-based real-time PCR assays for detection of DNA derived from genetically modified Bt plant products. Accred Qual Assur. 2015;20(2):85–96.
12. Li R, Wang Q, McHughen A. Chinese government reaffirms backing for GM products. Nat Biotechnol. 2015;33(10):1029.
13. Ramessar K, Capell T, Twyman RM, Christou P. Going to ridiculous lengths—European coexistence regulations for GM crops. Nat Biotechnol. 2010;28(2):133–6.
14. Li Y, Peng Y, Hallerman EM, Wu K. Biosafety management and commercial use of genetically modified crops in China. Plant Cell Rep. 2014;33(4):565–73.
15. Lu C. The first approved transgenic rice in China. GM Crops. 2010;1(3):113–5.
16. Chen M, Shelton A, Ye G-y. Insect-resistant genetically modified rice in China: from research to commercialization. Annu Rev Entomol. 2011;56:81–101.
17. Waltz E. USDA approves next-generation GM potato. Nat Biotechnol. 2015;33(1):12–3.
18. Holst-Jensen A, Bertheau Y, de Loose M, Grohmann L, Hamels S, Hougs L, Morisset D, Pecoraro S, Pla M, Van den Bulcke M. Detecting un-authorized genetically modified organisms (GMOs) and derived materials. Biotechnol Adv. 2012;30(6):1318–35.
19. Morisset D, Stebih D, Cankar K, Zel J, Gruden K. Alternative DNA amplification methods to PCR and their application in GMO detection: a review. Eur Food Res Technol. 2008;227 (5):1287–97.
20. Hemmer W. Foods derived from genetically modified organisms and detection methods. Basel: Agency BATS; 1997.
21. Xu X, Li Y, Zhao H, Wen S-y, Wang S-q, Huang J, Huang K-l, Luo Y-b. Rapid and reliable detection and identification of GM events using multiplex PCR coupled with oligonucleotide microarray. J Agric Food Chem. 2005;53(10):3789–94.
22. Schmidt A-m, Sahota R, Pope DS, Lawrence TS, Belton MP, Rott ME. Detection of genetically modified canola using multiplex PCR coupled with oligonucleotide microarray hybridization. J Agric Food Chem. 2008;56(16):6791–800.

23. Anklam E, Gadani F, Heinze P, Pijnenburg H, Van Den Eede G. Analytical methods for detection and determination of genetically modified organisms in agricultural crops and plant-derived food products. Eur Food Res Technol. 2002;214(1):3–26.
24. Taverniers I, Windels P, Vaïtilingom M, Milcamps A, Van Bockstaele E, Van den Eede G, De Loose M. Event-specific plasmid standards and real-time PCR methods for transgenic Bt11, Bt176, and GA21 maize and transgenic GT73 canola. J Agric Food Chem. 2005;53 (8):3041–52.
25. Berkhout RJ, Tieben LM, Smits HL, Bavinck JN, Vermeer BJ, Ter Schegget J. Nested PCR approach for detection and typing of epidermodysplasia verruciformis-associated human papillomavirus types in cutaneous cancers from renal transplant recipients. J Clin Microbiol. 1995;33(3):690–5.
26. Yunbo HKL. Detecting genetically modified soybean roundup ready ingredient in foodstuffs by nested PCR and semi-nested PCR. Chin J Agric Biotechnol. 2003;5:004.
27. Xu W-t, Huang K-l, Luo Y-b. YBR green I based PCR for detection of the bar and pat genes in genetically modified organisms. Food Sci. 2006;3:048.
28. Germini A, Zanetti A, Salati C, Rossi S, Forré C, Schmid S, Marchelli R. Development of a seven-target multiplex PCR for the simultaneous detection of transgenic soybean and maize in feeds and foods. J Agric Food Chem. 2004;52(11):3275–80.
29. Xu J, Miao H, Wu H, Huang W, Tang R, Qiu M, Wen J, Zhu S, Li Y. Screening genetically modified organisms using multiplex-PCR coupled with oligonucleotide microarray. Biosens Bioelectron. 2006;22(1):71–7.
30. Huang H-Y, Pan T-M. Detection of genetically modified maize MON810 and NK603 by multiplex and real-time polymerase chain reaction methods. J Agric Food Chem. 2004;52 (11):3264–8.
31. Feng J, Wang X, Li D, Tang S, You S, Wu X, Chen J. Study of multiplex-PCR for the detection of genetically modified contents in food. Inspection Quar Sci. 2006;4:005.
32. Jin Z, Pingjian D, Jianjun L. Multiplex qualitative PCR and real-time quantitative PCR for detection of genetically modified foods. Chin J Health Lab Technol. 2004;4:010.
33. Li X, Wu Y, Li J, Li Y, Long L, Li F, Wu G. Development and validation of a 48-target analytical method for high-throughput monitoring of genetically modified organisms. Sci Rep. 2015;5:7616.
34. Park S-B, Kim H-Y, Kim J-H. Multiplex PCR system to track authorized and unauthorized genetically modified soybean events in food and feed. Food Control. 2015;54:47–52.
35. Xu W, Yuan Y, Luo Y, Bai W, Zhang C, Huang K. Event-specific detection of stacked genetically modified maize Bt11× GA21 by UP-M-PCR and real-time PCR. J Agric Food Chem. 2008;57(2):395–402.
36. Xu W, Zhai Z, Huang K, Zhang N, Yuan Y, Shang Y, Luo Y. A novel universal primer-multiplex-PCR method with sequencing gel electrophoresis analysis. PLoS One. 2012;7(1):1.
37. Shang Y, Zhu P, Xu W, Guo T, Tian W, Luo Y, Huang K. Single universal primer multiplex ligation-dependent probe amplification with sequencing gel electrophoresis analysis. Anal Biochem. 2013;443(2):243–8.
38. Guo J, Chen L, Liu X, Gao Y, Zhang D, Yang L. A multiplex degenerate PCR analytical approach targeting to eight genes for screening GMOs. Food Chem. 2012;132(3):1566–73.
39. Fu W, Zhu P, Wang C, Huang K, Du Z, Tian W, Wang Q, Wang H, Xu W, Zhu S. A highly sensitive and specific method for the screening detection of genetically modified organisms based on digital PCR without pretreatment. Sci Rep. 2015;5. doi:10.1038/srep12715.
40. Holst-Jensen A, Rønning SB, Løvseth A, Berdal KG. PCR technology for screening and quantification of genetically modified organisms (GMOs). Anal Bioanal Chem. 2003;375 (8):985–93.
41. Tichoniuk M, Ligaj M, Filipiak M. Application of DNA hybridization biosensor as a screening method for the detection of genetically modified food components. Sensors. 2008;8(4):2118–35.

42. Waiblinger H-U, Grohmann L, Mankertz J, Engelbert D, Pietsch K. A practical approach to screen for authorised and unauthorised genetically modified plants. Anal Bioanal Chem. 2010;396(6):2065–72.
43. Querci M, Foti N, Bogni A, Kluga L, Broll H, Van den Eede G. Real-time PCR-based ready-to-use multi-target analytical system for GMO detection. Food Anal Methods. 2009;2 (4):325–36.
44. Higuchi R, Fockler C, Dollinger G, Watson R. Kinetic PCR analysis: real-time monitoring of DNA amplification reactions. Biotechnology. 1993;11:1026–30.
45. Zipper H, Brunner H, Bernhagen J, Vitzthum F. Investigations on DNA intercalation and surface binding by SYBR green I, its structure determination and methodological implications. Nucleic Acids Res. 2004;32(12):e103.
46. Clegg RM. Fluorescence resonance energy transfer. Curr Opin Biotechnol. 1995;6 (1):103–10.
47. Heid CA, Stevens J, Livak KJ, Williams PM. Real time quantitative PCR. Genome Res. 1996;6(10):986–94.
48. Taverniers I, Van Bockstaele E, De Loose M. Cloned plasmid DNA fragments as calibrators for controlling GMOs: different real-time duplex quantitative PCR methods. Anal Bioanal Chem. 2004;378(5):1198–207.
49. Holck A, Vaïtilingom M, Didierjean L, Rudi K. 5′-Nuclease PCR for quantitative event-specific detection of the genetically modified Mon810 MaisGard maize. Eur Food Res Technol. 2002;214(5):449–54.
50. Hernández M, Pla M, Esteve T, Prat S, Puigdomenech P, Ferrando A. A specific real-time quantitative PCR detection system for event MON810 in maize YieldGard® based on the 3′-transgene integration sequence. Transgenic Res. 2003;12(2):179–89.
51. Xu W-T, Huang K-L, Deng A-K, Liang Z-h, Luo Y-B. Variations of tissue DNA density and nuclear DNA content in soybean lines and their impacts on the GMO quantification. Food Control. 2007;18(10):1300–6.
52. Holland PM, Abramson RD, Watson R, Gelfand DH. Detection of specific polymerase chain reaction product by utilizing the $5' - 3'$ exonuclease activity of Thermus aquaticus DNA polymerase. Proc Natl Acad Sci. 1991;88(16):7276–80.
53. Bustin SA, Benes V, Garson JA, Hellemans J, Huggett J, Kubista M, Mueller R, Nolan T, Pfaffl MW, Shipley GL. The MIQE guidelines: minimum information for publication of quantitative real-time PCR experiments. Clin Chem. 2009;55(4):611–22.
54. Vogelstein B, Kinzler KW. Digital Pcr. Proc Natl Acad Sci. 1999;96(16):9236–41.
55. Burns MJ, Burrell AM, Foy CA. The applicability of digital PCR for the assessment of detection limits in GMO analysis. Eur Food Res Technol. 2010;231(3):353–62.
56. Morisset D, Stebih D, Milavec M, Gruden K, Zel J. Quantitative analysis of food and feed samples with droplet digital PCR. PLoS One. 2013;8:62583.
57. Corbisier P, Bhat S, Partis L, Xie VRD, Emslie KR. Absolute quantification of genetically modified MON810 maize (Zea mays L.) by digital polymerase chain reaction. Anal Bioanal Chem. 2010;396(6):2143–50.
58. Köppel R, Bucher T, Frei A, Waiblinger H-U. Droplet digital PCR versus multiplex real-time PCR method for the detection and quantification of DNA from the four transgenic soy traits MON87769, MON87708, MON87705 and FG72, and lectin. Eur Food Res Technol. 2015;241(4):521–7.
59. Dobnik D, Spilsberg B, Bogožalec Košir A, Holst-Jensen A, Žel J. Multiplex quantification of 12 European union authorized genetically modified maize lines with droplet digital polymerase chain reaction. Anal Chem. 2015;87(16):8218–26.
60. Halpin C. Gene stacking in transgenic plants–the challenge for 21st century plant biotechnology. Plant Biotechnol J. 2005;3(2):141–55.
61. Berdal KG, Holst-Jensen A. Roundup Ready® soybean event-specific real-time quantitative PCR assay and estimation of the practical detection and quantification limits in GMO analyses. Eur Food Res Technol. 2001;213(6):432–8.

62. Taverniers I, Windels P, Van Bockstaele E, De Loose M. Use of cloned DNA fragments for event-specific quantification of genetically modified organisms in pure and mixed food products. Eur Food Res Technol. 2001;213(6):417–24.
63. Terry CF, Harris N. Event-specific detection of roundup ready soya using two different real time PCR detection chemistries. Eur Food Res Technol. 2001;213(6):425–31.
64. Andersen CB, Holst-Jensen A, Berdal KG, Thorstensen T, Tengs T. Equal performance of TaqMan, MGB, molecular beacon, and SYBR green-based detection assays in detection and quantification of roundup ready soybean. J Agric Food Chem. 2006;54(26):9658–63.
65. Corbisier P, Trapmann S, Gancberg D, Hannes L, Van Iwaarden P, Berben G, Schimmel H, Emons H. Quantitative determination of Roundup Ready soybean (Glycine max) extracted from highly processed flour. Anal Bioanal Chem. 2005;383(2):282–90.
66. Zhang H, Yang L, Guo J, Li X, Jiang L, Zhang D. Development of one novel multiple-target plasmid for duplex quantitative PCR analysis of roundup ready soybean. J Agric Food Chem. 2008;56(14):5514–20.
67. La Paz JL, Esteve T, Pla M. Comparison of real-time PCR detection chemistries and cycling modes using Mon810 event-specific assays as model. J Agric Food Chem. 2007;55 (11):4312–8.
68. Hernández M, Rodríguez-Lázaro D, Zhang D, Esteve T, Pla M, Prat S. Interlaboratory transfer of a PCR multiplex method for simultaneous detection of four genetically modified maize lines: Bt11, MON810, T25, and GA21. J Agric Food Chem. 2005;53(9):3333–7.
69. Yang L, Xu S, Pan A, Yin C, Zhang K, Wang Z, Zhou Z, Zhang D. Event specific qualitative and quantitative polymerase chain reaction detection of genetically modified MON863 maize based on the 5′-transgene integration sequence. J Agric Food Chem. 2005;53(24):9312–8.
70. Pan A, Yang L, Xu S, Yin C, Zhang K, Wang Z, Zhang D. Event-specific qualitative and quantitative PCR detection of MON863 maize based upon the 3′-transgene integration sequence. J Cereal Sci. 2006;43(2):250–7.
71. Lee S-H, Min D-M, Kim J-K. Qualitative and quantitative polymerase chain reaction analysis for genetically modified maize MON863. J Agric Food Chem. 2006;54(4):1124–9.
72. Hemandez M, Esteve T, Prat S, Pla M. Development of real-time PCR system based on SYBR green I, Amplifluor and TaqMan technologies for specific quantitative detection of the transgenic maize event GA21. J Cereal Sci. 2004;39(1):99–107.
73. Nielsen CR, Berdal KG, Holst-Jensen A. Characterisation of the 5′ integration site and development of an event-specific real-time PCR assay for NK603 maize from a low starting copy number. Eur Food Res Technol. 2004;219(4):421–7.
74. Li X, Shen K, Yang L, Wang S, Pan L, Zhang D. Applicability of a novel reference molecule suitable for event-specific detections of maize NK603 based on both 5′ and 3′ flanking sequences. Food Control. 2010;21(6):927–34.
75. Collonnier C, Schattner A, Berthier G, Boyer F, Coué-Philippe G, Diolez A, Duplan M-N, Fernandez S, Kebdani N, Kobilinsky A. Characterization and event specific-detection by quantitative real-time PCR of T25 maize insert. J AOAC Int. 2005;88(2):536–46.
76. Yang R, Xu W, Luo Y, Guo F, Lu Y, Huang K. Event-specific qualitative and quantitative PCR detection of roundup ready event GT73 based on the 3′-integration junction. Plant Cell Rep. 2007;26(10):1821–31.
77. Demeke T, Ratnayaka I. Multiplex qualitative PCR assay for identification of genetically modified canola events and real-time event-specific PCR assay for quantification of the GT73 canola event. Food Control. 2008;19(9):893–7.
78. Triglia T, Peterson MG, Kemp DJ. A procedure for in vitro amplification of DNA segments that lie outside the boundaries of known sequences. Nucleic Acids Res. 1988;16(16):8186.
79. Zimmermann A, Lüthy J, Pauli U. Event specific transgene detection in Bt11 corn by quantitative PCR at the integration site. LWT-Food Sci Technol. 2000;33(3):210–6.
80. Kun Y, XueLong W, ChunXiu L, JinQing C. Isolation of the flanking sequences adjacent to transgenic T-DNA in Brassica napus genome by an improved inverse PCR method. Agric Sci Technol-Hunan. 2010;11(2):65–8, 139.

81. Liu YG, Mitsukawa N, Oosumi T, Whittier RF. Efficient isolation and mapping of Arabidopsis thaliana T-DNA insert junctions by thermal asymmetric interlaced PCR. Plant J. 1995;8(3):457–63.
82. Pfeifer GP, Steigerwald SD, Mueller PR, Wold B, Riggs AD. Genomic sequencing and methylation analysis by ligation mediated PCR. Science. 1989;246(4931):810–3.
83. Trinh Q, Xu W, Shi H, Luo Y, Huang K. An AT linker adapter polymerase chain reaction method for chromosome walking without restriction site cloning bias. Anal Biochem. 2012;425(1):62–7.
84. Trinh Q, Shi H, Xu W, Hao J, Luo Y, Huang K. Loop-linker PCR: an advanced PCR technique for genome walking. IUBMB Life. 2012;64(10):841–5.
85. Xu W, Shang Y, Zhu P, Zhai Z, He J, Huang K, Luo Y. Randomly broken fragment PCR with 5′ end-directed adaptor for genome walking. Sci Rep. 2013;3:3465.
86. Meyer R. Development and application of DNA analytical methods for the detection of GMOs in food. Food Control. 1999;10(6):391–9.
87. James D, Schmidt A-m, Wall E, Green M, Masri S. Reliable detection and identification of genetically modified maize, soybean, and canola by multiplex PCR analysis. J Agric Food Chem. 2003;51(20):5829–34.
88. Zimmermann A, Hemmer W, Liniger M, Lüthy J, Pauli U. A sensitive detection method for genetically modified MaisGard TM corn using a nested PCR-system. LWT-Food Sci Technol. 1998;31(7):664–7.
89. Vollenhofer S, Burg K, Schmidt J, Kroath H. Genetically modified organisms in food screening and specific detection by polymerase chain reaction. J Agric Food Chem. 1999;47(12):5038–43.
90. Weng H, Yang L, Liu Z, Ding J, Pan A, Zhang D. Novel reference gene, high-mobility-group protein I/Y, used in qualitative and real-time quantitative polymerase chain reaction detection of transgenic rapeseed cultivars. J AOAC Int. 2005;88(2):577–84.
91. Hernández M, Río A, Esteve T, Prat S, Pla M. A rapeseed-specific gene, acetyl-CoA carboxylase, can be used as a reference for qualitative and real-time quantitative PCR detection of transgenes from mixed food samples. J Agric Food Chem. 2001;49(8):3622–7.
92. Ding J, Jia J, Yang L, Wen H, Zhang C, Liu W, Zhang D. Validation of a rice specific gene, sucrose phosphate synthase, used as the endogenous reference gene for qualitative and real-time quantitative PCR detection of transgenes. J Agric Food Chem. 2004;52(11):3372–7.
93. Xu WT, Huang KL, Wang Y, Zhang HX, Luo YB. A cotton-specific gene, stearoyl-ACP desaturase, used as a reference for qualitative and real-time quantitative polymerase chain reaction detection of genetically modified organisms. J Sci Food Agric. 2006;86(7):1103–9.
94. Xu W, Bai W, Guo F, Luo Y, Yuan Y, Huang K. A papaya-specific gene, papain, used as an endogenous reference gene in qualitative and real-time quantitative PCR detection of transgenic papayas. Eur Food Res Technol. 2008;228(2):301–9.
95. Shang Y, Zhu P, Huang K, Liu W, Tian W, Luo Y, Xu W. A peach (Prunus persica)-specific gene, Lhcb2, used as an endogenous reference gene for qualitative and real-time quantitative PCR to detect fruit products. LWT-Food Sci Technol. 2014;55(1):218–23.
96. Yang L, Guo J, Pan A, Zhang H, Zhang K, Wang Z, Zhang D. Event-specific quantitative detection of nine genetically modified maizes using one novel standard reference molecule. J Agric Food Chem. 2007;55(1):15–24.
97. Wen-Tao X, Wei-Bin B, Yun-Bo L, Yan-Fang Y, Kun-Lun H. Research progress in techniques for detecting genetically modified organisms. Chin J Agric Biotechnol. 2009;6 (01):1–9.
98. Pi L, Li X, Cao Y, Wang C, Pan L, Yang L. Development and application of a multi-targeting reference plasmid as calibrator for analysis of five genetically modified soybean events. Anal Bioanal Chem. 2015;407(10):2877–86.
99. Stave JW. Detection of new or modified proteins in novel foods derived from GMO–future needs. Food Control. 1999;10(6):367–74.

100. Pardigol A, Guillet S, Pöpping B. A simple procedure for quantification of genetically modified organisms using hybrid amplicon standards. Eur Food Res Technol. 2003;216 (5):412–20.
101. Kuribara H, Shindo Y, Matsuoka T, Takubo K, Futo S, Aoki N, Hirao T, Akiyama H, Goda Y, Toyada M. Novel reference molecules for quantitation of genetically modified maize and soybean. J AOAC Int. 2002;85(5):1077–89.
102. Xu WT, Huang KL, Lu Y. Extraction methods of genome in processed rice and their influences on PCR. J Agric Biotechnol. 2007;15(1):97–101.
103. Lantz P-G, Al-Soud WA, Knutsson R, Hahn-Hägerdal B, Rådström P. Biotechnical use of polymerase chain reaction for microbiological analysis of biological samples. Biotechnol Annu Rev. 2000;5:87–130.
104. Shames B, Fox JG, Dewhirst F, Yan L, Shen Z, Taylor NS. Identification of widespread Helicobacter hepaticus infection in feces in commercial mouse colonies by culture and PCR assay. J Clin Microbiol. 1995;33(11):2968–72.
105. Al-Soud WA, Rådström P. Purification and characterization of PCR-inhibitory components in blood cells. J Clin Microbiol. 2001;39(2):485–93.
106. Sambrook J, Fritsch EF, Maniatis T. Molecular cloning, vol. 2. New York: Cold Spring Harbor Laboratory Press; 1989.
107. Wilson IG. Inhibition and facilitation of nucleic acid amplification. Appl Environ Microbiol. 1997;63(10):3741.
108. Vaneechoutte M, Van Eldere J. The possibilities and limitations of nucleic acid amplification technology in diagnostic microbiology. J Med Microbiol. 1997;46(3):188–94.
109. Abbaszadegan M, Huber MS, Gerba CP, Pepper IL. Detection of enteroviruses in groundwater with the polymerase chain reaction. Appl Environ Microbiol. 1993;59(5):1318–24.
110. Lantz P-G, Matsson M, Wadström T, Rådström P. Removal of PCR inhibitors from human faecal samples through the use of an aqueous two-phase system for sample preparation prior to PCR. J Microbiol Methods. 1997;28(3):159–67.
111. Cavallini A, Notarnicola M, Berloco P, Lippolis A, Di Leo A. Use of macroporous polypropylene filter to allow identification of bacteria by PCR in human fecal samples. J Microbiol Methods. 2000;39(3):265–70.
112. Zhang P, Gebhart CJ, Burden D, Duhamel GE. Improved diagnosis of porcine proliferative enteropathy caused by Lawsonia intracellularis using polymerase chain reaction-enzyme-linked oligosorbent assay (PCR-ELOSA). Mol Cell Probes. 2000;14(2):101–8.
113. Han X, Wang H, Chen H, Mei L, Wu S, Jia G, Cheng T, Zhu S, Lin X. Development and primary application of a fluorescent liquid bead array for the simultaneous identification of multiple genetically modified maize. Biosens Bioelectron. 2013;49:360–6.
114. Shao N, Jiang S-M, Zhang M, Wang J, Guo S-J, Li Y, Jiang H-W, Liu C-X, Zhang D-B, Yang L-T. MACRO: a combined microchip-PCR and microarray system for high-throughput monitoring of genetically modified organisms. Anal Chem. 2014;86(2):1269–76.
115. Huang S, Xu Y, Yan X, Shang Y, Zhu P, Tian W, Xu W. Development and application of a quantitative loop-mediated isothermal amplification method for detecting genetically modified maize MON863. J Sci Food Agric. 2015;95(2):253–9.
116. Huang X, Zhai C, You Q, Chen H. Potential of cross-priming amplification and DNA-based lateral-flow strip biosensor for rapid on-site GMO screening. Anal Bioanal Chem. 2014;406 (17):4243–9.
117. Xu C, Li L, Jin W, Wan Y. Recombinase Polymerase Amplification (RPA) of CaMV-35S promoter and nos terminator for rapid detection of genetically modified crops. Int J Mol Sci. 2014;15(10):18197–205.
118. Fu W, Wang H, Wang C, Mei L, Lin X, Han X, Zhu S. A high-throughput liquid bead array-based screening technology for Bt presence in GMO manipulation. Biosens Bioelectron. 2016;77:702–8.
119. Arugula MA, Zhang Y, Simonian AL. Biosensors as 21st century technology for detecting genetically modified organisms in food and feed. Anal Chem. 2013;86(1):119–29.

120. Feriotto G, Borgatti M, Mischiati C, Bianchi N, Gambari R. Biosensor technology and surface plasmon resonance for real-time detection of genetically modified roundup ready soybean gene sequences. J Agric Food Chem. 2002;50(5):955–62.
121. Feriotto G, Gardenghi S, Bianchi N, Gambari R. Quantitation of Bt-176 maize genomic sequences by surface plasmon resonance-based biospecific interaction analysis of multiplex polymerase chain reaction (PCR). J Agric Food Chem. 2003;51(16):4640–6.
122. Mariotti E, Minunni M, Mascini M. Surface plasmon resonance biosensor for genetically modified organisms detection. Anal Chim Acta. 2002;453(2):165–72.
123. Jang H-J, Cho I-H, Kim H-S, Jeon J-W, Hwang S-Y, Paek S-H. Development of a chemiluminometric immunosensor array for on-site monitoring of genetically modified organisms. Sens Actuators B. 2011;155(2):598–605.
124. Michelini E, Simoni P, Cevenini L, Mezzanotte L, Roda A. New trends in bioanalytical tools for the detection of genetically modified organisms: an update. Anal Bioanal Chem. 2008;392 (3):355–67.
125. Alishahi A, Farahmand H, Prieto N, Cozzolino D. Identification of transgenic foods using NIR spectroscopy: a review. Spectrochim Acta A Mol Biomol Spectrosc. 2010;75(1):1–7.
126. Xie L, Ying Y, Ying T. Quantification of chlorophyll content and classification of nontransgenic and transgenic tomato leaves using visible/near-infrared diffuse reflectance spectroscopy. J Agric Food Chem. 2007;55(12):4645–50.
127. Byrdwell WC, Neff WE. Analysis of genetically modified canola varieties by atmospheric pressure chemical ionization mass spectrometric and flame ionization detection. J Liq Chromatogr Relat Technol. 1996;19(14):2203–25.
128. Ricroch AE, Bergé JB, Kuntz M. Evaluation of genetically engineered crops using transcriptomic, proteomic, and metabolomic profiling techniques. Plant Physiol. 2011;155 (4):1752–61.
129. Hurburgh CR, Rippke GR, Heithoff C, Roussel SA, Hardy CL Detection of genetically modified grains by near-infrared spectroscopy. In: Proceedings PITTCON, 2000. p. 12–7.
130. Doveri S, Lee D. Development of sensitive crop-specific polymerase chain reaction assays using 5S DNA: applications in food traceability. J Agric Food Chem. 2007;55(12):4640–4.
131. Kues WA, Niemann H. Advances in farm animal transgenesis. Prev Vet Med. 2011;102 (2):146–56.
132. Vàzquez-Salat N, Salter B, Smets G, Houdebine L-M. The current state of GMO governance: are we ready for GM animals? Biotechnol Adv. 2012;30(6):1336–43.
133. Segal DJ, Meckler JF. Genome engineering at the dawn of the golden age. Annu Rev Genomics Hum Genet. 2013;14:135–58.
134. Voytas DF. Plant genome engineering with sequence-specific nucleases. Annu Rev Plant Biol. 2013;64:327–50.
135. Doudna JA, Charpentier E. The new frontier of genome engineering with CRISPR-Cas9. Science. 2014;346(6213):1258096.

Chapter 15
The Identification and Detection Technology of Research in Microorganisms Including Living or Dead Bacteria

Abstract Both dead and living bacteria can affect food quality. In recent years, the development of microbiological testing has undergone a series of progressions from traditional methods—including microbial cultivation, physiological and biochemical testing, instrument analysis, and immunology—to molecular biological detection. Reviewing its developments, it can be seen that the detection is becoming more rapid, sensitive, precise, integrated, and low cost. Of course, when all methods of microbial detection are compared, only living bacteria that detected by traditional methods could yield false negative results, especially if there are dormant bacteria. But with molecular methods, all microorganisms can be detected. With the development of technology, a lot of focus has been given to identification and detection technology for dead and living bacteria, which has massive potential and important significance in the area of microorganisms. There exists prospects of novel detection techniques for discrimination of viable and nonviable microorganisms, especially some more sensitive, precise, and comprehensive ones.

Keywords Viable • Nonviable • Bacteria • Nucleic acid analysis • Amplification

15.1 Introduction

Foods are polluted by pathogenic microorganisms or spoilage microorganisms in their processing, packing, storage, and transportation processes; this can easily cause poisoning and disease, and can cause sudden microbial food safety issues. According to the World Health Organization's (WHO's) statistics, about one third of people are infected with foodborne diseases in developed countries every year. In some developing countries, foodborne diseases are often the main cause of unnatural death in humans. Every year, millions of people die from foodborne diseases in the world. Thus, the detection of microorganisms is particularly important.

Cultural, immune, and nucleic acid–based methods all have disadvantages and advantages. For example, although the traditional culture approaches detect only living bacteria, they can accurately identify the existing targeted strains in real samples within 3–7 days [1]. Immunological diagnostic methods are quick but likely to produce nonspecific results [2]. Nucleic acid–based technologies, such as polymerase chain reaction (PCR) and real-time PCR [3], have been widely applied

W. Xu, *Functional Nucleic Acids Detection in Food Safety*,
DOI 10.1007/978-981-10-1618-9_15

in microbe detection. What is more, nucleic acid molecular biology technology has been widely applied in microbe testing and also has undergone innovative developments. Obvious advances in molecular methods have improved microbe analysis specificity and sensitivity while saving time. However, considering these advances, many nucleic acid targets are required to accomplish a detectable signal. Unfortunately, none of these approaches provide arapid and simple measure to distinguish viable and nonviable bacteria. In a lot of situations, detection of nonviable microbes would give rise to false positive results. Viable bacteria, rather than nonviable bacteria, can cause food corruption and pathogenicity. Precisely quantifying the numbers of living bacteria in food, the environment, and other fields has been a challenging problem.

Recently, identification and detection technology for viable and nonviable bacteria has drawn the attention of many researchers. It overcomes the disadvantage of all bacteria detection by the general molecular method and has massive potential and important significance in the area of microorganisms [4]. Most methods are still based on nucleic acid molecular detection technology, which is quick, specific, sensitive, able to be performed in real time, and accurate, and some methods use other treatments before the amplification.

In this chapter, the development of detection technology for microorganisms is introduced, especially emphasizing the identification of viable and nonviable bacteria.

15.2 Microorganism Detection Technology and Methods

15.2.1 Traditional Detection Methods

Traditional microbiological detection methods are based on isolation and culture techniques, which use culture media to isolate and culture and then biochemical identification or direct microscopy. Their advantage is relative accuracy, but they cannot detect those food microorganisms that are hardly culturing or uncultured. They also cannot achieve rapid and effective monitoring and preventive effects. Meanwhile, some factors, including cell damage caused by food processing and other food ingredients, also mean that traditional detection methods have certain restrictions.

For all their faults, the traditional isolation and culture techniques are still the gold standard for detection of microorganisms. They have many features, such as high accuracy, ability to discover new pathogens, ability to be used for other methods' further analysis (such as PCR or immunological analysis), and ability to be used as an effective control method. These features mean that they still occupy a prime position in the detection of microorganisms [1].

15.2.2 Instrumental Analytical Methods

Instrumental analytical methods can also be applied to microorganism detection. High-performance liquid chromatography (HPLC) and gas chromatography (GC) can be used to analyze the biochemical composition of different microorganisms and their metabolites for detecting and identifying microorganisms [5, 6]. The mass-spectrometric technique is based on specific markers existing in different microorganisms and uses the obtained specific mass spectrograms to identify microorganisms [7]. In the analysis of capillary electrophoresis (CE), different bacterial cells correspond with different characteristic peaks, thus analyzing and distinguishing different microorganisms [8]. Bionic instruments such as electronic nose instruments can use gas sensors and other gas-aborted means to analyze and process volatile odor substances produced by microorganisms and draw odor fingerprints for effectively studying microorganisms [9]. Moreover, automatic detectors of microorganisms have already been invented at home and abroad, including the Vitek system, AutoSceptor system, Biolog system, Sensititre system, and Midd system. The Vitek full-automatic microbial analytic system—without the process of isolating, culturing, and purifying microorganisms—is able to directly detect specific microbial species and flora from samples [10], and it has been approved as a legal analysis by the International Association for Analytical Chemistry. Its second-generation detection system—the Vitek immunodiagnostic testing system—offers some advantages of solid-phase absorption, enzyme-linked immunosorbent assay (ELISA), fluorescence detection, and latex agglutination in one unit, and it is a comprehensive microbial detection system.

The instrumental analytic technique is relatively simple, but there are also some shortcomings, such as high cost and low efficiency, so it is difficult to deal with food safety incidents occurring in a wide range of global settings.

15.2.3 Immunological Methods

Immunological methods are techniques dependent on the specific combinations between antigens and antibodies and thus qualitatively or quantitatively analyzing and assessing the samples. There are many types of immunological methods, including immunomagnetic separation (IMS) [11], immune diffusion technique (IDT) [12], immunofluorescence technique (IFT) [13], ELISA [14], enzyme-linked fluorescence immune assay (ELFIA) [15], and immunoblot technique (IBT) [16]. ELISA is the most common technology used for immunoassay, which combines specificity between antigens and antibodies with effective analysis of enzymes acting on the substrates. Then, according to the coloring reaction of enzymes acting on the substrates, when antigens and antibodies are combined, microorganisms can be qualitatively or quantitatively detected by colorimetric or fluorescent reactions. The coloring of enzymes reacting with substrates is proportional to the antigens' content in the samples. Depending on the operation, ELISA is divided into the double mono-cloned antigen method, the indirect method, and the competitive method.

Immunological methods have many advantages, such as relatively simple instruments and reagents, high specificity, ease of storing samples, and the fact that the results can be easily observed and can be quantitatively determined. Now, the data provided by the major database literature show that many pathogens have established serological detection methods, and those prove these methods have become more mature detection technologies in laboratories and widely used [15]. However, immunological methods also have some inherent technical limitations, such as small amounts of information, high selectivity for reagents, more complicated methods, difficulty of simultaneously analyzing multiple components, some cross-reactivity in structurally similar compounds, and limited sensitivity related to molecular biology methods. So, simple immunological technology cannot serve as the preferred method for microbiological safety.

15.2.4 Molecular Biological Detection Techniques

In recent years, molecular biological techniques have been widely used in microbial detection. The methods based on PCR have been most commonly used. PCR technology, invented by Kary Mullis in 1985, is one of the most rapidly developed and popularized molecular biological techniques. A variety of techniques have been derived on the basis of PCR technology, including nested PCR, multiplex PCR, PCR with denaturing gradient gel electrophoresis (PCR-DGGE) and PCR with thermal gradient gel electrophoresis (PCR-TGGE), random amplified polymorphic DNA technique, fluorescent quantitative PCR, and multiplex fluorescent quantitative PCR. These technologies have played important roles in microbial diversity analysis and in qualitative and quantitative analysis [17].

In addition, the molecular marker method, which is based on molecular hybridization technology, is used in microbial detection, such as fluorescence in situ hybridization and gene chip technology. The main characteristic of the fluorescence in situ hybridization technique is that it combines the accuracy of molecular biology and the visibility of microscopy, which can directly observe different microorganisms in complex environments. It also can provide much information—such as the morphology, quantity, and space distribution of bacterial colonies, and information on the cell environment—to evaluate the microbial community. Because it is sensitive, fast, safe, and specific, the fluorescence in situ hybridization technique has been widely used in microbial detection, including activated sludges, agricultural environments, marine environments, biological films, medical diagnosis, and soil [18–20]. Gene chip microarray protocols include a series of steps for DNA purification, specific amplification, fragmentation, labeling, concentration, quantitation, dilution, hybridization, washing, imaging, and data analysis. This technology is applied in the field of analysis and detection of microorganisms [21, 22].

15.2.4.1 Quantitative Technology for Nucleic Acid Analysis

DNA-based PCR approaches are mainly used because of their high sensitivity, specificity, and applicability for complex and processed samples. But traditional PCR is a qualitative method, which cannot assess microbe quantity for microbial detection. In other words, real-time PCR is necessary for precise microbe quantification, to evaluate safety and harmfulness [3]. Real-time PCR was first introduced in 1992, including the fluorescence probe system and the specific nucleic acid dye system. It is completed by monitoring the amplification of the target sequence using fluorescent signals, such as the TaqMan system, which relies on the release and detects the fluorescent signal following the cleavage of a fluorescent labeled probe. Real-time PCR is one of the rapid, sensitive, high-throughput, and quantitative methods that have been widely used to detect and identify microorganisms [23, 24]. Real-time PCR also can be used in microbial diversity analysis.

Digital PCR, a newly developed quantitative PCR–based detection method, has been regarded as the third generation of PCR technology for its better sensitivity and absolute quantitative detection [25]. After separating the original reaction volume by the microfluidic or oil-in-water method, huge numbers of separating reaction cells are generated, so that the original target is also divided into each cell. Thus, the copy number of the target for each partition is, according to the Poisson distribution, based on the ratio of positive partitions to total partitions. Finally, the copy number of the target fragments of the original volume can be calculated through the dilution percentage and Poisson distribution. After it was first reported in the 1990s, lots of research progress was achieved in the aspects of copy number determination, single-nucleotide polymorphisms, and absolute quantitative detection of low-abundance target samples. Compared with real-time PCR, digital PCR can determine the copy number of the target template in the original solution without generation of standard curves. This mostly avoids the uncertainties caused by standard curves and operations. So, digital PCR is regarded as an improved quantitative detection method compared with real-time PCR [26]. In the 2000s, Bio-Rad Company commercialized digital PCR with the facilities of the Bio-Rad QX100 digital PCR system. After development over nearly one decade, digital PCR is regarded as the best way of accuracy, stability, and cost performance that applied in bacteria [27].

15.2.4.2 High-Throughput Technology for Nucleic Acid Analysis

While traditional PCR has many good points, it has the shortcoming of the singleness of the product. Multiplex PCR is a kind of high-throughput PCR technology and is built upon ordinary PCR to improve the quantity. More than one region in a genome or different genes can be detected simultaneously in multiplex PCR, so there will be multiple targeted sequences amplified in an assay [28]. At present, multiplex PCR is widely used in microbial detection. Lindström established a multiplex PCR detection assay for botulinum neurotoxin, which could detect four kinds of botulinum toxin—*A*, *B*, *E*, and *F*—in 2001 [29]. With deep knowledge of botulinum neurotoxins, the researchers found that the previous test could distinguish toxins *B*, *E*, and *F* excellently, and only toxin *A1* was detected. Then Medici redesigned the primers for neurotoxin *A* of *Clostridium botulinum* and rebuilt the reaction system, which could distinguish toxins *A1*, *A2*, *A3*, and *A4* fully, in 2009 [30]. Ngamwongsatit established a multiplex PCR detection reaction system for *Bacillus cereus*, which could distinguish toxin genes (*hblCDA*, *nheABC*, *cytK*, *entFM*) in 2008 [31]. Monteiro developed a multiplex real-time PCR assay to detect six different genetic types of carbapenemases already identified in Enterobacteriaceae (KPC, GES, NDM, IMP, VIM, and OXA-48) [32]. Lee developed a multiplex PCR to detect *B. cereus*, *Salmonella* spp., *Escherichia coli* O157: H7, *Vibrio parahaemolyticus*, *Staphylococcus aureus*, and *Listeria monocytogenes* simultaneously [33].

Research has shown that the advantages of multiplex PCR are that it is rapid, reliable, low-cost, etc. But many primers can result in different amplification efficiency. In order to promote the application of multiplex PCR, much improvement has occurred through research. For example, universal primer multiplex PCR (UP-M-PCR) has been developed with a universal primer, which has common amplification efficiency in all primers and is more accurate and reliable [34]. The common single-primer multiplex PCR (CSP-M-PCR) method has been developed using a common single primer and applied to identify five species simultaneously [35]. These two novel multiplex PCR systems will be useful in many fields with sequencing gel electrophoresis analysis. The advantages of these two technologies are shown by two aspects: first, the disparity of different primers in multiplex PCR is avoided, and this generates higher sensitivity and specificity [36]; second, a universal adapter has been designed for each template, significantly reducing the complexity of the reaction system and inhibiting the form of primer dimers and non-specific products, avoiding self-inhibition and amplification disparity [37]. The UP-M-PCR system has been applied to simultaneous detection of *Salmonella* spp., *L. monocytogenes*, and *E. coli* [34], and for typing of toxicogenic *Pseudomonas aeruginosa* [38].

At the same time, along with the rise of large-scale genomics, a new generation of sequencing technology (called second-generation sequencing technology) has arisen at a historic moment. Second-generation sequencing has mostly adopted the large-scale matrix structure of microarray analysis technology using primers and DNA polymerase or ligase to extend and amplify. The method obtains the sequence information through continuous recording of the optical signal in the cycle and also can be applied in large-scale sequencing [39]. There are many sequencing instruments on the market, such as the Genome Analyzer and the HiSeq 2000 system from Illumina Company, the Genome Sequencer 20 System from 454 Life Science, and the AB SoLid sequencer from ABI Company. With the appearance of the first single-molecular DNA sequencer from Helicos Biosciences in 2008, genomics has ushered in a new development. Sequencing techniques can be used in microbial detection, including comparison of microbial genomics, taxonomy, macrogenomics, and single-celled bacteria sequencing [40]. The appearance and development of advanced molecular technologies provide a new approach for microbial detection.

15.2.4.3 Rapid Detection Technology for Nucleic Acid Analysis

PCR-based methods are suitable for various samples because of their high specificity and sensitivity. However, there are many limitations, depending on the instruments, including the long reaction times required and observation by agarose gel electrophoresis. Many isothermal methods based on nucleic acid amplification have now been developed. The isothermal amplification technologies have been proven to be a practical and rapid tool in the field for detecting microorganisms.

Nucleic acid sequence–based amplification (NASBA) is an RNA amplification technology that utilizes the promoter primer to produce DNA from the RNA templates by RNA polymerase and a reverse transcriptase. Each of the synthesized RNAs can be used in amplification steps again, and more RNA amplicons are accumulated. The products of NASBA can be detected by gel electrophoresis, fluorescence probes (real-time NASBA), and colorimetric assay (NASBA-ELISA) [41]. Many research studies have validated the technique applicability of NASBA on molecular detection of some pathogens such as *Salmonella enterica* [42], *S. aureus* [43], and *Vibrio cholerae* [44].

Strand displacement amplification (SDA) is an isothermal method of amplifying nucleic acid prior to detection by nicking the unmodified strand from a recognition site. The exonuclease deficient klenow (exo-klenow) can extend the 3′-end at the nick and displace the complementary DNA strand. Exponential amplification results from coupling sense and antisense reactions in which strands displaced from a sense reaction serve as a target for an antisense reaction and vice versa [45]. A multiplex SDA was subsequently developed for co-amplification of a target sequence (IS6110) of *Mycobacterium tuberculosis* and a target (16S ribosomal gene) of relevant *Mycobacterium* species [46]. This technique also can be used for RNA templates in an RT-SDA system by reverse transcriptase for the original process [41].

Rolling circle amplification (RCA) is a circle DNA replication technology that has traditionally used specific primer sequences to generate multiple copies of a sequence library by phi29 DNA polymerase [47]. Then more than one primer was used in RCA, and this was developed into multiply primed rolling circle amplification (multiply primed RCA). These methods allow replication of circular input DNA to achieve a 10,000-fold level. The continuous over length products can be identified with gel electrophoresis images [41]. RCA-based approaches have recently been attracting attention in fungal and bacteria testing studies [48, 49].

Loop-mediated isothermal amplification (LAMP) is the most mature technology, which was first described by Notomi in 2000. The amplification can be completed with four or six independent sequences under a constant temperature. The most important advantages of this method are its high efficiency and sensitivity. There is a large amount of products generated in the process, and the products can be determined by gel electrophoresis, SYBR Green I, and a turbidimeter, which contain a toxic dye, lead to contamination, and have low resolution, respectively [50]. LAMP has attracted the attention of researchers in microbe fields. For example, it has been established for the detection of *Shigella* [51], *Salmonella* [52], and *E. coli* [53]. Subsequently, Cheng developed a new quantitative LAMP based on EvaGreen dye to detect *L. monocytogenes* [54].

Helicase-dependent amplification (HDA) is a DNA isothermal technology that utilizes a helicase, rather than heat, to generate single-stranded templates for the purpose of amplification. A specific primer hybridizes to the single template and is then extended by DNA polymerases [55]. It has a simple reaction scheme and can be accomplished at constant temperature for the entire process, completely eliminating thermal cycling equipment dependency. The amplicons can be distinguished by gel electrophoresis, real-time format, and ELISA assay [41]. Many research studies have used the technique in HDA for molecular detection of bacteria [56].

Nicking endonuclease-mediated isothermal amplification (NEMA) is used widely because of its extraordinary advantages. This technology, based on the principle of strand displacement, cleaves only one strand of duplex DNA to amplify in an exponential form via nicking endonuclease activity [57]. Compared with LAMP, which is the most popular isothermal technology, NEMA requires only two pairs of ordinary primers, which greatly simplify primer designing work [50]. On the other hand, NEMA is more inhibited by aerosol pollution than LAMP, lowering its efficiency. Moreover, NEMA can amplify at most a 400 bp product, which is more universal than NASBA, RCA, HDA, and other isothermal technologies [42, 47, 55]. The most topical application of the approach is the detection of *M. tuberculosis* [57].

15.2.4.4 Diversity Analysis Technology for Nucleic Acid Analysis

Microbes can widely exist in different environments. The same microorganism in different environments also can present a high degree of genetic diversity. Thus, genotyping technology is necessary to precisely identify the microbes. In bacterial microorganism research studies, 16S ribosomal DNA (rDNA) is the most widely used marker, while in fungal microorganisms, 18S rDNA and ITS1/ITS2 genes are the famous markers. Microbial diversity detection not only is a highly efficient and comprehensive strategy for food microbial contamination detection but also aids understanding of the microbial mechanism by which the food is converted to be unsafe.

Restriction fragment length polymorphism (RFLP) is a method to exploit variations in homologous DNA sequences. In microbial diversity research studies, the 16S rDNA signal is first amplified using PCR, and the PCR products are then digested by restriction enzymes. Terminal RFLP (T-RFLP) is a method that results from the combination of RFLP and PCR. The PCR primers are both labeled at the 5′-end with phosphoramidite dyes, and the digested products are separated by electrophoresis using gel electrophoresis or capillary-based systems [58]. T-RFLP analysis has been applied to the analysis of various microorganisms and microbial communities [59].

Random amplified polymorphic DNA (RAPD) is a widely used and successful approach for the characterization of microbes. The advantage of RAPD is that no DNA sequence information is needed before experiments, and the main problem is that the repetition is not good. Single-strand conformation polymorphism (SSCP) analysis is an approach of analyzing the sequence variations among different homogeneous DNA fragments. Such sequence variations will result in different secondary structures of single-stranded DNA (ssDNA), which affect the migration speeds of the molecules during electrophoresis. In an analysis of microbial diversity, 16s rDNA is usually the target type of DNA. If the DNA fragments of interest are less than 400 bp in length, then SSCP generally offers suitable discrimination of single-nucleotide polymorphisms in the targeted genes [60].

DGGE and TGGE essentially consist of two procedures: PCR amplification of the target sequence and gradient gel electrophoresis. Separation is caused by the gradient decrease in the electrophoretic mobility of the target sequence in a polyacrylamide gel containing a linear gradient of denaturants in the case of DGGE, or a linear temperature gradient in the case of TGGE. The resolution of DGGE and TGGE is very high compared with other PCR electrophoresis–based methods. These methods are inexpensive, rapid, reproducible, and reliable [61], but the common restriction is that their community fingerprints cannot directly translate into taxonomic information [58].

Sanger sequencing is a method of DNA sequencing based on the selective incorporation of chain-terminating dideoxynucleotides by DNA polymerase during in vitro DNA replication. For the study of microbial diversity, the 16S rDNA of the metagenome should be amplified, and the clean PCR product should be cloned into a clone library. The clone library and Sanger sequencing–based method could provide the most comprehensive information on the microbial community, including detailed 16S rDNA sequences to indicate the species that constitute the community, the abundance and diversity of the community, and the relationships of different communities [62].

High-throughput sequencing based on the synthesis sequencing principle and the massively parallel signature sequencing strategy enable high-throughput sequencing to determine the sequences of millions of different DNA templates at the same time. High-throughput sequencing–based methods are highly sensitive and specific, inexpensive, and time-saving, compared with traditional molecular methods, for microbial diversity research. Amplicon sequencing and whole-metagenome shotgun sequencing are the two main methods used for microbial diversity studies. Amplicon sequencing is used for the detection of microbiota species composition to determine "species diversity" (however, the resolution of amplicon sequencing cannot reach species level at present), while whole-metagenome shotgun sequencing provides not only species diversity information but also genetic diversity information [63].

15.3 Identification and Detection Methods for Viable and Nonviable Bacteria

15.3.1 Reverse Transcription PCR

The words "alive" and "viable" describe different state of target cells. An acceptable definition of live cells is cells that are able to divide, metabolize, respire, and reproduce. Therefore, messenger RNA (mRNA) can be used as an appropriate landmark to detect viable microorganisms by PCR and to identify the effectiveness of PCR for detecting animate cells to accurately evaluate the status of bacteria from a sample. In that way, the detection of total RNA would be a cost-efficient and reproducible method, which could be applied to a wide range of microbial species [64].

mRNA might be a perfect referent for animate cells. Reverse transcription PCR (RT-PCR) can be utilized for analyzing mRNA to monitor gene expression in animate bacteria. While these research studies have really demonstrated the activities of animate bacteria in virus environments, their main purpose was not to distinguish the status as viable, half-viable, and nonviable. Some studies have particularly revealed the correlation between viability and microbial mRNA. RT-PCR has been used to diagnose living *Legionella pneumophila* and *V. cholerae*, and proved specific mRNA only in samples that contained viable bacteria detected by culturing. Equally, the viability of heat-killed cells was successfully proved with detection of heat-shock protein mRNA in living *Mycobacterium leprae* by Patel et al. Similarly, a specific method based on RT-PCR was built by Klein and Juneja to detect viable *L. monocytogenes*. In spite of the potential advantage, RT-PCR approaches have proved difficult to develop because of the complex assay, the practical problems of extracting detectable levels of complete mRNA from small amounts of bacteria, and a lack of basic information about the significance of detecting mRNA in stressed cells [65].

Additionally, a real-time RT-PCR (RRT-PCR) assay has been developed, and it is a new two-step RT-PCR multiplex assay based on targeted sequences for detection. RRT-PCR has been successfully utilized to detect various RNA viruses with hydrolysis probes by Holland et al. and Livak et al. RRT-PCR supplies the strengths of no sample handling and speed, thus reducing the probability of cross-contamination. Both RT-PCR and RRT-PCR require purification of the mRNA from objects. Because the half-life of most mRNA is 1.5–2 min, this is the biggest challenge to obtain intact mRNA from cells. Not many reports have described that prosthetic mRNA from microbes could be utilized for kinds of molecular analysis technology. Although a number of these extraction methods are time consuming, an effective approach has not been developed for the detection of microbes [64]. While advantages exist in which PCR can be modified to determine cell viability in different samples through RT-PCR or RRT-PCR, these methods can be difficult to perform because of the unsteadiness of RNA. At the same time, there are also some problems with extraction and preservation of intact RNA in physiological conditions or environment conditions [66].

15.3.2 Viability PCR

Real-time PCR (qPCR) combined with viability dyes have been widely used in different types of foodborne pathogens. DNA-binding dyes, such as ethidium monoazide bromide (EMA) or propidium monoazide (PMA), can permeate dead or membrane-compromised cells as pretreatment prior to the qPCR. This novel approach to detect viable and nonviable cells has introduced coupling PCR or qPCR with DNA-binding dyes, and is named viability PCR (v-PCR).

EMA treatment prior to DNA extraction is a simple approach to viable quantification that relies on preventing PCR amplification of genetic targets from nonviable cells. EMA is a DNA/RNA intercalating chromophore, which can penetrate damaged cytomembranes and embed in DNA upon photoactivation. As a result, it can inhibit subsequent molecular amplification including PCR, qPCR, and LAMP [67]. Nocker and Camper suggested that further optimization is necessary before EMA treatment can be incorporated into a viability assay prior to nucleic acid extraction. Although EMA can discriminate between viable and dead cells, high EMA concentrations would inhibit the qPCR process and cause negative impacts. Thus, it is important to confirm if the amplification of injured or senescent cells would be inhibited when EMA is incorporated into the experimental design, because these cells may be the most susceptible to treatment and could lead to false negative results [66]. Rapid quantification of viable and nonviable cells killed by heat was also developed using EMA in combination with LAMP [68] and real-time LAMP [69] assays.

Shi et al. have optimized the conditions of EMA treatment, the level of thermal disinfection, and the value of pH and osmotic pressure, and have well documented the possibilities of the EMA-qPCR assay [67]. Amplification will be inhibited when EMA enters nonviable bacteria and binds to the nucleic acid. Light can promote the combination of EMA and DNA. So, the optimum concentration of EMA is 10 μg ml^{-1}, and the photoactivation time is 20 min. Cells in various states can be bound in EMA totally by heating. The optimum treatment condition to yield inactivated cells could be considered as 85 °C for 35 min, which could be not detected by EMA-qPCR. High osmotic pressure ($\geq$4 %) could increase the inhibition of EMA-qPCR, and the inhibitory effect is positively correlated with the osmotic pressure. The counts of cells decreased sharply by EMA-qPCR when the osmotic pressure increased to 8 %. Cells were treated in different pH solutions and then subjected to EMA treatment, and the Ct values prominently increased from pH 1 to 5, with no difference after pH 5. It was also shown that the counts had already decreased significantly at pH 3. The sublethal acidification injury of the cells could be completely recovered, and EMA could not penetrate them with 40 min LB incubation.

The most important rule for distinguishing between viable and nonviable cells is membrane integrity. Viable cells with intact membranes can be distinguished by their ability to reject DNA-binding dyes. These dyes can easily infiltrate dead or membrane-compromised cells, and the most common dye is propidium iodide (PI). EMA-qPCR was reported to be an easy method to distinguish live and dead cells, but EMA had low ability to bind the DNA insolubly and strong ability to penetrate viable cells of other bacterial species [70]. In addition, propidium monoazide (PMA) has been successfully used to differentiate viable and nonviable bacteria in conjunction with qPCR. The principle is based on the integrity of bacterial cell membranes, as PMA can only penetrate cells with compromised membranes. So, PMA is able to enter the membrane of heat-killed bacterial cells and intercalate the DNA or bind to any free DNA in a sample and inhibit the activity of Taq polymerase [71]. Later, Lee et al. found that following heat treatment of cell suspensions, in the populations surviving with EMA and PMA, the percentage of erroneous survival was higher with PMA than with EMA treatment [72]. PMA was also coupled with multiplex PCR (mPCR) to detect enterotoxin genes (*nheA*, *entFM*, *hblD*, *cytK*) and emetic toxin (*ces*) in viable *B. cereus* [73]. PMA has also been combined with 454 pyrosequencing to detect bacteria [74].

V-PCR has shown potential for identification and detection of viable and nonviable microorganisms. Membrane-compromised microorganisms treated by PMA/EMA could inhibit PCR amplification. Moreover, PMA-qPCR has been more widely used in microorganisms than EMA-qPCR. The drawback of v-PCR is that the exclusion of dead cells can be incomplete, leading to false positive signals [75]. A new approach improves v-PCR by enabling it to also discriminate between cells that have an intact cell membrane and the ability to actively maintain bacterial homeostasis, and cells that have an intact membrane but are metabolically inactive [76].

15.3.3 Nuclease PCR

Microorganisms are susceptible to adverse environmental conditions. They can maintain low metabolic activity for a long time. The organisms may also remain in a dormant state, which has been named "viable but nonculturable" (VNC). These VNC organisms will not grow but can survive in the environment for a few weeks. V-PCR has been increasingly used for detection of viable and nonviable cells, but a hindrance still exists in v-PCR, which is the separation of viable and nonviable cells. Therefore, proper sample preparation methods have been sought for microorganisms to determine whether they are living, VNC, or dead. In the qPCR assay, when hybridization occurs, the dye-labeled probe is cleaved by the 5′-nuclease during the extension, generating increased fluorescence intensity. Using the 5′-nuclease PCR assay, the influence of nucleic acids from heat-killed bacteria on the detection of viable cells was investigated. An approach to the reduction of the background signal generated by DNA in heat-killed bacteria by using external DNases was evaluated [77]. The 5′-nuclease PCR assays were also applied to measure the effect of EMA, and the EVA could covalently bind to DNA inhibiting the 5′-nuclease PCR [78].

The quality of the nucleic acids from the dead cells is influenced by the nucleic acids themselves and by other treatment. The inherent bacterial DNases can affect the half-life of DNA. It is important to confirm the stability of nucleic acids to explain whether the positive PCR signal is rooted in living cells. The physical difference between viable and nonviable cells has not yet been widely applied in PCR. The nucleic acids in viable cells are protected by the intact membranes. But the nucleic acids are exposed in nonviable cells because of the damaged membranes. In others words, the exposure of DNA in nonviable cells is easily impaired by outside contamination. Deoxyribonuclease I (DNase I) is an endonuclease that can cleave single- and double-stranded DNA. DNase I has been widely used for removing genomic DNA in samples for RNA analyses. Free DNA in nonviable cells should not be protected from the DNase I, because of the damaged membranes. In theory, only DNA from viable cells will be present in the template after it is treated by DNase I for subsequent qPCR analyses. Villarreal et al. have reported on a DNase I and proteinase K–based treatment protocol developed and optimized for the detection, characterization, and analysis of live populations of bacteria present in drinking water biofilms [79].

15.3.4 Nanoparticle and Spectroscopy Technology

An important challenge in bacteriology is to identify whether a cell is living or dead. Especially when bacteria are starved or have been subjected to a severe environment, it is very difficult to distinguish whether a cell is alive but unable to grow, or dead. Many techniques have been developed to overcome this challenge, such as methods based on nucleic acid dyes to penetrate the incomplete membranes. Nevertheless, the problem of viability still exists, and a new approach to probe viable and nonviable cells, using nanoparticles and spectroscopy technology, has been developed. Atomic force microscopy (AFM) imaging of viable cells is becoming more commonplace, and AFM does not affect cell viability during and after the process. At the same time, there are plenty of probe–membrane interactions that lead to transfer of membrane components to the probe [80]. AFM has since been explored in viable cells at the nanoscale. This probed the mechanical properties of viable and nonviable cells via AFM indentation experiments and immobilized a single cell onto a surface for AFM, without the need for chemical fixation [81].

Both Fourier transform infrared (FTIR) and Raman spectroscopy have been applied to monitor biological activity by surface reactions. The Raman microspectrometer has also been used for in situ rapid discrimination of viable cells, when 58S substrates, 45S5 Bioglass, and bioinert silica were modified on the viable cells. Viable cells could be well monitored by a high-power 785 nm laser without cell damage, and cell death would induce strong changes in the Raman signature (1000–1150 cm^{-1} and 1550–1650 cm^{-1}) [82]. Both Raman and surface-enhanced Raman scattering (SERS) displayed highly specific spectral fingerprints of bacteria. Nevertheless, the sensitivity of SERS mapping should be developed with use of novel nanostructures. Novel SERS mapping could obtain a visual signal of live and dead bacteria and has been reported to recognize live and dead bacteria with silver nanoparticles (Bacteria@AgNPs) [83].

Laser-induced breakdown spectroscopy (LIBS) has the ability to provide ultrafast spectrochemical analysis of various samples. Recently, research has shown the advantage of LIBS with nanosecond or femtosecond laser pulses to distinguish microorganisms. *E. coli* are genetically well characterized, with divalent cations in the outer membrane. Moreover, the divalent cations can be detected by LIBS. These special characteristics make *E. coli* an excellent indicator organism to detect various samples, and it also can be an appropriate bacterial model to simulate the basic states of alive and dead [84].

15.3.5 Dielectrophoresis Technology

Dielectrophoresis (DEP), the development of microelectrodes, has been applied to characterize viable and non-viable cells by non-uniform alternating current (AC) electric fields. Using this novel physical method, viable yeast cells can effectively be distinguished from dead ones. Many elements should be considered in this technology, such as field strength, liquid conductivity, frequency range, and time. Viable and nonviable cells have different frequency responses, which can be selectively collected in the mixture. Microelectrodes can be used to make positive and negative DEP, which has been used to distinguish *Saccharomyces cerevisiae* [85]. Hydrodynamic DEP could be made on a high-throughput chip for separation of bacteria. The chip has three planar electrodes in every separation channel. In this channel, the target and non-target bacteria will move away or remain in the central streamline, respectively [86].

Insulator-based dielectrophoresis (iDEP) is a new kind of DEP. The innovation of iDEP is the use of insulators, avoiding the problems of electrodes. Viable and nonviable cells cannot be distinguished by electrokinetic mobility, but can be distinguished by dielectrophoretic mobility. Nonviable cells have lower dielectrophoretic mobility than viable cells. iDEP showed potency to separate viable and nonviable cells simultaneously for bacterial analysis [87]. iDEP trapping microchips with open-top microstructures are fit for capturing cells. Viable and nonviable cells have different frequency responses. At the frequency of 1 kHz, nonviable cells can be trapped from viable cells [88]. Contactless dielectrophoresis (cDEP) is a technology developed for cell screening. Viable cells can be screened from dead cells using their electrical characteristics. The cDEP method united with other advanced technologies can be used in lab-on-a-chip systems for identifying cells [89]. Reservoir-based dielectrophoresis (rDEP) has special advantages as compared with DEP, such as the occupation of zero channel space and the elimination of any mechanical or electrical parts inside microchannels. This approach has been used to selectively trap dead yeast cells and continuously separate them from live ones. It also can be combined with other components into lab-on-a-chip devices for applications in biomedical diagnostics [90].

15.3.6 Flow Cytometry

Flow cytometry (FCM) as a conventional means for cellular biology has been used since the 1970s. Recently, it has been applied to deal with bacteria. Because of their small size, they are very difficult to distinguish by FCM. However, FCM is still a potential method because of its high-throughput capacity and the ability to differentiate single cells. Some cells may show metabolic activity and capability to grow, while some may have intact membranes but no capability to grow. Other cells may be dead or dormant, have injured membranes, and be incapable of replication. Different characteristics between viable, dormant, and nonviable cells is the key to differentiation by FCM. Different characteristics such as cytoplasmic and membrane integrity, intracellular enzyme activity, respiration rates, and nucleic acid content, all can be key points. Nucleic acid binding dyes have been widely used to analyze cell viability, the most common being SYTO dyes and PI. When the viable and nonviable cells are combined with DNA dyes, they can be easily observed by UVA. EMA, annexin V, and amine reactive viability dyes (ViDs) also have been commonly used to exclude dead cells. At the same time, 3,3-′-dihexyloxacarbocyanine (DiOC6 (3)) was used to determine the membrane potential, 5-cyano-2,3-ditolyl tetrazolium chloride (CTC) dye was used to discriminate cell respiration, and carboxyfluorescein diacetate (cFDA) was used to monitor esterase activity [91]. Therefore, FCM provides a powerful and versatile approach to the measurement of viable and nonviable cells. When the cells were simultaneously treated with cFDA and PI, flow cytometry and cell sorting revealed striking physiological heterogeneity within the stressed *Bifidobacterium* population. In situ assessment of the physiological activity of stressed *Bifidobacterium* using multiparameter flow cytometry and cell sorting may provide a powerful and sensitive tool for assessment of the viability and stability of probiotics [92]. The LIVE/DEAD BacLight kit is becoming more and more popular in the various fields of bacteria. It can be combined with FCM to make a new approach to interpret LIVE/DEAD staining results. It has been applied in the detection of *E. coli*, *S. enterica* serovar Typhimurium, *Shigella flexneri*, and a community of freshwater bacteria, resulting in a clear and distinctive flow cytometric staining pattern [93].

15.3.7 Biosensors

Sensors can be classified into three types: physical sensors, chemical sensors, and biosensors. Biosensors bear remarkable advantages, such as specificity, sensitivity, real-time sensing, time-efficiency, and in situ monitoring. Their collaboration with nanomaterials has been applied in the detection of nucleic acids, enzymes, antibodies, bacteriophages, or even whole cells. Biosensors have various classifications, and one of them is electrochemical biosensors. Electrochemical biosensors also can be classified as potentiometric, impedimetric, amperometric, and conductimetric [94]. A highly specific and sensitive RNA biosensor has been developed to detect viable *E. coli*. Viable *E. coli* were identified and quantified by an mRNA (*clpB*) gene with a NASBA-based biosensor [95]. An impedance biosensor, with double interdigitated array microelectrode (IAM)–based flow cells, with two embedded IAMs on the top and bottom surfaces, has been developed for the detection of viable *E. coli* O157:H7. If bacteria grow, impedance would be measured inside the flow cell. A single IAM–based flow cell was less sensitive than a double IAM–based flow cell in the detection of *E. coli* O157:H7. Therefore, the double IAM–based flow cell can be used to sensitively detect viable bacterial [96]. A microfluidic platform using EMA has been used to detect and type viable and nonviable bacteria. It detected the viable cells by virtue of gold nanoparticle probes and typed the viable cells by an on-chip PCR. What is more, this is the first time viable bacteria have been detected and typed continuously on the same microfluidic platform [97].

Lateral flow tests (LFTs) or test strips have been developed as a diagnostic tool in the laboratory, and also are paper-based point-of-care biosensors. This technology has the advantage of being simple, accurate, quick, and cheap; it does not require technical expertise and complicated instruments. LFTs are usually labeled with colloidal gold, latex, carbon, up-converting phosphorus, single-strand nucleic acid, and so on. The most common state is LFT employing colloidal gold particles. Fluorescent immunoliposomes have more sensitivity than color labels for visual detection. Up-converting phosphorus may be 10- to 100-fold more sensitive than colored latex beads and colloidal gold. Europium chelate-loaded silica nanoparticles have been proven to improve sensitivity in LFTs compared with colloidal gold. LFTs present enormous potential and have been considered as an available method for field tests of bacteria, such as *Listeria*, *E. coli*, *S. aureus*, *Salmonella*, *Campylobacter*, and *Clostridium* [98]. Adenosine triphosphate (ATP)-based bioluminescence is the combination of LFT and mutant firefly luciferase, which is used for the detection of *Salmonella enteritidis*. In this approach, viable *S. enteritidis* could form a *Salmonella*–antibody–gold complex, which would be captured at the test line and lysed by heat treatment. At the same time, the viable cells would release ATP, measured by mutant luciferase [99].

15.4 Conclusion and Prospects

In recent years, the development of microbiological testing has undergone a series of progressions from traditional methods—including microbial cultivation, physiological and biochemical testing, instrument analysis, and immunology—to molecular biological detection, and nucleic acid molecular detection technology is playing a more and more important role, which traditional methods cannot match in microbial detection, with virtues of being quick, specific, sensitive, able to be performed in real time, and accurate in recent years. However, none of these methods can identify viable and nonviable bacteria effectively.

An important breakthrough to simultaneously differentiate viable and nonviable bacteria in a mixture assay is the integrity of cell membranes. Nucleic acid dyes, such as PMA and EMA, which can easily permeate membranes in dead cells, can be intercalated into DNA. This type of DNA cannot be the template in any amplification technology. Therefore, viable bacteria can be differentiated effectively. Novel molecular amplification methods have been widely applied in the detection of bacteria, but most of them have not been combined with nucleic acid dyes to distinguish viable, dormant, and nonviable bacteria. However when these techniques are used together, their combined strength can help overcome the weakness of the identification and detection methods for viable and nonviable bacteria, and extend into a new field. In addition, samples such as food, groundwater, or soil have many complex factors, which affect analyses and the results of post-processing. The special aptamer can associate with the target bacteria to form a complex that is completely undisturbed. Once the aptamer is associated with a biosensor, this will improve the sensitivity and specificity greatly.

Acknowledgments This work is supported by the Ministry of Science and Technology of Beijing (XX2014B069). Many thanks to Yuancong Xu, for her kindly help in manuscript conception and preparation.

References

1. Day J, Basavanna U, Sharma S. Development of a cell culture method to isolate and enrich Salmonella enterica serotype enteritidis from shell eggs for subsequent detection by real-time PCR. Appl Environ Microbiol. 2009;75(16):5321–7.
2. Charles RC, Sultana T, Alam MM, Yu Y, Wu-Freeman Y, Bufano MK, Rollins SM, Tsai L, Harris JB, Larocque RC. Identification of immunogenic Salmonella enterica serotype Typhi antigens expressed in chronic biliary carriers of S. Typhi in Kathmandu, Nepal. PLoS Negl Trop Dis. 2013;7(8):e2335.
3. Nadkarni MA, Martin FE, Jacques NA, Hunter N. Determination of bacterial load by real-time PCR using a broad-range (universal) probe and primers set. Microbiology. 2002;148 (1):257–66.
4. Shi H, Xu W, Trinh Q, Luo Y, Liang Z, Li Y, Huang K. Establishment of a viable cell detection system for microorganisms in wine based on ethidium monoazide and quantitative PCR. Food Control. 2012;27(1):81–6.

5. Fernández M, Linares DM, del Río B, Ladero V, Alvarez MA. HPLC quantification of biogenic amines in cheeses: correlation with PCR-detection of tyramine-producing microorganisms. J Dairy Res. 2007;74(03):276–82.
6. Sasser M. Identification of bacteria by gas chromatography of cellular fatty acids. 1990.
7. Claydon MA, Davey SN, Edwards-Jones V, Gordon DB. The rapid identification of intact microorganisms using mass spectrometry. Nat Biotechnol. 1996;14(11):1584–6.
8. Buszewski B, Szumski M, Kłodzińska E, Dahm H. Separation of bacteria by capillary electrophoresis. J Sep Sci. 2003;26(11):1045–9.
9. Marilley L, Ampuero S, Zesiger T, Casey MG. Screening of aroma-producing lactic acid bacteria with an electronic nose. Int Dairy J. 2004;14(10):849–56.
10. Huang X-z, Tan H-f, Chen H. Survey of Salmonella contamination in food and evaluation of Vitek microbial analytical system. J Trop Med. 2007;3:022.
11. Zhao Y, Ye M, Chao Q, Jia N, Ge Y, Shen H. Simultaneous detection of multifood-borne pathogenic bacteria based on functionalized quantum dots coupled with immunomagnetic separation in food samples. J Agric Food Chem. 2008;57(2):517–24.
12. Dudman W. Immune diffusion analysis of the extracellular soluble antigens of two strains of Rhizobium meliloti. J Bacteriol. 1964;88(3):782–94.
13. Peng Y-s, Peng K-y. A study on the possible utilization of immunodiffusion and immunofluorescence techniques as the diagnostic methods for American foulbrood of honeybees (Apis mellifera). J Invertebr Pathol. 1979;33(3):284–9.
14. Li J, Xia K, Yu C. Detection of Alicyclobacillus acidoterrestris in apple juice concentrate by enzyme-linked immunosorbent assay. Food Control. 2013;30(1):251–4.
15. Hou J-Y, Liu T-C, Lin G-F, Li Z-X, Zou L-P, Li M, Wu Y-S. Development of an immunomagnetic bead-based time-resolved fluorescence immunoassay for rapid determination of levels of carcinoembryonic antigen in human serum. Anal Chim Acta. 2012;734:93–8.
16. Wong K, Chua K, Lam S. Immunohistochemical detection of infected neurons as a rapid diagnosis of enterovirus 71 encephalomyelitis. Ann Neurol. 1999;45(2):271–2.
17. Justé A, Thomma B, Lievens B. Recent advances in molecular techniques to study microbial communities in food-associated matrices and processes. Food Microbiol. 2008;25(6):745–61.
18. Bottari B, Ercolini D, Gatti M, Neviani E. Application of FISH technology for microbiological analysis: current state and prospects. Appl Microbiol Biotechnol. 2006;73(3):485–94.
19. Lenaerts J, Lappin-Scott HM, Porter J. Improved fluorescent in situ hybridization method for detection of bacteria from activated sludge and river water by using DNA molecular beacons and flow cytometry. Appl Environ Microbiol. 2007;73(6):2020–3.
20. Bertaux J, Gloger U, Schmid M, Hartmann A, Scheu S. Routine fluorescence in situ hybridization in soil. J Microbiol Methods. 2007;69(3):451–60.
21. Brigham CJ, Speth DR, Rha C, Sinskey AJ. Whole-genome microarray and gene deletion studies reveal regulation of the polyhydroxyalkanoate production cycle by the stringent response in Ralstonia eutropha H16. Appl Environ Microbiol. 2012;78(22):8033–44.
22. Chandler DP, Knickerbocker C, Bryant L, Golova J, Wiles C, Williams KH, Peacock AD, Long PE. Profiling in situ microbial community structure with an amplification microarray. Appl Environ Microbiol. 2013;79(3):799–807.
23. Oblath EA, Henley WH, Alarie JP, Ramsey JM. A microfluidic chip integrating DNA extraction and real-time PCR for the detection of bacteria in saliva. Lab Chip. 2013;13 (7):1325–32.
24. Hierro N, Esteve-Zarzoso B, González Á, Mas A, Guillamón JM. Real-time quantitative PCR (QPCR) and reverse transcription-QPCR for detection and enumeration of total yeasts in wine. Appl Environ Microbiol. 2006;72(11):7148–55.
25. White RA, Blainey PC, Fan HC, Quake SR. Digital PCR provides sensitive and absolute calibration for high throughput sequencing. BMC Genomics. 2009;10(1):1.
26. Kim TG, Jeong S-Y, Cho K-S. Comparison of droplet digital PCR and quantitative real-time PCR for examining population dynamics of bacteria in soil. Appl Microbiol Biotechnol. 2014;98(13):6105–13.

27. Ottesen EA, Hong JW, Quake SR, Leadbetter JR. Microfluidic digital PCR enables multigene analysis of individual environmental bacteria. Science. 2006;314(5804):1464–7.
28. Chamberlain JS, Gibbs RA, Rainer JE, Nguyen PN, Thomas C. Deletion screening of the Duchenne muscular dystrophy locus via multiplex DNA amplification. Nucleic Acids Res. 1988;16(23):11141–56.
29. Lindström M, Keto R, Markkula A, Nevas M, Hielm S, Korkeala H. Multiplex PCR assay for detection and identification of Clostridium botulinum types A, B, E, and F in food and fecal material. Appl Environ Microbiol. 2001;67(12):5694–9.
30. De Medici D, Anniballi F, Wyatt GM, Lindström M, Messelhäußer U, Aldus CF, Delibato E, Korkeala H, Peck MW, Fenicia L. Multiplex PCR for detection of botulinum neurotoxin-producing clostridia in clinical, food, and environmental samples. Appl Environ Microbiol. 2009;75(20):6457–61.
31. Ngamwongsatit P, Buasri W, Pianariyanon P, Pulsrikarn C, Ohba M, Assavanig A, Panbangred W. Broad distribution of enterotoxin genes (hblCDA, nheABC, cytK, and entFM) among Bacillus thuringiensis and Bacillus cereus as shown by novel primers. Int J Food Microbiol. 2008;121(3):352–6.
32. Monteiro J, Widen RH, Pignatari AC, Kubasek C, Silbert S. Rapid detection of carbapenemase genes by multiplex real-time PCR. J Antimicrob Chemother. 2012;67(4):906–9.
33. Lee N, Kwon KY, Oh SK, Chang H-J, Chun HS, Choi S-W. A multiplex PCR assay for simultaneous detection of Escherichia coli O157: H7, Bacillus cereus, Vibrio parahaemolyticus, Salmonella spp., Listeria monocytogenes, and Staphylococcus aureus in Korean ready-to-eat food. Foodborne Pathog Dis. 2014;11(7):574–80.
34. Yuan Y, Xu W, Zhai Z, Shi H, Luo Y, Chen Z, Huang K. Universal primer-multiplex PCR approach for simultaneous detection of Escherichia coli, Listeria monocytogenes, and Salmonella spp. in food samples. J Food Sci. 2009;74(8):M446–52.
35. Xu W, Bai W, Luo Y, Yuan Y, Zhang W, Guo X, Huang K. A novel common single primer multiplex polymerase chain reaction (CSP-M-PCR) method for the identification of animal species in minced meat. J Sci Food Agric. 2008;88(15):2631–7.
36. Zhang C, Xu W, Zhai Z, Luo Y, Yan X, Zhang N, Huang K. Universal primer-multiplex-polymerase chain reaction (UP-M-PCR) and capillary electrophoresis–laser-induced fluorescence analysis for the simultaneous detection of six genetically modified maize lines. J Agric Food Chem. 2011;59(10):5188–94.
37. Xu W, Zhai Z, Huang K, Zhang N, Yuan Y, Shang Y, Luo Y. A novel universal primer-multiplex-PCR method with sequencing gel electrophoresis analysis. PLoS One. 2012;7(1):e22900.
38. Shi H, Trinh Q, Xu W, Zhai B, Luo Y, Huang K. A universal primer multiplex PCR method for typing of toxinogenic Pseudomonas aeruginosa. Appl Microbiol Biotechnol. 2012;95(6):1579–87.
39. Machida M, Asai K, Sano M, Tanaka T, Kumagai T, Terai G, Kusumoto K-I, Arima T, Akita O, Kashiwagi Y. Genome sequencing and analysis of Aspergillus oryzae. Nature. 2005;438(7071):1157–61.
40. Calistri A, Palù G. Unbiased next-generation sequencing and new pathogen discovery: undeniable advantages and still-existing drawbacks. Clin Infect Dis:ciu913. 2015.
41. Gill P, Ghaemi A. Nucleic acid isothermal amplification technologies—a review. Nucleosides Nucleotides Nucleic Acids. 2008;27(3):224–43.
42. Simpkins S, Chan A, Hays J, Cook N. An RNA transcription-based amplification technique (NASBA) for the detection of viable Salmonella enterica. Lett Appl Microbiol. 2000;30(1):75–9.
43. O'Grady J, Lacey K, Glynn B, Smith TJ, Barry T, Maher M. tmRNA–a novel high-copy-number RNA diagnostic target–its application for Staphylococcus aureus detection using real-time NASBA. FEMS Microbiol Lett. 2009;301(2):218–23.
44. Fykse EM, Nilsen T, Nielsen AD, Tryland I, Delacroix S, Blatny JM. Real-time PCR and NASBA for rapid and sensitive detection of Vibrio cholerae in ballast water. Mar Pollut Bull. 2012;64(2):200–6.

45. Walker GT, Fraiser MS, Schram JL, Little MC, Nadeau JG, Malinowski DP. Strand displacement amplification—an isothermal, in vitro DNA amplification technique. Nucleic Acids Res. 1992;20(7):1691–6.
46. Walker GT, Nadeau JG, Spears PA, Schram JL, Nycz CM, Shank DD. Multiplex strand displacement amplification (SDA) and detection of DNA sequences from Mycobacterium tuberculosis and other mycobacteria. Nucleic Acids Res. 1994;22(13):2670–7.
47. Detter JC, Jett JM, Lucas SM, Dalin E, Arellano AR, Wang M, Nelson JR, Chapman J, Lou Y, Rokhsar D. Isothermal strand-displacement amplification applications for high-throughput genomics. Genomics. 2002;80(6):691–8.
48. Javaheri Tehrani S, Aliabadian M, Fata A, Najafzadeh MJ. Rolling circle amplification (RCA): an approach for quick detection and identification of fungal species. J Mycol Res. 2014;1 (1):55–62.
49. Zhu D, Yan Y, Lei P, Shen B, Cheng W, Ju H, Ding S. A novel electrochemical sensing strategy for rapid and ultrasensitive detection of Salmonella by rolling circle amplification and DNA–AuNPs probe. Anal Chim Acta. 2014;846:44–50.
50. Mori Y, Kanda H, Notomi T. Loop-mediated isothermal amplification (LAMP): recent progress in research and development. J Infect Chemother. 2013;19(3):404–11.
51. Song T, Toma C, Nakasone N, Iwanaga M. Sensitive and rapid detection of Shigella and enteroinvasive Escherichia coli by a loop-mediated isothermal amplification method. FEMS Microbiol Lett. 2005;243(1):259–63.
52. Wang L, Shi L, Alam M, Geng Y, Li L. Specific and rapid detection of foodborne Salmonella by loop-mediated isothermal amplification method. Food Res Int. 2008;41(1):69–74.
53. Wang F, Jiang L, Ge B. Loop-mediated isothermal amplification assays for detecting Shiga toxin-producing Escherichia coli in ground beef and human stools. J Clin Microbiol. 2012;50 (1):91–7.
54. Cheng N, Xu Y, Yan X, Shang Y, Zhu P, Tian W, Liang Z, Xu W. An advanced visual qualitative and Eva Green-based quantitative isothermal amplification method to detect Listeria monocytogenes. J Food Saf. 2015.
55. Vincent M, Xu Y, Kong H. Helicase-dependent isothermal DNA amplification. EMBO Rep. 2004;5(8):795–800.
56. Gill P, Alvandi A-H, Abdul-Tehrani H, Sadeghizadeh M. Colorimetric detection of Helicobacter pylori DNA using isothermal helicase-dependent amplification and gold nanoparticle probes. Diagn Microbiol Infect Dis. 2008;62(2):119–24.
57. Kong H, Higgins LS, Dalton MA, Kucera RB, Schildkraut I, Wilson GG. N. bstnbi nicking endonuclease and methods for using endonucleases in single-stranded displacement amplification. US Patent 20,030,211,506. 2003.
58. Giraffa G, Neviani E. DNA-based, culture-independent strategies for evaluating microbial communities in food-associated ecosystems. Int J Food Microbiol. 2001;67(1):19–34.
59. Schütte UM, Abdo Z, Bent SJ, Shyu C, Williams CJ, Pierson JD, Forney LJ. Advances in the use of terminal restriction fragment length polymorphism (T-RFLP) analysis of 16S rRNA genes to characterize microbial communities. Appl Microbiol Biotechnol. 2008;80(3):365–80.
60. Vaneechoutte M. DNA fingerprinting techniques for microorganisms. Mol Biotechnol. 1996;6 (2):115–42.
61. Muyzer G. DGGE/TGGE a method for identifying genes from natural ecosystems. Curr Opin Microbiol. 1999;2(3):317–22.
62. Sanger F, Coulson AR. A rapid method for determining sequences in DNA by primed synthesis with DNA polymerase. J Mol Biol. 1975;94(3):441–8.
63. Claesson MJ, Wang Q, O'Sullivan O, Greene-Diniz R, Cole JR, Ross RP, O'Toole PW. Comparison of two next-generation sequencing technologies for resolving highly complex microbiota composition using tandem variable 16S rRNA gene regions. Nucleic Acids Res. 2010;38(22):e200.
64. Bej AK, Ng W-Y, Morgan S, Jones DD, Mahbubani MH. Detection of viable Vibrio cholerae by reverse-transcriptase polymerase chain reaction (RT-PCR). Mol Biotechnol. 1996;5 (1):1–10.

65. Sheridan G, Masters C, Shallcross J, Mackey B. Detection of mRNA by reverse transcription-PCR as an indicator of viability in Escherichia coli cells. Appl Environ Microbiol. 1998;64 (4):1313–8.
66. Gedalanga PB, Olson BH. Development of a quantitative PCR method to differentiate between viable and nonviable bacteria in environmental water samples. Appl Microbiol Biotechnol. 2009;82(3):587–96.
67. Shi H, Xu W, Luo Y, Chen L, Liang Z, Zhou X, Huang K. The effect of various environmental factors on the ethidium monazite and quantitative PCR method to detect viable bacteria. J Appl Microbiol. 2011;111(5):1194–204.
68. Wang L, Zhong Q, Liao Z. Specific detection of Vibrio parahaemolyticus in viable but non-culturable state by EMA-LAMP technique. Food Eng Biotechnol. 2013;50(34):169–73.
69. Wu GP, Chen SH, Levin RE. Application of ethidium bromide monoazide for quantification of viable and dead cells of Salmonella enterica by real-time loop-mediated isothermal amplification. J Microbiol Methods. 2015;117:41–8.
70. Nocker A, Cheung C-Y, Camper AK. Comparison of propidium monoazide with ethidium monoazide for differentiation of live vs. dead bacteria by selective removal of DNA from dead cells. J Microbiol Methods. 2006;67(2):310–20.
71. Vesper S, Mckinstry C, Hartmann C, Neace M, Yoder S, Vesper A. Quantifying fungal viability in air and water samples using quantitative PCR after treatment with propidium monoazide (PMA). J Microbiol Methods. 2008;72(2):180–4.
72. Lee J-L, Levin RE. A comparative study of the ability of EMA and PMA to distinguish viable from heat killed mixed bacterial flora from fish fillets. J Microbiol Methods. 2009;76(1):93–6.
73. Forghani F, Langaee T, Eskandari M, Seo K-H, Chung M-J, Oh D-H. Rapid detection of viable Bacillus cereus emetic and enterotoxic strains in food by coupling propidium monoazide and multiplex PCR (PMA-mPCR). Food Control. 2015;55:151–7.
74. Nocker A, Richter-Heitmann T, Montijn R, Schuren F, Kort R. Discrimination between live and dead cells in bacterial communities from environmental water samples analyzed by 454 pyrosequencing. Int Microbiol. 2010;13(2):59–65.
75. Elizaquível P, Aznar R, Sánchez G. Recent developments in the use of viability dyes and quantitative PCR in the food microbiology field. J Appl Microbiol. 2014;116(1):1–13.
76. Codony F, Agustí G, Allué-Guardia A. Cell membrane integrity and distinguishing between metabolically active and inactive cells as a means of improving viability PCR. Mol Cell Probes. 2015;29(3):190–2.
77. Nogva HK, Bergh A, Holck A, Rudi K. Application of the 5′-nuclease PCR assay in evaluation and development of methods for quantitative detection of Campylobacter jejuni. Appl Environ Microbiol. 2000;66(9):4029–36.
78. Nogva HK, Dromtorp S, Nissen H, Rudi K. Ethidium monoazide for DNA-based differentiation of viable and dead bacteria by 5′-nuclease PCR. Biotechniques. 2003;34(4):804–13.
79. Villarreal JV, Jungfer C, Obst U, Schwartz T. DNase I and proteinase K eliminate DNA from injured or dead bacteria but not from living bacteria in microbial reference systems and natural drinking water biofilms for subsequent molecular biology analyses. J Microbiol Methods. 2013;94(3):161–9.
80. Schaus SS, Henderson ER. Cell viability and probe-cell membrane interactions of XR1 glial cells imaged by atomic force microscopy. Biophys J. 1997;73(3):1205.
81. Cerf A, Cau J-C, Vieu C, Dague E. Nanomechanical properties of dead or alive single-patterned bacteria. Langmuir. 2009;25(10):5731–6.
82. Notingher I, Jones J, Verrier S, Bisson I, Embanga P, Edwards P, Polak J, Hench L. Application of FTIR and Raman spectroscopy to characterisation of bioactive materials and living cells. J Spectrosc. 2003;17(2–3):275–88.
83. Zhou H, Yang D, Ivleva NP, Mircescu NE, Schubert S, Niessner R, Wieser A, Haisch C. Label-free in situ discrimination of live and dead bacteria by surface-enhanced Raman scattering. Anal Chem. 2015;87(13):6553–61.

84. Sivakumar P, Fernández-Bravo A, Taleh L, Biddle J, Melikechi N. Detection and classification of live and dead Escherichia coli by laser-induced breakdown spectroscopy. Astrobiology. 2015;15(2):144–53.
85. Markx GH, Talary MS, Pethig R. Separation of viable and non-viable yeast using dielectrophoresis. J Biotechnol. 1994;32(1):29–37.
86. Doh I, Cho Y-H. A continuous cell separation chip using hydrodynamic dielectrophoresis (DEP) process. Sensors Actuators A Phys. 2005;121(1):59–65.
87. Lapizco-Encinas BH, Simmons BA, Cummings EB, Fintschenko Y. Dielectrophoretic concentration and separation of live and dead bacteria in an array of insulators. Anal Chem. 2004;76(6):1571–9.
88. Jen C-P, Chen T-W. Selective trapping of live and dead mammalian cells using insulator-based dielectrophoresis within open-top microstructures. Biomed Microdevices. 2009;11 (3):597–607.
89. Shafiee H, Sano MB, Henslee EA, Caldwell JL, Davalos RV. Selective isolation of live/dead cells using contactless dielectrophoresis (cDEP). Lab Chip. 2010;10(4):438–45.
90. Patel S, Showers D, Vedantam P, Tzeng T-R, Qian S, Xuan X. Microfluidic separation of live and dead yeast cells using reservoir-based dielectrophoresis. Biomicrofluidics. 2012;6 (3):034102.
91. Czechowska K, Johnson DR, van der Meer JR. Use of flow cytometric methods for single-cell analysis in environmental microbiology. Curr Opin Microbiol. 2008;11(3):205–12.
92. Amor KB, Breeuwer P, Verbaarschot P, Rombouts FM, Akkermans AD, De Vos WM, Abee T. Multiparametric flow cytometry and cell sorting for the assessment of viable, injured, and dead Bifidobacterium cells during bile salt stress. Appl Environ Microbiol. 2002;68 (11):5209–16.
93. Berney M, Hammes F, Bosshard F, Weilenmann H-U, Egli T. Assessment and interpretation of bacterial viability by using the LIVE/DEAD BacLight Kit in combination with flow cytometry. Appl Environ Microbiol. 2007;73(10):3283–90.
94. Yang X, Kirsch J, Simonian A. Campylobacter spp. detection in the 21st century: a review of the recent achievements in biosensor development. J Microbiol Methods. 2013;95(1):48–56.
95. Baeumner AJ, Cohen RN, Miksic V, Min J. RNA biosensor for the rapid detection of viable Escherichia coli in drinking water. Biosens Bioelectron. 2003;18(4):405–13.
96. Varshney M, Li Y. Double interdigitated array microelectrode-based impedance biosensor for detection of viable Escherichia coli O157: H7 in growth medium. Talanta. 2008;74(4):518–25.
97. Chang W-H, Wang C-H, Lin C-L, Wu J-J, Lee MS, Lee G-B. Rapid detection and typing of live bacteria from human joint fluid samples by utilizing an integrated microfluidic system. Biosens Bioelectron. 2015;66:148–54.
98. Ngom B, Guo Y, Wang X, Bi D. Development and application of lateral flow test strip technology for detection of infectious agents and chemical contaminants: a review. Anal Bioanal Chem. 2010;397(3):1113–35.
99. Urata M, Iwata R, Noda K, Murakami Y, Kuroda A. Detection of living Salmonella cells using bioluminescence. Biotechnol Lett. 2009;31(5):737–41.

Chapter 16
Characterization of Microbial Diversity in Food Chain: A Molecular Review

Abstract Three kinds of traditional molecular methods are available for the detection of microbial community diversity: nucleic-acid-hybridization-based methods, PCR-electrophoresis-based methods, and Sanger-sequencing-based methods. Distinct from Sanger sequencing, high-throughput sequencing is based on the sequencing-by-synthesis principle and the massively parallel signature sequencing strategy, allowing the simultaneous determination of the sequences of millions of different DNA templates. Thus, the metagenome of an entire microbial community can be sequenced simultaneously, without the need to isolate unique sequences through clone library construction and subculturing. High-throughput-sequencing-based methods are highly sensitive and specific, are inexpensive, and require relatively little time compared with traditional molecular methods for microbial diversity research. In the last section of this chapter, we introduce several traditionally used gene targets for microbial identification.

Keywords Microbial diversity • PCR • Hybridization • High-throughput sequencing • Food safety

16.1 Introduction

16.1.1 Definitions and Terms

In a narrow sense, microbial diversity is defined as the memberships and relative abundance of microorganisms in an ecological community, of which the species richness and species evenness are two basic parameters. In a broad sense, microbial diversity encompasses three levels of meaning: species diversity, genetic diversity, and ecosystem diversity.

Microbiota and microflora are two terms used to denote microbial communities, of which the former is more widely used at present. Species richness is the number of different species represented in a microbial community. Species richness is simply a count of species, and it does not account for the abundances of the species or their relative abundance distributions. Species diversity considers both species richness and species evenness. The structure of a microbiota refers to the

W. Xu, *Functional Nucleic Acids Detection in Food Safety*,
DOI 10.1007/978-981-10-1618-9_16

comprehensive composition of a microbial community, including both community membership and species diversity.

In 1960, RH Whittaker proposed the concepts of alpha diversity, beta diversity, and gamma diversity [1]. Whittaker's idea was that the total species diversity in a landscape (gamma diversity) is determined according to two different features: the mean species diversity of individual sites or habitats on the local scale (alpha diversity) and the differentiation among those habitats (beta diversity). In a microbial community, the alpha diversity is typically defined as the within-sample species diversity, whereas beta diversity is defined as the between-sample species diversity. Alpha diversity can be calculated using various indices, usually the Shannon–Wiener diversity index or the Simpson diversity index. Beta diversity is generally represented by the Bray–Curtis dissimilarity index in current microbiota studies.

16.1.2 Significance of Microbial Diversity Research in Food Safety Detection

Despite significant efforts to combat this problem, there remains a considerable burden of foodborne illness, in which microbes play a prominent role [2]. Microorganisms can enter the food chain at various stages, are highly versatile, and can adapt to their environment, facilitating their survival, growth, and production of toxic compounds. In addition to environmental factors, the microorganisms in a microbial community must also adapt to each other. The relationships between different microorganisms in the same community can take the form of competition, antagonism, alternation, or symbiosis. Microbes perform most of their functions in the form of integral microbiota; thus, compared with individual species detection, the detection of microbial communities could provide more information for food safety analysis.

The detection of microbial community diversity is more efficient than the separate detection of individual species. Most diversity detection methods are high in throughput, which means they can detect multiple species simultaneously, saving time and costs. Because the abundances of different species are shown in the same profile in most cases, the results could provide more precise information for comparing the abundances of different species, without between-batch deviations.

Furthermore, most methods for the detection of microbial community diversity are untargeted methods and thus are well suited for the detection of microbial contamination in food. In targeted detection methods, the potential contaminant strains must be specified before detection is performed. However, other strains not in the list of potential contaminants, which will escape targeted detection, also have the possibility to contaminate food. Untargeted methods can detect all microbial strains present in a community and thus are more comprehensive.

The application of methods for the detection of microbial community diversity has advanced our knowledge of the role played by microbes in the process of food spoilage and corruption. For example, Powell and Tamplin used a combination of culture-based and DNA-based microbial diversity detection methods to explore the microbial communities found on Atlantic salmon fillets packed in a modified atmosphere of carbon dioxide and nitrogen [3]. They found that *Shewanella* spp. or *Carnobacterium* spp. were the most abundant species, and a number of other genera were also present in lesser numbers in these communities after 15 days. This work also highlighted the potentially important role of lactic acid bacteria in fresh Atlantic salmon stored under modified atmosphere conditions.

Thus, microbial diversity detection not only is a highly efficient and comprehensive strategy for the detection of microbial contamination in food but also can help to gain a better understanding of the microbial mechanisms by which food can become unsafe.

16.2 A Brief Review of Traditional Molecular Methods for Microbial Diversity Research

Traditional molecular methods can be classified into three types: nucleic-acid-hybridization-based methods, PCR-electrophoresis-based methods, and Sanger-sequencing-based methods.

16.2.1 Nucleic-Acid-Hybridization-Based Methods

In molecular biology, hybridization is a phenomenon in which single-stranded deoxyribonucleic acid (DNA) or ribonucleic acid (RNA) molecules anneal to complementary DNA or RNA molecules through Watson–Crick base pairing [4]. The process of hybridization is highly specific; thus, specific target genes can be detected using synthetic probes.

Fluorescence in situ hybridization (FISH) is a technique for detecting in situ hybridization in samples using nonradioactive fluorescence signals, without the isolation of DNA or RNA. The tissues or cells of interest are first fixed on a slide or fiber membrane. After denaturation of the double-stranded DNA, fluorescently labeled probes can specifically hybridize the target genes, and the results can be observed using a fluorescence microscope or a laser scanning confocal microscope. In most such studies, fluorescently labeled 16S rDNA or 16S rRNA oligonucleotide sequences are used as probes. Because FISH provides a detailed picture of a microenvironment without the need for any selective purification or amplification steps, it has been extensively used in the field of environmental diversity research [5].

The performance of FISH technology can be summarized as follows: (1) False negatives—although 16S rDNA is highly expressed in most bacteria, a variation in abundance is observed not only among different strains but also among different growth phases of the same strains. Moreover, the low permeability of the cell walls of certain bacteria can prevent the entry of probes into the cells. These factors can result in false negatives when FISH technology is used. (2) False positives—the design of a FISH probe depends on the database of bacterial nucleic acid sequences that is used. The present databases are incomplete and inaccurate. Furthermore, some bacteria exhibit their own fluorescence. These factors can result in false positives when FISH technology is used. (3) The advantages of fluorescently labeled probes—such probes are safe, offer good resolution, and require no additional testing procedures. Moreover, different fluorescent dyes can be used at the same time for the simultaneous detection of different bacterial species.

16.2.2 PCR-Electrophoresis-Based Methods

In the application of detection methods to environmental samples, the interference components are usually complex, and the abundance of the target nucleic acid molecules is usually low. For the detection of microbial community diversity in such a complex environment, the sensitivity of hybridization-based methods would not be satisfactory. PCR-based methods, which can amplify the signals of interest, offer an advantage in terms of sensitivity. However, PCR amplification can bias the inferred composition of the microbial community of interest.

16.2.2.1 RT-PCR

Real-time polymerase chain reaction (RT-PCR) is a method in which fluorescent dyes are added to a PCR reaction system. The PCR reaction process is monitored throughout, and finally, the abundances of the target genes can be quantified based on standard curves. Different PCR primers can be designed to target different microbial taxa; thus, the microbial composition and abundance of a sample can be analyzed using a series of primers.

The primer design, which determines the specificity, is the most critical step of RT-PCR detection. Among the various known biomarker genes, 16S rDNA is the most popular target for primer design. However, in most cases, it is difficult to differentiate closely related species of the same genus using 16S rDNA. If species- or strain-specific primers are needed, the primers must be screened among from a large number of potential marker genes.

Although feasible in some cases, RT-PCR is not a recommended method for general microbial diversity research for the following reasons: (1) For specific primer design, the targeted taxa must be clearly known. However, in most cases of microbial diversity research, which taxa are present in the community of interest is unknown before a study is performed. (2) For complex microbial communities, the need to consider a different RT-PCR reaction for each taxon will result in a large workload.

16.2.2.2 RFLP and T-RFLP

Restriction fragment length polymorphism (RFLP) is a method of exploiting variations in homologous DNA sequences. In RFLP analysis, restriction enzymes are used to digest a DNA sample, and gel electrophoresis is then applied to separate the restriction fragments according to their lengths. Then, a Southern blot procedure is performed to transfer the electrophoresis profile to a membrane. The membrane is hybridized to a labeled DNA probe that is complementary to the target DNA to determine the length of the fragments. An RFLP occurs when the length of a detected fragment varies between different microbial strains. In microbial diversity studies, the 16S rDNA signal is first amplified using PCR, and the PCR products are then digested by restriction enzymes.

Terminal restriction fragment length polymorphism (T-RFLP) is a method that results from the combination of RFLP and PCR. PCR products are obtained using oligonucleotide primers that are typically designed to anneal at consensus sequences in bacterial 16S rDNA genes. The PCR primers are both labeled at the 5′ end with phosphoramidite dyes, and the digested products are separated via electrophoresis using gel electrophoresis or capillary-based systems [6]. Thus, only the terminal restriction fragments can be detected because the nonterminal fragments are not labeled with dyes. T-RFLP analysis is one of the most frequently used high-throughput fingerprinting methods. Because of its relative simplicity, T-RFLP analysis has been applied for the analysis of various microorganisms and microbial communities [7].

16.2.2.3 RAPD

Among PCR-based typing techniques, random amplified polymorphic DNA (RAPD) is a widely used and successful method for the characterization of bacteria. In this method, genomic DNA regions are amplified via PCR using oligonucleotide primers that are typically ten nucleotides long. After amplification, strain-specific patterns are visualized through electrophoresis; similar profiles indicate similar genomes or metagenomes of bacterial communities.

The advantages of RAPD are that no DNA sequencing information is needed prior to an experiment and that this method does not require Southern hybridization and are probe independent; as a result, RAPD technology is simple, fast, and inexpensive. The major shortcoming of this technique is that it does not offer good repeatability.

16.2.2.4 PCR-SSCP

Single-stranded conformation polymorphism (SSCP) analysis is a method of analyzing the sequence variations among different homogeneous DNA fragments. Such sequence variations will result in different secondary structures of single-stranded DNA, which affect the migration speeds of the molecules during electrophoresis. In an analysis of microbial diversity, 16S rDNA is usually the target type of DNA. If the DNA fragments of interest are less than 400 bp in length, then SSCP generally offers suitable discriminatory ability and reproducibility. The discriminatory ability is determined by the numbers and positions of single-nucleotide polymorphisms in the targeted genes [8].

16.2.2.5 PCR-DGGE and PCR-TGGE

Denaturing gradient gel electrophoresis (DGGE) and thermal gradient gel electrophoresis (TGGE) essentially consist of two procedures: the PCR amplification of the target genes and gradient gel electrophoresis. Separation is caused by the gradient decrease in the electrophoretic mobility of the target genes in a polyacrylamide gel containing a linear gradient of denaturants, in the case of DGGE, or a linear temperature gradient, in the case of TGGE [6].

The resolution of DGGE and TGGE is very high compared with other PCR-electrophoresis-based methods. Theoretically, DGGE and TGGE can distinguish single-base substitutions. Thus, they are very powerful tools for the analysis of a microbial community composed of closely related species. Although these techniques are reliable, reproducible, rapid, and inexpensive [9], their main limitation is that the community fingerprints they generate do not directly translate into taxonomic information [6]. Two potential strategies could be used to overcome this limitation: first, the retrieval of the DNA from the bands on the gel, followed by Sanger sequencing; second, if the main composition of the microbial community is known and the researchers wish to analyze the changes in abundance among different samples, a series of known pure cultured strains is used as markers.

16.2.2.6 PCR-PFGE

Pulsed-field gel electrophoresis (PFGE) is a technique that is used to separate large DNA molecules by applying an electric field that periodically changes in direction to a gel matrix. This electrophoresis method was developed in 1984 by Schwartz and Cantor [10]. Whereas traditional gel electrophoresis can only resolve DNA fragments that are no longer than 15 kb, as DNA molecules larger than 15–20 kb migrating through a gel will essentially move together in a size-independent manner, PFGE can be used to resolve DNA molecules of 10 kb to 10 Mb in length. In PCR-based microbial diversity studies, a longer length of the target genes allows more information to be acquired for use in taxonomic analysis. When long fragments are selected as the PCR targets, PFGE should be used for electrophoresis. The disadvantage of PFGE is that it is time-consuming, typically requiring 2 or 3 days for completion.

16.2.3 Clone-Library- and Sanger-Sequencing-Based Methods

Sanger sequencing is a method of DNA sequencing based on the selective incorporation of chain-terminating dideoxynucleotides by DNA polymerase during in vitro DNA replication [11]. The template for Sanger sequencing should be a unique sequence; therefore, the metagenome of an environmental microbial community, which will contain different genomes from the different member strains, cannot be directly subjected to Sanger sequencing. For microbial diversity research, the 16S rDNA of the metagenome should be amplified, and the clean PCR products should then be cloned to a clone library. Through subculturing of the clones, unique sequences can be acquired and then individually subjected to Sanger sequencing.

Clone-library- and Sanger-sequencing-based methods are able to provide the most comprehensive information on microbial communities, including the detailed 16S rDNA sequences that represent the species that comprise a community, the abundance and diversity of a community, and the relationships between different communities. However, these methods are time-consuming and expensive, and therefore, they have now been almost completely replaced by high-throughput-sequencing-based methods.

16.3 High-Throughput-Sequencing-Based Methods for Microbial Diversity Research and Their Application in Food Safety Detection

Distinct from Sanger sequencing, high-throughput sequencing is based on the sequencing-by-synthesis principle and the massively parallel signature sequencing strategy, thereby enabling the simultaneous determination of the sequences of millions of different DNA templates. Thus, the metagenome of a microbial community can be sequenced simultaneously, without the need to isolate unique sequences via clone library construction and subculturing. High-throughput-sequencing-based methods are highly sensitive and specific, are inexpensive, and require relatively little time compared with traditional molecular methods for microbial diversity research.

Amplicon sequencing and whole-metagenome shotgun sequencing are the two major methods used for microbial diversity studies. Amplicon sequencing is used for the detection of the species composition of a microbiota to determine the "species diversity" (however, the resolution of amplicon sequencing cannot reach the species level at present), whereas whole-metagenome shotgun sequencing provides not only species diversity information but also genetic diversity information.

16.3.1 Amplicon Sequencing

In amplicon sequencing, a phylogenetically informative marker is targeted for high-throughput sequencing. Such a marker should ideally be present in all expected organisms. In research on bacterial microorganisms, 16S rDNA is the most widely used marker, whereas for fungal microorganisms, 18S rDNA and the ITS1/ITS2 genes are the markers of interest. The chosen phylogenetically informative marker is amplified by PCR and then subjected to high-throughput sequencing. The raw sequencing data are filtered according to the sequence quality and length, and the remaining reads are then clustered and annotated to certain phylogenetic taxa.

16.3.1.1 Metagenomic DNA Extraction

In methods of nucleic acid detection, the extraction of nucleic acid molecules, whether DNA or RNA, is the first step and one of the most critical. In metagenomic methods, this step is of greater importance. The DNA isolation procedure may introduce species abundance bias into the taxonomic profile results because of the different sensitivities of Gram-positive and Gram-negative bacteria to the lysis reagent. Research conducted by Brooks et al. has revealed that different DNA extraction kits can produce dramatically different results, but bias is introduced regardless of the choice of kit [12]. To date, it is difficult to say which methods or kits are the best, and the most suitable choice may depend on the characteristics of different types of samples. Differential centrifugation is typically applied to isolate bacterial cells from environmental samples. During the cell lysis procedure, the use of lysozyme is recommended.

16.3.1.2 The Design of Amplicon Sequencing Experiments

Selection of the Sequencing Platform

Currently, the Roche 454 GS FLX system, the Illumina HiSeq and MiSeq systems, and the Life Technologies Ion Torrent system are the major high-throughput-sequencing platforms. In the area of amplicon sequencing, the HiSeq and MiSeq sequencing platforms have seen broader application, according to the published literature. The length of the target sequence and the number of reads are two important bases for choosing a sequencing platform. The Roche 454 GS FLX sequencing platform allows a longer length for each sequence but a lesser number of reads. The Illumina HiSeq PE250 allows a sequencing length of less than 500 nt but provides 10–100 times as many reads as the GS FLX. Furthermore, sequencing using a HiSeq or MiSeq system is much cheaper and faster than using the GS FLX.

Selection of the Target Region for Sequencing

The length of 16S rDNA is approximately 1540 bp; it contains nine variable regions and ten conserved regions. Unfortunately, because of the sequencing length restriction of high-throughput sequencing, it is impossible to achieve full-length sequencing of 16S rDNA in amplicon sequencing (however, by means of shotgun sequencing and read reconstruction, this task becomes possible; this method will be introduced in the following section). Thus, one to three adjacent variable regions should be selected as the sequencing target.

Claesson et al. reported inconsistencies when targeting different variable regions of this gene [13]. The reason for this may be that some variable regions are less phylogenetically informative than others, or it may be due to PCR amplification bias. No 16S rDNA primer is truly a universal primer, and different primers have mismatches with different microbes. The sequencing of different variable regions using different primers results in different PCR biases. Claesson et al. suggested the use of the V4+V5 variable regions (with the primer pair 563F/926R) for amplicon sequencing on a Roche 454 GS FLX platform [13]. Kozich et al. suggested the use of the V4 variable region for amplicon sequencing on a MiSeq platform [14]. However, for different types of microbial communities, the optimal target variable regions may be different. For example, we determined the distributions of variation (i.e., phylogenetic information) along the entire length of the 16S rDNA sequences of 85 *Lactobacillus* strains and 33 *Bifidobacterium* strains. The results showed there have more variation in V1–V3 and in V9 than in other variable regions in these lactic acid bacteria. Lundin et al. found that 5000 denoised sequences and 1000 denoised sequences per sample are sufficient to explain the trends in alpha diversity (Shannon index) and 90 % of the trends in β diversity (Bray–Curtis index) among a set of samples [15].

16.3.1.3 The Bioinformatic Analysis of Amplicon Sequencing Data

Data Preprocessing

The raw data from a sequencing platform typically consist of a mixture of reads from different samples, with the primer sequence and a sample-indicating barcode in each sequencing read, and the sequencing quality of the reads is uneven. Thus, before bioinformatic analysis, data preprocessing must be performed, primarily consisting of the classification of the reads based on the sample-indicating barcodes, the removal of the barcodes and primer sequences, and the filtering of low-quality reads.

Clustering and Annotation

A first and crucial step in microbial community analysis is the binning of 16S sequences into groups of sequences with a predetermined degree of similarity [16]. This clustering procedure can be performed using software such as CD-HIT [17], Mothur [18], QIIME [19], or DOTUR [20]. The resulting clusters of sequences are called operational taxonomic units (OTUs). Each OTU contains a series of sequences with a degree of similarity higher than some threshold (usually 97 % or 99 %). However, the concept of OTUs is different from the traditional definition of species, which means that one OTU may contain sequences from different species and that sequences from the same species may be distributed among different OTUs. Although the OTUs do not correspond to systematic taxonomy units, the systematic information is important for the analysis of a microbial community. Thus, the most abundant sequence in each OTU is selected as representative of that OTU for systematic annotation through alignment with a database such as that of the Ribosomal Database Project (RDP) or Greengenes.

Community Structure Analysis

The purpose of microbial community structure analysis is to analyze whether there is a significant difference between the microbial communities present in different samples to determine whether the factors under investigation cause changes in the microbial community from an overall perspective. The calculation of diversity indices, principal component analysis (PCA), principal coordinate analysis (PCoA), non-metric multidimensional scaling (NMDS), and clustering analysis are the most commonly used methods for the comprehensive structural analysis of microbial communities.

As mentioned in the first section of this chapter, for a microbial community, the alpha diversity is typically defined as the within-sample species diversity and calculated using the Shannon–Wiener diversity index or the Simpson diversity index, whereas the beta diversity is defined as the between-sample species diversity and is represented by the Bray–Curtis dissimilarity index in most current microbiota studies. These diversity indices for gut microbial communities can be calculated using software such as QIIME, Mothur, or PAST.

The PCA, PCoA, and NMDS approaches yield plots of different samples in a two-dimensional coordinate system to clearly illustrate the similarity relationship among the samples: a higher similarity (dissimilarity) between two microbial communities is indicated by a lesser (greater) distance between the points at the coordinates indicating those two communities in such a plot. The dissimilarity between samples can be calculated in terms of the Bray–Curtis dissimilarity, the unweighted UniFrac measure, or the weighted UniFrac measure. The plots can be generated by various statistical and bioinformatic software packages. Cluster analysis is another method of revealing the similarity relationships between samples. This method yields a phylogenetic tree of samples.

Key Taxon Analysis

If a significant difference between samples is revealed by the structural analysis, then the next question will be to determine the specific taxa for which changes in these taxa cause changes to the structure of the microbial community. The LEfSe online software is designed to help researches to find biomarkers that explain most of the effects that differentiate phenotypes of interest in biomarker discovery for comparative and hypothesis-driven investigations. The visualization of the discovered biomarkers on taxonomic trees provides an effective method of summarizing the results in a biologically meaningful way, as such visualizations both statistically and visually capture the hierarchical relationships inherent in 16S-based taxonomies/phylogenies or in ontologies of pathways and biomolecular functions [21].

16.3.2 New Advancements in Overcoming the Resolution Limitation of Sequencing-Based Methods

As mentioned in the previous section, current amplicon sequencing methods for 16S rDNA can yield only a genus-level profile of a microbial community because of the short-read length of high-throughput sequencing; this phenomenon can be called the "resolution limitation" of sequencing-based methods. The high-resolution identification of bacterial taxa, to at least the species level, is necessary to fully understand the composition and diversity of microbial communities. For food safety detection, gene-level resolution cannot distinguish the virulence of bacterial contaminants.

Because the resolution limitation is primarily due to the conflict between the existence of too many bacterial species to be identified and the availability of too little information to distinguish among them, two strategies have been developed to achieve high-resolution profiles by means of high-throughput sequencing: first, reducing the number of potential species to be analyzed, and second, increasing the amount of available sequencing information.

16.3.2.1 Limiting the Potential Species to a Certain Range

In certain closed environments, the membership of the microbial community remains almost constant. Thus, the number of potential species can be reduced by limiting the species of interest to a certain range that contains all species ever to be reported as present in that community. The critical step of this method is to construct a reference database that contains all potential species present in the community but does not include any other species.

An exemplary study was conducted by Fettweis et al. [22]. They developed a non-redundant, comprehensive Vaginal 16S rDNA Reference Database using vaginal-health-related bacterial species. They also constructed a rapid species-level classification method, STIRRUPS, for 16S rDNA partial sequencing data. Using these tools, they analyzed two datasets: one mock community containing six vaginal bacterial strains and one sequencing dataset constructed from 1000 mid-vaginal samples. In both datasets, they classified more than 95 % of the processed reads to species-level taxa.

16.3.2.2 Shotgun Sequencing of 16S rDNA

Through shotgun sequencing of 16S rDNA, short reads covering a longer fragment or the entire range of 16S rDNA could be acquired. However, the question of how to reconstruct these short reads into a long sequence poses a conundrum because of the conservative regions of 16S rDNA. Several research groups have made efforts to solve this conundrum, and their methods could provide a higher resolution in microbial community analysis compared with amplicon sequencing.

Miller et al. reported a novel iterative mapping method called expectation maximization iterative reconstruction of genes from the environment (EMIRGE) [23], using the expectation maximization algorithm [24]. This method constructs whole-length 16S rDNA sequences from shotgun sequencing reads of a microbial community. The raw reads and quality values of 16S rDNA sequences and short-read DNA sequencing projects are taken as inputs, and several iterative read-mapping cycles are performed to gradually discover and correct the most probable consensus sequences based on the mapped reads. This method can be used on dataset of different sizes acquired from microbial communities of widely varying diversity.

In 2012, Fan et al. described a method of reconstructing nearly whole-length 16S rDNA reads from metagenomic shotgun pyrosequencing data [25]. These authors developed a process of data filtering and stringent assembly through the simulation of bacterial communities with different complexity. They then applied this method to assess the species in two marine sponge microbial communities and compared the outcomes to those of PCR-based methods. The results showed that approximately 30 % of the abundant phylotypes reconstructed from the metagenomic reads failed to be amplified by PCR [25].

In 2013, Ong et al. reported a shotgun short-read sequencing method [26] for reconstructing 16S rDNA amplicon sequences. This method is able to capture a greater portion of 16S rDNA gene sequences with less bias. This method also allows precise assignment at deeper taxonomic levels. In the same year, Amir et al. presented Convex Optimization for Microbial Profiling by Aggregating

Short Sequence reads (COMPASS), a method for integrating high-throughput shotgun sequencing reads from a large genomic region or multiple regions, thereby enabling high-resolution microbial profiling [27]. Unlike de novo assembly methods, this method relies on a database of bacterial sequences, under the assumption that the sequences of most bacterial species in the community are represented in this database. A theoretical analysis of COMPASS yielded the sufficient conditions for performance in the limit of a sufficient number of sequencing reads and bounds on the reconstruction error in the case of finite reads. As intended, COMPASS is completely agnostic to chimeric reads and robust to sequencing errors.

16.3.2.3 Using Whole-Genome Shotgun Sequencing for Species Profiling

Because of its universality in prokaryotes and the availability of large, curated reference databases, 16S rDNA is a powerful phylogenetic marker [28]. However, the use of 16S rDNA has inherent disadvantages, as introduced in the previous section, including biases introduced by differences in amplification efficiency [29], inconsistencies when targeting different variable regions of the gene [13], and an inherent resolution limitation that prevents certain related species from being distinguished even when the entire length of 16S rDNA is used.

In contrast to this 16S-rDNA-targeted amplicon approach, the shotgun sequencing of an entire metagenome results in millions of short reads that are sampled from the genomes of the microbial community at random. However, present approaches that use metagenomic shotgun sequencing data primarily for taxonomic composition analysis rely on the availability of reference genomes. Hence, they cannot resolve taxa for which no representative sequence information is available, although members of these unrepresented taxa may constitute a large fraction or even the majority of microbial communities.

To solve this limitation, Sunagawa et al. [28] developed a method using universal, single-copy marker genes that provide prokaryotic species boundaries at higher resolution than 16S rDNA, to estimate the relative abundances of known and currently unknown microbial community members using metagenomic data at a species-level resolution. This method uses both reference genomes and metagenomes to cluster marker gene sequences into mOTUs. Based on covariance data collected across multiple samples, mOTUs of common species origin are combined into mOTU linkage groups (mOTU-LGs). The method was used to analyze 252 human fecal samples. The results revealed that on average, 43 % of the species abundance and 58 % of the richness could not be identified using current reference-genome-based methods [28].

In 2014, Nielsen et al. developed a method to achieve the de novo segregation of complex metagenomic data into particular biological entities, such as specific bacterial or viral strains. Their approach, which is based on the binning of co-abundant genes across a set of metagenomic samples, enables the complete discovery of novel bacterial organisms, viruses, and coinherited genetic entities and helps in the assembly of microbial genomes without the requirements of reference sequences. The authors demonstrated their method on data from 396 human fecal samples to study their gut microbiota and identify 7381 co-abundance gene groups, including 741 metagenomic species. They then used these groups to assemble 238 high-quality microbial genomes and identify affiliations between the metagenomic species and hundreds of viruses or genetic entities. Their method provides the ways for comprehensive profiling of the diversity within complex metagenomic samples [30].

16.4 Gene Targets in Microbial Diversity Studies

16.4.1 The Principle of Gene Target Selection

Whether in traditional molecular methods, amplicon sequencing methods, or even shotgun-sequencing-based methods, the selection of the gene targets for microbial diversity research is the critical step that initially determines the final resolution with which different taxa in microbial communities can be distinguished. Essentially, when more gene targets are used and when those gene targets are greater in length, more information will be obtained, allowing a higher resolution to be achieved. Sunagawa et al. have presented a vivid example of this basic principle [28]. They used ten marker genes, each of which was longer than 16S rDNA, to obtain a higher resolution profile of the gut microbiota.

However, the number of gene targets and the length of each gene cannot be extended ad infinitum. At present, the read length in high-throughput sequencing is limited to 500 bp for Illumina sequencing platforms. Thus, the sequencing regions of gene targets can be no longer than 500 bp, unless shotgun sequencing and assembly are used. Furthermore, in metagenomic studies, in which all of the genes from a given microbial community are mixed, it is difficult (although not impossible, as proven by Sunagawa et al. using shotgun sequencing) to link different gene targets with each other and with particular bacterial taxa. For each potential target gene, there should be a variable region of sufficient length to provide taxonomic information, and in amplicon sequencing, there should also be a conservative region of sufficient length for primer design.

16.4.2 Traditionally Used Gene Targets

Specific genotypic markers are essential to analyze the phylogenetic lineages for correlation, as well as to find what lineages a particular bacterial species belong to. Various housekeeping genes are targeted in different microbial taxa to get precise genetic pattern. Several genetic markers are available for the rapid cataloguing of gene families with conserved sequences falling within definite microbial clades.

16S rDNA is a component of the small 30S subunit of prokaryotic ribosomes. The genes that code for it are referred to as 16S rDNA and are used in reconstructing phylogenies because of the slow rate of evolution of this region of the gene [31]. Woese and Fox recommended the use of 16S rDNA in reconstructing phylogenies in 1990 [32]. As introduced in Sect. 16.3.1, there are nine variable regions in the 16S rDNA gene, and different variable regions present different taxonomic information. In eukaryotic microorganisms such as fungi, 18S rDNA, intergenic spacer region 1 (ITS1), and intergenic spacer region 2 (ITS2) are often used as taxonomic gene markers.

rpoA and *rpoB* are the genes that encode for different subunits of the bacterial RNA polymerase. *rpoA* gene sequences yield a robust classification into species groups, similar to the classification based on 16S rDNA gene sequence analysis [33]. The comparison of *rpoB* sequences has also been explored as a means for the phylogenetic analysis of archaea and bacteria [34].

The *recA* gene encodes a 38 kDa protein that is essential for the repair and maintenance of DNA. Many classifications of Gram-negative and Gram-positive bacteria have been reported based on the results obtained from *recA* gene sequences [35]. Comparative phylogenetic analysis of *recA* and 16S rDNA gene sequences reveals highly similar branching patterns, suggesting that it may be advantageous to use this gene sequence for molecular systematics as well as species-level identification [36].

gyrA and *gyrB* encode for subunit A and subunit B, respectively, of DNA gyrase, which is one of two essential members of the type II topoisomerase family. The presence of highly conserved motifs in these gene sequences provides a useful tool for the design of universal primers for studies of bacterial identification and diversity. Because greater genetic variation is observed among the protein-coding genes, they can be used for the identification and classification of closely related taxa. Variations of *gyrA* gene sequences have been found to enable discrimination among members of the *Bacillus subtilis* group, whereas *gyrB* is useful for discriminating among members of the *Bacillus cereus* group [37].

16.4.3 Prospects for Identifying New Gene Targets for Microbial Diversity Analysis

With the development of high-throughput-sequencing technologies and the accumulation of bacterial genome information, whole-genome-scale screening for marker genes in microbial diversity analyses could become feasible. The entire genome sequences of potentially present bacteria could be downloaded from a public database such as GenBank, and analysis software could then be used to perform whole-genome alignment. The potential target regions should have characteristics such as those of 16S rDNA, i.e., both a conservative region and a variable region should be present, of which the conservative region would be used for primer design, and the variable region would be used for taxonomical analysis. In particular, virulence genes may be promising targets for microbial diversity analysis because the sequence variations in virulence genes are directly related to the infection capabilities and toxin production behavior of pathogenic bacteria. Thus, if virulence genes were to be used as target genes in diversity analyses, the resulting profiles could provide additional relevant information for food safety risk analysis.

Acknowledgments This work is supported by the Ministry of Science and Technology of Beijing (XX2014B069). Many thanks to Mingzhang Guo, for his kind help in manuscript conception and preparation.

References

1. Whittaker RH. Vegetation of the Siskiyou mountains, Oregon and California. Ecol Monogr. 1960;30(3):279–338.
2. Havelaar AH, Brul S, de Jong A, de Jonge R, Zwietering MH, Ter Kuile BH. Future challenges to microbial food safety. Int J Food Microbiol. 2010;139 Suppl 1:S79–94.
3. Powell SM, Tamplin ML. Microbial communities on Australian modified atmosphere packaged Atlantic salmon. Food Microbiol. 2012;30(1):226–32.
4. Felsenfeld G, Miles HT. The physical and chemical properties of nucleic acids. Annu Rev Biochem. 1967;36:407–48.
5. Moter A, Gobel UB. Fluorescence in situ hybridization (FISH) for direct visualization of microorganisms. J Microbiol Methods. 2000;41(2):85–112.
6. Giraffa G, Neviani E. DNA-based, culture-independent strategies for evaluating microbial communities in food-associated ecosystems. Int J Food Microbiol. 2001;67(1-2):19–34.
7. Schutte UM, Abdo Z, Bent SJ, Shyu C, Williams CJ, Pierson JD, Forney LJ. Advances in the use of terminal restriction fragment length polymorphism (T-RFLP) analysis of 16S rRNA genes to characterize microbial communities. Appl Microbiol Biotechnol. 2008;80(3):365–80.
8. Vaneechoutte M. DNA fingerprinting techniques for microorganisms. A proposal for classification and nomenclature. Mol Biotechnol. 1996;6(2):115–42.
9. Muyzer G. DGGE/TGGE a method for identifying genes from natural ecosystems. Curr Opin Microbiol. 1999;2(3):317–22.
10. Schwartz DC, Cantor CR. Separation of yeast chromosome-sized DNAs by pulsed field gradient gel electrophoresis. Cell. 1984;37(1):67–75.
11. Sanger F, Coulson AR. A rapid method for determining sequences in DNA by primed synthesis with DNA polymerase. J Mol Biol. 1975;94(3):441–8.
12. Brooks JP, Edwards DJ, Harwich Jr MD, Rivera MC, Fettweis JM, Serrano MG, Reris RA, Sheth NU, Huang B, Girerd P, Strauss 3rd JF, Jefferson KK, Buck GA. The truth about metagenomics: quantifying and counteracting bias in 16S rRNA studies. BMC Microbiol. 2015;15:66.
13. Claesson MJ, Wang Q, O'Sullivan O, Greene-Diniz R, Cole JR, Ross RP, O'Toole PW. Comparison of two next-generation sequencing technologies for resolving highly complex microbiota composition using tandem variable 16S rRNA gene regions. Nucleic Acids Res. 2010;38(22):e200.

14. Kozich JJ, Westcott SL, Baxter NT, Highlander SK, Schloss PD. Development of a dual-index sequencing strategy and curation pipeline for analyzing amplicon sequence data on the MiSeq Illumina sequencing platform. Appl Environ Microbiol. 2013;79(17):5112–20.
15. Lundin D, Severin I, Logue JB, Ostman O, Andersson AF, Lindstrom ES. Which sequencing depth is sufficient to describe patterns in bacterial alpha- and beta-diversity? Environ Microbiol Rep. 2012;4(3):367–72.
16. Wang X, Cai Y, Sun Y, Knight R, Mai V. Secondary structure information does not improve OTU assignment for partial 16s rRNA sequences. ISME J. 2012;6(7):1277–80.
17. Li W, Godzik A. Cd-hit: a fast program for clustering and comparing large sets of protein or nucleotide sequences. Bioinformatics. 2006;22(13):1658–9.
18. Schloss PD, Westcott SL, Ryabin T, Hall JR, Hartmann M, Hollister EB, Lesniewski RA, Oakley BB, Parks DH, Robinson CJ, Sahl JW, Stres B, Thallinger GG, Van Horn DJ, Weber CF. Introducing mothur: open-source, platform-independent, community-supported software for describing and comparing microbial communities. Appl Environ Microbiol. 2009;75 (23):7537–41.
19. Caporaso JG, Kuczynski J, Stombaugh J, Bittinger K, Bushman FD, Costello EK, Fierer N, Pena AG, Goodrich JK, Gordon JI, Huttley GA, Kelley ST, Knights D, Koenig JE, Ley RE, Lozupone CA, McDonald D, Muegge BD, Pirrung M, Reeder J, Sevinsky JR, Turnbaugh PJ, Walters WA, Widmann J, Yatsunenko T, Zaneveld J, Knight R. QIIME allows analysis of high-throughput community sequencing data. Nat Methods. 2010;7(5):335–6.
20. Schloss PD, Handelsman J. Introducing DOTUR, a computer program for defining operational taxonomic units and estimating species richness. Appl Environ Microbiol. 2005;71(3):1501–6.
21. Segata N, Izard J, Waldron L, Gevers D, Miropolsky L, Garrett WS, Huttenhower C. Metagenomic biomarker discovery and explanation. Genome Biol. 2011;12(6):R60.
22. Fettweis JM, Serrano MG, Sheth NU, Mayer CM, Glascock AL, Brooks JP, Jefferson KK, Buck GA. Species-level classification of the vaginal microbiome. BMC Genomics. 2012;13 Suppl 8:S17.
23. Miller CS, Baker BJ, Thomas BC, Singer SW, Banfield JF. EMIRGE: reconstruction of full-length ribosomal genes from microbial community short read sequencing data. Genome Biol. 2011;12(5):R44.
24. Dempster AP, Laird NM, Rubin DB. Maximum likelihood from incomplete data via the EM algorithm. J R Stat Soc Ser B (Methodol). 1977;39:1–38.
25. Fan L, McElroy K, Thomas T. Reconstruction of ribosomal RNA genes from metagenomic data. PLoS One. 2012;7(6):e39948.
26. Ong SH, Kukkillaya VU, Wilm A, Lay C, Ho EX, Low L, Hibberd ML, Nagarajan N. Species identification and profiling of complex microbial communities using shotgun Illumina sequencing of 16S rRNA amplicon sequences. PLoS One. 2013;8(4):e60811.
27. Amir A, Zeisel A, Zuk O, Elgart M, Stern S, Shamir O, Turnbaugh PJ, Soen Y, Shental N. High-resolution microbial community reconstruction by integrating short reads from multiple 16S rRNA regions. Nucleic Acids Res. 2013;41(22):e205.
28. Sunagawa S, Mende DR, Zeller G, Izquierdo-Carrasco F, Berger SA, Kultima JR, Coelho LP, Arumugam M, Tap J, Nielsen HB, Rasmussen S, Brunak S, Pedersen O, Guarner F, de Vos WM, Wang J, Li J, Dore J, Ehrlich SD, Stamatakis A, Bork P. Metagenomic species profiling using universal phylogenetic marker genes. Nat Methods. 2013;10(12):1196–9.
29. Engelbrektson A, Kunin V, Wrighton KC, Zvenigorodsky N, Chen F, Ochman H, Hugenholtz P. Experimental factors affecting PCR-based estimates of microbial species richness and evenness. ISME J. 2010;4(5):642–7.
30. Nielsen HB, Almeida M, Juncker AS, Rasmussen S, Li J, Sunagawa S, Plichta DR, Gautier L, Pedersen AG, Le Chatelier E, Pelletier E, Bonde I, Nielsen T, Manichanh C, Arumugam M, Batto JM, Quintanilha Dos Santos MB, Blom N, Borruel N, Burgdorf KS, Boumezbeur F, Casellas F, Dore J, Dworzynski P, Guarner F, Hansen T, Hildebrand F, Kaas RS, Kennedy S, Kristiansen K, Kultima JR, Leonard P, Levenez F, Lund O, Moumen B, Le Paslier D, Pons N, Pedersen O, Prifti E, Qin J, Raes J, Sorensen S, Tap J, Tims S, Ussery DW, Yamada T,

Renault P, Sicheritz-Ponten T, Bork P, Wang J, Brunak S, Ehrlich SD. Identification and assembly of genomes and genetic elements in complex metagenomic samples without using reference genomes. Nat Biotechnol. 2014;32(8):822–8.

31. Woese CR, Fox GE. Phylogenetic structure of the prokaryotic domain: the primary kingdoms. Proc Natl Acad Sci U S A. 1977;74(11):5088–90.
32. Woese CR, Kandler O, Wheelis ML. Towards a natural system of organisms: proposal for the domains Archaea, Bacteria, and Eucarya. Proc Natl Acad Sci U S A. 1990;87(12):4576–9.
33. Naser SM, Thompson FL, Hoste B, Gevers D, Dawyndt P, Vancanneyt M, Swings J. Application of multilocus sequence analysis (MLSA) for rapid identification of Enterococcus species based on rpoA and pheS genes. Microbiology. 2005;151(Pt 7):2141–50. doi:151/7/2141 [pii] 10.1099/mic.0.27840-0.
34. Klenk HP, Zillig W. DNA-dependent RNA polymerase subunit B as a tool for phylogenetic reconstructions: branching topology of the archaeal domain. J Mol Evol. 1994;38(4):420–32.
35. Karlin S, Weinstock GM, Brendel V. Bacterial classifications derived from recA protein sequence comparisons. J Bacteriol. 1995;177(23):6881–93.
36. Eisen JA. The RecA protein as a model molecule for molecular systematic studies of bacteria: comparison of trees of RecAs and 16S rRNAs from the same species. J Mol Evol. 1995;41(6):1105–23.
37. Wang LT, Lee FL, Tai CJ, Kasai H. Comparison of gyrB gene sequences, 16S rRNA gene sequences and DNA-DNA hybridization in the Bacillus subtilis group. Int J Syst Evol Microbiol. 2007;57(Pt 8):1846–50.

Chapter 17
Identification and Assessment of Heavy Metal Pollution Using Nucleic Acid-Mediated Technologies

Abstract Heavy metal ions in polluted drinking water or in plants have attracted increasing public attention because of their extremely harmful effects on human beings and ecological equilibria. Therefore, identifying and assessing heavy metal pollution in different substrates are important. Traditional heavy metal ion detection methods, such as atomic absorption spectroscopy, electron capture devices, inductively coupled plasma optical emission spectroscopy, and mass spectrometry, which can be coupled to chromatographic techniques, require the use of large instruments. These techniques have high sensitivity to and specificity for different ions. However, the large, expensive instruments required by these techniques hinder the rapid, cost-effective detection of heavy metal ions in real applications. Thus, these methods are not suitable for real-time detection, and numerous studies have focused on making heavy metal ion detection more convenient and feasible. Accordingly, detection methods based on nucleic acid signal transmission have been developed to achieve rapid or real-time detection. In this review, the necessity of heavy metal ion detection methods, the advantages and drawbacks of traditional detection methods, and the latest nucleic acid-based detection methods are discussed. Moreover, the prospective applications of the different nucleic acid-based detection methods are presented.

Keywords Heavy metal ion • Analytical methods • Functional nucleic acid • Biosensor

17.1 Necessity of Heavy Metal Ion Detection

Heavy metals are highly toxic, low-density chemical elements [1]. They are non-biodegradable and ubiquitous in nature [2, 3] and pose significant risks to human health and the environment [3–6]. These elements accumulate in the biosphere and enter living organisms through the food chain [7], thereby adversely affecting human health [8, 9]. Their toxicity is due to their bonding to protein thiol groups [10] and disruption of biochemical cycles in cells [4, 6]. Some heavy metals, such as iron, cobalt, zinc, copper, manganese, etc., are required by living organisms in small quantities; however, higher concentrations of these elements lead to toxic effects [11]. Heavy metal ions such as cadmium, lead, arsenic, chromium, and

W. Xu, *Functional Nucleic Acids Detection in Food Safety*,
DOI 10.1007/978-981-10-1618-9_17

mercury are considered to be hazardous compounds even at low concentrations [10, 12].

Because some heavy metal ions are very important for human biological functions, they are safe and even essential at very low concentrations. However, most heavy metal ions have obvious side effects, even at very low concentrations. Heavy metals are ubiquitous in nature and can enter the human body by various routes to disrupt cellular functions, making them toxic. They are then excreted by the liver, kidney, or spleen [11]. Unlike other molecules, these metals are excreted very slowly. These metals are stored in excretory organs for long time periods because of their interactions with cysteine-rich proteins (e.g., metallothioneins), leading to organ damage by changing cellular functioning.

The side effects and toxicity mechanisms of different heavy metals (their negative effects on the liver, kidney, and spleen) are discussed in the following sections.

17.1.1 Arsenic Toxicity

Arsenic can be released to the environment by wooden electricity poles that are treated with arsenic-based preservatives, pesticides, fertilizers, and pyrite (FeS) and arsenopyrite (FeAsS) oxidation and in untreated effluents. It is widespread in the environment. Arsenic toxicity could be due to enzyme inhibition and oxidative stress induction. In particular, arsenic inhibits enzyme oxidation, reduction, and hydrolysis reactions. When arsenic first enters liver cells, they release several enzymes, such as cytochrome P450 or mixed function oxidase (MFO). Some food substrates can cause the toxic molecules to remain in the body for a long time. After the enzymes are released, arsenic can conjugate to them and become incorporated into normal human pathways, and the toxic substance is converted into a water-soluble product that is excreted in bile or urine. Arsenic induces oxidative stress by altering the signal cascades involved in gene expression, inflammatory responses, and cell apoptosis [13, 14].

17.1.2 Cadmium Toxicity

Cadmium is thought to be the most essential element for human nutrition. However, it is also a carcinogen that potentially inhibits the enzymes involved in DNA mismatch repair. Because cadmium inhibits the mismatch repair (MMR) system [15, 16], the cellular error is amplified, and gene mutations increase, thereby leading to cancer. Although cadmium itself does not generate free radicals, it can promote the cellular release of copper and iron, which can induce oxidative stress and produce hydrogen peroxide.

17.1.3 Mercury Toxicity

Mercury has a large affinity for sulfur and therefore interacts with thiol-containing molecules such as glutathione (GSH) [17], cysteine, and metallothionein (MT) and alters the levels of superoxide dismutase (SOD) [17], glutathione peroxidase (GPx) [18], and glutathione reductase (GR) in the kidneys [11, 19–23]. Mercury exposure could result in the inhibition of cytochrome C oxidase (CCO), succinate dehydrogenase (SDH), and SOD, which are involved in the tricarboxylic acid cycle. Mercury can significantly affect the mitochondrial electron transport chain in cells by altering the K^+ and Ca^{2+} concentrations. Therefore, more free radicals can be generated, leading to oxidative stress.

17.1.4 Lead Toxicity

The toxicity mechanism of lead is the same as that of mercury. Lead inhibits enzymes, such as δ-aminolevulinic acid dehydratase (ALAD) [12] and glutathione reductase (GR), by binding to them. ALAD inhibition can lead to mitochondria-associated degradation (MAD) due to oxidative stress caused by free radical release.

17.1.5 Silver Toxicity

Silver is used in various commercial products, such as costumes, food storage products, and water filters. However, it has been proven that silver has both toxic cytological and physiological effects [24]. Exposing cells to silver can lead to the inhibition of GSH and other antioxidant marker enzymes. Silver exposure can also result in a decrease in the SOD levels and subsequent increase in ROS generation. In the presence of molecular oxygen, Ag^+ ions interact with ascorbate-forming free radicals. In addition, silver exposure can lead to carcinogenic malondialdehyde production. Malondialdehyde damages both purine and pyrimidine bases in DNA [25]. Ag^+ binding can lead to fibroblast membrane damage, which affects ATP synthesis and results in DNA damage.

17.1.6 Copper Toxicity

Copper is also believed to be carcinogenic because it can potentially cause DNA damage by interacting with lipid hydroxyperoxides to form malondialdehyde and 4-hydroxynonenal. It can also induce oxidative stress by generating ROS.

17.2 Traditional Analytical Methods

Most traditional analytical methods for detecting heavy metal ions employ large, expensive instruments. In this section, non-chromatographic, gas chromatography, and liquid chromatography methods are reviewed.

17.2.1 *Non-Chromatographic Methods*

Non-chromatographic methods typically involve cold vapor atomic absorption spectrometry (CVAAS), in which each heavy metal ion exhibits a different specific peak. This technique is commonly used to identify heavy metal ions and is highly sensitive. Cold vapor atomic fluorescence spectrometry (CVAFS) and inductively coupled plasma mass spectrometry (ICP-MS) have also been used.

In metal ion detection, some ion species must be selectively reduced. The reduction process involves selecting an appropriate reducing agent based on the behavior of different heavy metal ions and their formation mechanisms (e.g., Hg^{2+} is reduced to Hg^0 by $SnCl_2$, whereas methylmercury is not). However, selective reduction cannot distinguish between metal ions in different organic compounds; it can only distinguish between inorganic and organic ions.

Prior to analysis by non-chromatographic methods, the different types of ions, such as various organic and inorganic ions and selectively immobilized ionic compounds (generally applicable for air and gas samples), must be identified.

17.2.2 *Gas Chromatography Methods*

Gas chromatography is the most widely used technique for measuring different metal ion concentrations. A recent review describes the most important advantages of this technique [26].

The most obvious advantage of gas chromatography is that the concentrations of generated ionic compounds can be directly measured. In addition, in many cases, individual ions of organic compounds can be measured. The main drawback is that inorganic and organic ions must often be derivatized before the measurement due to their low volatility in the column and complex nature. The three most commonly used derivatizations are (1) butylation using a Grignard reagent (butyl magnesium chloride tetrahydrofuran), (2) hydride and sodium borohydride formation, and (3) ethylation with $NaBEt_4$.

The three main detection systems that are widely used after separation by gas chromatography are electronic capture devices (ECD), atomic spectroscopy (AS) detectors, and coupled gas chromatography–mass spectrometry (MS) systems.

17.2.3 High-Performance Liquid Chromatography Methods

In the last few years, high-performance liquid chromatography (HPLC) has been the most widely used traditional method for detecting heavy metal ions. In particular, to determine mercury contents, HPLC has been combined with (1) voltammetry, (2) UV–vis spectroscopy, (3) cold atomic absorption spectrometry, and (4) plasma (microwave, ICP-AES, ICP-MS) detection systems for different applications.

- Liquid chromatography is most commonly combined with electrochemical detection systems, which depend on the column used and the specific ion binding character, to achieve highly sensitive and specific ion detection with a detection limit of 1–2 ng/mL [26].
- UV–vis spectroscopy can be used to identify various organic and inorganic ionic compounds. However, its relatively low sensitivity is the main drawback of this approach. The detection limit is 7–95 ng/mL [26].
- Cold vapor atomic absorption spectroscopy is a widely used technique due to its high sensitivity and ability to detect the target ion in the presence of other ions, even when its content is very low. The reported detection limit ranges from 0.1 to 0.2 ng/mL [26].
- High-performance liquid chromatography combined with plasma techniques is highly promising. However, the plasma can be easily extinguished by liquid aerosols, greatly limiting its use.

17.3 Nucleic Acid-Based Detection Methods

Biosensors, which are made from biological materials, produce detectable signals in response to physical or chemical stimuli. Sensors used to be fabricated primarily from electronic materials, but the development of biosensors has stimulated the use of new materials. Sensors contain at least two components for target recognition and signal transduction. Any chemical or biological entity, such as a small organic molecule, polypeptide, protein, nucleic acid, carbohydrate, or whole cell, can be used for target recognition. The target recognition component must have a high affinity for the target, low detection limit, high specificity with low interference, wide dynamic range, rapid response time, high stability, and good versatility for detecting a wide range of similar analytes. Antibodies were thought to be promising target recognition candidates because they satisfy most of these criteria. However, antibody generation is too complex, and their range of suitable analytes is narrow, meaning different antibodies are required for the detection of different targets. In contrast, single-stranded DNA and RNA have high specificity for binding to their complementary strands and are useful for nucleic acid detection. Single-stranded nucleic acid ligands, including DNAzymes and aptamers (the two main types), can be obtained by the systematic evolution of ligands by exponential enrichment (SELEX).

Although these ligands are similar to antibodies, they have several competitive advantages over antibodies in heavy metal ion detection [27, 28]. First, because they are produced by in vitro selection, they can be designed to bind any selected target molecule. Antibodies, on the other hand, do not bind small molecules (e.g., Mg^{2+} or Pb^{2+} ions that are not chelated) or molecules with poor immunogenicity or high toxicity. It is difficult to review all the ligands generated by SELEX methods because more than 100 of these ligands were isolated by NeXstar Pharmaceuticals, Inc., and the University of Colorado by 1999 [29]. The number of all generated ligands is far greater than the number of scientists worldwide. Ellington and colleagues created an online database for finding more detailed information on nucleic acids [30]. Many reports of RNA adaptations have appeared in the literature, and the binding properties of RNA are generally similar to those of DNA. In this section, different heavy metal ion detection methods based on functional nucleic acids are summarized.

17.3.1 Nucleic Acid Enzyme (NAE)-Based Methods

In NAE sensors, the target molecule is usually a cofactor that accelerates the catalytic reaction or a reaction inhibitor. Accordingly, the selection pressure can be varied during the selection process to ensure that the desired reaction occurs only in the presence of the target analyte. To improve the target specificity, the sequences that compete with the NAE can be removed during the selection process [31]. Due to the chemical nature of NAE catalytic reactions, metal ions are usually required as auxiliary factors; therefore, metal detection by NAE sensors is highly successful [32–41]. Compared to protein enzymes, DNAzymes, which are widely used for DNA, protein, and metal ion detection, are cheaper to synthesize and more difficult to hydrolyze, meaning they are more stable [42–44]. In addition, they can bind multiple substrates, although they have less functionality than proteins [45]. For example, a Pb^{2+}-specific DNAzyme can catalyze the cleavage of an RNA substrate and its cofactor [41]. It has also been shown that aptamers exhibit high affinity and specificity for target cancer cells, proteins, and small molecules [46–48]. Moreover, aptamers can be easily labeled by techniques such as fluorescent and electroactive nanoparticle labeling [49–51] and are easier to obtain than antibodies [52]. Notably, two extremely simple and specific aptamers based on mismatched base pairs, i.e., thymine–Hg^{2+}–thymine (T–Hg^{2+}–T) [53, 54] and cytosine–Ag^{+}–cytosine (C–Ag^{+}–C) [55], have been employed in Hg^{2+} and Ag^{+} detection assays.

17.3.1.1 Principles of NAE Metal Ion Detection

Most NAE systems use Mg^{2+} as the metal cofactor, because it has been shown to be highly specific in several in vitro screening studies. Figure 17.1 shows some examples of detection systems for other metals endemic to the region. In Fig. 17.1, the enzyme strands, substrate strands, and cleavage sites are shown in green, black, and red, respectively. A Pb^{2+}-specific ribozyme called leadzyme, which was isolated by Pan and Uhlenbeck, is presented in Fig. 17.1a [56, 57]. This RNA pattern is the smallest catalytic pattern and requires Pb^{2+} to be active. The Pb^{2+}-dependent 8–17 DNA enzyme is shown in Fig. 17.1b. Interestingly, this motif has been independently selected by several research groups under different conditions, indicating that the sequence is a dominant choice from the selection pool [58–64]. In a comprehensive study by Lee and colleagues on this particular DNAzyme, it was found that the recurrence of this motif is due to its high catalytic efficiency, ability to cleave nucleotides, small catalytic core, and many nucleotide mutations [64]. The substrate cleavage site might be on an RNA strand or at a single RNA linkage site (i.e., rA) as shown in Fig. 17.1b. The DNAzyme exhibited high activity in the presence of Pb^{2+} [59]. When the metal concentration was low (in the nanomolar to micromolar range), only Pb^{2+} could activate the enzyme. Due to its unique and attractive properties, this type of DNA has been widely studied to elucidate its biochemical [59, 60, 64] and biophysical [65–67] properties. In another example, Joyce and colleagues developed a DNAzyme that could be used for target recognition in the presence of Zn^{2+} by introducing imidazole-modified deoxyuridine into an extended DNA fragment, imparting chemical functionality to it (Fig. 17.2c). The DNAzyme activity in the presence of Zn^{2+} was 100 times higher than that in the presence of Cd^{2+}, which is competitive

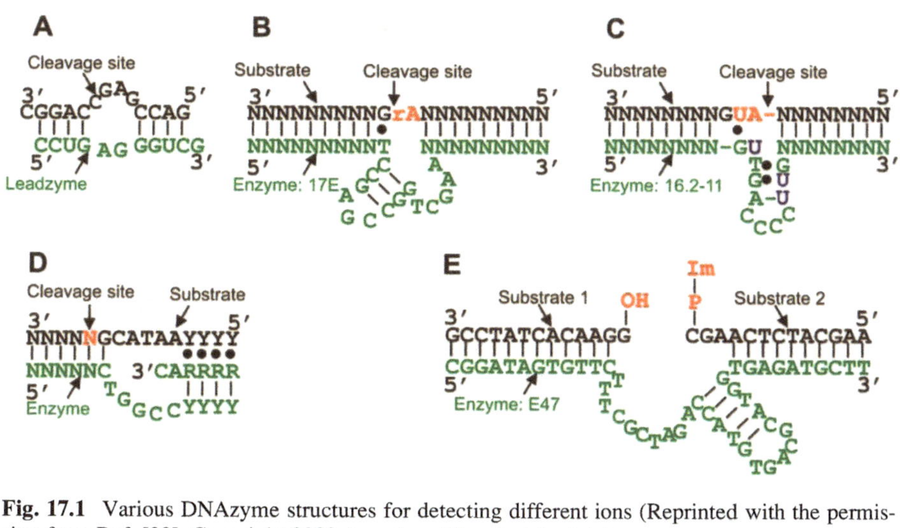

Fig. 17.1 Various DNAzyme structures for detecting different ions (Reprinted with the permission from Ref. [33]. Copyright 2009 American Chemical Society)

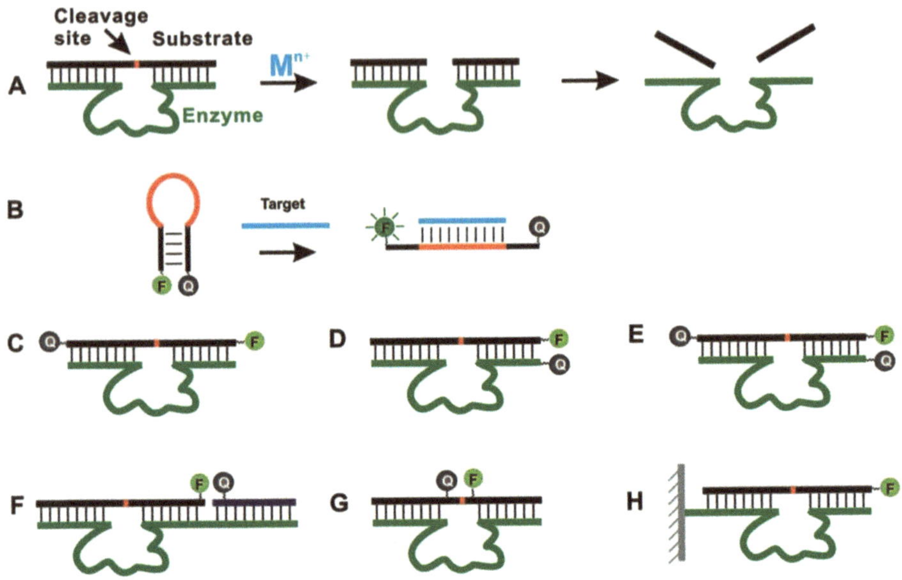

Fig. 17.2 Principal structures for NAE-based fluorescence detection (Reprinted with the permission from Ref. [33]. Copyright 2009 American Chemical Society)

with Zn^{2+}. In addition to RNA cleavage, DNA cleavage was achieved by Breaker and colleagues using a copper ion cofactor and DNAzyme (Fig. 17.2d). The Cu^{2+} catalytic oxidative cleavage of DNA is highly specific [68–72]. The activity of the related DNA ligase enzyme in Fig. 17.1e has also been shown to be specific to Cu^{2+} and Zn^{2+} at higher metal concentrations [73]. Li and coworkers have developed other metal-specific DNAzymes, such as DNA phosphorylation [74] and signaling RNA-cleaving [32, 75] DNAzymes. NAEs that are extremely sensitive to other environmental factors, such as the pH [76] and temperature [77], have also been developed.

17.3.1.2 NAE-Based Fluorescence Detection

Fluorescence detection techniques usually have high sensitivity. High fluorescence can be achieved at concentrations below 1 nM using ordinary equipment. When a more advanced instrument is employed, single-molecule fluorescence can be detected. Therefore, material consumption by fluorescence sensors is very low. In addition, some portable fluorescence sensors are commercially available, enabling their use in the field. External fluorescent groups are needed for fluorescent sensors based on natural nucleic acids because they are non-fluorescent.

In a typical NAE catalytic pyrolysis reaction (Fig. 17.2a), the substrates and DNAzyme are firstly hybridized. In the presence of the target metal ion, the annealed substrate is cleaved into two pieces, and then the substrate–enzyme complex is dehybridized. Consequently, the distance between the two substrate fragments and the distance between these fragments and the enzyme increase, and a fluorescence energy transfer signal similar to that in a molecular beacon can be exploited. A molecular beacon is a hairpin-shaped single-stranded DNA molecule with fluorescence and fluorescence-quenching groups attached to opposite ends of the hairpin (Fig. 17.2b) [78–81]. The hairpin loop section is developed to be reversed complementary to the target gene; thus, target binding to this section forces the hairpin to open and triggers an increase in the fluorescence. Molecular beacons have been successfully employed in many nucleic acid detection studies [78–82]. Unlike molecular beacons, which require the fluorescence group and quenching agent to be located at opposite ends of the hairpin, catalytic beacons can be labeled at several different locations. For example, a fluorophore and quencher can be placed at opposite ends of the substrate chain (Fig. 17.2c). This strategy was used in some early works in which a radioactive isotope was replaced by a fluorescence group for kinetics monitoring [83–89]. Figure 17.2d shows another option in which the fluorophore and quencher are on the same side but on different strands [41, 89]. Because of the proximity of the fluorophore and quencher, the background is lower. Other similar modifications are shown in Fig. 17.2f with the quencher placed at the end of a DNA enzyme template near the fluorophore on the substrate.

17.3.1.3 Catalytic Beacons with Labels in Different Positions

Although most sensors used for detection cannot compete with instrumental analysis methods, UO_2^{2+} has been used in catalytic beacon sensors to achieve a performance comparable to that of many analytical instruments [90]. In one system, this ion binded to the asymmetric substrate of an enzyme system with two quenchers (Fig. 17.3b), leading to a 15-fold increase in the fluorescence at room temperature. The UO_2^{2+} detection limit was 45 pmol, which is much lower than that

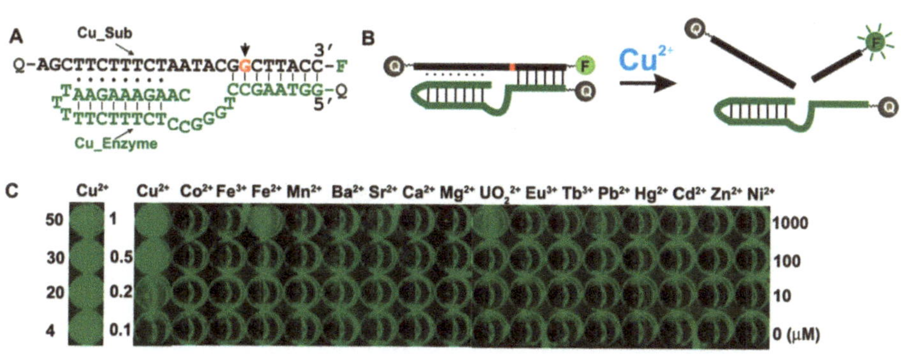

Fig. 17.3 Cu^{2+} catalytic beacon sensor (Reprinted with the permission from Ref. [16]. Copyright 2007 American Chemical Society)

of most analytical instruments (e.g., the uranium detection limit in ICP-MS is 420 pM) [91]. The US Environmental Protection Agency (EPA) has set the maximum uranium level in water to 130 nM, which is within the dynamic range of the sensor. This sensor selectivity to uranium was also more than one million times that to other metal ions.

One of the biggest challenges in designing fluorescent metal sensors is developing suitable sensors for paramagnetic metal ions (i.e., metal ions with unpaired electrons), such as copper ions, due to their intrinsic fluorescence-quenching properties. In this method, a major challenge is to convert the metal binding into a fluorescence signal when the two fragments are separated. Catalytic beacons are very appropriate for this type of system. Breaker and colleagues identified a series of Cu^{2+}-dependent DNAzymes for DNA cleavage. Their common structure is shown in Fig. 17.1d [69, 70, 92]. Figure 17.4a shows the DNAzyme sequences, and the red guanine symbol indicates the cleavage site in this catalytic beacon design (schematic shown in Fig. 17.4b). At the Cu^{2+} detection limit of 35 nM, which is much lower than the US EPA limit of 20 μM, the fluorescence was ~13 times higher. The fluorescence was only enhanced in the presence of 1 mM UO_2^{2+} and Fe^{2+} (Fig. 17.4c), and the sensor selectivity to Cu^{2+} was greater than 2000 times that to the other ions.

17.3.2 Colorimetric Sensors for Heavy Metal Ion Detection

17.3.2.1 Colorimetric Sensors Using Cleaving DNAzymes

Colorimetric methods are advantageous due to their simplicity and the low cost of monitoring the results by the naked eye, i.e., these methods do not require expensive instruments. Gold nanoparticles are usually used for monitoring because of their desirable optical properties, which depend on the distance, and their high extinction coefficients. Liu and Lu reported the colorimetric detection of Pb^{2+} by the sensor illustrated schematically in Fig. 17.4a [40]. The substrate strand of the DNAzyme was functionalized with gold nanoparticles that were arranged in a "head-to-tail" structure. In the absence of lead, the DNA-functionalized gold nanoparticles aggregated due to the hybridization of the gold nanoparticle substrate chain and enzyme strand, resulting in a blue color. In contrast, in the presence of Pb^{2+}, the DNA-functionalized gold nanoparticles were removed by substrate cleavage, resulting in a color change from blue to red. It should be noted that the steric effects of the "tail" prevented the complete annealing of the gold nanoparticles. To overcome this limitation, the "tail" must be designed to allow the Au nanoparticles to aggregate at room temperature during the annealing process (Fig. 17.4b) [36]. In addition, several simple colorimetric Pb^{2+} sensors have been designed to detect stimuli-induced responses of nanoparticle aggregates [35], even in the absence of cross-linked DNA-functionalized gold nanoparticles [93]. It should be noted that a

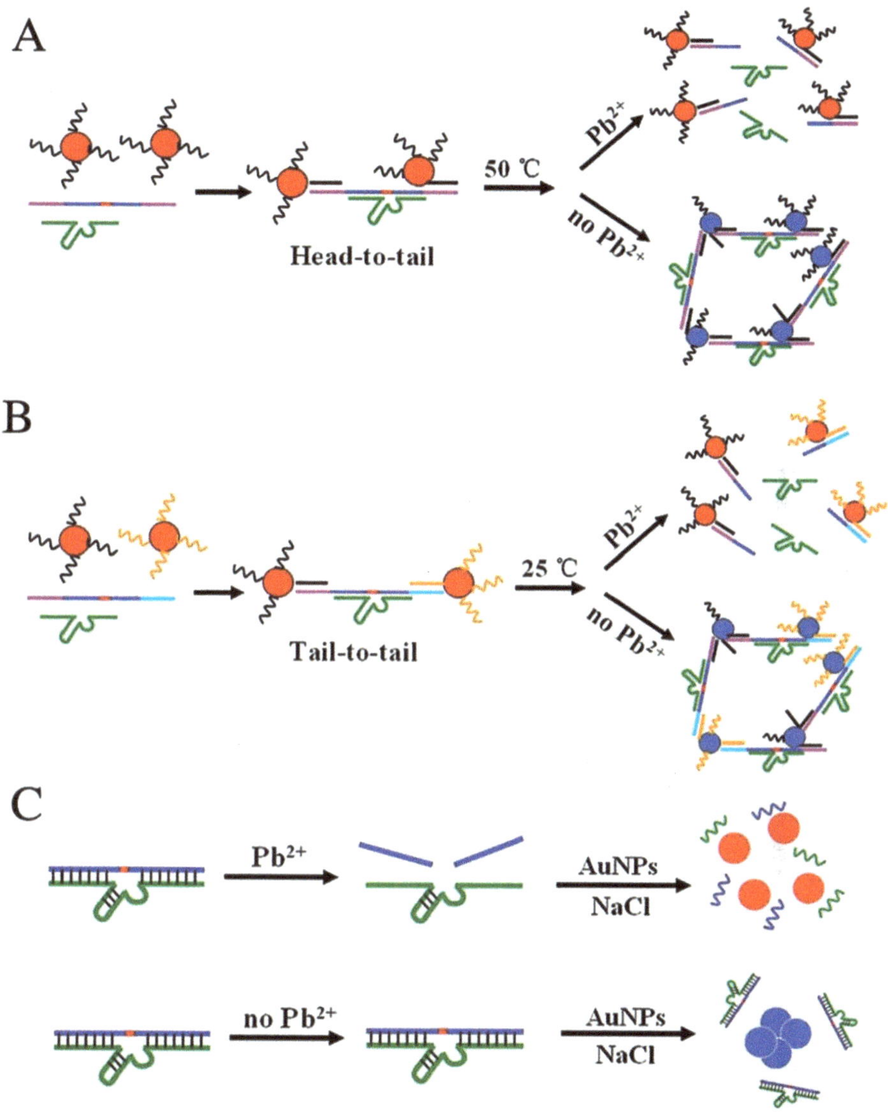

Fig. 17.4 Colorimetric Pb^{2+} detection by AuNPs that are aligned (**a** head-to-tail and (**b**) tail-to-tail, (**c**) label-free colorimetric Pb^{2+} detection (Reproduced from Ref. [40] by permission of The Royal Society of Chemistry)

label-free colorimetric Pb^{2+} sensor was shown to detect unmodified gold nanoparticles (Fig. 17.4c) [94]. The salt-induced aggregation of the gold nanoparticles resulted in a color change from red to blue, but the aggregates could be disassembled by the lead-induced cleavage of a single-stranded DNA enzyme complex (Fig. 17.5).

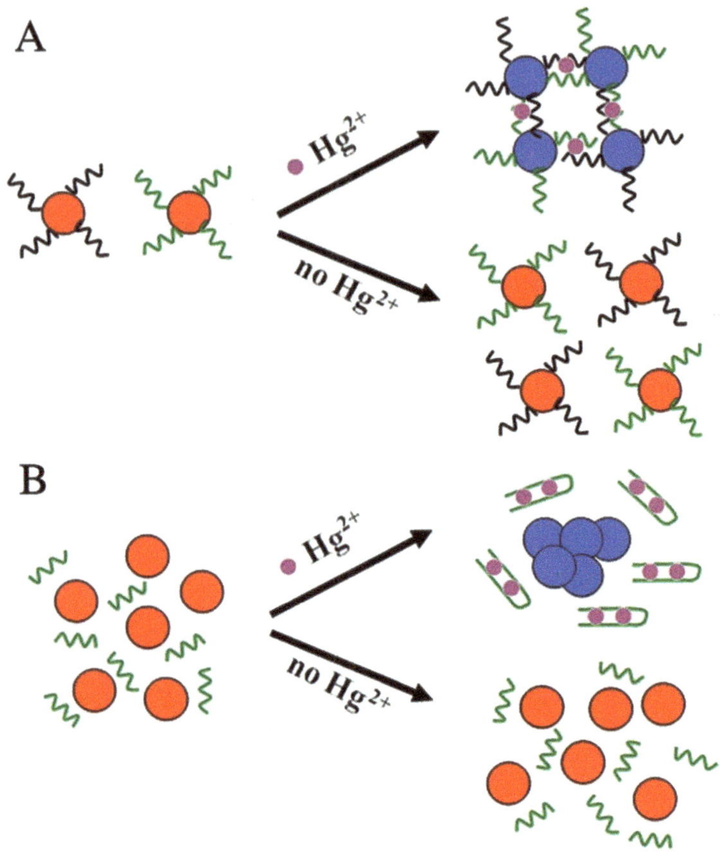

Fig. 17.5 (**a**) DNA-labeled and (**b**) label-free colorimetric Hg^{2+} sensors (Reproduced from Ref. [40] by permission of The Royal Society of Chemistry)

17.3.2.2 Colorimetric Sensors Based on Base Pair Mismatches

Colorimetric Hg^{2+} sensors can utilize a DNA marker, or they can be label-free. Li et al. developed a representative DNA-labeled colorimetric sensor, which is shown in Fig. 17.6a [95]. Gold nanoparticles were modified with two incomplete but complementary single-stranded DNA molecules with T-T base pairs. The presence of Hg^{2+} led to gold nanoparticle aggregation due to the formation of an Hg^{2+}-mediated T-T base pair. The Hg^{2+} concentration could be determined by monitoring the DNA–gold nanoparticle aggregates in the molten solution. The sensor detection limit was 0.1 mM. However, this sensor is not suitable for field analysis because the temperature must be precisely controlled. By introducing a suitable number of oligonucleotide connections and controlling the mismatch, a new colorimetric sensor that performs efficiently at room temperature was developed [96]. Hg^{2+} can also be detected by various label-free colorimetric sensors [97–101].

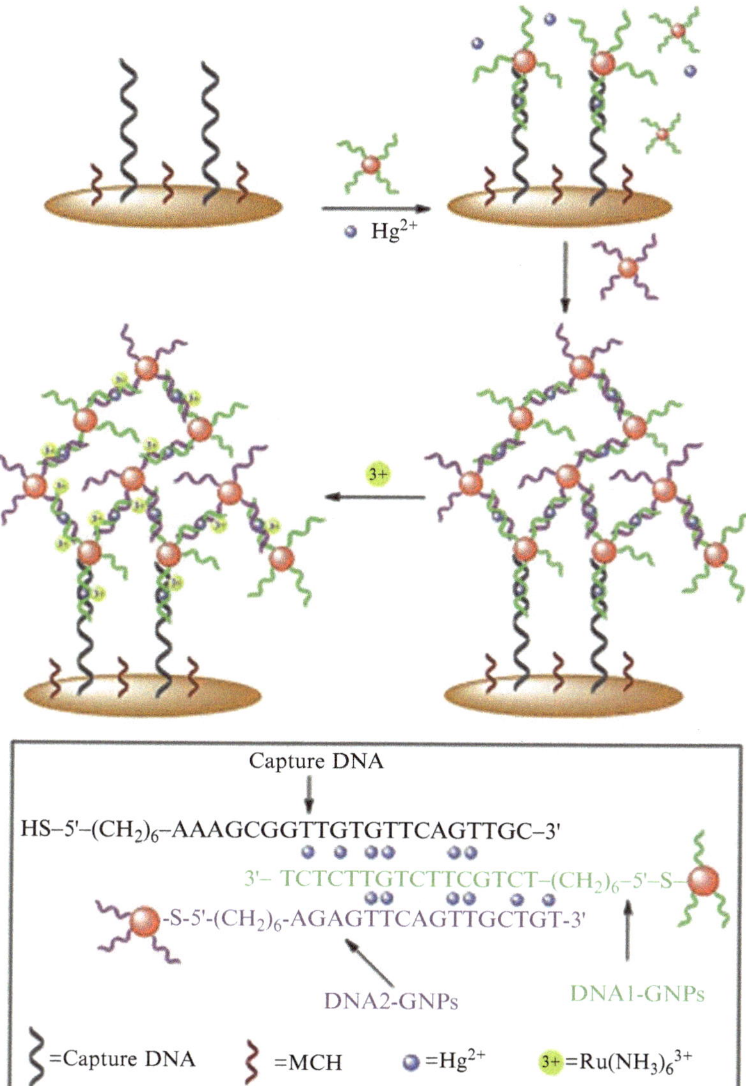

Fig. 17.6 Principle of the three-dimensional fishnet electrochemical Hg^{2+} sensor (Reproduced from Ref. [111] by permission of The Royal Society of Chemistry)

In a typical label-free colorimetric sensor (Fig. 17.6b) [101], ssDNA prevents AuNP aggregation in the absence of Hg^{2+}. In the presence of Hg^{2+}, however, the formation of Hg^{2+}-mediated T-T base pairs leads to the formation of DNA hairpin structures with gold nanoparticles on their surfaces. The gold nanoparticles can then aggregate, resulting in a color change from red to blue. The detection limit of these sensors can be as low as 10 nM.

In addition, gold nanoparticles were combined with hemin/G-quadruplexes to develop a label-free colorimetric Hg^{2+} sensor [102, 103]. In this system, 2,20-azino-bis(3-ethylbenzothiazoline-6-sulfonic acid) diammonium salt (ABTS) and 3,30,5,50-tetramethylbenzidine (TMB) were employed as catalytic substrates to achieve detection limits of 50 nM and 100 nM, respectively. Interestingly, when an imprinted membrane was used as the detection platform, the probe could be concentrated to a small point, resulting in a detection limit of 0.1 nM [104].

17.3.3 Electrochemical Heavy Metal Ion Sensors

Electrochemical sensors have high sensitivity, simple designs, sensitive electrochemical labels that detect changes in the distance between the double-stranded DNA chains, and a nanomolar detection limit at gold electrode surfaces [105–110]. Tang et al. designed an Hg^{2+} chemical sensor with a three-dimensional, supramolecular, network structure consisting of gold nanoparticles and DNA (Fig. 17.6). The detection limit of this sensor was 7.38 pM [111]. To fabricate this sensor, DNA was immobilized on a gold electrode and functioned as a platform for combining Hg^{2+} and DNA-modified AuNPs. T-Hg^{2+}-T interactions between the immobilized DNA and DNA1–AuNPs led to the formation of an internal binding layer via DNA hybridization. Then, the DNA1–AuNP complexes could bind to DNA2–AuNPs to form an outer layer, eventually leading to the formation of supramolecular structures. Carbon nanotubes can also be used as carriers in highly sensitive electrochemical Hg^{2+} sensors [111, 112]. It should be noted that the detection limit reached the femtomolar level when graphene oxide and quercetin–Cu^{2+} complexes were used as the carrier and DNA intercalator, respectively [113].

17.3.4 Graphene-Based Heavy Metal Ion Sensors

Graphene (G) is widely utilized because of its unique physical and electronic properties. Graphene oxide (GO) can be inexpensively reduced to produce reduced GO, which can lead to a decrease in its conductivity.

Due to small sp^2 carbon domains embedded in the sp^3 matrix, these materials can be studied by visible-near-infrared fluorescence spectroscopy [114]. To date, several spectroscopic techniques have been used to detect metal ions [115–117]. Here, the development of a heavy metal ion detection method is discussed.

Fu and coworkers developed a platform for effective Pb^{2+} sensing in which the fluorescence of GO was quenched by AuNPs [118]. The GO fluorescence quenching might be due to fluorescence resonance energy transfer (FRET) between GO and the AuNPs [119, 120]. Pb^{2+} could accelerate the rate of AuNP leaching from the graphene surface, which led to "turn-on" fluorescence. This Pb^{2+} sensing strategy based on graphene–Au nanocomposites is illustrated in Fig. 17.7a. Adding

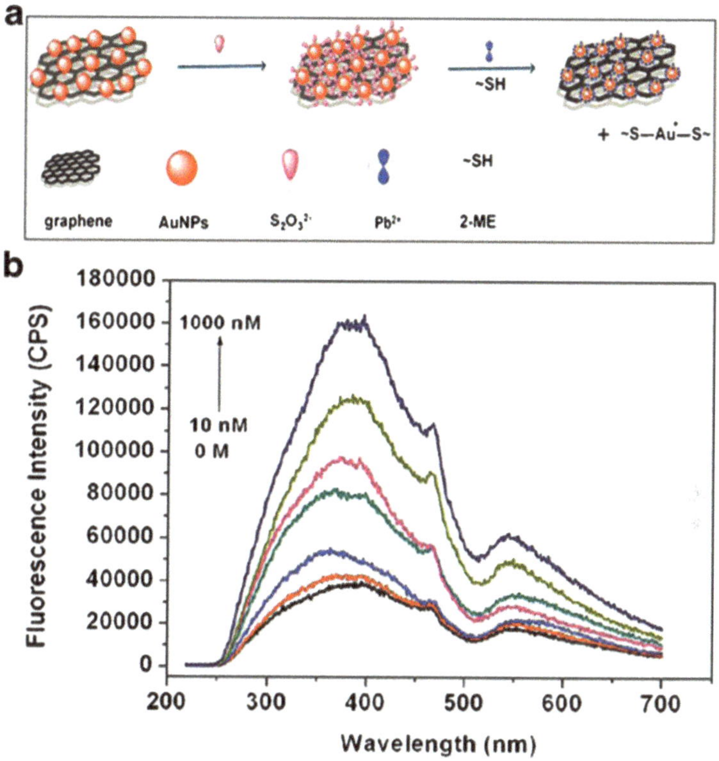

Fig. 17.7 Schematic of and results for Pb^{2+} detection using a GO-based sensor (Reprinted with the permission from Ref. [118]. Copyright 2012 American Chemical Society)

Pb^{2+} and 2-mercaptoethanol (2-ME) to an AuNP solution led to the rapid dissolution of the AuNPs and subsequent formation of Au(2-ME)2 complexes. Consequently, the graphene fluorescence increased due to the gradual dissolution of the AuNPs. The fluorescence properties depended strongly on the size of the GO precursor. GO with a lateral size of <1 μm was used because the "turn-on" fluorescence was highly efficient. Figure 17.7b shows the increase in the fluorescence intensity with increasing Pb^{2+} concentration. The relative fluorescence intensity had a strong linear dependence on the logarithm of the Pb^{2+} concentration in the range of 50–1000 nM. The relative standard deviation of the relative fluorescence intensity was less than 0.3 %, indicating that this method was reproducible. This sensor platform had a high selectivity for Pb^{2+} over common coexisting metal ions (Al^{3+}, Ca^{2+}, Cd^{2+}, K^+, Mg^{2+}, Zn^{2+}, Li^+, Co^{2+}, and Ni^{2+}) and a low detection limit of 10 nM.

Kundu et al. [121] reported a fascinating tool for selective Au^{3+} sensing in aqueous media using enhanced GO fluorescence. They proposed that the enhanced fluorescence properties of GO in acidic media were due to passivation by hydrogen bonding between polyvinyl alcohol (PVA) and GO hydroxy groups, as shown

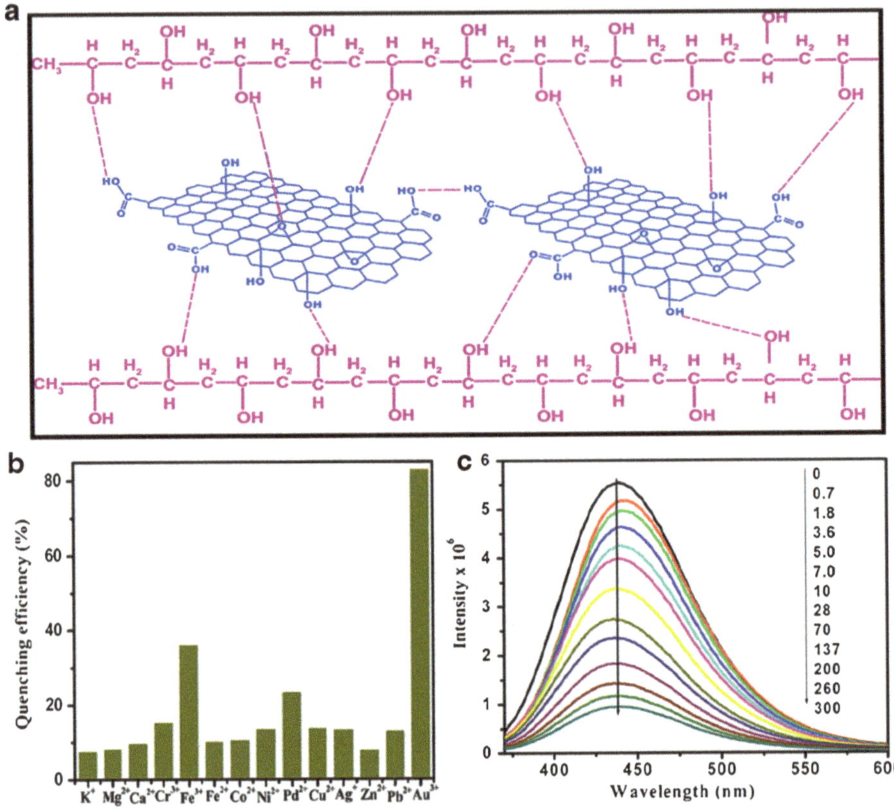

Fig. 17.8 **a** Schematic of the supramolecular interactions between GO and PVA **b** specificity and **c** detection results for the corresponding sensor (Reprinted with the permission from Ref. [121]. Copyright 2012 American Chemical Society)

in Fig. 17.8a. They studied the effect of different metal ions, including Au^{3+}, Cu^{2+}, Ag^{+}, Mg^{2+}, Ca^{2+}, Zn^{2+}, K^{+}, Pb^{2+}, Co^{2+}, Ni^{2+}, Pd^{2+}, Fe^{2+}, Fe^{3+}, and Cr^{3+}, on the GO/PVA optical properties (Fig. 17.8b). Au^{3+} chloride clearly quenched the GO/PVA hybrid fluorescence ($\approx$83 %), whereas the other metal salts had a significantly smaller effect on the hybrid fluorescence. As the Au^{3+} solution concentration increased, the fluorescence quenching increased (Fig. 17.8c). Therefore, the highly fluorescent GO/PVA hybrid had a low Au^{3+} detection limit of 1.4 μM and a high Au^{3+} selectivity in aqueous media because Au^{3+} has a higher reduction potential than the other metal ions.

Electrochemical sensor analytical techniques, particularly stripping voltammetry, are very attractive for on-site heavy metal ion monitoring and for addressing other environmental needs [122–125]. Electrochemical sensors are fast, accurate, portable, inexpensive, and inherently sensitive and selective toward electroactive species [126, 127].

Li et al. [128] reported that not only did Nafion–graphene (Nafion–G) composite film-based electrochemical sensors exhibit improved sensitivity for metal ion (Pb^{2+} and Cd^{2+}) detection but interference effects were also reduced due to synergy between the graphene nanosheets and Nafion. The stripping current signal was thus greatly enhanced on the graphene electrodes. The high sensitivity resulted from the enhanced rGO electron conduction combined with the high Nafion cation exchange capacity [129–131]. Linear calibration curves were obtained over the range of 0.5–50 mg/L for Pb^{2+} and 1.5–30 mg/L for Cd^{2+}. Willemse et al. [132] also reported a highly sensitive electrochemical platform consisting of a Nafion–G nanocomposite solution and an in situ plated mercury film electrode for detecting Zn^{2+}, Cd^{2+}, Pb^{2+}, and Cu^{2+} by square-wave anodic stripping voltammetry (SWASV). The Zn^{2+}, Cd^{2+}, Pb^{2+}, and Cu^{2+} ions could be accumulated on the Nafion–G composite electrode surface (the optimum deposition potentials were −1.3 V for Zn^{2+}, Cd^{2+}, and Pb^{2+} and −1.0 V for Cu^{2+}), resulting in high sensitivity. Zn^{2+}, Cd^{2+}, and Pb^{2+} could be easily identified using the Nafion–G–Hg sensors. Copper could not be detected simultaneously with Zn^{2+}, Cd^{2+}, and Pb^{2+} due to intermetallic interference between Cu and Zn. For individual metal ion detection, the lower detection limits were 0.07 μg/L (0.338 nM) for Pb^{2+}, 0.08 μg/L (1.23 nM) for Zn^{2+}, 0.13 μg/L (2.03 nM) for Cu^{2+}, and 0.08 μg/L (0.71 nM) for Cd^{2+}. A linear individual metal detection response was obtained over the range from 1 to 7 μg/L for Zn^{2+}, Cd^{2+}, and Pb^{2+} and from 20 to 180 μg/L for Cu^{2+}. For simultaneous metal detection, the detection limits were 0.07 μg/L (0.338 nM), 0.13 μg/L (1.16 nM), and 0.14 μg/L (2.15 nM) for Pb^{2+}, Cd^{2+}, and Zn^{2+}, respectively, and their correlation coefficients were 0.990, 0.983, and 0.999, respectively. The accuracy of the Nafion–G-modified electrode platform in real applications was comparable to that of inductively coupled plasma mass spectrometry (ICP-MS). Nafion–G was sensitive enough to detect Zn^{2+}, Cd^{2+}, and Pb^{2+}, and ICP-MS only exhibited a higher sensitivity for Cd^{2+}. In summary, a highly sensitive sensing platform based on Nafion–G nanocomposite films was used to detect Zn^{2+}, Cd^{2+}, Pb^{2+}, and Cu^{2+} individually and simultaneously by SWASV. The excellent lower detection limits and high sensitivity for Zn^{2+}, Cd^{2+}, Pb^{2+}, and Cu^{2+} were due to the high conductivity of graphene combined with the high cation exchange capacity of Nafion.

17.4 Raman Spectroscopy-Based Methods for Detecting Heavy Metal Ions

Surface-enhanced Raman spectroscopy (SERS) has become one of the most promising analytical tools in recent years. Due to its advantages, such as high sensitivity, high specificity, and facile operation, SERS is particularly suitable for environmental analysis. In this section, the development of detection methods based on surface-enhanced Raman spectroscopy is summarized.

Surface-enhanced Raman spectroscopy has become increasingly attractive for contaminant detection because of its unique advantages. Although some comments on the application of surface-enhanced Raman spectroscopy in environmental analysis have appeared in the literature [133–135], many new achievements have been reported in the past few years. These achievements, which include the detection of surface organic contaminants, heavy metal ions, and pathogens, are described below.

After the SERS of silver colloids was used to successfully detect As^{5+}, several studies of arsenic detection by SERS were performed [136]. Yang et al. fabricated closely spaced polymer-coated Ag nanoparticles by the LB technique and exploited their SERS response to arsenic for As detection [137]. The detection sensitivity of the poly(vinyl ketone)-coated nano-octahedra was in the ppb range. Since then, Meng et al. prepared a silver nanofilm SERS substrate by a mirror reaction and chemical deposition for arsenic detection [138]. They found that the SERS response of As^{5+} and As^{3+} species on this substrate exhibited a good linear relationship with their concentrations, enabling the determination of the detection limit, which was 5–10 ppb. In addition, a highly sensitive, highly selective SERS platform based on reduced glutathione (GSH)/thiol pyridine (4-MPY)-modified nanosilver was developed for As^{3+} detection [136]. In this system, the GSH on the Ag nanoparticle surfaces binded to As^{3+} via an As–O linkage, and 4-MPY acted as a Raman reporter. Using this method, As^{3+} could be quickly and accurately detected with a sensitivity of less than 1 ppb.

The mercury ion (Hg^{2+}) is considered to be one of the most toxic metal pollutants, and its detection by SERS is challenging. Therefore, the "turn-on" or "turn-off" mechanism was recently developed for the indirect SERS detection of Hg^{2+} [139–144]. For example, Wang et al. used surface-enhanced Raman spectroscopy combined with microfluidic water droplets for highly sensitive Hg^{2+} detection [144]. The water droplets, which contained Hg^{2+} and a SERS reporter molecular substrate (rhodamine B), adsorbed on AuNPs as individual reaction vessels. An initial SERS signal for rhodamine B adsorption on the Au nanoparticles was recorded. The SERS signal decreased as Hg^{2+} displaced rhodamine B at the AuNP surface due to its higher affinity for the surface. By monitoring the decrease in the SERS signal, the amount of Hg^{2+} could be quantified at approximately 100 % of the detection limit. This "turn-off" mechanism was exploited to develop various SERS platforms based on the interactions between the 4-thiol group of SERS substrates and nanosilver, aptamer-modified Au/Ag core-shell NPs, and rhodamine 6G adsorption on Au–Ag alloy [141] for highly efficient Hg^{2+} detection. Despite these advances, the "turn-on" mechanism is a more promising alternative for improving the selectivity of mercury ion detection by surface-enhanced Raman spectroscopy. For example, Chung et al. used the selective Hg^{2+} mediation of thymine base pair mismatches (strong T–Hg^{2+}–T complexes) to "turn on" the SERS response of DNA-modified microspheres for Hg^{2+} detection [145]. This strategy is related to the later development of a single Au nanowire thin-film SERS platform by Kim et al. for Hg^{2+} detection. The Hg^{2+} detection limit was 100 when the sensor was stably stored. Similarly, other single metal ions, such as

Pb^{2+}, Cu^{2+}, Cd^{2+}, and Cr^{3+}, can also be indirectly detected by the SERS response of DNAzyme and 2-mercaptobenzimidazole, trimercaptotriazine, and citric acid ligand platforms [139, 140]. The specific interaction between the metal ion and ligand, which also acts as a Raman reporter, is detected by changes in the SERS signal. The potential of background signals from other species in the sample can be reduced by using the ligand as an internal standard.

17.5 Quantum Dot-Based Methods for Detecting Heavy Metal Ions

Since 2002, quantum dots (QDs)26 and a wide range of ions have been investigated for the spectral detection of transition metals [146] and heavy metal ions [147, 148]. Here, the development of quantum dot sensors for the most relevant toxic heavy metal ions is summarized.

The simplest quantum dot sensors utilize quantum dots as a passive fluorescent marker. Many systems that utilize quantum dots modified by thiolate ligands (e.g., thioglycolic acid, MAA, L-cysteine, cystine acid by different water soluble, and N-acetyl-L-cysteine (NAC)) have been developed for heavy metal ion detection. In these systems, the ligand solubilizes the quantum dots in the aqueous solution and also responds to the metal ions. Water-soluble, thioacetic acid-modified CdS and CdS/ZnS quantum dots (CdS MAA and CdS/ZnS MAA) were used as fluorescent probes for mercury ions, which quenched their fluorescence. Under optimal conditions, the degree of fluorescence quenching was linearly proportional to the Hg^{2+} concentration in the working range, and the detection limits were 4.2 nM and 2.2 nM for CdS MAA [149] and CdS/ZnS MAA [150], respectively. To develop a less toxic sensor, ZnS quantum dots were modified with NAC (NAC-modified ZnS) in aqueous media [151]. The synthesized nanoparticles were utilized to detect trace mercury in water samples with a detection limit of 5 nM.

In many sensing structures, quantum dots are not used as passive tags, and some energy can flow between the system components because nanoscale charge and energy transfer processes can be changed, initiated, or destroyed by small perturbations at the quantum dot surface. Here, we focus on the interactions between the assembly system and heavy metal ions that drive the association or dissociation of the acceptors or change the donor–acceptor distance. The signal is modulated by the subsequent changes in the charge or energy transfer efficiency.

Banerjee et al. reported a novel chemical Cd^{2+} sensor based on quantum dots functionalized with 1,10-diaza-18 crown-6 rings [152]. First, core-shell CdS:Mn/ZnS quantum dots were synthesized in an AOT/heptane/water mixture. Then, a Cd^{2+}-selective ligand (i.e., 1,10-diaza-18-crown-6) was covalently coupled to the quantum dot surface via a CS_2 linkage. The modified quantum dots were in the "off" state because the quantum dot fluorescence was quenched by electron transfer to the auxiliary ligand. This electron transfer process could be blocked by the formation of a Cd^{2+}-1,10-diaza-18-crown-6 complex, which resulted in enhanced quantum dot fluorescence (the quantum dots were in the “on” state). This system had a higher selectivity for Cd^{2+} than for other metal ions. The sensor selectivity was due to the fact that the binding constant of Cd^{2+} in the aza-crown ether was higher than those of the other ions. Another metal ion detection method based on charge transfer processes involves the combination of quantum dots and organic dyes. S^{2-} was added to the solution to quench the quantum dot luminescence peak (“off” state), and the addition of Cd^{2+} led to its recovery (“on” state). Interestingly, S^{2-} and Cd^{2+} did not significantly affect the fluorescence emission peaks, which enabled the development of the ratiometric sensor. The Cd^{2+} detection limit of this luminescence detection method was 12 nm. Although the sensitivity of this system is satisfactory for real applications, it cannot sufficiently discriminate between chemically similar metal ions.

The development of ion sensors based on changes in quantum dot photoluminescence is a very active research field. In these systems, the QDs often act as donors, transferring their absorbed light energy to acceptors such as organic fluorophores or other nanoparticles. According to Forster’s theory, the energy transfer rate depends on the following factors: the donor emission and acceptor absorption, the donor quantum yield, the relative orientation of the donor–acceptor transition dipoles, and the distance between the donor and acceptor [153, 154]. The use of quantum dots as donor species in fluorescence resonance systems for heavy metal ion detection has been reported in several studies. In these systems, the ions are detected by changes in the FRET efficiency between the quantum dot and acceptor.

17.6 Amplification-Based Methods for Detecting Heavy Metal Ions

To improve the sensitivity of heavy metal detection methods, amplification techniques are used. In this section, isothermal amplification and sensitive detection tools are reviewed.

17.6.1 Strand Displacement Amplification

Strand displacement amplification (SDA) is an isothermal, primer-driven reaction that can produce interesting target sequences [155]. Li et al. demonstrated the use of SDA for Hg^{2+} detection for the first time (Fig. 17.9). Ninety-seven single-stranded DNA sequences were synthesized with three domains: (I) an Hg^{2+} recognition sequence, (II) a sequence with an Nt.BbvCI nicking enzyme recognition site, and (III) an HRP-mimicking sequence. In the presence of Hg^{2+} ions, the 3′ end domain

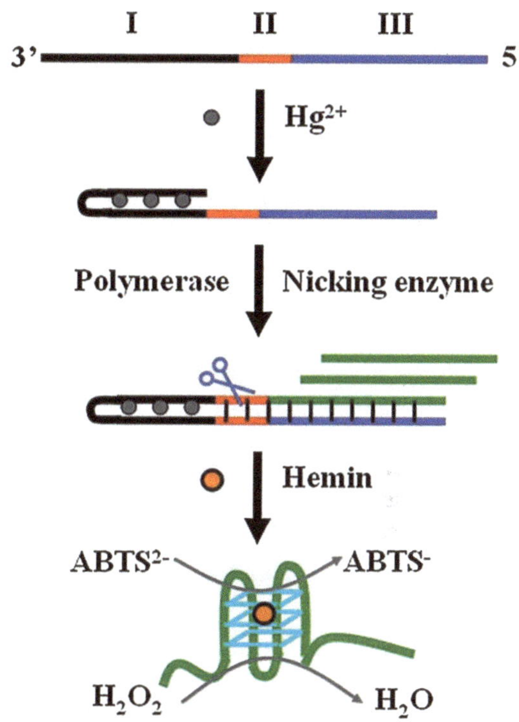

Fig. 17.9 Schematic illustration of an SDA-based colorimetric assay for Hg^{2+} detection (Reprinted from Ref. [96] by permission of Taylor & Francis Ltd)

folded to form a hairpin structure, acting as a primer extension reaction to produce a DNA double chain. Then, the Nt.BbvCI nicking enzyme cut the strand at the domain II recognition site to start a new expansion cycle. The HRP-mimicking sequences could self-assemble into a hemin/G-quadruplex structure, which enabled reactant oxidation at the DNAzyme catalytic active site. Thus, mercury could be detected indirectly by a colorimetric method with a detection limit of 1 nM by exploiting the SDA technique [96]. The sensitivity of this method can be improved by combining it with a fluorescence system. Yin et al. monitored changes in the fluorescence signal of a silver cluster due to its interactions with an SDA product for Hg^{2+} detection. The detection limit of the system was 80 pM [156].

17.6.2 Nicking Enzyme Signal Amplification

Nicking enzymes can recognize specific nucleotide sequences to cleave double-stranded DNA only at specific positions along the chain [157]. This unique feature of nicking enzyme signal amplification (NESA) has been exploited to develop nucleic acid biosensors and protein and small molecule assays [158]. While expansion and molecular beacon nodes provide inherent stability and are straightforward,

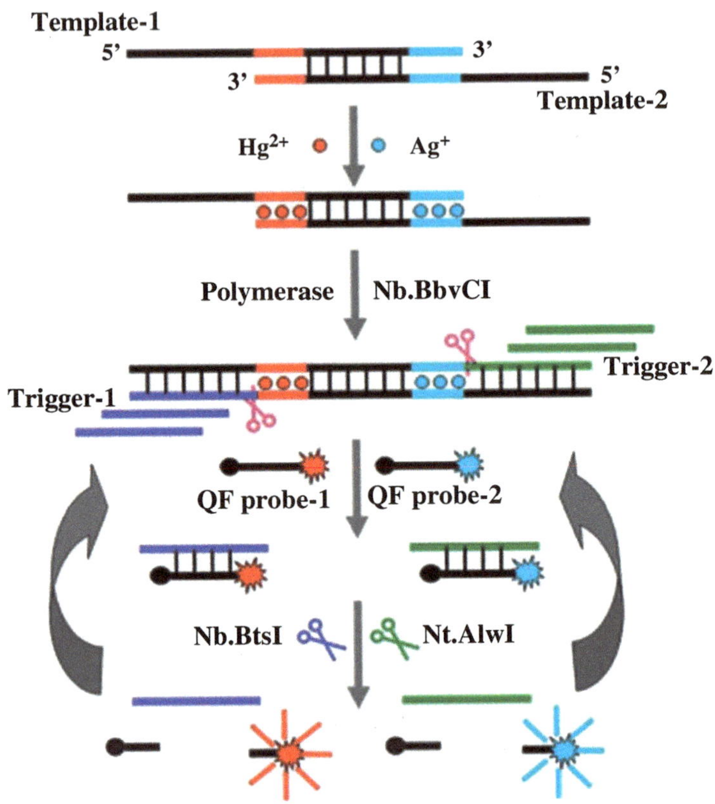

Fig. 17.10 Simultaneous detection of Hg^{2+} and Ag^{+} by a fluorescent sensor based on an SDA-triggered NESA reaction (Reproduced from Ref. [161] by permission of The Royal Society of Chemistry)

the sensitivity of this method is unsatisfactory, and the SDA target must have nicking enzyme recognition sites [159]. These limitations can be overcome depending on the material employed [160]. Recently, an SDA fluorescence sensor was developed for the simultaneous detection of Hg^{2+} and Ag^{+} (Fig. 17.10). In this system, the Hg^{2+} and Ag^{+} ions formed T–Hg^{2+}–T and C–Ag^{+}–C complexes, respectively, resulting in two different triggers that quenched the fluorescence of quencher–fluorophore (QF) probes by forming hybridized DNA duplexes with them. The fluorescence signals of the QF probes, which had different emission spectra, were enhanced after cutting them with the nicking enzyme. This sensor was extremely sensitive to Hg^{2+} and Ag^{+} with detection limits of 2 pM and 16 pM, respectively. In addition, this NESA-based sensor was combined with quantum dots to achieve an Hg^{2+} detection limit of 0.8 nM [161].

17.6.3 Hybridization Chain Reaction

Hybridization chain reactions (HCRs) are the most widely used nonenzymatic strategy for amplifying target molecules [162]. In this method, ssDNA molecules can be assembled into an orderly unit [163, 164]. A label-free fluorescent sensor based an HCR that produced a double-stranded DNA structure with hundreds of repeat units was developed to detect Hg^{2+} [165].

Recently, Huang et al. reported the use of graphene oxide and HCR fluorescence sensors for Hg^{2+} detection (Fig. 17.11) [166]. In these sensors, the dye and auxiliary DNA used for the ion detection were labeled with hairpin probes. The probes adsorbed on the graphene oxide surface, effectively quenching their fluorescence. The presence of Hg^{2+} ions initiated the HCR between the two hairpin probes, and the resulting double-stranded DNA product was released from the graphene oxide surface, leading to the recovery of the fluorescence signal. The detection limit was 0.3 nM [166].

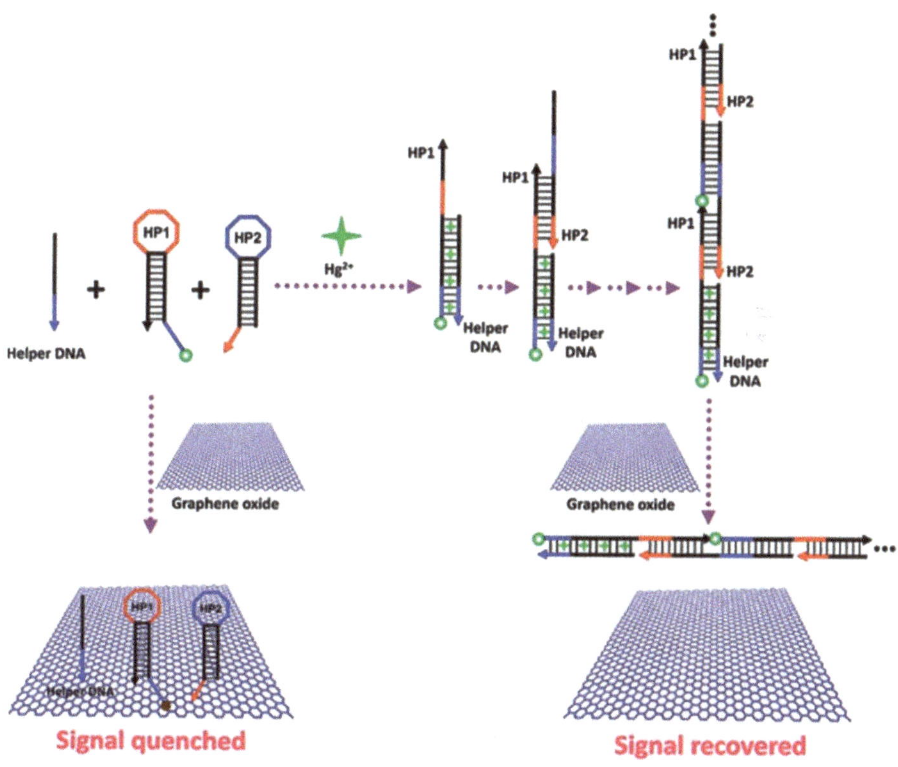

Fig. 17.11 Illustration of graphene oxide- and HCR-based fluorescent Hg^{2+} sensors (Reprinted with the permission from Ref. [166]. Copyright 2014 American Chemical Society)

17.6.4 Digital PCR (dPCR)

Digital PCR, a newly developed quantitative PCR-based detection method, is considered to be the third generation of PCR technology because of its higher sensitivity and ability to directly quantify the detected species [167–170]. After separating the original reaction volume in a microfluidic system or oil-in-water emulsion [168, 170], many separate reaction cells containing the original target are generated. The target copy number of each partition follows a Poisson distribution, which can be fitted to the ratio of positive partitions to total partitions. Since dPCR was first introduced in the 1990s [167], much progress has been made in determining the copy number [171], identifying single-nucleotide polymorphisms and detecting and quantifying targets with low abundances [172–174]. Compared to real-time PCR, digital PCR can determine the copy number of a target template in the original solution without the generation of standard curves [175]. This feature mostly avoids the uncertainties associated with standard curves and their generation procedures. Thus, digital PCR is considered to be a more accurate quantitative detection method than real-time PCR.

An example sensing system is illustrated in Fig. 17.12. This system consists of two main components, the biosensing (Fig. 17.12a, b) and digital PCR (Fig. 17.12c,

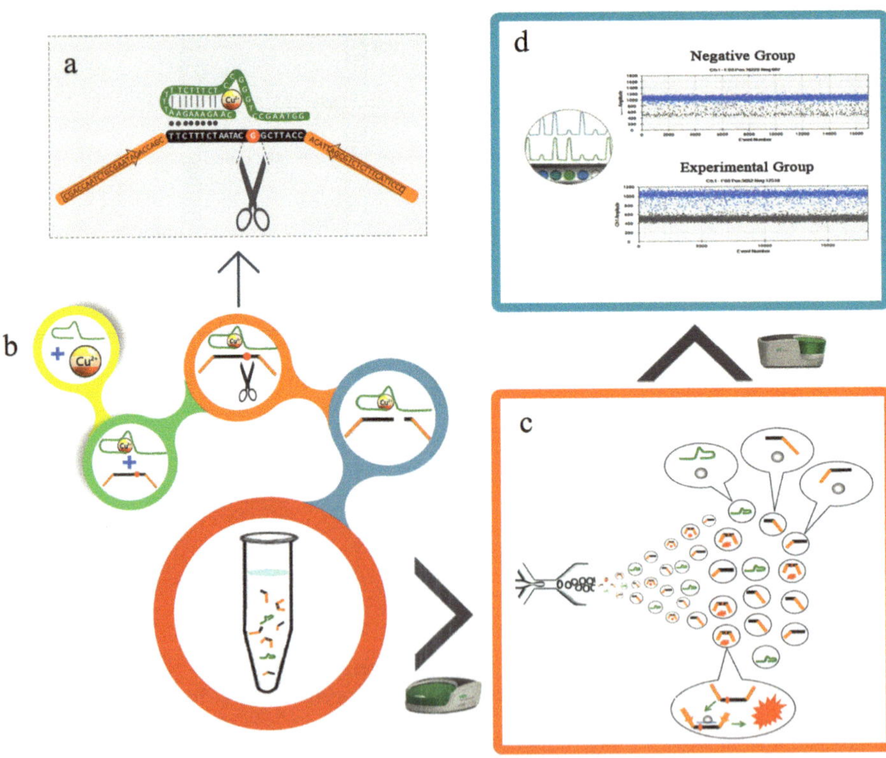

Fig. 17.12 Illustration of the working principle of a DNAzyme biosensor detection system

d) steps. The biosensor sequence designation is shown in Fig. 17.12a. In the example system, the substrate sequence contained both the primer locations and DNAzyme cleavage site. In the first step in Fig. 17.12b, the single substrate strand was hydrolyzed by the DNAzyme in the presence of Cu^{2+} at the appropriate temperature. The DNAzyme sequence was based on original research on Cu^{2+} biosensors [72]. After the biosensing step, three types of single strands, the DNAzyme and the cut and uncut substrate strands, were obtained. All of these products were loaded on the digital PCR platform after serial dilutions in which the different sequences were loaded into different droplets (Fig. 17.12c). Only the uncut substrates exhibited a positive signal. For the dPCR data analysis, the fluorescence threshold was set to distinguish between the negative and positive droplets. Therefore, during the droplet reading step after the PCR amplification (Fig. 17.12d), the droplets containing the DNAzyme or cleaved substrate sequences exhibited negative signals, which are represented as grey dots in the final heat map. The blue dots in the heat map represent the positive signals of the droplets containing the uncut substrate sequence. Finally, the cleavage rate was calculated using Eq. 3 and used to evaluate the final results. The final quantitative results showed that the limit of quantitation (LOQ) was as low as 0.5 pmol, whereas the sensitivity was found to be 50 fmol.

To achieve ultrasensitive, absolute quantitative Hg^{2+} detection, we combined a mismatch Hg^{2+} biosensor and emulsion PCR. This detection system is illustrated in Fig. 17.13. In this system, three main steps, the isothermal biosensing, emulsion

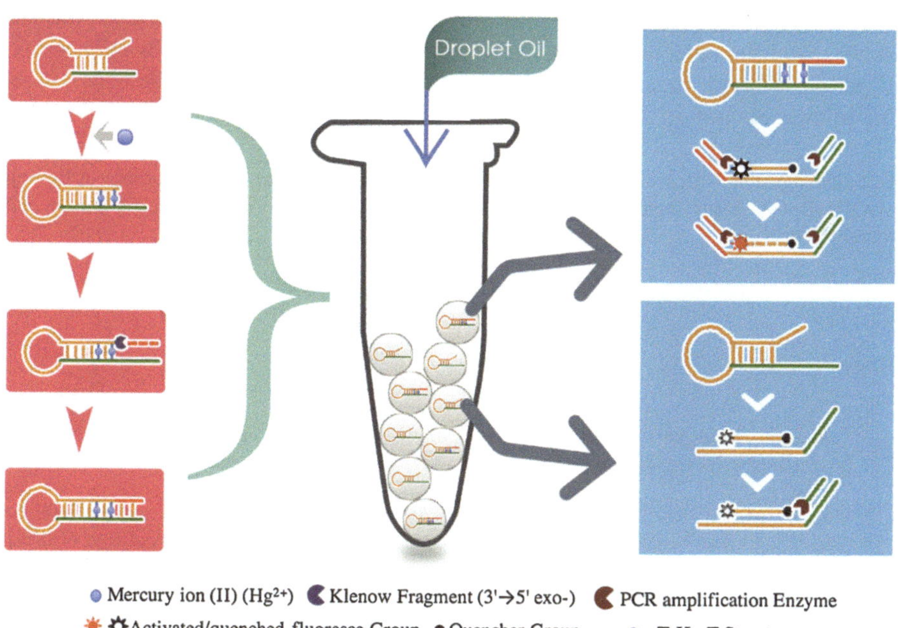

Fig. 17.13 Mechanism of an Hg^{2+} detection system based on a mismatch biosensor and emulsion PCR

PCR, and droplet reading steps, were performed sequentially. In the first step, the Hg^{2+} signal was converted to a nucleic acid signal by a linear relationship. Because the T–Hg–T structure was reported to be stable at low temperatures, the Klenow fragment (3′→5′ exo-), which exhibits its highest activity at 37 °C, was chosen as the isothermal amplification enzyme. The T–Hg–T structure was generated after denaturation and slow cooling to 37 °C. Then, the isothermal enzyme was used to generate blunt ends. During the emulsion PCR step, droplets containing positive or negative sequences were generated. Only the droplets containing blunt terminal sequences generated positive signals in the droplet reading step. The final heat map was generated based on the analysis of each droplet. A unique threshold was set to distinguish between the negative and positive droplets. The absolute copy number of the target sequence was calculated by fitting the ratio of positive droplets to total droplets to a Poisson distribution. The final results indicated that the limit of quantitation (LOQ) was as low as 40 fmol, and the limit of detection (LOD) was 10 fmol.

17.7 Prospects

The development of heavy metal ion detection methods based on nucleic acids has attracted increasing attention. Highly sensitive and selective ion detection has been achieved with various advanced materials. Therefore, the development of advanced materials and downstream detection methods are considered to be essential for heavy metal ion detection methods. To achieve more sensitive, rapid, and specific detection of heavy metal ions, research should focus on downstream detection tools due to their economic and practical feasibility.

Acknowledgments This work is supported by the Ministry of Science and Technology of Beijing (XX2014B069). Many thanks to Pengyu Zhu, for his kindly help in manuscript conception and preparation.

References

1. Tekaya N, Saiapina O, Ben Ouada H, et al. Ultra-sensitive conductometric detection of pesticides based on inhibition of esterase activity in Arthrospira platensis. Environ Pollut. 2013;178:182–8.
2. Singh A, Sharma RK, Agrawal M, Marshall FM. Health risk assessment of heavy metals via dietary intake of foodstuffs from the wastewater irrigated site of a dry tropical area of India. Food Chem Toxicol. 2010;48:611–9.
3. Rogers KR. Biosensors for environmental applications. Biosens Bioelectron. 1995;10:533–41.
4. Turdean GGL. Design and development of biosensors for the detection of heavy metal toxicity. Int J Electrochem. 2011;2011:1–15.

5. Gao C, Yu XY, Xiong SQ, et al. Electrochemical detection of arsenic(III) completely free from noble metal: Fe3O4 microspheres-room temperature ionic liquid composite showing better performance than gold. Anal Chem. 2013;85:2673–80.
6. Tag K, Riedel K, Bauer H-J, et al. Amperometric detection of Cu2+ by yeast biosensors using flow injection analysis (FIA). Sensors Actuators B Chem. 2007;122:403–9.
7. Arao T, Ishikawa S, Murakami M, et al. Heavy metal contamination of agricultural soil and countermeasures in Japan. Paddy Water Environ. 2010;8:247–57.
8. Guascito MR, Malitesta C, Mazzotta E, Turco A. Inhibitive determination of metal ions by an amperometric glucose oxidase biosensor. Sensors Actuators B Chem. 2008;131:394–402.
9. Li M, Gou H, Al-Ogaidi I, Wu N. Nanostructured sensors for detection of heavy metals: a review. ACS Sustain Chem Eng. 2013;1:713–23.
10. Bagal-Kestwal D, Karve MS, Kakade B, Pillai VK. Invertase inhibition based electrochemical sensor for the detection of heavy metal ions in aqueous system: application of ultra-microelectrode to enhance sucrose biosensor's sensitivity. Biosens Bioelectron. 2008;24:657–64.
11. Valko M, Morris H, Cronin MTD. Metals, toxicity and oxidative stress. Curr Med Chem. 2005;12:1161–208.
12. Patrick L. Lead toxicity part II: the role of free radical damage and the use of antioxidants in the pathology and treatment of lead toxicity. Altern Med Rev. 2006;11:114–27.
13. Han YH, Kim SZ, Kim SH, Park WH. Suppression of arsenic trioxide-induced apoptosis in HeLa cells by N-acetylcysteine. Mol Cells. 2008;26:18–25.
14. Ho S-Y, Wu W-J, Chiu H-W, et al. Arsenic trioxide and radiation enhance apoptotic effects in HL-60 cells through increased ROS generation and regulation of JNK and p38 MAPK signaling pathways. Chem Biol Interact. 2011;193:162–71.
15. Jin YH, Clark AB, Slebos RJC, et al. Cadmium is a mutagen that acts by inhibiting mismatch repair. Nat Genet. 2003;34:326–9.
16. Wieland M, Levin MK, Hingorani KS, et al. Mechanism of cadmium-mediated inhibition of Msh2-Msh6 function in DNA mismatch repair. Biochemistry. 2009;48:9492–502.
17. Kobal AB, Horvat M, Prezelj M, et al. The impact of long-term past exposure to elemental mercury on antioxidative capacity and lipid peroxidation in mercury miners. J Trace Elem Med Biol. 2004;17:261–74.
18. Srikanth K, Ahmad I, Rao JV, et al. Modulation of glutathione and its dependent enzymes in gill cells of Anguilla anguilla exposed to silica coated iron oxide nanoparticles with or without mercury co-exposure under in vitro condition. Comp Biochem Physiol C Toxicol Pharmacol. 2014;162:7–14.
19. Jarup L. Hazards of heavy metal contamination. Br Med Bull. 2003;68:167–82.
20. Falco G, Llobet JM, Bocio A, Domingo JL. Daily intake of arsenic, cadmium, mercury, and lead by consumption of edible marine species. J Agric Food Chem. 2006;54:6106–12.
21. Monteiro DA, Rantin FT, Kalinin AL. Inorganic mercury exposure: toxicological effects, oxidative stress biomarkers and bioaccumulation in the tropical freshwater fish matrinxã, Brycon amazonicus (Spix and Agassiz, 1829). Ecotoxicology. 2010;19:105–23.
22. Linšak Ž, Linšak DT, Špirić Z, et al. Effects of mercury on glutathione and glutathione-dependent enzymes in hares (Lepus europaeus Pallas). J Environ Sci Health A-Toxic/Hazard Subst Environ Eng. 2013;48:1325–32.
23. Patra RC, Rautray AK, Swarup D. Oxidative stress in lead and cadmium toxicity and its amelioration. Vet Med Int. 2011;2011:457327.
24. Gong H, Li X. Y-type, C-rich DNA probe for electrochemical detection of silver ion and cysteine. Analyst. 2011;136:2242–6.
25. AshaRani PV, Low Kah Mun G, Hande MP, Valiyaveettil S. Cytotoxicity and genotoxicity of silver nanoparticles in human cells – SOM. ACS Nano. 2009;3:279–90.
26. Cano-Pavon JM, De Torres AG, Sánchez-Rojas F, Cañada-Rudner P. Analytical methods for mercury speciation in environmental and biological samples-An overview. Int J Environ Anal Chem. 1999;75:93–106.

27. Bunka DHJ, Stockley PG. Aptamers come of age – at last. Nat Rev Microbiol. 2006;4:588–96.
28. O'Sullivan CK. Aptasensors – the future of biosensing? Anal Bioanal Chem. 2002;372:44–8.
29. Brody EN, Willis MC, Smith JD, et al. The use of aptamers in large arrays for molecular diagnostics. Mol Diagnosis. 1999;4:381–8.
30. Lee JF, Hesselberth JR, Meyers LA, Ellington AD. Aptamer database. Nucleic Acids Res. 2004;32:D95–100.
31. Bruesehoff PJ, Li J, Augustine AJ, Lu Y. Improving metal ion specificity during in vitro selection of catalytic DNA. Comb Chem High Throughput Screen. 2002;5:327–35.
32. Shen Y, Mackey G, Rupcich N, et al. Entrapment of fluorescence signaling DNA enzymes in sol-gel-derived materials for metal ion sensing. Anal Chem. 2007;79:3494–503.
33. Liu J, Cao Z, Lu Y. Functional nucleic acid sensors. Chem Rev. 2009;109(5):1948–98.
34. Liu J, Lu Y. Fluorescent DNAzyme biosensors for metal ions based on catalytic molecular beacons. Methods Mol Biol. 2006;335:275–88.
35. Liu J, Lu Y. Stimuli-responsive disassembly of nanoparticle aggregates for light-up colorimetric sensing. J Am Chem Soc. 2005;127:12677–83.
36. Liu J, Lu Y. Accelerated color change of gold nanoparticles assembled by DNAzymes for simple and fast colorimetric Pb2+ detection. J Am Chem Soc. 2004;126:12298–305.
37. Liu J, Lu Y. Optimization of a Pb2+-directed gold nanoparticle/DNAzyme assembly and its application as a colorimetric biosensor for Pb2+. Chem Mater. 2004;16:3231–8.
38. Liu J, Lu Y. Colorimetric biosensors based on DNAzyme-assembled gold nanoparticles. J Fluoresc. 2004;14:343–54.
39. Liu J, Lu Y. Improving fluorescent DNAzyme biosensors by combining inter- and intramolecular quenchers. Anal Chem. 2003;75:6666–72.
40. Zhu G, Zhang CY. Functional nucleic acid-based sensors for heavy metal ion assays. Analyst. 2014;139(24):6326–42.
41. Li J, Lu Y. A highly sensitive and selective catalytic DNA biosensor for lead ions. J Am Chem Soc. 2000;122:10466–7.
42. Niazov T, Pavlov V, Xiao Y, et al. DNAzyme-functionalized Au nanoparticles for the amplified detection of DNA or telomerase activity. Nano Lett. 2004;4:1683–7.
43. Cheng Y, Huang Y, Lei J, et al. Design and biosensing of Mg2+-dependent DNAzyme-triggered ratiometric electrochemiluminescence. Anal Chem. 2014;86:5158–63.
44. Zhang Y, Wang L-J, Zhang C-Y. Highly sensitive detection of telomerase using a telomere-triggered isothermal exponential amplification-based DNAzyme biosensor. Chem Commun (Camb). 2014;50:1909–11.
45. Lu Y. New transition-metal-dependent DNAzymes as efficient endonucleases and as selective metal biosensors. Chem A Eur J. 2002;8:4588–96.
46. Zhang ZZ, Zhang CY. Highly sensitive detection of protein with aptamer-based target-triggering two-stage amplification. Anal Chem. 2012;84:1623–9.
47. Herr JK, Smith JE, Medley CD, et al. Aptamer-conjugated nanoparticles for selective collection and detection of cancer cells. Anal Chem. 2006;78:2918–24.
48. Zhang CY, Johnson LW. Single quantum-dot-based aptameric nanosensor for cocaine. Anal Chem. 2009;81:3051–5.
49. Xue L, Zhou X, Xing D. Highly sensitive protein detection based on aptamer probe and isothermal nicking enzyme assisted fluorescence signal amplification. Chem Commun (Camb). 2010;46:7373–5.
50. Yang K, Zhang C. Simple detection of nucleic acids with a single-walled carbon-nanotube-based electrochemical biosensor. Biosens Bioelectron. 2011;28:257–62.
51. Zhou X, Tang Y, Xing D. One-step homogeneous protein detection based on aptamer probe and fluorescence cross-correlation spectroscopy. Anal Chem. 2011;83:2906–12.
52. Sassolas A, Blum LJ, Leca-Bouvier BD. Homogeneous assays using aptamers. Analyst. 2011;136:257–74.

53. Ono A, Togashi H. Highly selective oligonucleotide-based sensor for mercury(II) in aqueous solutions. Angew Chem Int Ed. 2004;43:4300–2.
54. Miyake Y, Togashi H, Tashiro M, et al. MercuryII-mediated formation of thymine-HgII-thymine base pairs in DNA duplexes. J Am Chem Soc. 2006;128:2172–3.
55. Ono A, Cao S, Togashi H, et al. Specific interactions between silver(I) ions and cytosine-cytosine pairs in DNA duplexes. Chem Commun. 2008;39:4825–7.
56. Pan T, Uhlenbeck OC. In vitro selection of RNAs that undergo autolytic cleavage with Pb2+. Biochemistry. 1992;31:3887–95.
57. Pan T, Uhlenbeck OC. A small metalloribozyme with a two-step mechanism. Nature. 1992;358:560–3.
58. Santoro SW, Joyce GF. A general purpose RNA-cleaving DNA enzyme. Proc Natl Acad Sci U S A. 1997;94:4262–6.
59. Brown AK, Li J, Pavot CM-B, Lu Y. A lead-dependent DNAzyme with a two-step mechanism. Biochemistry. 2003;42:7152–61.
60. Peracchi A, Bonaccio M, Clerici M. A mutational analysis of the 8-17 deoxyribozyme core. J Mol Biol. 2005;352:783–94.
61. Faulhammer D, Famulok M. The Ca2+ ion as a cofactor for a novel RNA-cleaving deoxyribozyme. Angew Chem Int Ed Engl. 1996;35:2837–41.
62. Peracchi A. Preferential activation of the 8-17 deoxyribozyme by Ca(2+) ions. Evidence for the identity of 8-17 with the catalytic domain of the Mg5 deoxyribozyme. J Biol Chem. 2000;275:11693–7.
63. Li J, Zheng W, Kwon AH, Lu Y. In vitro selection and characterization of a highly efficient Zn (II)-dependent RNA-cleaving deoxyribozyme. Nucleic Acids Res. 2000;28:481–8.
64. Cruz RPG, Withers JB, Li Y. Dinucleotide junction cleavage versatility of 8-17 deoxyribozyme. Chem Biol. 2004;11:57–67.
65. Liu J, Lu Y. FRET study of a trifluorophore-labeled DNAzyme. J Am Chem Soc. 2002;124:15208–16.
66. Kim HK, Liu J, Li J, et al. Metal-dependent global folding and activity of the 8-17 DNAzyme studied by fluorescence resonance energy transfer. J Am Chem Soc. 2007;129:6896–902.
67. Kim H-K, Rasnik I, Liu J, et al. Dissecting metal ion-dependent folding and catalysis of a single DNAzyme. Nat Chem Biol. 2007;3:763–8.
68. Santoro SW, Joyce GF, Sakthivel K, et al. RNA cleavage by a DNA enzyme with extended chemical functionality. J Am Chem Soc. 2000;122:2433–9.
69. Carmi N, Balkhi SR, Breaker RR. Cleaving DNA with DNA. Proc Natl Acad Sci U S A. 1998;95:2233–7.
70. Carmi N, Shultz LA, Breaker RR. In vitro selection of self-cleaving DNAs. Chem Biol. 1996;3:1039–46.
71. Carmi N, Breaker RR. Characterization of a DNA-cleaving deoxyribozyme. Bioorg Med Chem. 2001;9:2589–600.
72. Liu J, Lu Y. A DNAzyme catalytic beacon sensor for paramagnetic Cu 2+ ions in aqueous solution with high sensitivity and selectivity. J Am Chem Soc. 2007;129:9838–9.
73. Cuenoud B, Szostak JW. A DNA metalloenzyme with DNA ligase activity. Nature. 1995;375:611–4.
74. Wang W, Billen LP, Li Y. Sequence diversity, metal specificity, and catalytic proficiency of metal-dependent phosphorylating DNA enzymes. Chem Biol. 2002;9:507–17.
75. Chiuman W, Li Y. Revitalization of six abandoned catalytic DNA species reveals a common three-way junction framework and diverse catalytic cores. J Mol Biol. 2006;357:748–54.
76. Liu Z, Mei SHJ, Brennan JD, Li Y. Assemblage of signaling DNA enzymes with intriguing metal-ion specificities and pH dependences. J Am Chem Soc. 2003;125:7539–45.
77. Nelson KE, Bruesehoff PJ, Lu Y. In vitro selection of high temperature Zn2+-dependent DNAzymes. J Mol Evol. 2005;61:216–25.
78. Dubertret B, Calame M, Libchabera J. Single-mismatch detection using gold-quenched fluorescent oligonucleotides. Nat Biotechnol. 2001;19:365–70.

79. Tyagi S, Bratu DP, Kramer FR. Multicolor molecular beacons for allele discrimination. Nat Biotechnol. 1998;16:49–53.
80. Tyagi S, Marras SA, Kramer FR. Wavelength-shifting molecular beacons. Nat Biotechnol. 2000;18:1191–6.
81. Tyagi S, Kramer FR. Molecular beacons: probes that fluoresce upon hybridization. Nat Biotechnol. 1996;14:303–8.
82. Cao Z, Suljak SW, Tan W. Molecular beacon aptamers for protein monitoring in real-time and in homogeneous solutions. Curr Proteomics. 2005;2:31–40.
83. Stojanovic MN, de Prada P, Landry DW. Homogeneous assays based on deoxyribozyme catalysis. Nucleic Acids Res. 2000;28:2915–8.
84. Vitiello D, Pecchia DB, Burke JM. Intracellular ribozyme-catalyzed trans-cleavage of RNA monitored by fluorescence resonance energy transfer. RNA. 2000;6:628–37.
85. Jenne A, Hartig JS, Piganeau N, et al. Rapid identification and characterization of hammerhead-ribozyme inhibitors using fluorescence-based technology. Nat Biotechnol. 2001;19:56–61.
86. Singh KK, Parwaresch R, Krupp G. Rapid kinetic characterization of hammerhead ribozymes by real-time monitoring of fluorescence resonance energy transfer (FRET). RNA. 1999;5:1348–56.
87. Sekar RB, Periasamy A. Fluorescence resonance energy transfer (FRET) microscopy imaging of live cell protein localizations. J Cell Biol. 2003;160:629–33.
88. Walter NG, Burke JM. Real-time monitoring of hairpin ribozyme kinetics through base-specific quenching of fluorescein-labeled substrates. RNA. 1997;3:392–404.
89. Perkins TA, Wolf DE, Goodchild J. Fluorescence resonance energy transfer analysis of ribozyme kinetics reveals the mode of action of a facilitator oligonucleotide. Biochemistry. 1996;35:16370–7.
90. Liu J, Brown AK, Meng X, et al. A catalytic beacon sensor for uranium with parts-per-trillion sensitivity and millionfold selectivity. Proc Natl Acad Sci U S A. 2007;104:2056–61.
91. Boulyga SF, Heumann KG. Determination of extremely low 236U/238U isotope ratios in environmental samples by sector-field inductively coupled plasma mass spectrometry using high-efficiency sample introduction. J Environ Radioact. 2006;88:1–10.
92. Carmi N, Breaker RR. Characterization of a DNA-cleaving deoxyribozyme. Bioorganic Med Chem. 2001;9:2589–600.
93. Zhao W, Lam JCF, Chiuman W, et al. Enzymatic cleavage of nucleic acids on gold nanoparticles: a generic platform for facile colorimetric biosensors. Small. 2008;4:810–6.
94. Wei H, Li BL, Li J, et al. DNAzyme-based colorimetric sensing of lead (Pb2+) using unmodified gold nanoparticle probes. Nanotechnology. 2008;19:095501.
95. Lee JS, Han MS, Mirkin CA. Colorimetric detection of mercuric ion (Hg2+) in aqueous media using DNA-functionalized gold nanoparticles. Angew Chem Int Ed. 2007;46:4093–6.
96. Li D, Wieckowska A, Willner I. Optical analysis of Hg2+ ions by oligonucleotide-gold-nanoparticle hybrids and DNA-based machines. Angew Chem Int Ed. 2008;47:3927–31.
97. Chen L, Lu W, Wang X, Chen L. A highly selective and sensitive colorimetric sensor for iodide detection based on anti-aggregation of gold nanoparticles. Sensors Actuators B Chem. 2013;182:482–8.
98. Xu X, Wang J, Jiao K, Yang X. Colorimetric detection of mercury ion (Hg2+) based on DNA oligonucleotides and unmodified gold nanoparticles sensing system with a tunable detection range. Biosens Bioelectron. 2009;24:3153–8.
99. Chai F, Wang C, Wang T, et al. Colorimetric detection of Pb2+ using glutathione functionalized gold nanoparticles. ACS Appl Mater Interfaces. 2010;2:1466–70.
100. Liu C-W, Hsieh Y-T, Huang C-C, et al. Detection of mercury(II) based on Hg2+ -DNA complexes inducing the aggregation of gold nanoparticles. Chem Commun (Camb). 2008;22:2242–4.

101. Xia F, Zuo X, Yang R, et al. Colorimetric detection of DNA, small molecules, proteins, and ions using unmodified gold nanoparticles and conjugated polyelectrolytes. Proc Natl Acad Sci U S A. 2010;107:10837–41.
102. Li T, Dong S, Wang E. Label-free colorimetric detection of aqueous mercury ion (Hg2+) using Hg2+-modulated G-quadruplex-based DNAzymes. Anal Chem. 2009;81:2144–9.
103. Li T, Li B, Wang E, Dong S. G-quadruplex-based DNAzyme for sensitive mercury detection with the naked eye. Chem Commun (Camb). 2009;24:3551–3.
104. Li J, Yao J, Zhong W. Membrane blotting for rapid detection of mercury(II) in water. Chem Commun (Camb). 2009;137:4962–4.
105. Wang J, Liu B. Highly sensitive and selective detection of Hg(2+) in aqueous solution with mercury-specific DNA and Sybr Green I. Chem Commun (Camb). 2008;39:4759–61.
106. Zhang Z, Yin J, Wu Z, Yu R. Electrocatalytic assay of mercury(II) ions using a bifunctional oligonucleotide signal probe. Anal Chim Acta. 2013;762:47–53.
107. Zhang Z, Tang A, Liao S, et al. Oligonucleotide probes applied for sensitive enzyme-amplified electrochemical assay of mercury(II) ions. Biosens Bioelectron. 2011;26:3320–4.
108. Wu D, Zhang Q, Chu X, et al. Ultrasensitive electrochemical sensor for mercury (II) based on target-induced structure-switching DNA. Biosens Bioelectron. 2010;25:1025–31.
109. Liu SJ, Nie HG, Jiang JH, et al. Electrochemical sensor for mercury(II) based on conformational switch mediated by interstrand cooperative coordination. Anal Chem. 2009;81:5724–30.
110. Han D, Kim Y-R, Oh J-W, et al. A regenerative electrochemical sensor based on oligonucleotide for the selective determination of mercury(II). Analyst. 2009;134:1857–62.
111. Tang X, Liu H, Zou B, et al. A fishnet electrochemical Hg 2+ sensing strategy based on gold nanoparticle-bioconjugate and thymine–Hg 2+–thymine coordination chemistry. Analyst. 2012;137:309–11.
112. Gong JL, Sarkar T, Badhulika S, Mulchandani A. Label-free chemiresistive biosensor for mercury (II) based on single-walled carbon nanotubes and structure-switching DNA. Appl Phys Lett. 2013. doi:10.1063/1.4773569.
113. Li H, Xue Y, Wang W. Femtomole level photoelectrochemical aptasensing for mercury ions using quercetin-copper(II) complex as the DNA intercalator. Biosens Bioelectron. 2014;54:317–22.
114. Eda G, Lin YY, Mattevi C, et al. Blue photoluminescence from chemically derived graphene oxide. Adv Mater. 2010;22:505–9.
115. Bingol H, Kocabas E, Zor E, Coskun A. A novel benzothiazole based azocalix[4]arene as a highly selective chromogenic chemosensor for Hg2+ ion: a rapid test application in aqueous environment. Talanta. 2010;82:1538–42.
116. Wilson D, Del Valle M, Alegret S, et al. Potentiometric electronic tongue-flow injection analysis system for the monitoring of heavy metal biosorption processes. Talanta. 2012;93:285–92.
117. Chandrasoma A, Hamid AAA, Bruce AE, et al. An infrared spectroscopic based method for mercury(II) detection in aqueous solutions. Anal Chim Acta. 2012;728:57–63.
118. Fu X, Lou T, Chen Z, et al. "Turn-on" fluorescence detection of lead ions based on accelerated leaching of gold nanoparticles on the surface of graphene. ACS Appl Mater Interfaces. 2012;4:1080–6.
119. Jung JH, Cheon DS, Liu F, et al. A graphene oxide based immuno-biosensor for pathogen detection. Angew Chem Int Ed Engl. 2010;49:5708–11.
120. Huang X, Lan T, Zhang B, Ren J. Gold nanoparticle–enzyme conjugates based FRET for highly sensitive determination of hydrogen peroxide, glucose and uric acid using tyramide reaction. Analyst. 2012;137:3659–66.
121. Kundu A, Layek RK, Kuila A, Nandi AK. Highly fluorescent graphene oxide-poly(vinyl alcohol) hybrid: an effective material for specific Au 3+ ion sensors. ACS Appl Mater Interfaces. 2012;4:5576–82.

122. Zhang W, Wei J, Zhu H, et al. Self-assembled multilayer of alkyl graphene oxide for highly selective detection of copper(ii) based on anodic stripping voltammetry. J Mater Chem. 2012;22:22631.
123. Zhou N, Chen H, Li J, Chen L. Highly sensitive and selective voltammetric detection of mercury(II) using an ITO electrode modified with 5-methyl-2-thiouracil, graphene oxide and gold nanoparticles. Microchim Acta. 2013;180:493–9.
124. Wu S, He Q, Tan C, et al. Graphene-based electrochemical sensors. Small. 2013;9:1160–72.
125. Wang B, Chang YH, Zhi LJ. High yield production of graphene and its improved property in detecting heavy metal ions. Xinxing Tan Cailiao/New Carbon Mater. 2011;26:31–5.
126. Chow E, Goading JJ. Peptide modified electrodes as electrochemical metal ion sensors. Electroanalysis. 2006;18:1437–48.
127. Çeken B, Kandaz M, Koca A. Electrochemical metal-ion sensors based on a novel manganese phthalocyanine complex. Synth Met. 2012;162:1524–30.
128. Li J, Guo S, Zhai Y, Wang E. High-sensitivity determination of lead and cadmium based on the Nafion-graphene composite film. Anal Chim Acta. 2009;649:196–201.
129. Chen TY, Leddy J. Ion exchange capacity of Nafion and Nafion composites. Langmuir. 2000;16:2866–71.
130. Schrenk MJ, Villigram RE, Torrence NJ, et al. Effects of mixture casting Nafion?? with quaternary ammonium bromide salts on the ion-exchange capacity and mass transport in the membranes. J Memb Sci. 2002;205:3–10.
131. Iwai Y, Hiroki A, Tamada M, Yamanishi T. Radiation deterioration in mechanical properties and ion exchange capacity of Nafion N117 swelling in water. J Memb Sci. 2008;322:249–55.
132. Willemse CM, Tlhomelang K, Jahed N, et al. Metallo-Graphene nanocomposite electrocatalytic platform for the determination of toxic metal ions. Sensors. 2011;11:3970–87.
133. Fan M, Andrade GFS, Brolo AG. A review on the fabrication of substrates for surface enhanced Raman spectroscopy and their applications in analytical chemistry. Anal Chim Acta. 2011;693:7–25.
134. Álvarez-Puebla RA, Liz-Marzán LM. Environmental applications of plasmon assisted Raman scattering. Energy Environ Sci. 2010;3:1011.
135. Halvorson RA, Vikesland PJ. Surface-enhanced Raman spectroscopy (SERS) for environmental analyses. Environ Sci Technol. 2010;44:7749–55.
136. Li J, Chen L, Lou T, Wang Y. Highly sensitive SERS detection of As3+ ions in aqueous media using glutathione functionalized silver nanoparticles. ACS Appl Mater Interfaces. 2011;3:3936–41.
137. Mulvihill M, Tao A, Benjauthrit K, et al. Surface-enhanced Raman spectroscopy for trace arsenic detection in contaminated water. Angew Chem Int Ed. 2008;47:6456–60.
138. Han MJ, Hao J, Xu Z, Meng X. Surface-enhanced Raman scattering for arsenate detection on multilayer silver nanofilms. Anal Chim Acta. 2011;692:96–102.
139. Yin J, Wu T, Song J, et al. SERS-active nanoparticles for sensitive and selective detection of cadmium ion (Cd2+). Chem Mater. 2011;23:4756–64.
140. Wang Y, Irudayaraj J. A SERS DNAzyme biosensor for lead ion detection. Chem Commun (Camb). 2011;47:4394–6.
141. Li P, Liu H, Yang L, Liu J. Sensitive and selective SERS probe for Hg(II) detection using aminated ring-close structure of Rhodamine 6G. Talanta. 2013;106:381–7.
142. Ma Y, Liu H, Qian K, et al. A displacement principle for mercury detection by optical waveguide and surface enhanced Raman spectroscopy. J Colloid Interface Sci. 2012;386:451–5.
143. Li K, Liang A, Jiang C, et al. A stable and reproducible nanosilver-aggregation-4-mercaptopyridine surface-enhanced Raman scattering probe for rapid determination of trace Hg^{2+}. Talanta. 2012;99:890–6.

144. Choo J, Lim C, Chen L, et al. Surface-enhanced Raman scattering in nanoliter droplets: towards high-sensitivity detection of mercury (II) ions. Anal Bioanal Chem. 2009;394:1827–32.
145. Han D, Lim SY, Kim BJ, et al. Mercury(ii) detection by SERS based on a single gold microshell. Chem Commun (Camb). 2010;46:5587–9.
146. Chen J, Zheng A, Gao Y, et al. Functionalized CdS quantum dots-based luminescence probe for detection of heavy and transition metal ions in aqueous solution. Spectrochim Acta A Mol Biomol Spectrosc. 2008;69:1044–52.
147. Chen J, Gao Y, Xu Z, et al. A novel fluorescent array for mercury (II) ion in aqueous solution with functionalized cadmium selenide nanoclusters. Anal Chim Acta. 2006;577:77–84.
148. Ali EM, Zheng Y, Yu HH, Ying JY. Ultrasensitive Pb2+ detection by glutathione-capped quantum dots. Anal Chem. 2007;79:9452–8.
149. Koneswaran M, Narayanaswamy R. Mercaptoacetic acid capped CdS quantum dots as fluorescence single shot probe for mercury(II). Sensors Actuators B Chem. 2009;139:91–6.
150. Deka S, Quarta A, Lupo MG, et al. CdSe/CdS/ZnS double shell nanorods with high photoluminescence efficiency and their exploitation as biolabeling probes. J Am Chem Soc. 2009;131:2948–58.
151. Duan J, Jiang X, Ni S, et al. Facile synthesis of N-acetyl-l-cysteine capped ZnS quantum dots as an eco-friendly fluorescence sensor for Hg 2+. Talanta. 2011;85:1738–43.
152. Banerjee S, Kar S, Santra S. A simple strategy for quantum dot assisted selective detection of cadmium ions. Chem Commun (Camb). 2008;1:3037–9.
153. Algar WR, Tavares AJ, Krull UJ. Beyond labels: a review of the application of quantum dots as integrated components of assays, bioprobes, and biosensors utilizing optical transduction. Anal Chim Acta. 2010;673:1–25.
154. Medintz IL, Clapp AR, Mattoussi H, et al. Self-assembled nanoscale biosensors based on quantum dot FRET donors. Nat Mater. 2003;2:630–8.
155. Van Ness J, Van Ness LK, Galas DJ. Isothermal reactions for the amplification of oligonucleotides. Proc Natl Acad Sci U S A. 2003;100:4504–9.
156. Yin J, He X, Jia X, et al. Highly sensitive label-free fluorescent detection of Hg2+ ions by DNA molecular machine-based Ag nanoclusters. Analyst. 2013;138:2350–6.
157. Zou B, Ma Y, Wu H, Zhou G. Ultrasensitive DNA detection by cascade enzymatic signal amplification based on Afu flap endonuclease coupled with nicking endonuclease. Angew Chem – Int Ed. 2011;50:7395–8.
158. Xue L, Zhou X, Xing D. Sensitive and homogeneous protein detection based on target-triggered aptamer hairpin switch and nicking enzyme assisted fluorescence signal amplification. Anal Chem. 2012;84:3507–13.
159. Li JJ, Chu Y, Lee BYH, Xie XS. Enzymatic signal amplification of molecular beacons for sensitive DNA detection. Nucleic Acids Res. 2008. doi:10.1093/nar/gkn033.
160. Zhu G, Yang K, Zhang C, Yang. Sensitive detection of methylated DNA using the short linear quencher-fluorophore probe and two-stage isothermal amplification assay. Biosens Bioelectron. 2013;49:170–5.
161. Ma J, Chen Y, Hou Z, et al. Selective and sensitive mercuric (ii) ion detection based on quantum dots and nicking endonuclease assisted signal amplification. Biosens Bioelectron. 2013;43:84–7.
162. Dirks RM, Pierce NA. Triggered amplification by hybridization chain reaction. Proc Natl Acad Sci U S A. 2004;101:15275–8.
163. Huang J, Wu Y, Chen Y, et al. Pyrene-excimer probes based on the hybridization chain reaction for the detection of nucleic acids in complex biological fluids. Angew Chem Int Ed. 2011;50:401–4.
164. Xu Q, Zhu G, Zhang CY. Homogeneous bioluminescence detection of biomolecules using target-triggered hybridization chain reaction-mediated ligation without luciferase label. Anal Chem. 2013;85:6915–21.

165. Xu N, Wang Q, Lei J, et al. Label-free triple-helix aptamer as sensing platform for "signal-on" fluorescent detection of thrombin. Talanta. 2015;132:387–91.
166. Huang J, Gao X, Jia J, et al. Graphene oxide-based amplified fluorescent biosensor for Hg2+ detection through hybridization chain reactions. Anal Chem. 2014;86:3209–15.
167. Vogelstein B, Kinzler KW. Digital PCR. Proc Natl Acad Sci. 1999;96:9236–41.
168. Sanders R, Huggett JF, Bushell CA, et al. Evaluation of digital PCR for absolute DNA quantification. Anal Chem. 2011;83:6474–84.
169. Miotto E, Saccenti E, Lupini L, et al. Quantification of circulating miRNAs by droplet digital PCR: comparison of EvaGreen- and TaqMan-based chemistries. Cancer Epidemiol Biomarkers Prev. 2014;23:2638–42.
170. Pinheiro LB, Coleman VA, Hindson CM, et al. Evaluation of a droplet digital polymerase chain reaction format for DNA copy number quantification. Anal Chem. 2012;84:1003–11.
171. Hindson BJ, Ness KD, Masquelier DA, Colston BW. High-throughput droplet digital PCR system for absolute quantitation of DNA copy number. Anal Chem. 2011;83:8604–10.
172. Burns MJ, Burrell AM, Foy CA. The applicability of digital PCR for the assessment of detection limits in GMO analysis. Eur Food Res Technol. 2010;231:353–62.
173. Demeke T, Gräfenhan T, Holigroski M, et al. Assessment of droplet digital PCR for absolute quantification of genetically engineered OXY235 canola and DP305423 soybean samples. Food Control. 2014;46:470–4.
174. Corbisier P, Bhat S, Partis L, et al. Absolute quantification of genetically modified MON810 maize (Zea mays L.) by digital polymerase chain reaction. Anal Bioanal Chem. 2010;396:2143–50.
175. Morisset D, Štebih D, Milavec M, et al. Quantitative analysis of food and feed samples with droplet digital PCR. PLoS One. 2013;8:e62583.

Chapter 18
Genotoxicity Detection at the Molecular Level in Food Safety Assessment: Conventional Methods and Developments

Abstract Genotoxicity detection is an important part of food safety assessment because many food contaminants are genotoxic. Through the regulation of gene expression, causing DNA damage, inducing gene mutation, and influencing DNA methylation, such contaminants pose a severe threat to the health of human and animals and can even induce cancer. The development of PCR technology makes it possible and easier to detect genotoxicity at the molecular level. Of the various genotoxic pathways, gene mutation, DNA methylation, and miRNA regulation are issues of particular concern. This chapter therefore focuses on the technologies for the detection of gene mutation, DNA methylation, and miRNA and reviews the conventional methods and developments in these fields. Direct DNA sequencing, PCR-SSCP, PCR-RFLP, ASO-PCR, biosensors, SNP detection, and other methods are discussed in the context of gene mutation detection. With regard to DNA methylation detection, techniques for both genome-wide methylation and gene-specific methylation are reviewed. In the discussion of miRNA detection, we introduce two of the most easily accessible and widely used methods for this purpose, namely, poly(A)-tailing PCR and stem-loop PCR, as well as innovative techniques such as biosensors. Finally, the prospects under development in these fields are discussed.

Keywords Genotoxicity detection • Food contaminants • Gene mutation detection • DNA methylation detection • miRNA detection

18.1 Introduction

In each step of food production, genotoxic food contaminants may be introduced. The plants and animals that are used as raw materials may be contaminated by mycotoxins, such as aflatoxins and ochratoxins; heavy metals, such as lead, mercury, and cadmium; and overused or illegally used pesticides and drugs. Products or by-products of improper processing, such as acrylamide in baked goods and benzopyrene in roast meat, can also raise health concerns.

For example, aflatoxins are secondary metabolites produced by many strains of *Aspergillus* via a polyketide pathway [1]. The most widespread member of this family of toxins is AFB1, which is also the most toxic in terms of both acute and

W. Xu, *Functional Nucleic Acids Detection in Food Safety*,
DOI 10.1007/978-981-10-1618-9_18

chronic toxicity; it plays a synergistic role with the hepatitis B virus, causing hepatocellular carcinoma through the integration of the hepatitis B virus x gene, interference with nucleotide excision repair, alteration of the methylation of genes, and DNA mutation [2]. The genotoxicity of heavy metals to living organisms involves the interaction of heavy metals with genetic material through binding to DNA bases [3], eventually causing gene mutation. It has recently been reported that acrylamide-induced carcinogenicity in the mouse lung is related to mutagenicity and that A→T and G→C are the predominant types of mutations [4]. Another mycotoxin, ochratoxin A, can induce global DNA demethylation and influence the expression of several critical tumor-related genes by inducing methylation modifications [5].

These genotoxic food contaminants, through the regulation of gene expression, cause DNA damage, induce gene mutation, and influence DNA methylation, thereby severely endangering the health of humans and animals and even causing cancer. Therefore, genotoxicity detection has become an important aspect of food safety assessment. This chapter focuses on the technologies that are available for the detection of gene mutation, DNA methylation, and miRNA and reviews the conventional methods and developments in these fields.

18.2 Technologies for Gene Mutation Detection: Conventional Methods and Developments

Genetic mutations include point mutations, insertions, deletions, translocations, abnormal gene structures, and polymorphisms, among others. With the rapid development of molecular biology, PCR and its derivative technologies have become widely used in gene mutation detection because of their rapidity and accuracy. We review the conventional methods and developments in gene mutation detection, with a particular focus on those based on polymerase chain reaction, as follows.

18.2.1 Direct DNA Sequencing

The direct sequencing of PCR products is the most direct and comprehensive of all gene mutation detection methods. It can not only determine the locations of mutation but can also determine the mutation patterns.

DNA sequencing refers to deciphering the continuum of nucleotide bases of a given DNA segment. In 1971, Frederick Sanger made this task possible by inventing the first rapid DNA sequencing technology, that is, Sanger sequencing [6]. There are two approaches: the first is based on enzymatic synthesis from a single-stranded DNA template with chain termination using dideoxynucleotides (ddNTPs), and the second is Maxim-Gilbert sequencing, which involves the chemical degradation of end-radio-labeled DNA fragments. Both methods rely on four-lane, high-resolution polyacrylamide gel electrophoresis to separate the labeled fragments and allow the base sequences to be read in a staggered, ladder-like fashion [7].

Second-generation technologies, namely, the various implementations of cyclic-array sequencing, have truly revolutionized the field of DNA sequencing. The concept of this technology can be summarized as the sequencing of a dense array of DNA features via iterative cycles of enzymatic manipulation and imaging-based data collection [8]. Two reports in 2005 described the first integrated implementations of cyclic-array strategies that were both practical and cost-competitive with conventional sequencing [9, 10]. Second-generation read lengths range from 30 to 450 bases; however, the reliability of these reads is lower than that of Sanger sequencing, introducing an inherent error rate [6].

Real-time DNA sequencing from single polymerase molecules represents the third generation of DNA sequencing methods. Eid J. et al. presented this method in 2009 [11], and it possesses several advantages over the first and second generations of DNA sequencing approaches, including longer sequencing read lengths and greater flexibility. It is also lower in cost than first-generation methods and has a higher throughput than second-generation methods.

However, the instruments required for sequencing and the high cost limit the application of DNA sequencing in gene mutation detection, and thus, simpler and more convenient methods have emerged.

18.2.2 PCR-SSCP

Rapid determination of whether sequence differences exist between certain segments of DNA can be achieved through a simple method known as polymerase chain reaction single-strand conformation polymorphism (PCR-SSCP), first reported by Orita M. et al. in 1989 [12]. Since that time, SSCP has been extensively applied to identify and characterize genetic polymorphisms and mutations in host genes such as p53 [13, 14], especially in combination with sequence analysis. The principle of SSCP is that even single-base differences can cause different secondary conformations and therefore result in different migration rates of the DNA strands in nondenaturing polyacrylamide gel. Radioactive nucleotides can be incorporated into DNA strands via PCR, making it possible to detect the DNA through autoradiography [15]. Recently, in 2015, Geng R. Q. et al. used PCR-SSCP and DNA sequencing to detect genetic mutations of the EGLN1 gene in different sheep breeds [16].

The primary advantages of this method are its simple operation, its high sensitivity, and its ability to be used for the analysis of various types of mutations, such as deletion, insertion, and point mutations, making it one of the simplest methods of testing for gene mutations. The shortcoming of this technology is that it is possible to miss detecting certain mutations because different sequences of DNA, the PAGE electrophoresis conditions, and the size of the PCR products all influence the detection accuracy.

18.2.3 PCR-RFLP

For determining known sequence variants in nucleic acids, restriction fragment length polymorphism (RFLP) is an option [17]. This convenient method consists of three steps: first, the segment to be searched for a mutation is amplified from genomic DNA or cDNA by PCR; second, the PCR products are digested by an appropriate restriction enzyme; and afterward, the digested products are separated by agarose gel electrophoresis. In the 1990s, the PCR-RFLP technique was successfully applied for the determination of K-ras mutations in BAL specimens, tumor biopsies, and microdissected tumor cell populations [18, 19]. Schuermann, M. applied this technique for the sensitive detection of mutated p53 alleles in several of the "hot spot" positions that are most frequently affected in lung cancer [20]. In recent years, Ouyang L. et al. applied the PCR-RFLP method to determine codon 248 mutation in exon 7 of the p53 tumor suppressor gene; MspI restriction mapping confirmed the G-C mutation in codon 248, which is closely related to cancer susceptibility [21]. Anitha S. et al. examined mutations in p53 via PCR-RFLP analysis, confirming that the G-to-T transversion at the third position of codon 249 (AGG) of the p53 gene, substituting arginine with serine, is the most common aflatoxin-induced mutation linked to HCC [22].

However, this facile and highly sensitive method suffers from the disadvantage that simple PCR and RFLP analysis fail to detect the polymorphisms resulting from single-base substitutions because such substitutions do not create or remove any restriction sites [17]. To overcome this disadvantage, an artificially created restriction site should be applied, which can be achieved by means of a mismatch primer [23]. This mismatch primer contains a one- or two-base mismatch near its 3′ end such that, in the amplified product, a restriction site for the corresponding endonuclease is added or removed in the case of a base substitution [24, 25].

Another development is the combination of PCR-RFLP with capillary electrophoresis (CE). CE methods have been reported for separating PCR-RFLP fragments from mRNA, DNA from bacteriophages, and oncogenes [26, 27]. CE offers high throughput and high resolution, automatic operation, and online detection with automatic data acquisition, and these features have stimulated its particular application for the analysis of DNA mutations for genetic analysis and medical diagnosis [28].

18.2.4 ASO-PCR

Another simple, rapid, and nonradioactive method for the detection of mutations is the combination of PCR with allele-specific oligonucleotides (ASOs) [29]. The main distinguishing feature of this technique is the primer used in the amplification: only a perfectly matched oligonucleotide can fulfill the task. Oligonucleotide primers are designed complementary to either the wild-type or mutant sequence, and both are used in conjunction with a common primer. Because DNA polymerase lacks a 3′ exonuclease activity, it is unable to repair a single-base mismatch between the primer and the template at the 3′ end of a DNA primer. Thus, under appropriately stringent conditions, only target DNA that is exactly complementary to the primer will be amplified [29].

This method was first described for the determination of mutations in the β-antitrypsin gene in 1985 [30]. Since that time, it has been widely applied in studies of a variety of genes, such as point mutations in the ras oncogene [31] and polymorphisms of apolipoprotein E [32]. Liu Z. et al. detected ras oncogene mutations in diethylnitrosamine-induced hepatic tumors in medaka, a teleost fish, using this method [33]. Diethylnitrosamine is a kind of nitrosamine, which can be produced in meat processing. Nitrosamines are very powerful carcinogens that are formed by nitric acid salts or nitrites bonded to amines. In the cited study, a single-point mutation was detected in codon 12 position two in 8/25 (32 %) of the examined tumors, and ras mutations were induced by diethylnitrosamine and were present in over 30 % of the fish that developed tumors.

ASO-PCR is also widely used in the medical and health fields. Ashraf S. et al. reported that the ASO-PCR method appears to be useful for the detection of IDH2 mutations and the routine diagnostic screening of prognostically relevant alterations in acute myeloid leukemia and may be conveniently incorporated into the diagnostic workup [34]. In 2015, Jara-Acevedo M. et al. determined the frequency of peripheral blood involvement in the KIT D816V mutation, as assessed using two highly sensitive PCR methods, ASO-PCR and a peptide nucleic acid-mediated PCR approach [35], and concluded that the ASO-PCR method was more sensitive ($P < 0.0001$).

18.2.5 Biosensors

A biosensor typically consists of a bio-recognition component, a biotransducer component, and an electronic system, which includes a signal amplifier, a processor, and a display. Because of their rapidity and high sensitivity, biosensors have become a promising technique that is being increasingly and widely used in the detection of gene mutation.

In 2001, Patolsky et al. assembled a primer thiolated oligonucleotide, complementary to the target DNA up to one base before the mutation site, on an electrode or a gold–quartz piezoelectric crystal [36]. After hybridization of the target DNA, the resulting assembly was reacted with a biotinylated nucleotide, complementary to the mutation site, in the presence of polymerase. The labeled nucleotide was coupled only to the double-stranded assembly that included the mutant site. Subsequent binding of avidin–alkaline phosphatase to the assembly, and the biocatalyzed precipitation of an insoluble product on the transducer, provided a means of confirming and amplifying the detection of the mutant. Wang's group developed a reusable quartz crystal microbalance (QCM) biosensor based on the toehold-mediated strand displacement reaction (SDR) for the highly specific detection of single-base DNA mutations [37]. In this sensor, the capture DNA immobilized on the gold electrode surface of the QCM hybridizes with the conjugate of biotinylated reporter DNA and streptavidin to form the recognition layer. A toehold domain in the reporter DNA is specifically designed to bind the complementary mutant DNA and trigger the SDR, eventually resulting in the release of a complex of the reporter probe and the mutant DNA from the gold surface.

18.2.6 Others

In addition, there are many more convenient methods that can be used in the detection of gene mutation, including allele-specific PCR (AS-PCR) [38, 39], chemical mismatch cleavage (CMC) [40, 41], heteroduplex analysis (HA) [42, 43], denaturing gradient gel electrophoresis (DGGE) [44, 45], and denaturing high-performance liquid chromatography (DHPLC) [46, 47].

18.2.7 SNP Detection Methods

Single-nucleotide polymorphisms (SNPs) refer to groups of single-nucleotide changes that appear at the same position in polymorphic DNA sequences, including the transformation of a single base, transversion, deletion, or insertion. SNPs are a special kind of genetic mutation, distinguished from rare variations by the requirement that the least abundant allele must have a frequency of 1 % or more [48]. SNPs are the most frequent type of variation observed in the human genome, and they provide powerful tools for a variety of medical genetic studies [49].

Bi-allelic SNPs comprise four distinct types, including one transition, C→T (G→A), and three transversions, C→A (G→T), C→G (G→C), and T→A (A→T). One quarter of SNPs occur at CpG loci in the form of C→T conversion, associated with the methylation of the site [48]. Sachidanandam R. et al. reported a map of human genome sequence variations containing 1.42 million single-nucleotide polymorphisms [50]. Inherited differences in DNA sequences contribute to phenotypic variations, influencing an individual's anthropometric characteristics, risk of disease, and response to the environment. Therefore, SNP mapping and detection should assist in defining haplotype variations across the genome and in identifying biomedically important genes for diagnosis and therapy.

SNP detection technology can be classified into traditional electrophoresis technology and modern high-throughput detection technology. The traditional methods include PCR-RFLP [51, 52], PCR-SSCP [53, 54], ASO-PCR [55, 56], and DGGE [57, 58], among others (all of the listed methods have been introduced above). However, their application is restricted because they can only be used for small-scale SNP typing and detection; moreover, automation is difficult to achieve.

Modern high-throughput SNP detection techniques can be applied to large samples and can enable the genotyping of multiple SNP loci [59]. These techniques include the TaqMan probe method [60, 61], the ligase detection reaction (LDR) method [62, 63], pyrophosphate sequencing [64, 65], DHPLC [66, 67], biosensors [68], and the DNA chip method [69, 70], among others.

18.3 Technologies for DNA Methylation Detection: Conventional Methods and Developments

DNA methylation is an important mode of epigenetic modification and has great significance in biochemistry, medicine, and genomics. It plays a key role in gene expression control. DNA methylation is a type of epigenetic change; that is, it modifies DNA without changing its sequence [71]. Generally, epigenetic changes mainly comprise DNA methylation, histone modifications, and noncoding RNA regulation. Many epigenetic studies, including experiments focusing on genomic imprinting [72], the manner of monoallelic expression [73], and the use of high-throughput sequencing technology [74], have recently been widely performed to gain insight into gene regulation patterns. In the DNA methylation process, under catalysis by methyltransferases, the fifth carbon of the C in dinucleotide 5′-CG-3′ is selectively modified with a methyl group to form 5-mC. Methyl groups are added by a family of DNA methyltransferase enzymes. DNMT1 catalyzes the majority of maintenance methylation (the addition of methyl groups to hemimethylated DNA strands immediately following DNA replication), whereas DNMT3a and DNMT3b are believed to be more strongly involved in de novo methylation [75]. Depending on whether the level of genomic modification is increased or decreased, DNA methylation can be classified as hyper- or hypomethylation, respectively. Hypermethylation can result in gene-inactivating deletions and translocations [76].

DNA methylation occurs in all living creatures and takes several forms. In prokaryotes, it usually occurs on cytosine (C) in CCA/TGG and GATC, whereas in eukaryotes, it occurs on purine and guanine but mostly on C [77]. Approximately 60–90 % of 5′-CG-3′ dinucleotides in eukaryotes are methylated. DNA methylation occurs most often at cytosine–guanine sites (CpG dinucleotides) of gene promoters. CpG islands are regions with a high frequency of CpG sites that are frequently located in promoter regions. The most widely used definition of a CpG island is that provided by Gardiner-Garden and Frommer [78], who specified that a CpG island must be greater than 200 bp in length, with a GC content of >50 % and an observed/expected CpG ratio of ≥0.6. Approximately 40 % of tissue-specific genes contain CpG islands [79], and methylation may play a role in the tissue-specific expression of some of these genes [80]. CpG island methylation has also been shown to increase with age in certain tissues, such as the colonic mucosa [81–83]. Aberrant methylation of CpG islands is the earliest molecular alteration that occurs during carcinogenesis; thus, it is regarded as a biomarker for the early diagnosis of cancers.

Many studies have confirmed that DNA methylation plays a crucial role in the regulation of gene transcription [84, 85]. The aberrant methylation of gene promoters may cause the silencing of important genes, further influencing their related transcriptional pathways and eventually leading to cell dysfunction and even disease. DNA methylation plays an essential role in gene regulation; it affects genomic imprinting [86], the silencing of microRNA genes [87], Alzheimer's disease [88], and X chromosome inactivation [89]. Aberrant DNA methylation reduces or increases the activity of genes at the cellular level. These changes are often inherited during cell division, leading to diseases at the organism level, such as cardiovascular diseases [90], schizophrenia [91], and various cancers [92], including lung cancer [93], breast cancer [94], colon cancer [95], urological cancer [96], ovarian cancer [97], sarcomas [98], Hodgkin's lymphoma [99], and leukemia [100]. Tumor cells often exhibit methylation patterns that are substantially different from those of cells in healthy tissues (e.g., cancer is typically characterized by both global and site-specific gene hypomethylation and hypermethylation) [101]. Regarding brain tumors, it has even been shown that they are more often linked to aberrant methylation than to DNA copy changes [101].

Because so many complicated diseases have been correlated with methylation status, which may be affected by harmful components in food and induce cancer, investigating the potential and practical methods of detecting DNA methylation and aberrant DNA methylation within CpG-rich sequences is of great significance.

The development of techniques for DNA methylation analysis is fundamental and pivotal for epigenomic research. Over approximately 30 years of diligent study, a variety of specific and sensitive methods have emerged to detect these small alterations. These methods can be divided into two groups based on the target of investigation, namely, direct and indirect approaches. Direct approaches primarily focus on global methylation analysis, the purpose of which is to quantify the total amount of 5-mC in the genome; typical methods include liquid chromatography/electrospray ionization mass spectrometry (LC-ESI-MS) [102], LC-MS/MS [103], and SssI methylase analysis [104]. By contrast, indirect approaches rely on the specific recognition of DNA methylation loci. These approaches include methylation-specific PCR (MS-PCR), amplification of inter-methylated sites [105], methylation-sensitive restriction fingerprinting [106], split β-lactamase sensors [107], methylation-specific multiplex ligation-dependent probe amplification [108], and methylation-specific single-strand conformation analysis [109].

18.3.1 Detection of Genome-Wide Methylation

LC and capillary electrophoresis (CE) are common direct approaches to DNA methylation analysis and have been frequently used to analyze hydrolyzed DNA. The LC method was first utilized by Kuo and his team in 1980 [110]; after hydrolyzation with DNase 1, nuclease P1, and alkaline phosphatase, the deoxyribonucleosides were directly separated via reversed-phase high-performance liquid chromatography with detection based on ultraviolet absorption at 254 and 280 nm. This highly sensitive and selective dual-wavelength quantitation greatly enhanced the precision and accuracy of the chromatographic analysis. Kuo and his team have used this method for the quantitation of 5-mC in 5 μg of calf thymus and salmon sperm DNA, in which 5-mC comprises only 1–2 % of the total bases. This method was further enhanced with the incorporation of PCR to develop high-performance LC-PCR (HPLC-PCR) [111]. The latest reported technique uses octyl-modified quaternized cellulose in CE [112]. The method was developed on the basis of the hydrolysis of DNA with formic acid and the separation of nucleic acid bases via capillary electrophoresis. Cytosine and 5-methylcytosine were separated with a resolution of approximately 4.0 within less than 10 min. The detection limits ($S/N = 3$) were 1.1 and 1.5 μg/mL for cytosine and 5-methylcytosine, respectively. With the application of MS, the selectivity and sensitivity of this technology were significantly improved. The detection range and limit of detection (LOD) for 5-mC were 0.2–100 ng mL^{-1} and 45 pg mL^{-1}, respectively, and for 5-hmC analysis, they were 0.1–50 ng mL^{-1} and 57 pg mL^{-1}, respectively [113].

18.3.2 Detection of Gene-Specific Methylation

In comparison with the above direct approaches, indirect methods offer numerous other approaches that rely on the specific recognition of DNA methylation loci, followed by the amplification of recognition processes or the isolation of recognized DNA strands based on changes in their properties. These methods often provide higher specificity and sensitivity in the analysis of DNA methylation. In this section, we introduce typical methods of this type, especially those based on further developments of the most widely used methods of methylation locus recognition, namely, bisulfite conversion, biological recognition, and chemical cleavage. Most of these methods have been developed based on polymerase chain reaction and signal amplification by probe hybridization [114]. In fact, DNA methylation is not preserved during PCR reactions; most PCR-based methods use genomic DNA templates that have been treated with sodium bisulfite or enzymes to differentiate methylated from unmethylated bases.

18.3.2.1 Techniques Based on Bisulfite Treatment

Bisulfite treatment is a widely used, effective method for sorting 5-mC from unmethylated bases. During bisulfite treatment, the 5-mC in the gene remains unchanged, whereas C is specifically converted into uracil (U), which will be further transformed into thymine (T) after PCR amplification [115]. As a result, the methylation information of the gene is translated into sequence information. The features of the gene, such as melting temperature and specific recognition interactions, are altered by this change in sequence; thus, the differences in these features are the key to DNA methylation analysis. Bisulfite converts unmethylated cytosine residues into uracil residues, while the methylated cytosine residues remain their status [115].

Bisulfite conversion of unmethylated cytosines followed by DNA sequencing requires no live cells and only limited amounts of DNA, which also need not be of high quality; therefore, DNA methylation is less sensitive to specimen handling compared with RNA or proteins [116]. As a result, bisulfite sequencing, which, importantly, cannot distinguish between 5-mC and 5-hmC, can even be performed on DNA isolated from small amounts of fixed tissue [117]. Indeed, bisulfite sequencing and methylation-specific PCR (MSP) are accepted as the gold standard methods for single-locus DNA methylation analysis. The former method provides information regarding the methylation status at single-base resolution, which sometimes requires the cloning of PCR amplicons to obtain qualitative and quantitative results. The latter method enables simple and rapid detection of DNA methylation through the use of two sets of primers, designed for methylated and unmethylated DNA, although it lacks quantitative accuracy.

Bisulfite Sequencing

The most straightforward method of distinguishing 5-mC from other bases is gene sequencing. Frommer et al. achieved a major breakthrough in the analysis of DNA methylation by employing bisulfite treatment, in which only unmethylated cytosine is converted into uracil [118]. The DNA was first treated with sodium bisulfite and then amplified via PCR. Through the gene sequencing of the resultant product, the 5-mC status was obtained. This technique is suitable for multiplex DNA methylation determination, although the need for massive-scale cloning and sequencing processes causes sample preparation to be tedious and time-consuming. P. Subiyantoro and his team verified DNA hypermethylation, especially in the P14ARF part of the CDKN2A gene in chromosome 9, using bisulfite sequencing [119].

Methylation-Specific PCR (MSP)

In 1996, Herman developed an advanced sequencing-based technique named methylation-specific PCR (MSP), which employs the bisulfite treatment of DNA as a starting point for methylation analysis [120]. Unlike the previous bisulfite sequencing technique, two specific promoters were predesigned, one for the methylated sequence and the other for the unmethylated sequence of interest. After treatment with sodium bisulfite, the target sequence was amplified via PCR using these two specific promoters. If the target sequence was methylated, then the PCR products contained amplified 5-mC. This technique avoids the need for a complicated sequencing process. However, the sequence and the methylation loci must be known in advance because of the need to predesign the primer; thus, the technique can be used to perform only a qualitative determination. Later, the performance of MSP was improved using a variety of PCR-based methods, resulting in approaches such as real-time PCR [97], quantitative MSP [121], methylation-sensitive single-nucleotide primer extension [122], and sensitive melting analysis after real-time MSP [123].

Pannone et al. [124] have used MSP to analyze, in a series of primary OSCCs with matched normal oral mucosa under chronic exposure to alcohol and smoking, the methylation status of a panel of genes, including hMLH1, CDH1, CDKN2a, MGMT, RAR-β2, SFRP-1, SFRP-2, SFRP-4, SFRP-5, WIF-1, and DKK-3, with the purpose of defining an epigenetic fingerprint for precancer lesions and oral cancers. Using the methylation-specific PCR (MSP) method, Raluca Dumache [125] analyzed the methylation status of the p16INK4a gene from serum and its corresponding tumor in 42 bladder cancer patients, as well as 35 samples from cancer-free individuals as controls, to investigate whether the hypermethylation of the p16INK4a gene can be used as a serum biomarker for the early detection of bladder cancer.

18.3.2.2 Techniques Based on Biological Recognition

Although bisulfite treatment has been widely used to distinguish 5-mC from other bases, it suffers from issues of incomplete conversion, false-positive results, difficult operation, and high time consumption. In comparison, biological identification approaches, such as enzymatic digestion or bioaffinity reactions, can rapidly recognize methylation loci with high specificity under mild reaction conditions; as a result, 5-mC analysis methods based on biological recognition have become a new focus of attention [115]. At the beginning of the twenty-first century, techniques based on enzymes were widely applied for the detection of DNA methylation. The typical method is restriction-landmark genomic scanning (RLGS) [126]. In 2004, an advanced enzyme digestion technique named methylation-sensitive restriction endonuclease-PCR/Southern (MSRE-PCR/Southern) was developed. In 2007, methylation-sensitive restriction endonuclease was utilized for online monitoring of the activity of MTase and evaluation of the methylation status [127]. The activity of MTase and the DNA methylation status were monitored based on distinct fluorescence signals. However, the need for the fluorescence-labeling procedure and its cost are drawbacks of this technique. As an alternative approach, certain biomolecules, such as MBD, anti-5-methylcytosine IgG1 antibody, and zinc finger (ZF) proteins, exhibit an inherent capability of methylation locus specificity and can be directly used for DNA methylation analysis [128–130]. In 2004, Shiraishi proposed an MBD-based method for screening and identifying 5-mC in genomes [131]. In 2006, an enhanced technique was developed by combining methylated DNA precipitation with methylation-sensitive restriction enzyme (COMPARE-MS).

As an advanced enzyme digestion technique, methylation-sensitive restriction endonuclease-PCR/Southern (MSRE-PCR/Southern) was developed in 2004 for detecting DNA methylation. DNA samples were treated with Hpa II and Msp I restriction enzymes. Both the Hpa II and Msp I enzymes can specifically recognize the 5′-CCGG-3′ sequence. Hpa II and Msp I cannot digest unmethylated C in 5′-mCCGG-3′, whereas Msp I can cleave methylated C in 5′-CmCGG-3′. DNA samples were digested with Hpa II and Msp I enzymes and analyzed via PCR or Southern blot. The DNA methylation statuses at multiple methylation loci were obtained. The technique is simple and economical, but it is not suitable for complex gene samples.

Hua D investigated the methylation levels of multiple genes in hepatocellular carcinoma (HCC) and identified a combination of methylation markers that could be useful for the diagnosis of HCC by utilizing methylation-sensitive restriction endonuclease-PCR/Southern [132]. In 2007, the methylation-sensitive restriction endonuclease-PCR/Southern approach was applied to investigate the epigenetic regulatory mechanism of BMP-6 gene expression in the breast cancer cell lines MDA-MB-231, MCF-7, and T47D with regard to the methylation status in the 5′ flanking region of the human BMP-6 gene [133].

18.3.2.3 Chemical Oxidation-Based Methods

Recently, several new chemical methods based on chemical oxidation cleavage have been proposed for the discrimination of 5-mC. Inspired by the distinctive capability of OsO4 to discriminate 5-mC from C, Yamata et al. developed a method employing 2-methyl-1,4-naphthoquinone chromophore to identify 5-mC through photosensitized oxidation [134]. Later, an advanced method using NaIO4/LiBr was reported [135]. The C5-C6 double bond of 5-mC in DNA was selectively oxidized by NaIO4/LiBr and subsequently cut by means of hot piperidine treatment. The chemically cleaved DNA segments were analyzed via denaturing polyacrylamide gel electrophoresis (PAGE), and the DNA methylation status was easily obtained. The chemical cleavage method shows high efficiency for methylation locus recognition, but its lack of sequence selectivity and unsatisfactory sensitivity hinder its application in gene-specific methylation analysis.

18.3.2.4 Biosensors

Biosensors are also seeing increasingly wide use in the detection of DNA methylation because of their convenience. By using DNA-modified gold nanoparticles (AuNPs) coupled with enzyme linkage reactions, Liu's team proposed a simple colorimetric assay method for monitoring DNA methylation by sensitively detecting the activity of methylases [136]. The duplex DNA molecules on the surfaces of the AuNPs are first methylated with DNA adenine methylation methyltransferase (Dam MTase) and then cut using methylation-sensitive restriction endonuclease Dpn I. The removal of these duplexes from the AuNP surfaces through the methylation/cleavage process destabilizes the nanoparticles, leading to aggregation of the AuNPs and a red-to-blue color change. Consequently, the enzyme activity of Dam MTase can be assayed and thus DNA methylation can be monitored.

Duan et al. developed a novel method of DNA methylation detection based on single-base extension reactions and cationic conjugated polymers (CCPs) [68]. Treatment of DNA with bisulfite followed by PCR amplification converts unmethylated DNA into a C/T polymorphism. After single-nucleotide base extension with Taq polymerase, fluorescein-labeled dGTP (dGTP-Fl) is incorporated into the methylation-specific probe for methylated DNA but not into the probe for nonmethylated DNA. Upon the addition of cationic polyfluorene (PFP1a), the strong electrostatic interactions between DNA and PFP1a bring the fluorescein close to the PFP1a, and efficient fluorescence resonance energy transfer from the PFP1a to the fluorescein occurs for the methylated DNA.

18.4 Technologies for miRNA Detection: Conventional Methods and Developments

Animals and plants contain miRNAs, which are a type of small molecule with approximately 22 nucleotides acting as a crucial mediator of gene expression. By primarily binding to complementary sequences in the 3′-UTR region of specific mRNAs, they contribute to the cleavage of mRNA or to the repression of translation, thereby regulating the expression of target genes [137].

Food contaminants have been reported to induce genotoxicity through the abnormal expression of miRNAs. The genotoxic mycotoxin AFB1 can induce liver tumorigenesis, and Zhu et al. recently found that miRNAs are involved in this inducement. Based on miRNA profiling in HepG2 cells, they suggested that one mechanism of the genotoxicity of AFB1 is the downregulation of the Wnt/β-catenin signaling pathway through the upregulation of miR-34a [138], in accordance with an in vivo study conducted by Liu's group. Liu et al. demonstrated that rno-miR-34a-5p responds sensitively to AFB1 exposure and facilitates p53 repair of DNA damage by impacting the cell cycle. Thus, circulating rno-miR-34a-

5p may be a sensitive indicator for the inducement of hepatic genotoxicity by AFB1 in rats [139]. Therefore, the detection of this important kind of gene regulator should not be ignored in genotoxicity detection.

18.4.1 Common Methods Based on PCR

The most easily accessible and widely used technologies based on PCR for miRNA detection are poly(A)-tailing PCR and stem-loop PCR.

18.4.1.1 Poly(A)-Tailing PCR

Poly(A)-tailing PCR is a fast and convenient method reported by Shi and Chiang in 2005 [140]. The principle and procedure of poly(A)-tailing PCR are illustrated in Fig.18.1. In brief, in the presence of poly(A) polymerase and ATP, a poly(A) tail is added to mature miRNA, thus forming a longer detection target. Then, with the assistance of a poly(T) adapter, reverse transcription is performed to obtain the first strand cDNAs of the polyadenylated miRNAs. The final step is PCR amplification, during which an miRNA-specific primer is used as the forward primer, whereas the reverse primer is complementary to the poly(T) adapter.

18.4.1.2 Stem-Loop PCR

The other widely used PCR-based method for miRNA detection is stem-loop PCR, which is efficient and specific for the detection of mature miRNA, as reported by Chen et al., also in 2005 [141]. Stem-loop PCR consists of two steps. The first is reverse transcription. A stem-loop primer binds to the target miRNA at the 3′ region, and in the presence of reverse transcriptase, the cDNA product becomes linked to a corresponding stem-loop. Then, conventional PCR is applied for the quantification of the target miRNA, during which an miRNA-specific forward primer and a stem-loop-specific reverse primer are used. The principle and procedure are illustrated in Fig.18.2.

Mou et al. evaluated these two methods for the detection of several miRNAs and found the stem-loop method can detect highly and moderately abundant miRNAs but fails to detect those of low abundance, whereas poly(A)-tailing PCR is applicable for miRNA detection regardless of amount but is not suitable for miRNAs with a hairpin structure [142]. Following a comparison with other miRNA detection methods, Chen et al. concluded that a stem-loop RT-based TaqMan miRNA assay can be regarded as the gold standard method because of its large dynamic range and high specificity and sensitivity [143].

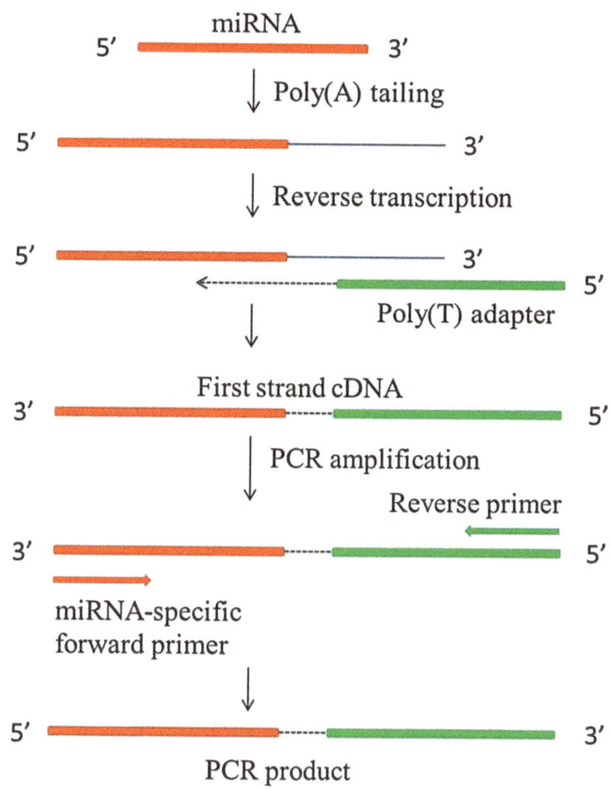

Fig. 18.1 Principle and procedure of poly(A)-tailing PCR [140] (Reproduced from Shi R et al. 2005. © 2009 BioTechniques. Used by Permission)

18.4.2 Biosensors

Innovative methods based on nanoparticles, surface plasmon resonance, locked nucleic acids, and luciferase imaging have emerged in recent years [144], among which electrochemical biosensors are of particular interest because of their higher selectivity and lower cost. Gao et al. reported a DNA-gold nanoparticle (DNA-GNP)-based lateral flow nucleic acid biosensor as a highly sensitive tool for the visual detection of miRNA with a limit of detection of 60 pM [145]. The operational basis of this device is the sandwich-type hybridization reactions among GNP-labeled DNA probes, miRNA, and biotin-modified DNA probes.

Ge and colleagues demonstrated an ultrasensitive detection platform for miRNA by combining tetrahedral DNA nanostructure probes with hybridization chain reaction (HCR) amplification in 2014 [146]. Taking advantage of DNA nanotechnology, they employed a three-dimensional tetrahedral DNA nanostructure as the scaffold on which to immobilize DNA recognition probes to increase reactivity and accessibility, whereas DNA nanowire tentacles were used for efficient signal amplification through the capture of multiple catalytic enzymes in a highly ordered way. The detection limit of this platform for miRNA is 10 aM. These innovative

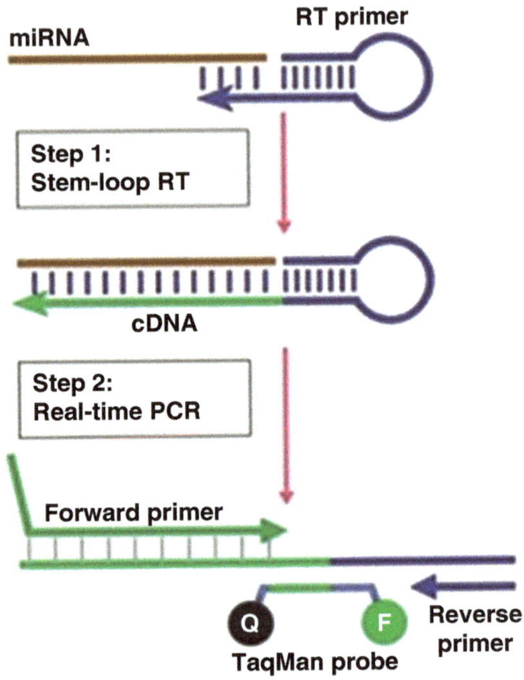

Fig. 18.2 Principle and process of stem-loop PCR (Reproduced from [141] with permission of Oxford)

methods vividly represent the prevailing trend of development in miRNA detection, namely, the pursuit of higher sensitivity, higher speed, and lower cost.

Duan and colleagues developed a hybridization-triggered fluorescence strategy for label-free, microarray-based high-throughput miRNA expression profiling in 2011[147]. The design is based on "base stacking hybridization" phenomenon that has been used to describe the additional stability associated with DNA hybridization reactions wherein two or more DNA oligonucleotides hybridize in a contiguous tandem to a longer complementary DNA single strand. Furthermore, they demonstrated that the proposed microarray assay is well suited for unbiased profiling of both normal and methylated small RNA species in 2012 [148]. Notably, as compared to many delicate techniques, the enzyme-free and label-free approach requires much less reagent and manipulation, which can produce more efficient workflow and highly reproducible results.

18.5 Prospects

Benefiting from rapid developments in molecular biology, technologies for the detection of gene mutations, DNA methylation, and miRNA are becoming increasingly sophisticated and specific. Single-cell sequencing and big data analysis are among the most remarkable prospects for these technologies. Single-cell sequencing has been used in the detection of gene mutations and DNA methylation. Yao et al. recently utilized single-cell sequencing to identify mutations in mitochondrial DNA in different human blood cells and confirmed the superiority of this approach in DNA mutation detection from extremely small amounts of DNA extracted from limited amounts of tissue and/or degraded samples [149]. Methods for the detection of DNA methylation at the single-cell level are also seeing rapid progress. Guo et al. developed a single-cell reduced representation bisulfite sequencing technique. This method provides digitized methylation information regarding ~1 million CpG sites within an individual diploid mouse or human cell at single-base resolution [150]. The same group applied this method to detect the methylation status within the genome of an individual mouse embryonic stem cell and found that the demethylation dynamics of maternal and paternal genomes after fertilization can be traced within the individual pronuclei of mouse zygotes [151].

Big data analysis (high-throughput technology), which is also known as omics, has been widely applied and is becoming a popular trend [152]. Regarding miRNA omics, studies of the genotoxicity of several mycotoxins based on miRNA profiling have been published [138, 153]. In addition to the sequencing approach used in these studies, another high-throughput miRNA detection method is based on miRNA microarrays [154]. In addition, a combination of more than one method may be necessary to achieve reliable detection of the miRNAs of interest.

Other aspects of genotoxicity detection, such as DNA adduct, base pair mismatches, and the regulation of other noncoding RNAs, have not been addressed in this chapter. However, the PCR-based detection techniques and development trends related to these topics are similar to those for the detection of gene mutation, DNA methylation, and miRNA.

Acknowledgments This work is supported by the Ministry of Science and Technology of Beijing (XX2014B069). Many thanks to Boyang Zhang, for his kindly help in manuscript conception and preparation.

References

1. Goto T, Wicklow DT, Ito Y. Aflatoxin and cyclopiazonic acid production by a Sclerotium-producing Aspergillus tamarii strain. Appl Environ Microb. 1996;62(11):4036–8.
2. Raiola A, Tenore GC, Manyes L, Meca G, Ritieni A. Risk analysis of main mycotoxins occurring in food for children: an overview. Food Chem Toxicol: Int J Publ Br Ind Biol Res Assoc. 2015;84:169–80. doi:10.1016/j.fct.2015.08.023.
3. Khan A, Khan S, Khan MA, Qamar Z, Waqas M. The uptake and bioaccumulation of heavy metals by food plants, their effects on plants nutrients, and associated health risk: a review. Environ Sci Pollut Res Int. 2015;22(18):13772–99. doi:10.1007/s11356-015-4881-0.
4. Manjanatha MG, Guo LW, Shelton SD, Doerge DR. Acrylamide-induced carcinogenicity in mouse lung involves mutagenicity: cII gene mutations in the lung of big blue mice exposed to acrylamide and glycidamide for up to 4 weeks. Environ Mol Mutagen. 2015;56(5):446–56. doi:10.1002/em.21939.

5. Li X, Gao J, Huang K, Qi X, Dai Q, Mei X, Xu W. Dynamic changes of global DNA methylation and hypermethylation of cell adhesion-related genes in rat kidneys in response to ochratoxin A. World Mycotoxin J. 2015;8(4):465–76. doi:10.3920/Wmj2014.1795.
6. McCarthy A. Third generation DNA sequencing: pacific biosciences' single molecule real time technology. Chem Biol. 2010;17(7):675–6. doi:10.1016/j.chembiol.2010.07.004.
7. Graham CA, Hill AJ. Introduction to DNA sequencing. Methods Mol Biol. 2001;167:1–12. doi:10.1385/1-59259-113-2:001.
8. Shendure J, Ji H. Next-generation DNA sequencing. Nat Biotechnol. 2008;26(10):1135–45. doi:10.1038/nbt1486.
9. Shendure J, Porreca GJ, Reppas NB, Lin X, McCutcheon JP, Rosenbaum AM, Wang MD, Zhang K, Mitra RD, Church GM. Accurate multiplex polony sequencing of an evolved bacterial genome. Science. 2005;309(5741):1728–32. doi:10.1126/science.1117389.
10. Margulies M, Egholm M, Altman WE, Attiya S, Bader JS, Bemben LA, Berka J, Braverman MS, Chen YJ, Chen Z, Dewell SB, Du L, Fierro JM, Gomes XV, Godwin BC, He W, Helgesen S, Ho CH, Irzyk GP, Jando SC, Alenquer ML, Jarvie TP, Jirage KB, Kim JB, Knight JR, Lanza JR, Leamon JH, Lefkowitz SM, Lei M, Li J, Lohman KL, Lu H, Makhijani VB, McDade KE, McKenna MP, Myers EW, Nickerson E, Nobile JR, Plant R, Puc BP, Ronan MT, Roth GT, Sarkis GJ, Simons JF, Simpson JW, Srinivasan M, Tartaro KR, Tomasz A, Vogt KA, Volkmer GA, Wang SH, Wang Y, Weiner MP, Yu P, Begley RF, Rothberg JM. Genome sequencing in microfabricated high-density picolitre reactors. Nature. 2005;437(7057):376–80. doi:10.1038/nature03959.
11. Eid J, Fehr A, Gray J, Luong K, Lyle J, Otto G, Peluso P, Rank D, Baybayan P, Bettman B, Bibillo A, Bjornson K, Chaudhuri B, Christians F, Cicero R, Clark S, Dalal R, Dewinter A, Dixon J, Foquet M, Gaertner A, Hardenbol P, Heiner C, Hester K, Holden D, Kearns G, Kong XX, Kuse R, Lacroix Y, Lin S, Lundquist P, Ma CC, Marks P, Maxham M, Murphy D, Park I, Pham T, Phillips M, Roy J, Sebra R, Shen G, Sorenson J, Tomaney A, Travers K, Trulson M, Vieceli J, Wegener J, Wu D, Yang A, Zaccarin D, Zhao P, Zhong F, Korlach J, Turner S. Real-time DNA sequencing from single polymerase molecules. Science. 2009;323 (5910):133–8. doi:10.1126/science.1162986.
12. Orita M, Suzuki Y, Sekiya T, Hayashi K. Rapid and sensitive detection of point mutations and DNA polymorphisms using the polymerase chain-reaction. Genomics. 1989;5(4):874–9. doi:10.1016/0888-7543(89)90129-8.
13. Bosari S, Marchetti A, Buttitta F, Graziani D, Borsani G, Loda M, Bevilacqua G, Coggi G. Detection of p53 mutations by single-strand conformation polymorphisms (SSCP) gel electrophoresis. A comparative study of radioactive and nonradioactive silver-stained SSCP analysis. Diagn Mol Pathol: Am J Surg Pathol B. 1995;4(4):249–55.
14. Kurvinen K, Hietanen S, Syrjanen K, Syrjanen S. Rapid and effective detection of mutations in the P53 gene using nonradioactive Single-Strand Conformation Polymorphism (Sscp) Technique Applied on Phastsystem(Tm). J Virol Methods. 1995;51(1):43–53. doi:10.1016/0166-0934(94)00099-3.
15. Han M, Robinson MA. PCR-SSCP analysis of polymorphism: a simple and sensitive method for detecting differences between short segments of DNA. Methods Mol Biol. 2003;226:327–34. doi:10.1385/1-59259-384-4:327.
16. Geng RQ, Wang LP. The PCR-SSCP and DNA sequencing methods detecting genetic mutations of EGLN1 gene in different sheep breeds. Indian J Anim Res. 2015;49(1):44–7. doi:10.5958/0976-0555.2015.00009.6.
17. Nakashima H, Akahoshi M, Tanaka Y. Mutation detection using RT-PCR-RFLP. Methods Mol Biol. 2003;226:319–22. doi:10.1385/1-59259-384-4:319.
18. Sugio K, Kishimoto Y, Virmani AK, Hung JY, Gazdar AF. K-Ras mutations are a relatively late event in the pathogenesis of lung carcinomas. Cancer Res. 1994;54(22):5811–5.
19. Mills NE, Fishman CL, Scholes J, Anderson SE, Rom WN, Jacobson DR. Detection of K-Ras oncogene mutations in bronchoalveolar lavage fluid for lung-cancer diagnosis. J Natl Cancer I. 1995;87(14):1056–60. doi:10.1093/jnci/87.14.1056.

20. Schuermann M. Detection of K-ras and p53 mutations by "mutant-enriched" PCR-RFLP. Methods Mol Med. 2003;75:325–33.
21. Ouyang L, Ge C, Wu H, Li S, Zhang H. PCR-RFLP to detect codon 248 mutation in exon 7 of p53 tumor suppressor gene. Biochem Mol Biol Educ: Bimonthly Publ Int Union Biochem Mol Biol. 2009;37(2):106–9. doi:10.1002/bmb.20255.
22. Anitha S, Raghunadharao D, Waliyar F, Sudini H, Parveen M, Rao R, Kumar PL. The association between exposure to aflatoxin, mutation in TP53, infection with hepatitis B virus, and occurrence of liver disease in a selected population in Hyderabad, India. Mutat Res-Gen Toxicol Environ. 2014;766:23–8. doi:10.1016/j.mrgentox.2013.12.011.
23. Hingorani AD, Brown MJ. A simple molecular assay for the C-1166 variant of the angiotensin-II type-1 receptor gene. Biochem Biophys Res Co. 1995;213(2):725–9. doi:10.1006/bbrc.1995.2190.
24. Shibata S, Asano Y, Yokoyama T, Shimoda K, Nakashima H, Okamura S, Niho Y. Analysis of the granulocyte colony-stimulating factor receptor gene structure using PCR-SSCP in myeloid leukemia and myelodysplastic syndrome. Eur J Haematol. 1998;60(3):197–201.
25. Iwata I, Nagafuchi S, Nakashima H, Kondo S, Koga T, Yokogawa Y, Akashi T, Shibuya T, Umeno Y, Okeda T, Shibata S, Kono S, Yasunami M, Ohkubo H, Niho Y. Association of polymorphism in the NeuroD/BETA2 gene with type 1 diabetes in the Japanese. Diabetes. 1999;48(2):416–9. doi:10.2337/diabetes.48.2.416.
26. Butler JM, Wilson MR, Reeder DJ. Rapid mitochondrial DNA typing using restriction enzyme digestion of polymerase chain reaction amplicons followed by capillary electrophoresis separation with laser-induced fluorescence detection. Electrophoresis. 1998;19 (1):119–24. doi:10.1002/elps.1150190120.
27. Butler JM, Reeder DJ. Detection of DNA polymorphisms using PCR-RFLP and capillary electrophoresis. In: Capillary electrophoresis of nucleic acids. New York: Humana Press; 2001. p. 49–56.
28. Mitchelson KR. Overview: the application of capillary electrophoresis for DNA polymorphism analysis. In: Capillary electrophoresis of nucleic acids. New York: Humana Press; 2001. p. 3–26.
29. Green EK. Allele-specific oligonucleotide PCR. In: PCR mutation detection protocols. New York: Humana Press; 2002. p. 47–50.
30. Saiki RK, Scharf S, Faloona F, Mullis KB, Horn GT, Erlich HA, Arnheim N. Enzymatic amplification of beta-globin genomic sequences and restriction site analysis for diagnosis of sickle-cell anemia. Science. 1985;230(4732):1350–4. doi:10.1126/science.2999980.
31. Ehlen T, Dubeau L. Detection of Ras point mutations by polymerase chain-reaction using mutation-specific, inosine-containing oligonucleotide primers. Biochem Biophys Res Co. 1989;160(2):441–7. doi:10.1016/0006-291x(89)92452-2.
32. Green EK, Bain SC, Day PJR, Barnett AH, Charleson F, Jones AF, Walker MR. Detection of human apolipoprotein-E3, apolipoprotein-E2, and apolipoprotein-E4 genotypes by an allele-specific oligonucleotide-primed polymerase chain-reaction assay – development and validation. Clin Chem. 1991;37(7):1263–8.
33. Liu Z, Kullman SW, Bencic DC, Torten M, Hinton DE. ras oncogene mutations in diethylnitrosamine-induced hepatic tumors in medaka (Oryzias latipes), a teleost fish. Mutat Res. 2003;539(1–2):43–53.
34. Ashraf S, Noguera NI, Di Giandomenico J, Zaza S, Hasan SK, Lo-Coco F. Rapid detection of IDH2 (R140Q and R172K) mutations in acute myeloid leukemia. Ann Hematol. 2013;92 (10):1319–23. doi:10.1007/s00277-013-1868-0.
35. Jara-Acevedo M, Teodosio C, Sanchez-Munoz L, Alvarez-Twose I, Mayado A, Caldas C, Matito A, Morgado JM, Munoz-Gonzalez JI, Escribano L, Garcia-Montero AC, Orfao A. Detection of the KIT D816V mutation in peripheral blood of systemic mastocytosis: diagnostic implications. Mod Pathol. 2015;28(8):1138–49. doi:10.1038/modpathol.2015.72.
36. Patolsky F, Lichtenstein A, Willner I. Detection of single-base DNA mutations by enzyme-amplified electronic transduction. Nat Biotechnol. 2001;19(3):253–7. doi:10.1038/85704.

37. Wang D, Chen G, Wang H, Tang W, Pan W, Li N, Liu F. A reusable quartz crystal microbalance biosensor for highly specific detection of single-base DNA mutation. Biosens Bioelectron. 2013;48:276–80. doi:10.1016/j.bios.2013.04.035.
38. Staiger AM, Ott MM, Parmentier S, Rosenwald A, Ott G, Horn H, Griese EU. Allele-specific PCR is a powerful tool for the detection of the MYD88 L265P mutation in diffuse large B cell lymphoma and decalcified bone marrow samples. Br J Haematol. 2015;171(1):145–8. doi:10.1111/bjh.13369.
39. Wang T, Liu JH, Zhang J, Wang L, Chen C, Dai PG. A multiplex allele-specific real-time PCR assay for screening of ESR1 mutations in metastatic breast cancer. Exp Mol Pathol. 2015;98(2):152–7. doi:10.1016/j.yexmp2015.03.004.
40. Sheridan E, Hancock BW, Goyns MH. High-incidence of mutations of the P53 gene detected in ovarian-tumors by the use of chemical mismatch cleavage. Cancer Lett. 1993;68(1):83–9. doi:10.1016/0304-3835(93)90223-V.
41. Belyakova AA, Yakubovskaya MG, Gasanova VK, Belitsky GA, Dolinnaya NG. Specificity and sensitivity of point mutation detection in tumor suppressor genes via chemical cleavage of mismatches. EJC Suppl. 2007;5(4):84–5. doi:10.1016/S1359-6349(07)70421-4.
42. Glavac D, Dean M. Applications of heteroduplex analysis for mutation detection in disease genes. Hum Mutat. 1995;6(4):281–7. doi:10.1002/humu.1380060402.
43. Velasco E, Infante M, Duran M, Esteban-Cardenosa E, Lastra E, Garcia-Giron C, Miner C. Rapid mutation detection in complex genes by heteroduplex analysis with capillary array electrophoresis. Electrophoresis. 2005;26(13):2539–52. doi:10.1002/elps.200410425.
44. Macek M, Mercier B, Mackova A, Miller PW, Hamosh A, Ferec C, Cutting GR. Sensitivity of the denaturing gradient gel electrophoresis technique in detection of known mutations and novel Asian mutations in CFTR gene. Hum Mutat. 1997;9(2):136–47. doi:10.1002/(Sici)1098-1004(1997)9:2<136::Aid-Humu6>3.0.Co;2-7.
45. Schwarzova L, Stekrova J, Florianova M, Novotny A, Schneiderova M, Lnenicka P, Kebrdlova V, Kotlas J, Vesela K, Kohoutova M. Novel mutations of the APC gene and genetic consequences of splicing mutations in the Czech FAP families. Fam Cancer. 2013;12(1):35–42. doi:10.1007/s10689-012-9569-8.
46. Ravnik-Glavac M, Atkinson A, Glavac D, Dean M. DHPLC screening of cystic fibrosis gene mutations. Hum Mutat. 2002;19(4):374–83. doi:10.1002/humu.10065.
47. Kantorova B, Malcikova J, Smardova J, Pavlova S, Trbusek M, Tom N, Plevova K, Tichy B, Truong S, Diviskova E, Kotaskova J, Oltova A, Patten N, Brychtova Y, Doubek M, Mayer J, Pospisilova S. TP53 mutation analysis in chronic lymphocytic leukemia: comparison of different detection methods. Tumor Biol. 2015;36(5):3371–80. doi:10.1007/s13277-014-2971-0.
48. Brookes AJ. The essence of SNPs. Gene. 1999;234(2):177–86. doi:10.1016/S0378-1119(99)00219-X.
49. Wang DG, Fan JB, Siao CJ, Berno A, Young P, Sapolsky R, Ghandour G, Perkins N, Winchester E, Spencer J, Kruglyak L, Stein L, Hsie L, Topaloglou T, Hubbell E, Robinson E, Mittmann M, Morris MS, Shen NP, Kilburn D, Rioux J, Nusbaum C, Rozen S, Hudson TJ, Lipshutz R, Chee M, Lander ES. Large-scale identification, mapping, and genotyping of single-nucleotide polymorphisms in the human genome. Science. 1998;280(5366):1077–82. doi:10.1126/science.280.5366.1077.
50. Sachidanandam R, Weissman D, Schmidt SC, Kakol JM, Stein LD, Marth G, Sherry S, Mullikin JC, Mortimore BJ, Willey DL, Hunt SE, Cole CG, Coggill PC, Rice CM, Ning ZM, Rogers J, Bentley DR, Kwok PY, Mardis ER, Yeh RT, Schultz B, Cook L, Davenport R, Dante M, Fulton L, Hillier L, Waterston RH, McPherson JD, Gilman B, Schaffner S, Van Etten WJ, Reich D, Higgins J, Daly MJ, Blumenstiel B, Baldwin J, Stange-Thomann NS, Zody MC, Linton L, Lander ES, Altshuler D, Grp ISMW. A map of human genome sequence variation containing 1.42 million single nucleotide polymorphisms. Nature. 2001;409(6822):928–33. doi:10.1038/35057149.

51. Fukuen S, Fukuda T, Maune H, Ikenaga Y, Yamamoto I, Inaba T, Azuma J. Novel detection assay by PCR-RFLP and frequency of the CYP3A5 SNPs, CYP3A5*3 and *6, in a Japanese population. Pharmacogenetics. 2002;12(4):331–4.
52. Chang HW, Cheng YH, Chuang LY, Yang CH. SNP-RFLPing 2: an updated and integrated PCR-RFLP tool for SNP genotyping. BMC Bioinforma. 2010;11:173. doi:10.1186/1471-2105-11-173.
53. Martins-Lopes P, Zhang H, Koebner R. Detection of single nucleotide mutations in wheat using single strand conformation polymorphism gels. Plant Mol Biol Rep. 2001;19 (2):159–62. doi:10.1007/Bf02772158.
54. Sasaki T, Tahira T, Suzuki A, Higasa K, Kukita Y, Baba S, Hayashi K. Precise estimation of allele frequencies of single-nucleotide polymorphisms by a quantitative SSCP analysis of pooled DNA. Am J Hum Genet. 2001;68(1):214–8. doi:10.1086/316928.
55. Kim S, Misra A. SNP genotyping: technologies and biomedical applications. Annu Rev Biomed Eng. 2007;9:289–320. doi:10.1146/annurev.bioeng.9.060906.152037.
56. Lustosa Souza CR, Azevedo Shimmoto MM, Vicari P, Mecabo G, Arruda MM, Figueiredo MS. Klotho gene polymorphisms and their association with sickle cell disease phenotypes. Rev Bras Hematol Hemoter. 2015;37(4):275–6. doi:10.1016/j.bjhh.2015.02.009.
57. Petersen DC, Laten A, Zeier MD, Grimwood A, van Rensburg EJ, Hayes VM. Novel mutations and SNPs identified in CCR2 using a new comprehensive denaturing gradient gel electrophoresis assay. Hum Mutat. 2002;20(4):253–9. doi:10.1002/humu.10111.
58. Chen XW, Yue LJ, Li CG, Li CR, Zhang M, Shi HS. Polymorphisms of thymidylate synthase gene detected by RT-PCR-denaturing gradient gel electrophoresis in children with acute leukemia. Zhongguo Dang Dai Er Ke Za Zhi Chin J Contemp Pediatr. 2009;11(4):251–4.
59. Ding C, Jin S. High-throughput methods for SNP genotyping. In: Single nucleotide polymorphisms. New York: Humana Press; 2009. p. 245–54.
60. Birdsell DN, Vogler AJ, Buchhagen J, Clare A, Kaufman E, Naumann A, Driebe E, Wagner DM, Keim PS. TaqMan real-time PCR assays for single-nucleotide polymorphisms which identify Francisella tularensis and its subspecies and subpopulations. Plos One. 2014;9(9). doi:ARTN e107964 10.1371/journal.pone.0107964.
61. Martinez-Cruz B, Ziegle J, Sanz P, Sotelo G, Anglada R, Plaza S, Comas D, Genographic C. Multiplex single-nucleotide polymorphism typing of the human Y chromosome using TaqMan probes. Investigative Genet. 2011;2:13. doi:10.1186/2041-2223-2-13.
62. Asari M, Omura T, Maseda C, Shiono H, Tasaki Y, Matsubara K, Shimizu K. Enhanced discrimination of single nucleotide polymorphisms using 3′ nucleotide differences in ligase detection reaction probes. Mol Cell Probes. 2010;24(6):381–6. doi:10.1016/j.mcp.2010.08.005.
63. Chen Y, Yang ML, Xiang Y, Yuan R, Chai YQ. Ligase chain reaction amplification for sensitive electrochemiluminescent detection of single nucleotide polymorphisms. Anal Chim Acta. 2013;796:1–6. doi:10.1016/j.aca.2013.07.057.
64. Alderborn A, Kristofferson A, Hammerling U. Determination of single-nucleotide polymorphisms by real-time pyrophosphate DNA sequencing. Genome Res. 2000;10(8):1249–58. doi:10.1101/Gr.10.8.1249.
65. Lavebratt C, Sengul S. Single nucleotide polymorphism (SNP) allele frequency estimation in DNA pools using Pyrosequencing (TM). Nat Protoc. 2006;1(6):2573–82. doi:10.1038/nprot.2006.442.
66. Underwood JL, Stamper RL, Kukreja RC, Samson V. dHPLC analysis of single nucleotide polymorphism in the MYOC/TIGR gene. Mol Biol Cell. 2001;12:125a–6.
67. Hoogendoorn B, Norton N, Kirov G, Williams N, Hamshere ML, Spurlock G, Austin J, Stephens MK, Buckland PR, Owen MJ, O'Donovan MC. Cheap, accurate and rapid allele frequency estimation of single nucleotide polymorphisms by primer extension and DHPLC in DNA pools. Hum Genet. 2000;107(5):488–93. doi:10.1007/s004390000397.

68. Duan XR, Liu LB, Feng FD, Wang S. Cationic conjugated polymers for optical detection of DNA methylation, lesions, and single nucleotide polymorphisms. Acc Chem Res. 2010;43 (2):260–70. doi:10.1021/ar9001813.
69. Syvanen AC. Toward genome-wide SNP genotyping. Nat Genet. 2005;37:S5–10. doi:10.1038/ng1558.
70. Deng HY, Zhang XE, Mang Y, Zhang ZP, Zhou YF, Liu Q, Lu HB, Fu ZJ. Oligonucleotide ligation assay-based DNA chip for multiplex detection of single nucleotide polymorphism. Biosens Bioelectron. 2004;19(10):1277–83. doi:10.1016/j.bios.2003.11.022.
71. Rush LJ, Plass C. Restriction landmark genomic scanning for DNA methylation in cancer: past, present, and future applications. Anal Biochem. 2002;307(2):191–201.
72. Reik W, Collick A, Norris ML, Barton SC, Surani MA. Genomic imprinting determines methylation of parental alleles in transgenic mice. Nature. 1987;328(6127):248–51. doi:10.1038/328248a0.
73. Onyango P, Jiang S, Uejima H, Shamblott MJ, Gearhart JD, Cui H, Feinberg AP. Monoallelic expression and methylation of imprinted genes in human and mouse embryonic germ cell lineages. Proc Natl Acad Sci U S A. 2002;99(16):10599–604. doi:10.1073/pnas.152327599.
74. Tost J, Dunker J, Gut IG. Analysis and quantification of multiple methylation variable positions in CpG islands by Pyrosequencing. Biotechniques. 2003;35(1):152–6.
75. Bestor TH. The DNA methyltransferases of mammals. Hum Mol Genet. 2000;9 (16):2395–402. doi:10.1093/hmg/9.16.2395.
76. Dong Y, Zhao H, Li H, Li X, Yang S. DNA methylation as an early diagnostic marker of cancer (review). Biomed Rep. 2014. doi:10.3892/br.2014.237.
77. Tucker KL. Methylated cytosine and the brain: a new base for neuroscience. Neuron. 2001;30 (3):649–52.
78. Gardinergarden M, Frommer M. Cpg islands in vertebrate genomes. J Mol Biol. 1987;196 (2):261–82. doi:10.1016/0022-2836(87)90689-9.
79. Larsen F, Gundersen G, Lopez R, Prydz H. CpG islands as gene markers in the human genome. Genomics. 1992;13(4):1095–107.
80. Yin H, Blanchard KL. DNA methylation represses the expression of the human erythropoietin gene by two different mechanisms. Blood. 2000;95(1):111–9.
81. Issa JP, Ottaviano YL, Celano P, Hamilton SR, Davidson NE, Baylin SB. Methylation of the oestrogen receptor CpG island links ageing and neoplasia in human colon. Nat Genet. 1994;7 (4):536–40. doi:10.1038/ng0894-536.
82. Ahuja N, Li Q, Mohan AL, Baylin SB, Issa JP. Aging and DNA methylation in colorectal mucosa and cancer. Cancer Res. 1998;58(23):5489–94.
83. Toyota M, Issa JP. CpG island methylator phenotypes in aging and cancer. Semin Cancer Biol. 1999;9(5):349–57. doi:10.1006/scbi.1999.0135.
84. Di Ruscio A, Ebralidze AK, Benoukraf T, Amabile G, Goff LA, Terragni J, Figueroa ME, De Figueiredo Pontes LL, Alberich-Jorda M, Zhang P, Wu M, D'Alo F, Melnick A, Leone G, Ebralidze KK, Pradhan S, Rinn JL, Tenen DG. DNMT1-interacting RNAs block gene-specific DNA methylation. Nature. 2013;503(7476):371–6. doi:10.1038/nature12598.
85. Xie W, Schultz MD, Lister R, Hou Z, Rajagopal N, Ray P, Whitaker JW, Tian S, Hawkins RD, Leung D, Yang H, Wang T, Lee AY, Swanson SA, Zhang J, Zhu Y, Kim A, Nery JR, Urich MA, Kuan S, Yen CA, Klugman S, Yu P, Suknuntha K, Propson NE, Chen H, Edsall LE, Wagner U, Li Y, Ye Z, Kulkarni A, Xuan Z, Chung WY, Chi NC, Antosiewicz-Bourget JE, Slukvin I, Stewart R, Zhang MQ, Wang W, Thomson JA, Ecker JR, Ren B. Epigenomic analysis of multilineage differentiation of human embryonic stem cells. Cell. 2013;153 (5):1134–48. doi:10.1016/j.cell.2013.04.022.
86. Li E, Beard C, Jaenisch R. Role for DNA methylation in genomic imprinting. Nature. 1993;366(6453):362–5. doi:10.1038/366362a0.
87. Lujambio A, Calin GA, Villanueva A, Ropero S, Sanchez-Cespedes M, Blanco D, Montuenga LM, Rossi S, Nicoloso MS, Faller WJ, Gallagher WM, Eccles SA, Croce CM,

Esteller M. A microRNA DNA methylation signature for human cancer metastasis. Proc Natl Acad Sci U S A. 2008;105(36):13556–61. doi:10.1073/pnas.0803055105.
88. Mastroeni D, Grover A, Delvaux E, Whiteside C, Coleman PD, Rogers J. Epigenetic changes in Alzheimer's disease: decrements in DNA methylation. Neurobiol Aging. 2010;31 (12):2025–37. doi:10.1016/j.neurobiolaging.2008.12.005.
89. Mohandas T, Sparkes RS, Shapiro LJ. Reactivation of an inactive human X chromosome: evidence for X inactivation by DNA methylation. Science. 1981;211(4480):393–6.
90. Udali S, Guarini P, Moruzzi S, Choi SW, Friso S. Cardiovascular epigenetics: from DNA methylation to microRNAs. Mol Aspects Med. 2013;34(4):883–901. doi:10.1016/j.mam.2012.08.001.
91. Wockner LF, Noble EP, Lawford BR, Young RM, Morris CP, Whitehall VL, Voisey J. -Genome-wide DNA methylation analysis of human brain tissue from schizophrenia patients. Transl Psychiatry. 2014;4:e339. doi:10.1038/tp.2013.111.
92. Baylin SB. DNA methylation and gene silencing in cancer. Nat Clin Pract Oncol. 2005;2 Suppl 1:S4–11. doi:10.1038/ncponc0354.
93. Son JW, Jeong KJ, Jean WS, Park SY, Jheon S, Cho HM, Park CG, Lee HY, Kang J. Genome-wide combination profiling of DNA copy number and methylation for deciphering biomarkers in non-small cell lung cancer patients. Cancer Lett. 2011;311 (1):29–37. doi:10.1016/j.canlet.2011.06.021.
94. Szyf M. DNA methylation signatures for breast cancer classification and prognosis. Genome Med. 2012;4(3):26. doi:10.1186/gm325.
95. Hong L, Ahuja N. DNA methylation biomarkers of stool and blood for early detection of colon cancer. Genet Test Mol Biomarkers. 2013;17(5):401–6. doi:10.1089/gtmb.2012.0478.
96. Andres G, Ashour N, Sanchez-Chapado M, Ropero S, Angulo JC. The study of DNA methylation in urological cancer: present and future. Actas Urol Esp. 2013;37(6):368–75. doi:10.1016/j.acuro.2013.03.001.
97. Bondurant AE, Huang Z, Whitaker RS, Simel LR, Berchuck A, Murphy SK. Quantitative detection of RASSF1A DNA promoter methylation in tumors and serum of patients with serous epithelial ovarian cancer. Gynecol Oncol. 2011;123(3):581–7. doi:10.1016/j.ygyno.2011.08.029.
98. Renner M, Wolf T, Meyer H, Hartmann W, Penzel R, Ulrich A, Lehner B, Hovestadt V, Czwan E, Egerer G, Schmitt T, Alldinger I, Renker EK, Ehemann V, Eils R, Wardelmann E, Buttner R, Lichter P, Brors B, Schirmacher P, Mechtersheimer G. Integrative DNA methylation and gene expression analysis in high-grade soft tissue sarcomas. Genome Biol. 2013;14 (12):r137. doi:10.1186/gb-2013-14-12-r137.
99. Ammerpohl O, Haake A, Pellissery S, Giefing M, Richter J, Balint B, Kulis M, Le J, Bibikova M, Drexler HG, Seifert M, Shaknovic R, Korn B, Kuppers R, Martin-Subero JI, Siebert R. Array-based DNA methylation analysis in classical Hodgkin lymphoma reveals new insights into the mechanisms underlying silencing of B cell-specific genes. Leukemia. 2012;26(1):185–8. doi:10.1038/leu.2011.194.
100. Schoofs T, Berdel WE, Muller-Tidow C. Origins of aberrant DNA methylation in acute myeloid leukemia. Leukemia. 2014;28(1):1–14. doi:10.1038/leu.2013.242.
101. Zardo G, Tiirikainen MI, Hong C, Misra A, Feuerstein BG, Volik S, Collins CC, Lamborn KR, Bollen A, Pinkel D, Albertson DG, Costello JF. Integrated genomic and epigenomic analyses pinpoint biallelic gene inactivation in tumors. Nat Genet. 2002;32(3):453–8. doi:10.1038/ng1007.
102. Friso S, Choi SW, Dolnikowski GG, Selhub J. A method to assess genomic DNA methylation using high-performance liquid chromatography/electrospray ionization mass spectrometry. Anal Chem. 2002;74(17):4526–31.
103. Hu CW, Lee H, Chen JL, Li YJ, Chao MR. Optimization of global DNA methylation measurement by LC-MS/MS and its application in lung cancer patients. Anal Bioanal Chem. 2013;405(27):8859–69. doi:10.1007/s00216-013-7305-3.

104. Zhang Z, Chen CQ, Manev H. Enzymatic regional methylation assay for determination of CpG methylation density. Anal Chem. 2004;76(22):6829–32. doi:10.1021/ac049114+.
105. Frigola J, Ribas M, Risques RA, Peinado MA. Methylome profiling of cancer cells by amplification of inter-methylated sites (AIMS). Nucleic Acids Res. 2002;30(7):e28.
106. Huang TH, Laux DE, Hamlin BC, Tran P, Tran H, Lubahn DB. Identification of DNA methylation markers for human breast carcinomas using the methylation-sensitive restriction fingerprinting technique. Cancer Res. 1997;57(6):1030–4.
107. Porter JR, Stains CI, Segal DJ, Ghosh I. Split beta-lactamase sensor for the sequence-specific detection of DNA methylation. Anal Chem. 2007;79(17):6702–8. doi:10.1021/ac071163+.
108. Nygren AO, Ameziane N, Duarte HM, Vijzelaar RN, Waisfisz Q, Hess CJ, Schouten JP, Errami A. Methylation-specific MLPA (MS-MLPA): simultaneous detection of CpG methylation and copy number changes of up to 40 sequences. Nucleic Acids Res. 2005;33(14): e128. doi:10.1093/nar/gni127.
109. Dobrovic A, Bianco T, Tan LW, Sanders T, Hussey D. Screening for and analysis of methylation differences using methylation-sensitive single-strand conformation analysis. Methods. 2002;27(2):134–8.
110. Kuo KC, McCune RA, Gehrke CW, Midgett R, Ehrlich M. Quantitative reversed-phase high performance liquid chromatographic determination of major and modified deoxyribonucleosides in DNA. Nucleic Acids Res. 1980;8(20):4763–76.
111. Deng D, Deng G, Lu Y. Analysis of the methylation in CpG island by denaturing high-performance liquid chromatography. Zhonghua Yi Xue Za Zhi. 2001;81(3):158–61.
112. Huang DH, Yang Q, Jin SX, Deng QC, Zhou P. Determination of global DNA methylation level by capillary electrophoresis using octyl-modified quaternized cellulose as an electrolyte additive. Anal Bioanal Chem. 2014;406(12):2771–7. doi:10.1007/s00216-014-7698-7.
113. Zhang L, Zhang L, Zhou K, Ye X, Zhang J, Xie A, Chen L, Kang JX, Cai C. Simultaneous determination of global DNA methylation and hydroxymethylation levels by hydrophilic interaction liquid chromatography-tandem mass spectrometry. J Biomol Screen. 2012;17 (7):877–84. doi:10.1177/1087057112447946.
114. Ramanathan A, Numata K, Wu ST, Li SX, Dawsey MW, Mao JP, Kawa SR, Riris H. Airborne measurement of atmospheric methane concentration using pulsed lidar. Proc Spie. 2012;8511. doi:Artn 85110e 10.1117/12.929990.
115. Zhang L, Xu Y-Z, Xiao X-F, Chen J, Zhou X-Q, Zhu W-Y, Dai Z, Zou X-Y. Development of techniques for DNA-methylation analysis. TrAC Trends Anal Chem. 2015;72:114–22. doi:10.1016/j.trac.2015.03.025.
116. Heyn H, Esteller M. DNA methylation profiling in the clinic: applications and challenges. Nat Rev Genet. 2012;13(10):679–92. doi:10.1038/nrg3270.
117. Schillebeeckx M, Schrade A, Lobs AK, Pihlajoki M, Wilson DB, Mitra RD. Laser capture microdissection-reduced representation bisulfite sequencing (LCM-RRBS) maps changes in DNA methylation associated with gonadectomy-induced adrenocortical neoplasia in the mouse. Nucleic Acids Res. 2013;41(11):e116. doi:10.1093/nar/gkt230.
118. Frommer M, McDonald LE, Millar DS, Collis CM, Watt F, Grigg GW, Molloy PL, Paul CL. A genomic sequencing protocol that yields a positive display of 5-methylcytosine residues in individual DNA strands. Proc Natl Acad Sci U S A. 1992;89(5):1827–31.
119. Subiyantoro P. Methylation detection of oral cancer using bisulfite sequencing. Int J Oral Maxillofac Surg. 2015;44:e291–2. doi:10.1016/j.ijom.2015.08.331.
120. Herman JG, Graff JR, Myohanen S, Nelkin BD, Baylin SB. Methylation-specific PCR: a novel PCR assay for methylation status of CpG islands. Proc Natl Acad Sci U S A. 1996;93 (18):9821–6.
121. Hibi K, Goto T, Shirahata A, Saito M, Kigawa G, Nemoto H, Sanada Y. Detection of TFPI2 methylation in the serum of colorectal cancer patients. Cancer Lett. 2011;311(1):96–100. doi:10.1016/j.canlet.2011.07.006.

122. Gonzalgo ML, Jones PA. Rapid quantitation of methylation differences at specific sites using methylation-sensitive single nucleotide primer extension (Ms-SNuPE). Nucleic Acids Res. 1997;25(12):2529–31.
123. Kristensen LS, Mikeska T, Krypuy M, Dobrovic A. Sensitive Melting Analysis after Real Time- Methylation Specific PCR (SMART-MSP): high-throughput and probe-free quantitative DNA methylation detection. Nucleic Acids Res. 2008;36(7):e42. doi:10.1093/nar/gkn113.
124. Pannone G, Santoro A, Cagiano S, Loreto C, Mattoni M, Lo Muzio L, Papagerakis S, Papagerakis P, Rubini C, Bufo P. The role of epigenetic alteration in oral cancer and in oral mucosa under alcohol and cigarette smoking. A study of a panel of eleven genes by methylation specific PCR (MSP). Oral Oncol. 2013;49:S8. doi:10.1016/j.oraloncology.2013.03.018.
125. Dumache R, Puiu M, Negru S, Minciu R, Bucur A, Bumbacila B. Molecular diagnosis of bladder cancer from body fluids by methylation-specific PCR. Toxicol Lett. 2012;211:S53. doi:10.1016/j.toxlet.2012.03.212.
126. Costello JF, Smiraglia DJ, Plass C. Restriction landmark genome scanning. Methods. 2002;27(2):144–9. doi:Pii S1046-2023(02)00067-1 10.1016/S1046-2023(02)00067-1.
127. Li J, Yan H, Wang K, Tan W, Zhou X. Hairpin fluorescence DNA probe for real-time monitoring of DNA methylation. Anal Chem. 2007;79(3):1050–6. doi:10.1021/ac061694i.
128. Hiraoka D, Yoshida W, Abe K, Wakeda H, Hata K, Ikebukuro K. Development of a method to measure DNA methylation levels by using methyl CpG-binding protein and luciferase-fused zinc finger protein. Anal Chem. 2012;84(19):8259–64. doi:10.1021/ac3015774.
129. Yegnasubramanian S, Lin X, Haffner MC, DeMarzo AM, Nelson WG. Combination of methylated-DNA precipitation and methylation-sensitive restriction enzymes (COMPARE-MS) for the rapid, sensitive and quantitative detection of DNA methylation. Nucleic Acids Res. 2006;34(3):e19. doi:10.1093/nar/gnj022.
130. Wang X, Song Y, Song M, Wang Z, Li T, Wang H. Fluorescence polarization combined capillary electrophoresis immunoassay for the sensitive detection of genomic DNA methylation. Anal Chem. 2009;81(19):7885–91. doi:10.1021/ac901681k.
131. Shiraishi M, Sekiguchi A, Oates AJ, Terry MJ, Miyamoto Y, Sekiya T. Methyl-CpG binding domain column chromatography as a tool for the analysis of genomic DNA methylation. Anal Biochem. 2004;329(1):1–10. doi:10.1016/j.ab.2004.02.024.
132. Hua D, Hu Y, Wu YY, Cheng ZH, Yu J, Du X, Huang ZH. Quantitative methylation analysis of multiple genes using methylation-sensitive restriction enzyme-based quantitative PCR for the detection of hepatocellular carcinoma. Exp Mol Pathol. 2011;91(1):455–60. doi:10.1016/j.yexmp.2011.05.001.
133. Zhang M, Wang Q, Yuan W, Yang S, Wang X, Yan JD, Du J, Yin J, Gao SY, Sun BC, Zhu TH. Epigenetic regulation of bone morphogenetic protein-6 gene expression in breast cancer cells. J Steroid Biochem Mol Biol. 2007;105(1–5):91–7. doi:10.1016/j.jsbmb.2007.01.002.
134. Yamada H, Tanabe K, Ito T, Nishimoto S. The pH effect on the naphthoquinone-photosensitized oxidation of 5-methylcytosine. Chemistry. 2008;14(33):10453–61. doi:10.1002/chem.200800840.
135. Bareyt S, Carell T. Selective detection of 5-methylcytosine sites in DNA. Angew Chem. 2008;47(1):181–4. doi:10.1002/anie.200702159.
136. Liu T, Zhao J, Zhang D, Li G. Novel method to detect DNA methylation using gold nanoparticles coupled with enzyme-linkage reactions. Anal Chem. 2009;82(1):229–33.
137. Bartel DP. MicroRNAs: genomics, biogenesis, mechanism, and function. Cell. 2004;116(2):281–97. doi:10.1016/S0092-8674(04)00045-5.
138. Zhu L, Gao J, Huang K, Luo Y, Zhang B, Xu W. miR-34a screened by miRNA profiling negatively regulates Wnt/beta-catenin signaling pathway in Aflatoxin B1 induced hepatotoxicity. Sci Rep. 2015;5:16732. doi:10.1038/srep16732.

139. Liu C, Yu H, Zhang Y, Li D, Xing X, Chen L, Zeng X, Xu D, Fan Q, Xiao Y, Chen W, Wang Q. Upregulation of miR-34a-5p antagonizes AFB1-induced genotoxicity in F344 rat liver. Toxicon: Off J Int Soc Toxinology. 2015;106:46–56. doi:10.1016/j.toxicon.2015.09.016.
140. Shi R, Chiang VL. Facile means for quantifying microRNA expression by real-time PCR. Biotechniques. 2005;39(4):519–25.
141. Chen C, Ridzon DA, Broomer AJ, Zhou Z, Lee DH, Nguyen JT, Barbisin M, Xu NL, Mahuvakar VR, Andersen MR, Lao KQ, Livak KJ, Guegler KJ. Real-time quantification of microRNAs by stem-loop RT-PCR. Nucleic Acids Res. 2005;33(20):e179. doi:10.1093/nar/gni178.
142. Mou G, Wang K, Xu D, Zhou G. Evaluation of three RT-qPCR-based miRNA detection methods using seven rice miRNAs. Biosci Biotechnol Biochem. 2013;77(6):1349–53. doi:10.1271/bbb.130192.
143. Chen C, Tan R, Wong L, Fekete R, Halsey J. Quantitation of microRNAs by real-time RT-qPCR. Methods Mol Biol. 2011;687:113–34. doi:10.1007/978-1-60761-944-4_8.
144. Dong H, Lei J, Ding L, Wen Y, Ju H, Zhang X. MicroRNA: function, detection, and bioanalysis. Chem Rev. 2013;113(8):6207–33. doi:10.1021/cr300362f.
145. Gao XF, Xu H, Baloda M, Gurung AS, Xu LP, Wang T, Zhang XJ, Liu GD. Visual detection of microRNA with lateral flow nucleic acid biosensor. Biosens Bioelectron. 2014;54:578–84. doi:10.1016/j.bios.2013.10.055.
146. Ge ZL, Lin MH, Wang P, Pei H, Yan J, Sho JY, Huang Q, He DN, Fan CH, Zuo XL. Hybridization chain reaction amplification of MicroRNA detection with a tetrahedral DNA nanostructure-based electrochemical biosensor. Anal Chem. 2014;86(4):2124–30. doi:10.1021/ac4037262.
147. Duan D, Zheng K-X, Shen Y, Cao R, Jiang L, Lu Z, Yan X, Li J. Label-free high-throughput microRNA expression profiling from total RNA. Nucleic Acids Res. 2011;39(22):e154.
148. Shen Y, Zheng K-X, Duan D, Jiang L, Li J. Label-free MicroRNA profiling not biased by 3′ end 2′-O-methylation. Anal Chem. 2012;84(15):6361–5.
149. Yao YG, Kajigaya S, Young NS. Mitochondrial DNA mutations in single human blood cells. Mutat Res. 2015;779:68–77. doi:10.1016/j.mrfmmm.2015.06.009.
150. Guo HS, Zhu P, Guo F, Li XL, Wu XL, Fan XY, Wen L, Tang FC. Profiling DNA methylome landscapes of mammalian cells with single-cell reduced-representation bisulfite sequencing. Nat Protoc. 2015;10(5):645–59. doi:10.1038/nprot.2015.039.
151. Guo HS, Zhu P, Wu XL, Li XL, Wen L, Tang FC. Single-cell methylome landscapes of mouse embryonic stem cells and early embryos analyzed using reduced representation bisulfite sequencing. Genome Res. 2013;23(12):2126–35. doi:10.1101/gr.161679.113.
152. Alyass A, Turcotte M, Meyre D. From big data analysis to personalized medicine for all: challenges and opportunities. BMC Med Genom. 2015;8. doi:ARTN 33 10.1186/s12920-015-0108-y.
153. Dai Q, Zhao J, Qi XZ, Xu WT, He XY, Guo MZ, Dweep H, Cheng WH, Luo YB, Xia K, Gretz N, Huang KL. MicroRNA profiling of rats with ochratoxin A nephrotoxicity. BMC Genom. 2014;15. doi:Artn 333 10.1186/1471-2164-15-333.
154. Nelson PT, Baldwin DA, Scearce LM, Oberholtzer JC, Tobias JW, Mourelatos Z. - Microarray-based, high-throughput gene expression profiling of microRNAs. Nat Methods. 2004;1(2):155–61. doi:10.1038/NMETH717.

Chapter 19
A Molecular Review of the Detection of Specific Nucleic Acids by Amplification and Hybridization Characterization of Microbial Diversity in the Food Chain: A Molecular Review

Abstract The ideal target DNA or RNA region used in nucleic acid detection in food safety should be appropriate in length, suitable for the design of primers and probes, and include minimal repeat sequences. However, in some cases, specific nucleic acid targets as short as approximately 20 nt or consisting of short tandem repeats must be determined, with microRNA and telomeres as two actual examples. In this chapter, specific nucleic acid detection methods involving amplification and hybridization will be reviewed. For the detection of miRNA, microarray chip methods, Northern blot methods, real-time PCR-based methods, in situ hybridization-based methods, and biosensor-based methods have been developed to overcome the too short length of miRNA. For the detection of telomere length, terminal restriction fragment analysis, quantitative PCR-based telomere detection, single telomere length analysis (STELA), and fluorescence in situ hybridization-based methods have been used. The advantages and disadvantages of these methods are analyzed in this review.

Keywords miRNA • Telomere • Triplex DNA • Hybridization in situ • PCR

19.1 Introduction

The ideal target DNA or RNA region for nucleic acid detection in food safety should be appropriate in length, suitable for primer and probe design, and with minimal repeat sequences. Generally, a sequence target that is too short in length results in difficult within-region primer or probe design as well as difficult signal detection. A sequence target that is too long makes it difficult to maintain integrity during the signal amplification. Highly repetitive sequences usually result in complicated secondary structures, and the primers designed for the detection of tandem repeats can easily form primer dimers.

In some cases, however, we must address specific nucleic acid targets as short as approximately 20 nt or consisting of short tandem repeats, such as microRNA and telomeres. MicroRNA (miRNA) is an important type of nucleic acid molecule, usually 19–25 nt. In animals, the expression profiles of specific marker miRNAs

W. Xu, *Functional Nucleic Acids Detection in Food Safety*,
DOI 10.1007/978-981-10-1618-9_19

can indicate the risk or progress of certain diseases. In plants, miRNAs are involved in the resistance of plantlets and the ripening of fruits. In bacteria, miRNA-like small noncoding RNAs are important virulence factors for controlling infection and toxin production in pathogens. These miRNAs usually act upstream of gene expression regulation pathways, and thus, compared with the conventional detection of mRNA, the detection of miRNA could offer more predictiveness and sensitivity. A telomere is a region of repetitive nucleotide sequences at each end of a chromosome that protects the end of the chromosome from deterioration or from fusion with neighboring chromosomes. The length of the telomere, i.e., the number of repeats, is often detected in biomedical research to indicate the degree of aging of a cell. However, the short length of miRNAs and the short tandem repeats characteristic of telomeres introduce many difficulties in their detection.

19.2 Detection of Short Single-Stranded Nucleotides: miRNA

MicroRNAs (miRNAs) are a type of small, endogenous, conserved noncoding RNA. The length is 19–25 nt. They are processed from primary miRNAs (pri-miRNAs). The RNA-pol II promoter activates their transcription from DNA [1]. Pri-miRNAs are processed into mature miRNAs by two shear reactions. In animals [1], after the first shear in the cell nucleus, stem-loop structure pre-miRNAs of approximately 70 nt are produced. The second shear occurs in the cell cytoplasm. Pre-miRNAs are cut into 19–25 nt mature miRNAs. Two kinds of RNase III, Drosha and Dicer, catalyze the shear. Drosha in the cell nucleus is coordinated with DGCR8 for the first. When the two shears are complete, the single-stranded miRNA will integrate into the RNA-induced silencing complex (RISC) and recognize the target gene. The expression of the target gene is thus regulated, while the other strand, miRNA*, may degrade. In plants, the processing of mature miRNAs is different from the processing in animals, as Drosha is not found in plants [2]. However, in the model plant, *Arabidopsis thaliana*, the Dice family is found, including DCL1, DCL2, DCL3, and DCL4. It is proposed that the role of Drosha may be replaced by Dice. In addition, the maturation of miRNAs in plants may occur in the cell nucleus.

To date, hundreds of miRNAs have been found in plants and animals, and their roles in biological processes have been researched. In recent years, more and more researchers have found that miRNAs are involved in many biological processes [3–5], such as proliferation, apoptosis, differentiation, and development, in regulatory roles. Changes in miRNAs can result in the abnormal expression of target genes by interacting with complementary sites in the 3′UTR of the target mRNA. The expression levels of miRNAs are very different in different tissues.

Moreover, in animals, many studies have been performed showing that the abnormal expression of miRNAs can induce all kinds of diseases [6–8], including leukemia, colon cancer, lung cancer, and diabetes, such as the downregulation of miR-15 and miR-16 in leukemia [9, 10] and the downregulation of miR-143 and miR-145 in colorectal cancer. MiRNAs have been considered as biomarkers for certain diseases. In plants, miRNAs can regulate flower development, leaf development, and embryonic patterning [11, 12]. In insects, miRNAs also regulate the processes of apoptosis, growth, and fat metabolism [4, 13]. Thus, the detection of miRNA is an urgent need.

MiRNA microarrays, high-throughput sequencing, Northern blotting, and real-time PCR are the most common methods for the detection of miRNAs.

19.2.1 MiRNA Microarray

19.2.1.1 Principles of miRNA Microarrays

Numerous miRNAs have been found in all kinds of species. To detect changes in massive miRNA expression, DNA microarray technology is applied. The DNA microarray is a high-efficiency and high-throughput detection technology for thousands of DNA at once and has been applied in studies of many biomolecules.

The principle of the DNA microarray is hybridization and sequencing. In other words, it can obtain the sequence of DNA by hybridization with a probe of known DNA sequence. First, a series of known biomolecules are arranged in an orderly lattice on the substrate surface, such as glass, nylon, silicon, or metal. Then, under the same conditions, these biomolecules react with their "ligand molecules." Detection is conducted using radiography technology, fluorescence labeling, or chemiluminescence. Finally, based on the computer software analysis, the result is shown. In a biological organism, the reading, transcription, and translation of genes are all completed by molecular recognition. Therefore, massive molecule sequences can be detected via complementary base pairing with a known nucleic acid probe. High-throughput analysis was performed for the unknown genes. This method can also be applied for the rapid, accurate, and high-throughput analysis of proteins, chemical compounds, cells, and other tissues. Based on this reasoning, the DNA microarray technique was further applied to miRNA analysis (miRNA microarray). Moreover, miRNA microarrays have been used extensively in miRNA expression profiling studies on specific tissues and cells [5, 14, 15].

There are two types of miRNA microarray platform [16]. One is the fabricated miRNA microarray, where the known oligonucleotide probe is set in the nylon membrane surface with an amino group at the 5′ end. In general, there are three copies of every target miRNA in the probe. In addition, experiments have proven that the overlay of multiple copies is needless, and the signal of the hybridization and complementary sequences within the targeted RNA molecules is low, which may be because the sequences far away from the substrate have more freedom and easier conjunction with the target miRNAs.

The other is the in situ-*synthesized* microRNA microarray. For detection, the oligonucleotide probe is synthesized from the 3′ end to the 5′ end and set in the substrate surface by an amino group at the 3′ end. Thus, the direction of the probe is very important.

19.2.1.2 Advantages and Disadvantages

The biggest advantage of miRNA microarrays is the high-throughput detection, which is very important for detecting miRNA changes in samples. There are two reasons. First, there are more than 1000 miRNAs in an organism, which are conservative across different species. Additionally, more than 5000 genes are regulated by miRNAs. Thus, high-throughput detection at once is necessary for detecting the overall changes in miRNAs. Second, every miRNA has many target genes, and one gene can also be regulated by multiple miRNAs. Thus, the regulation process is complex.

The detection of miRNAs by a miRNA microarray is indirect, and the specificity and sensitivity of the analysis process are low. Therefore, false-positive results will occur, and thus the results need further verification by other methods, such as the Northern blot. Moreover, to reduce the false-positive results, the RNA sample should be high purity.

19.2.2 Northern Blot

19.2.2.1 Principle of Northern Blot

The Northern blot is a method for detecting the abundance and molecular weight of miRNAs in eukaryotes. The nucleic acid hybridization in the Southern blot and Northern blot can detect DNA and RNA content, respectively. The principle is the use of agarose gel electrophoresis (AGE) to separate DNA or RNA. DNA or RNA is transferred to the nylon membrane from agarose gels, fixed, and the nylon membrane blocked by pre-hybridization. Then, the nucleic acids are hybridized with the labeled specific probes, and the nonspecific probes are washed away. Finally, based on the different labeling forms of specific probes, a method such as autoradiography or chemiluminescence is used to detect the signal. There will be an image distribution [17, 18]. However, miRNAs are only 19–25 nt, and the molecular weight is too small for easy separation by AGE. Thus, denaturing polyacrylamide gel is chosen to separate miRNAs in the Northern blot [19–21].

The probe for the Northern blot can be labeled in different ways, including nick translation, random priming, end labeling, single-stranded DNA labeling, oligonucleotide labeling, and RNA probe labeling [22]. The markers are radiolabeling or nonradioactive labeling.

Nick Translation

When one chain of the double-stranded DNA has a gap, *E. coli* DNA polymerase I can connect the nucleotide to the end with a 3′-OH. *E. coli* DNA polymerase I also has 5–3′ exonuclease activity and thus can cut the nucleotide from the incision at the 5′ end. Because it can both cut the nucleotide at the notch and connect the nucleotide at the notch to the 3′ end, the notch can transfer along the DNA strand. Thus, the original nonradioactive oligonucleotide should be replaced by a radioactive oligonucleotide as the radioisotopes are added into the new DNA chains. The best length for nick translation is 30–50 nucleotides.

Random Priming

Random priming is binding on the oligonucleotide primers to the DNA template. The DNA probe is synthesized by the Klenow enzyme. The length of the synthesis product is 400–600 nucleotides.

Random priming has the following advantages:

1. The Klenow enzyme has no 5–3′ exonuclease activity. The action is steady, and numerous effective probes can be synthesized.
2. The activity is not rigorously specific to the DNA template. Trace DNA templates can also activate.
3. The activity ratio of the reaction product is higher, at up to 4×109 cpm/μg probes.
4. The arbitrarily primed polymerase chain reaction can be directly processed in low melting point agarose.

End Labeling

End labeling is catalyzing the DNA probe to label the 3′ end using Klenow fragments.

Single-Stranded DNA Labeling

In general, when distant DNA is detected by a double-stranded DNA probe, there are many mismatches between the probe sequence and the unknown sequence. Moreover, the matched pairs of the two probes' complementary strands are stable, decreasing the detection efficiency. Therefore, single-stranded DNA labeling is studied. There are two synthetic methods for single-stranded DNA probes: by Klenow fragment using the M13 carrier derived sequence as a template or by reverse transcriptase using RNA as the template.

Oligonucleotide Labeling

An oligonucleotide labeling probe can be used to detect the point mutation of a single nucleotide in a target gene. There are two common oligonucleotide labeling probes. One is an oligonucleotide probe library consisting of a single known sequence oligonucleotide probe. The probe can match the target sequence.

The other is an oligonucleotide probe library consisting of many degenerate oligonucleotide probes. For the probe to hybridize with the target sequence, the hybridization conditions should be designed accurately.

19.2.2.2 Advantages and Disadvantages

The Northern blot can provide accurate hybridization information for miRNAs, especially using an isotope-labeled probe. This method also can detect low-content miRNAs and can verify the results of an miRNA microarray and decrease the false-positive results. The Northern blot is also the classical method for miRNA detection. However, the Northern blot cannot detect numerous miRNAs at once. Therefore, for the detection of many miRNAs, an miRNA microarray and Northern blot should be used jointly.

19.2.3 Real-Time PCR

19.2.3.1 Principles of Real-Time PCR

Low sensitivity and specificity are problems for miRNA quantification because miRNA is difficult to amplify. For the highly sensitive detection of miRNAs at low levels of abundance, real-time polymerase chain reaction (real-time PCR) is employed. The method was established by Applied Biosystems Company in 1996 [23]. The principle of real-time PCR is to illuminate each sample with a beam of light of at least one specified wavelength and to detect the fluorescence emitted. The thermal cycler is necessary because it can rapidly heat and chill samples. In the process, the nucleic acids are amplified and the fluorescence was captured. By analysis, the relative expression level of nucleic acids is calculated.

Similar to the ordinary PCR, a series of temperature are designed in real-time PCR. In the process, four stages are included. First, DNA denaturation is proceeded at 95 °C for 3–5 min to separate the nucleic acid's double strand; second, DNA renaturation is proceeded at 50–60 °C, which is to bind of the primers to the DNA template; and third, DNA extension is proceeded at 68–72 °C. In this stage, the polymerization is facilitated; after the three stages, the melt curve is added at the fourth stage. The temperature is set from 65 to 95 °C. Step 1–3 is set for 25–50 cycles. In the process, the fluorescence is measured during a short temperature phase lasting only a few seconds in each cycle [25].

19.2.3.2 Primers for Real-Time PCR

The first application of real-time PCR was the detection of miRNA precursor expression [26]. The primer should be designed according to the following rules: (1) The forward and reverse primer should be located in the stem-loop of the

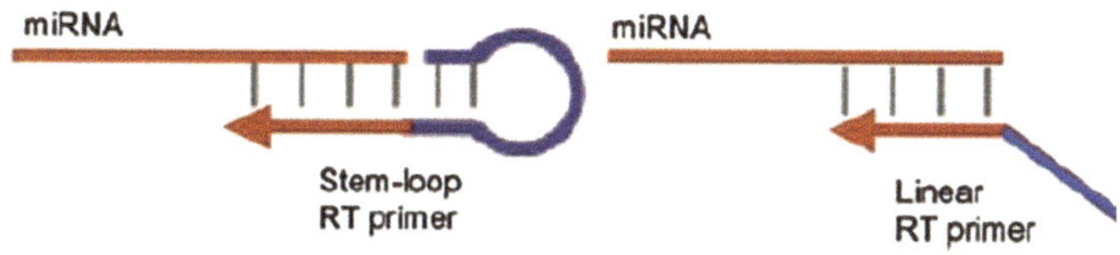

Fig. 19.1 Stem-loop RT-primer and linear RT-primer (Reprinted from Ref. [24] with permission of Oxford)

miRNA precursor. (2) Assume that the sequence of the 3′ or 5′ end of the miRNA precursor is the same as in the mature miRNA. (3) The length of the primer is 18–24 nt, and the annealing temperature is 49–59 °C. No GC should be present at the 3′ end of the primer designed for the miRNA precursor (Fig. 19.1).

Real-time PCR has also been used to detect mature miRNAs. Based on this application, more optimization methods and throughput screening have been developed. There are two kinds of RT-primer in the detection of mature miRNAs: the stem-loop RT-primer and the linear RT-primer. Experiments have confirmed that the stem-loop primer offers higher amplification efficiency and specialized amplification activity. Therefore, for the detection of a single miRNA by real-time PCR, the stem-loop primer is always chosen [24].

The stem-loop RT-primer for a mature miRNA consists of a universal sequence and a specific sequence. The specific sequence can effectively bind to the 3′ end of the mature miRNA (Fig. 19.2). Then, reverse transcription is performed using reverse transcriptase, and the RT product is quantified using conventional TaqMan PCR including a miRNA-specific forward primer, reverse primer, and dye-labeled TaqMan probes. The purpose of the tailed forward primer at 50 nt is to increase its melting temperature (Tm) depending on the sequence composition of the miRNA molecules [24].

19.2.4 MiRNA Hybridization In Situ

19.2.4.1 Principles of miRNA Hybridization In Situ

In situ hybridization (ISH) was invented by Joseph G. Gall and Mary-Lou Pardue. ISH is a type of hybridization for localizing a specific DNA or RNA sequence in tissues or cells by applying a labeled complementary DNA, RNA, or modified nucleic acid strand. It is benefit to understand the organization, regulation, and function of genes. Until now, DNA ISH and RNA ISH have been used extensively. For example, the structure of chromosomes can be determined using DNA ISH; in medical diagnostics, fluorescent DNA ISH (FISH) is used. Moreover, no matter in tissues or in cells, RNAs (mRNAs, lncRNAs, and miRNAs) can be located by RNA ISH [27].

The principle of ISH is the combination of single-stranded probe including specific sequence and the DNA or RNA and the capture of the fluorescent signal

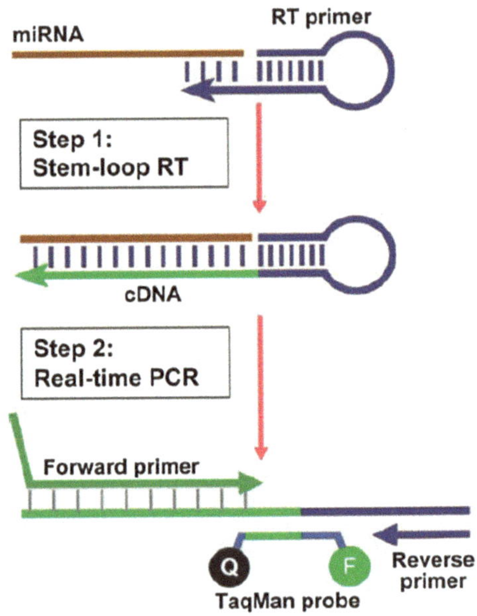

Fig. 19.2 Detection procedures for mature miRNAs (Reprinted from Ref. [24] with permission of Oxford)

from the detection system labeled on the probes at certain temperature and icon concentration. Based on this, the location of a specific DNA or RNA sequence in tissues or cells can be detected.

19.2.4.2 Characteristics of miRNA In Situ Hybridization

Compared with the original in situ hybridization, miRNA in situ hybridization has two distinctive characteristics. One is the use of an LNA probe from the company Exiqon in Denmark by the researchers. This probe is a highly sensitive locking oligonucleotide probe. Moreover, the hybridization efficiency is clearly higher than for an ordinary nucleic acid probe. Another is the use of a digoxigenin-labeled probe, which is nonradioactive and is safer, more convenient, and time-saving. Additionally, in terms of sensitivity and quality control, this probe offers a great advantage over a biotin-labeled probe. Most importantly, it can detect a single copy of a gene in the human genome [28, 29]. The basic steps of making a digoxigenin-labeled probe are as follows: (1) permeabilization of cells with proteinase K to open cell membranes (approximately 25 min, not needed for tissue sections or some early-stage embryos), (2) binding of mRNAs to the marked RNA probe (usually overnight), (3) antibody–phosphatase binding to the RN-probe (some hours), and (4) staining of the antibody (e.g., with alkaline phosphatase).

The protocol takes approximately 2–3 days and requires some time to set up. Some companies sell robots to automate the process. As a result, some laboratories have conducted large-scale screenings on thousands of genes. The results can usually be accessed via websites.

Now, some reports have presented the expressions of certain miRNAs in specific histological contexts [30–33]. For example, Ryan et al. [31] locate the accurate distribution of miR-181, miR-182, and miR-183 in the mouse retina. Darnell et al. [30] locate the accurate distribution of miR-206 in the chick embryo.

19.2.5 Biosensor-Based Detection

19.2.5.1 Principles of Biosensor-Based miRNA Detection

Biosensors are devices that recognize the concentrations of biological molecular and transform this information into physical signals. Typically, a biosensor consists of a bio-recognition component, a biotransducer component, and an electronic system. Due to their good conductivity, biocompatibility, and large surface area, nanoparticles, especially carbon- and metal-based nanoparticles, are excellent candidates as carriers to enhance signal transduction. For the detection of miRNA, the complementary strand of the target miRNA is usually connected to a biotransducer component. Then, this system can specifically recognize the target miRNA, and the Au nanoparticle can be detected to determine the concentration of the target miRNA. In 2012, Crew et al. first proved that the immobilization of miRNA on the Au nanoparticle is feasible and that the miRNA nanoparticle system could be transfected into animal cells [34]. Then, in the same year, Yin et al. first developed an "off–on" signaling genosensor platform for miRNA-21 detection. This biosensor consisted of a locked nucleic acid (LNA)-integrated nucleic acid hairpin probe, a biotin-labeled bridge Au nanoparticle barcode signal amplification unit, and enzymatic signal amplification [35].

In 2012, Alhasan et al. designed a novel high-throughput miRNA detection array using Au nanoparticles. They connected the miRNAs with a universal linker by enzymes. Then, the linker-miRNAs were hybridized with a miRNA microarray, and the nucleic acid-functionalized Au nanoparticles were used to recognize the linker sequences hybridized on the microarray [36].

In addition to Au nanoparticles, other biological and chemical nanoparticles have also been used to develop biosensors for miRNA detection. In 2010, Wanunu et al. developed a biosensor system to detect miRNAs. In this system, the target miRNA was first hybridized with a probe and bound to viral protein p19. Then, using a nanopore, the concentration of the miRNA–probe duplex was quantified [37]. In 2011, Peng et al. developed an ultrasensitive miRNA biosensor using RuO_2 nanoparticles. The tag miRNA is directly coupled to RuO_2 nanoparticles and then detected by a microarray. RuO_2 acts as a polymerization initiator and catalyst for the polymerization of 3,3′-dimethoxybenzidine, and the hybridized miRNA strands and free capture probe strands guide the deposition of poly 3,3′-dimethoxybenzidine [38]. In 2011, Yang and Vosch developed a silver nanocluster probe-based biosensor system for the rapid detection of miRNAs. They connected the complementary sequence of the target miRNA to the silver nanocluster, and in the presence of the target miRNA, the fluorescence properties of the DNA-linked nanosilver are significantly attenuated [39].

19.2.5.2 Advantages of Biosensor-Based miRNA Detection

Nanoparticle biosensor-based methods usually provide ultrasensitive, high-throughput detection of miRNAs. Their detection limitation is usually on the fM or pM level, due to the small size of the nanoparticle and excellent signal amplification system. Nanoparticles are in direct contact with the environment, which can accelerate signal transduction, enhance the ability of rapid analysis, and lower the detection limit. Moreover, the results of nanoparticle biosensor-based miRNA detection are usually optical and easy to read.

19.3 Detection of Short Tandem Repeat Sequences: Telomeres

A telomere is a region of repetitive nucleotide sequences at each end of a chromosome, which protects the end of the chromosome from deterioration or from fusion with the neighboring chromosomes. The observation that telomeres shorten with increasing age and are implicated in cellular aging has led to the proposal that telomere length is a biomarker of aging [40, 41], although some authors consider the evidence supporting the hypothesis that telomere length is a biomarker of aging to be equivocal, and more data are required from studies that assess telomere length and aging-related functional measures and collect mortality data [42].

Human telomere DNA consists of short tandem repeats of 5′-TTAGGG-3′, with lengths ranging from 4 to 15 kb, caused by the different degrees of cell division. The telomere ends consist of 50 to several hundred repeats of single-stranded TTAGGG DNA called 3′ overhang. This single-stranded overhang, combined with several protein factors, forms a T-loop structure to protect the end of the chromosome from deterioration or fusion. It is thought that both the length and the structure are related to the telomere functions. Telomeres are always in a dynamic balance. When telomeres are shortened, the telomere proteins will depolymerize and then the 3′ ends of the telomeres will be exposed for telomerase to lengthen the telomeres [43]. In most tumor cells, telomeres are very short, and tumor cells express telomerase to extend their telomeres, which also indicate that the telomere must have a minimal length to maintain the T-loop structure and their function [44].

Detection of the length of telomeric DNA is usually a critical step in studies of telomeres, in either single cells or populations of cells. The realization that the length of telomere repeats at individual chromosome ends is a critical variable in cell fate decisions and biological functions ranging from aging to carcinogenesis [45] has highlighted the need for techniques that can provide accurate information on the lengths of telomere repeats in different cell types [46]. However, the short tandem repeat characteristic of telomeres introduces many difficulties in the detection. Several methods have been developed for the detection of the length of telomere, including terminal restriction fragment analysis, PCR-based methods, single telomere length analysis, and fluorescence in situ hybridization-based methods.

19.3.1 Terminal Restriction Fragment Analysis of Telomere Length

19.3.1.1 Principle of Terminal Restriction Fragment Analysis

Terminal restriction fragment (TRF) analysis determines the length of the telomere DNA sequence by digesting regions of DNA sequences other than the telomere using restriction enzymes. There are plenty of restriction sites in chromosomal DNA sequences, and now that we know the sequences of telomeres are tandem repeats of TTAGGG, it is feasible to avoid restriction enzymes that recognize the telomere during the digestion of genomic DNA. Thus, the digestion of genomic DNA will result in the telomere sequences as the only long undigested fragments, and these fragments can then be detected using either Southern blotting or in-gel hybridization using a labeled probe specific for telomeric DNA. The qualitative analysis of telomere lengths from different samples can be performed by comparing their positions on the gel. The quantitative lengths of telomere sequences can be analyzed by comparing the position of the band or smear on the gel with a standard DNA ladder marker.

As longer telomeres can hybridize with more probes, the intensity of the bands may directly indicate the telomere abundance, and some correction should be conducted.

The TRF technology has been used for the detection of telomeres since the 1980s. And for a long period, it was the only method for telomere detection. In 1988, Moyzis et al. digested the genomic DNA using *Sau*3AI or *Rsa*I, and the size-fractionated genomic DNA was hybridized to ^{32}P-labeled (TTAGGG)n [47]. In 1989, Allshire et al. used a combination of different restriction enzymes and oligonucleotides to examine the telomeres of human chromosomes and found that in addition to the classical TTAGGG repeats in the telomere, there are also at least two other types of repeats. They also used TRF to analyze the lengths of telomeres in different types of tissue cells and reported that the differences in telomere lengths are due to the variable amounts of the TTAGGG repeat and not the other repeat types [48]. In 1990, Harley et al. determined the lengths of telomeres by TRF using the restriction enzymes *Msp*I and *Rsa*I and a $(TTAGGG)_3$ or $(TTAGGG)_4$ probe, and by these methods, they found that telomeres shorten during the aging of human fibroblasts [49].

In the subsequent years, several attempts have been made to improve the sensitivity of TRF. In 1997, Bryant et al. used a slot blot assay instead of the Southern blot to determine the relative content, instead of length, of telomere DNA. The relative contents of telomere DNA determined by this slot blot assay were directly proportional to the relative lengths of telomere DNA determined in parallel by Southern blot analysis [50]. Norwood and Dimitrov further optimized this method for the determination of telomeres from whole cells or smaller amounts of purified DNA [51].

19.3.1.2 The Advantages and Disadvantages of TRF

The main advantage of this technique is that the errors of TRF are relatively small [46]. For a long period, TRF has been used as the gold standard of telomere determination, and little has changed in the procedure since the first reports. However, due to the use of different restriction enzymes, the results from different research groups are difficult to standardize and compare. There are uncertain regions between the restriction sites and the start of the telomere, and when different restriction enzyme sets are selected, the lengths of uncertain region are different, resulting in systematic error.

The main disadvantage of TRF is its insensitivity to very short telomeres, which is often the case in studies of tumor cells. Moreover, the TRF technique requires a large amount of DNA and is time consuming. These disadvantages make TRF insufficient in many cases, especially when cell samples are rare and precious. The development of new techniques for telomere length detection is necessary.

19.3.2 Quantitative PCR-Based Methods for the Detection of Telomere Length

19.3.2.1 Principle of Quantitative PCR-Based Methods for the Detection of Telomere Length

Traditionally, the measurement of telomere length using PCR amplification techniques with primers designed on the TTAGGG repeat sequences was thought to be infeasible because only primer dimer-derived products were expected. However, in 2002, Richard M. Cawthon broke through these expectations. Cawthon designed the primers tel 1 (5′-GGTTTT TGAGGG TGAGGG TGAGGG TGAGGG TGAGGG T-3′) and tel 2 (5′-TCCCGAC TATCCC TATCCC TATCCC TATCCC TATCCC TA-3′). The tel 1 primer TGAGGG unit could match the telomere repeat sequence AATCCC with one mismatch, and the tel 2 primer TATCCC unit could also match the telomere repeat sequence TTAGGG with only one mismatch. The 3′ ends of the primers match the telomere sequence perfectly. The primer pairs could also anneal with each other. The strongest possible hybridizations of the primers to each other involve a repeated pattern of six bases containing four consecutive paired bases followed by two mismatched bases, and the 3′-terminal base of each primer cannot form a stable base pair with the base opposite it, thereby blocking the addition of bases by DNA polymerase [52].

In addition to these primers, Cawthon also developed a new detection strategy for the measurement of telomere lengths, instead of measuring the direct terminal quantity by quantitative PCR (Q-PCR). In this strategy, Cawthon introduced a reference DNA sample that could be from a single individual or a pooled sample from multiple individuals. To determine the difference in the quantity of genomes between the experimental and reference DNA, one single-copy gene was chosen as the target of quantitative PCR, and the experimental DNA was diluted to different levels until the single-copy gene in the experimental and reference samples was amplified equivalently with regard to the number of cycles of PCR needed to generate a given amount of PCR product during the exponential phase of PCR amplification. This level of dilution was denoted as "S." Then, another dilution level "T" of experimental DNA was determined to cause the telomere in the experimental and reference samples to be amplified equivalently. For each experimental sample, the ratio of these dilution factors is the relative telomere to single-copy gene (T/S) ratio [52], and the telomeres in different samples could be compared if they used the same reference DNA sample.

In 2009, Cawthon improved the PCR-based telomere length detection methods. In the previous methods, the single-copy gene PCR and telomere PCR were conducted in two separate tubes containing an equivalent sample, which could result in significant errors due to pipetting volume discrepancies between the telomere and single-copy gene reactions. Thus, Cawthon developed a novel monochrome multiplex quantitative PCR (MMQ-PCR) method to conduct the telomere and single-copy gene quantitative PCR reaction in the same tube [53]. Simply, when the telomere repeats were the high-abundance species and a single-copy gene was the low-abundance species, Cawthon used a single fluorescent DNA-intercalating dye and detected the T signal in the early cycles before the S signals rose above the baseline and then detected the S signals in the following cycles at a temperature where the telomere product was fully melted [53]. This MMQ-PCR method showed a strong correlation ($R = 0.844$) with the gold standard method TRF [53], and compared with previous Q-PCR methods, MMQ-PCR increases throughput and lowers costs, as half as many reactions are needed [53].

19.3.2.2 The Advantages and Disadvantages of Quantitative PCR-Based Telomere Detection

As quantitative PCR is a prevalent and convenient technique at this time, the PCR-based telomere detection methods have been widely applied. The advantages of these methods are the use of far less DNA sample and much less time, compared with TRF-Southern blot based methods. The main disadvantage of these methods is the requirement of too many assumptions to make them feasible, whereas in individual experiments, these assumptions may not be satisfied. For example, the reference single-copy gene may not really be a single copy due to chromosomal duplications or copy number variation. Some of these assumptions are difficult to validate [46].

19.3.3 Single Telomere Length Analysis

19.3.3.1 Principles of Single Telomere Length Analysis

The single telomere length analysis (STELA) method was first described by Baird et al. in 2003. The main principle of STELA is the use of the single-stranded TTAGGG DNA in the 3′ overhang [54]. Baird et al. used a series of "telorette" linkers to recognize the start site of the 3′ overhang. These linkers consist of a 20 nucleotide noncomplementary tail (5′-TGC TCC GTG CAT CTG GCA TC-3′) and one of the seven-base sequences homologous to TTAGGG (CCCTAAC, TAACCCT, CCTAACC, CTAACCC, AACCCTA, or ACCCTAA) at the 3′ end. These telorettes could anneal with the G-rich 3′ overhang and be ligated to the 5′ end of the complementary C-rich strand of the chromosome. Then, the 20 nucleotide noncomplementary tail was used as one of the primers, with one chromosome-specific upstream primer located in the subtelomeric region as the other primer. This primer pair could be used to amplify almost the whole length of the telomere except for the 3′ overhang [54]. Baird et al. used this method first to detect the human XpYp telomeres in fibroblasts [54], and then in their following studies, they designed chromosome-specific primers for the telomeres of 2p, 11q, 12q, and 17p within the same cell strains. Using STELA, they found that the telomere of 17p was more stable, with a striking paucity of large-scale length changes, exhibited the shortest recorded allelic distribution (300 bp) in senescent cells, and displayed a general, but not absolute, trend toward being the shortest telomere [55]. To reduce the artificial bias caused by PCR amplification, limited PCR cycles were conducted, and the STELA products were resolved by Southern blot using probes for the specific subtelomeric sequence. The positions and intensities of bands in the blot could be used to calculate the length of each single telomere.

19.3.3.2 The Advantages and Disadvantages of Single Telomere Length Analysis

The main advantages of STELA are its ability to generate highly accurate telomere length measurements and its low requirement for the amount of DNA samples. While the restriction site in TRF analysis is an unknown distance from the start of the telomere, the exact base of the subtelomeric primer used in STELA is known and stable between cells, samples, and individuals, and therefore telomere length is measured very accurately [46]. The main disadvantage of this method is that not all chromosome ends have a suitable sequence for the design of unique chromosome arm primers, and thus STELA is usually restricted to several well-characterized ends. Another disadvantage of STELA is that this method is not suitable for the measurement of very long telomeres.

19.3.4 Fluorescence In Situ Hybridization-Based Methods

19.3.4.1 Principles of Fluorescence In Situ Hybridization-Based Methods

Fluorescence in situ hybridization (FISH) is a technique for detecting in situ hybridization samples using nonradioactive fluorescence signals, without the isolation of DNA or RNA. The tissues or cells were first fixed on the slide or fiber membrane. After denaturation of the double-stranded DNA, the fluorescently labeled probes could specifically hybridize with the target genes, and the results could be observed and analyzed using a fluorescence microscope, laser scanning confocal microscope (Q-FISH), or flow cytometer (flow-FISH). The probes used in FISH for telomere length detection are usually fluorescently labeled $(CCCTAA)_3$ peptide nucleic acid (PNA) probes. In 1998, Martens et al. validated the use of Q-FISH for the quantitative measurement of telomeric repeats. They hybridized purified plasmid DNA with different sizes of (TTAGGG)n repeat inserts with Cy3-labeled $(CCCTAA)_3$ PNA probes, and they found that the intensity of the fluorescent spots was homogeneous for each plasmid and that the medians were directly proportional to the size of the (TTAGGG)n repeat inserts for different plasmids. The resulting calibration line could be used to express telomere fluorescence in telomere fluorescence units, with each unit corresponding to 1 kb of TTAGGG repeats [56]. In the same year, the same research group developed flow cytometry-based FISH using similar PNA probes, and the results of flow-FISH correlated with the results of conventional telomere length measurements by Southern blot analysis with a correlation coefficient R of 0.9. They used this method in telomere length detection in $CD8^+$ T-cell subsets [57].

19.3.4.2 The Advantages and Disadvantages of Fluorescence In Situ Hybridization-Based Methods

The PNA-based FISH methods can be used to determine very short telomere lengths, and the flow-FISH now offers the most extensive quantitative reference data related to telomere length currently available and is the first of the telomere length methods to have been validated for clinical diagnostic purposes [58]. The main disadvantage of FISH-based methods for detection of telomeres is that these methods are not suitable for measuring telomeres in terminally senescent cells, highly aberrant cells, or generally in cells that are unable to divide. Further, it may be challenging to test cell types with very low mitotic indices [46]. Moreover, compared with the immunophenotyping of cells, the flow-FISH technique requires accurate measurements of relatively weak fluorescence signals on a linear scale. The required calibrations and controls make the whole procedure quite time consuming and technically demanding [59].

19.4 Triplex DNAs for Detection

Triplex DNAs form in a sequence-specific manner when purine or pyrimidine bases lie in the major groove of the DNA double helix via Hoogsteen pairs with purines of the Watson–Crick base pairs. It was first mentioned by Felsenfeld et al. in 1957 [60] that a specific DNA triple helix was produced between one strand of poly(adenylic acid) and two strands of poly(uridylic acid) in the presence of divalent cations.

Since the pioneering work on oligoribonucleotide-based DNA triple helices, many studies have demonstrated the structure of triplex DNAs. Triplex DNAs mainly contain two motifs: the YR*Y motif and the YR*R motif. In the YR*Y motif, the triplex DNAs consist of one purine strand and two pyrimidine strands, where the homopyrimidine oligonucleotide binds to the purine strand of the duplex in a direct parallel way, with the noncanonical triads of AT*G and GC*T as well as the canonical triads of TA*T and CG*C [61]. It is worth mentioning that because the formation of the CG*C triad requires the protonation of the cytosine of the third strand, the triplexes are able to exist stably in acidic conditions. In contrast, in the YR*R motif, the triplex DNAs consist of one pyrimidine strand and two purine strands, where the homopurine strand binds to the purine strand of the duplex in an antiparallel way, with the versatile triads of CG*G, TA*A, TA*T, and CG*A [62]. In contrast to the YR*Y triplex, the stability of the YR*R triplex depends dramatically on the presence of bivalent metal cations [63]. Despite some differences, the YR*Y triplex and the YR*R triplex share certain fundamental features: (1) there must be a homopurine sequence in one strand of the duplex for the formation of the triplex and (2) the orientation is antiparallel in the two homologous strands.

In early studies, triplex-forming oligonucleotides (TFO) were described as tools to restrain gene expression via inhibiting transcription initiation through binding to the promoter sites to block transcription factor access and stop gene activation [64]. In 1988, Scott et al. [65] proved that 15 nt oligonucleotides cause sequence-specific double-strand breaks, equipped with a cleaving EDTA-Fe at the 5′ end, at one site in the plasmid DNA. Several studies have investigated TFOs as candidates to mediate genome modification, resulting in a change in target sequences [66–68] Then, the modified third strand, such as psoralen, was used for site-directed mutagenesis studies in mammalian cells and in yeast [68, 69]. In addition, psoralen-TFO mediation was extended to knock out chromosomal genes [70]. In 2000, Vasquez et al. argued that bare TFOs without psoralen or any other DNA reactive conjugate could induce specific mutations at sequence-specific genome sites in the somatic cells of adult mice, meaning that site-directed DNA-binding molecules can mediate gene modification and gene targeting in vivo upon systemic administration. Based on the idea that third-strand binding could trigger DNA repair, it is hypothesized that the formation of a triplex might induce recombination because of the production of repair-dependent DNA strand breaks. It was found that both non-psoralen TFOs [71] and psoralen-TFOs [72] could trigger recombination within the SV40 virus genome. Similar research using TFOs to trigger recombination has been demonstrated by Wilson, using a locus in CHO cells containing duplicated *APRT* genes as a target [73]. In 2008, Feng et al. [74] demonstrated that a conserved triple helix contributes to the catalysis of telomerase activity, occurring within a pseudoknot structure of RN, but not by binding the reverse transcriptase (TERT) protein. As telomerase may be a molecular fossil representing an evolutionary link between ribonucleoprotein (RNP) enzymes and RNA enzymes, the role of the triple-helix structure in telomerase catalysis has been acquired.

19.4.1 Aptamer-Based Sensing Platform Using a Triple-Helix Molecular Switch

Effective detection strategies for aptasensors are paid great attention in chemistry and biotechnology domain. Aptamers are artificial short single-stranded ligands of deoxyribonucleic acid (DNA) or ribonucleic acid (RNA), selected from a combinatorial random nucleic acid library of 10^{13}–10^{16} sequences using an in vitro process known as the systematic evolution of ligands by exponential enrichment (SELEX), which is a selection process involving iterative binding, separation, and amplification in vitro [75, 76]. Due to their inherent selectivity, sensitivity, and affinity toward targets and ease of preparation and use, aptamers are perfect recognition module for biosensor applications [77].

In aptamer-based biosensor design, the conformational alteration of specific aptamer usually plays a key role during the target recognition event. Labeling different aptamers with specific signaling reporter, conformational changes of specific aptamer are able to produce visible, fluorescent, or electrochemical signal after aptamer–target binding event.

Zheng et al. established two kinds of aptamer-based detection model using a triple-helix molecular switch (THMS). The triple-helix molecular switch is composed of two parts: one part is target-specific aptamer element in the center flanked by two arm components, and the other is single-stranded oligonucleotide which acted as signal transduction. After specific target recognition based on aptamer, THMS disassembles and releases signal transduction element to produce output signal. In the reference [78], the signal transduction element was named STP, doubly labeled with pyrene at the 5′ and 3′ end, respectively. In Fig. 19.3, although STP was hairpin-shaped structure when it separately existed, it still transformed to "open" configuration as binding two arm components of the aptamer element, leading to output signal. Utilizing this module, small molecules (e.g., adenosine

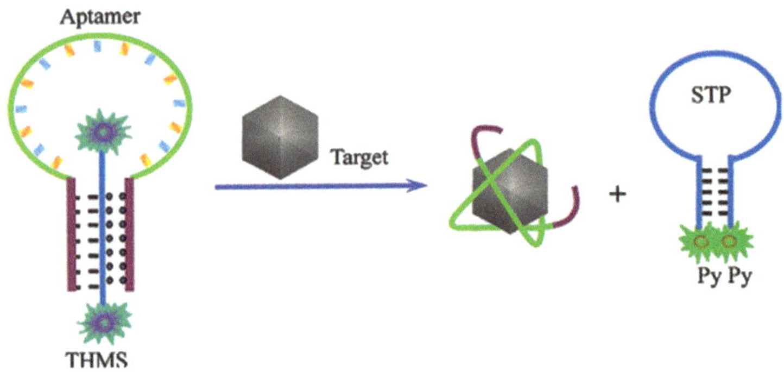

Fig. 19.3 Design scheme of THMS for signaling aptamer–target binding event. "-" and "•" represent Watson–Crick and Hoogsteen base pairings, respectively (Reprinted with the permission from Ref. [78]. Copyright 2011 American Chemical Society)

triphosphate) and protein (e.g., human R-thrombin and L-argininamide) were successfully detected.

In the reference [79], a universal detection method combined surface-enhanced Raman scattering (SERS) with triple-helix molecular switch (THMS) was established for the analysis of multiple targets. In Fig. 19.4, THMS was disassembled after the process of target recognition and then the signal transduction element was served as universal trigger (UT) to initiate hybridization chain reaction (HCR). By virtue of 4-aminobenzenethiol (4-ABT)-encoded gold nanoparticles (AuNPs), the SERS active hot spots were formed as output signal. Utilizing this module, small molecules (e.g., adenosine triphosphate), protein (e.g., human R-thrombin and L-argininamide), and cells (e.g., CEM cancer cells) were successfully determined, respectively.

Ye et al. established another THMS method to detect nucleic acids and proteins [80]. As shown in Fig. 19.5, the target-specific aptamer element was designed as an intramolecular hairpin structure immobilized on magnetic beads (MBs). The stem of the hairpin element formed triple helix, bonded to a single-stranded trigger DNA. When the target recognition event occurred, the THMS would produce output signal.

19.4.2 Other Applications Based on a Triplex–Helix Molecular Switch

19.4.2.1 DNA Colorimetric Logic Gates Based on a Triplex–Helix Molecular Switch

Molecular logic gates, binary switches of molecular level that generate the output state (0 or 1) based on input conditions (0 or 1), have recently captured great attention [81–88]. Gao et al. reported a new-style detection method which

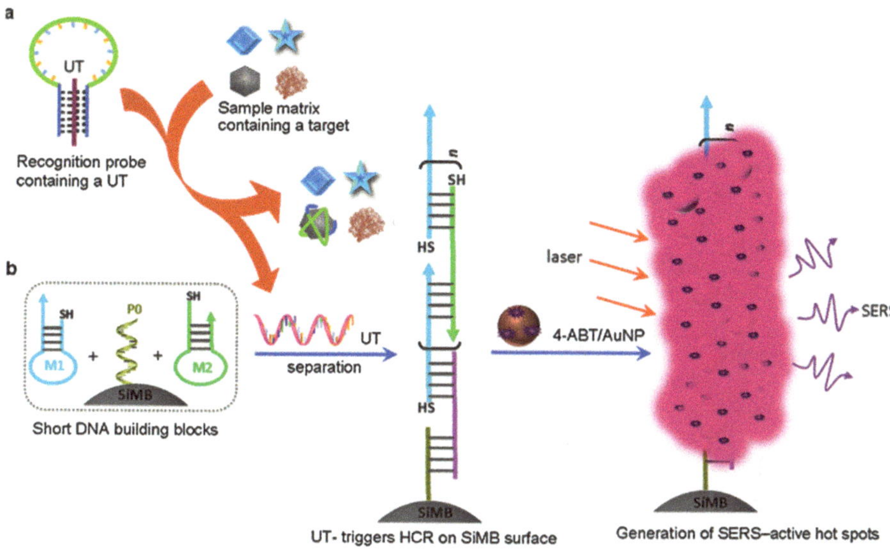

Fig. 19.4 (**a**) The molecular recognition probe is hybridized with a universal trigger (UT) to form a triple-helix structure via Watson–Crick (-) and Hoogsteen (•) base pairings. Following addition of the target, interaction between the recognition probe and its target perturbs the triple-helix structure and causes the release of the UT. (**b**) Upon the release of UT, HCR is triggered between M1 and M2 on the SiMB surface to form a long double-stranded DNA polymer, and finally, conjugation of the 4-ABT-functionalized AuNPs on the DNA polymer forms a SERS-active substrate (Reprinted with the permission from Ref. [79]. Copyright 2014 American Chemical Society)

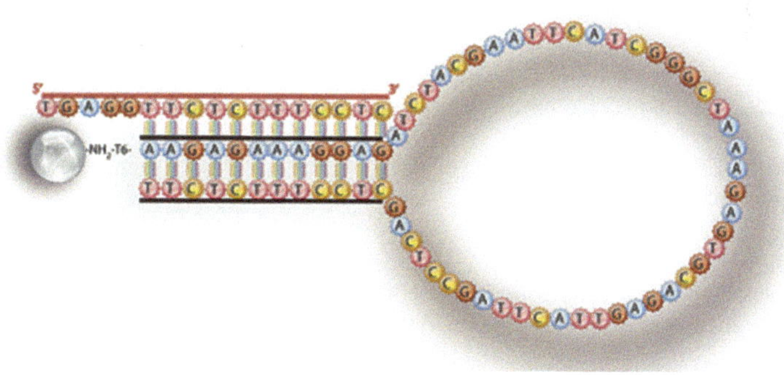

Fig. 19.5 The structure and sequences of the THM immobilized on MBs (Reproduced from Ref. [80] by permission of The Royal Society of Chemistry)

combined colorimetric logic gate with THMS [89]. In this design, G-quadruplex DNAzyme, which has a horseradish peroxidase (HRP) activity, was inhibited from THMS. As shown in Fig. 19.6, upon recognition with the target sequence, the THMS decomposed and self-assembles into activated DNAzyme to produce a naked-eye output signal for different molecular logic gates.

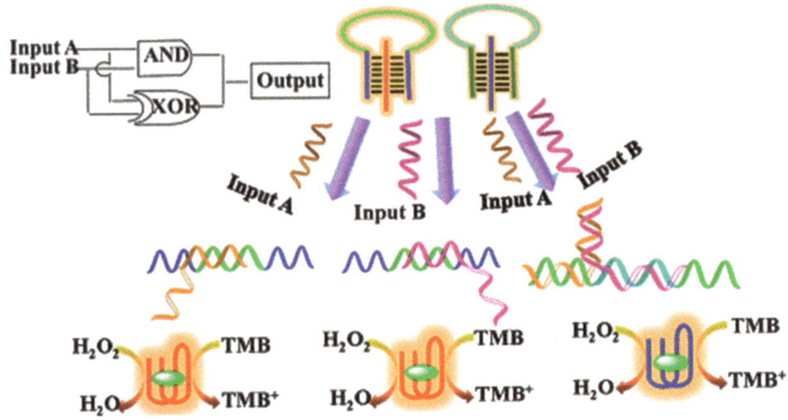

Fig. 19.6 Schematic representation of a tMB-based logic gate based on triplex–helix molecular switch (Reproduced from Ref. [80] by permission of The Royal Society of Chemistry)

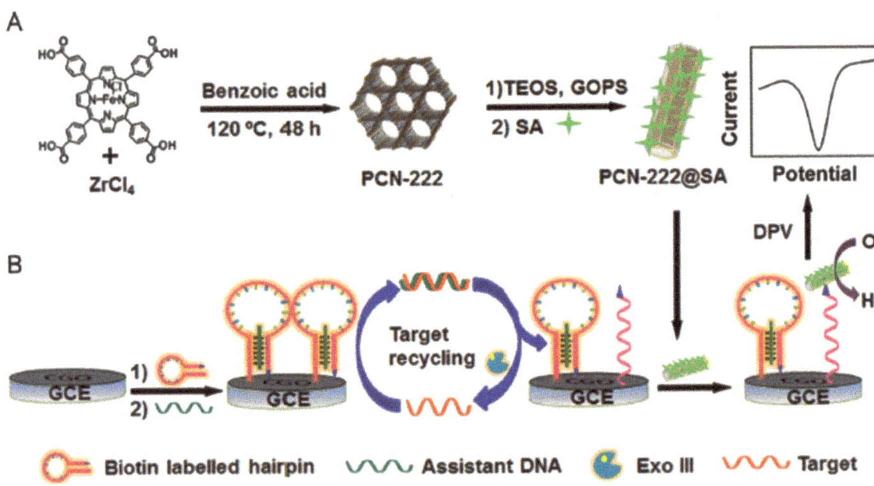

Fig. 19.7 (**a**) Synthesis of PCN-222@SA composite and (**b**) electrochemical strategy coupling with target recycling amplification for DNA sensing (Reprinted with the permission from Ref. [90]. Copyright 2014 American Chemical Society)

19.4.2.2 Porphyrinic Metal–Organic Framework as an Electrochemical Probe for DNA Sensing via Triple-Helix Molecular Switch

Ye et al. established an electrocatalysis detection method which combined a porphyrinic metal–organic framework (MOF) with THMS [90]. In this design, THMS was immobilized on the surface of a glassy carbon electrode, and zirconium–porphyrin MOF labeled with streptavidin (PCN-222@SA) was synthesized (Fig. 19.7). Upon the arrival of target DNA, THMS was broken and released the assistant DNA that hybridized with the target sequence. Upon biotin/streptavidin bio-recognition, the end biotin away from the electrode would integrate with PCN-222@SA to the output signal.

Acknowledgments This work is supported by the Ministry of Science and Technology of Beijing (XX2014B069). Many thanks to Mingzhang Guo, Jingjing Tian and Xiangli Shao, for their kindly help in manuscript conception and preparation.

References

1. Cullen BR. Transcription and processing of human microRNA precursors. Mol Cell. 2004;16(6):861–5.
2. Zeng Y. Principles of micro-RNA production and maturation. Oncogene. 2006;25(46):6156–62.
3. Chen CZ, Li L, Lodish HF, Bartel DP. MicroRNAs modulate hematopoietic lineage differentiation. Science. 2004;303(5654):83–6.
4. Xu P, Vernooy SY, Guo M, Hay BA. The Drosophila microRNA Mir-14 suppresses cell death and is required for normal fat metabolism. Curr Biol. 2003;13(9):790–5.
5. Barad O, Meiri E, Avniel A, Aharonov R, Barzilai A, Bentwich I, Einav U, Gilad S, Hurban P, Karov Y, Lobenhofer EK, Sharon E, Shiboleth YM, Shtutman M, Bentwich Z, Einat P. MicroRNA expression detected by oligonucleotide microarrays: system establishment and expression profiling in human tissues. Genome Res. 2004;14(12):2486–94.
6. Cummins JM, He Y, Leary RJ, Pagliarini R, Diaz Jr LA, Sjoblom T, Barad O, Bentwich Z, Szafranska AE, Labourier E, Raymond CK, Roberts BS, Juhl H, Kinzler KW, Vogelstein B, Velculescu VE. The colorectal microRNAome. Proc Natl Acad Sci U S A. 2006;103(10):3687–92.
7. Pfeffer S, Voinnet O. Viruses, microRNAs and cancer. Oncogene. 2006;25(46):6211–9.
8. Taft RJ, Pang KC, Mercer TR, Dinger M, Mattick JS. Non-coding RNAs: regulators of disease. J Pathol. 2010;220(2):126–39.
9. Calin GA, Croce CM. MicroRNAs and chromosomal abnormalities in cancer cells. Oncogene. 2006;25(46):6202–10.
10. Cimmino A, Calin GA, Fabbri M, Iorio MV, Ferracin M, Shimizu M, Wojcik SE, Aqeilan RI, Zupo S, Dono M, Rassenti L, Alder H, Volinia S, Liu CG, Kipps TJ, Negrini M, Croce CM. miR-15 and miR-16 induce apoptosis by targeting BCL2. Proc Natl Acad Sci U S A. 2005;102(39):13944–9.
11. Aukerman MJ, Sakai H. Regulation of flowering time and floral organ identity by a MicroRNA and its APETALA2-like target genes. Plant Cell. 2003;15(11):2730–41.
12. Palatnik JF, Allen E, Wu X, Schommer C, Schwab R, Carrington JC, Weigel D. Control of leaf morphogenesis by microRNAs. Nature. 2003;425(6955):257–63.
13. Brennecke J, Hipfner DR, Stark A, Russell RB, Cohen SM. bantam encodes a developmentally regulated microRNA that controls cell proliferation and regulates the proapoptotic gene hid in Drosophila. Cell. 2003;113(1):25–36.
14. Nelson PT, Baldwin DA, Scearce LM, Oberholtzer JC, Tobias JW, Mourelatos Z. Microarray-based, high-throughput gene expression profiling of microRNAs. Nat Methods. 2004;1(2):155–61.
15. Babak T, Zhang W, Morris Q, Blencowe BJ, Hughes TR. Probing microRNAs with microarrays: tissue specificity and functional inference. RNA. 2004;10(11):1813–9.

16. Beuvink I, Kolb FA, Budach W, Garnier A, Lange J, Natt F, Dengler U, Hall J, Filipowicz W, Weiler J. A novel microarray approach reveals new tissue-specific signatures of known and predicted mammalian microRNAs. Nucleic Acids Res. 2007;35(7):e52.
17. Liu C, Yu J, Yu SR, Cai L, Liu W, Yang K, He X, Chen S. MicroRNA-21 acts as an oncomir through multiple targets in human hepatocellular carcinoma. J Hepatol. 2010;53(1):98–107.
18. Yu J, Wang F, Yang GH, Wang FL, Ma YN, Du ZW, Zhang JW. Human microRNA clusters: genomic organization and expression profile in leukemia cell lines. Biochem Biophys Res Commun. 2006;349(1):59–68.
19. Pfeffer S, Zavolan M, Grasser FA, Chien M, Russo JJ, Ju J, John B, Enright AJ, Marks D, Sander C, Tuschl T. Identification of virus-encoded microRNAs. Science. 2004;304 (5671):734–6.
20. Takamizawa J, Konishi H, Yanagisawa K, Tomida S, Osada H, Endoh H, Harano T, Yatabe Y, Nagino M, Nimura Y, Mitsudomi T, Takahashi T. Reduced expression of the let-7 microRNAs in human lung cancers in association with shortened postoperative survival. Cancer Res. 2004;64(11):3753–6.
21. Valoczi A, Hornyik C, Varga N, Burgyan J, Kauppinen S, Havelda Z. Sensitive and specific detection of microRNAs by northern blot analysis using LNA-modified oligonucleotide probes. Nucleic Acids Res. 2004;32(22):e175.
22. Sambrook J, Fritsch EF, Maniatis T. Molecular cloning: a laboratory manual. Immunology. 1983;49(1):895–909.
23. Heid CA, Stevens J, Livak KJ, Williams PM. Real time quantitative PCR. Genome Res. 1996;6(10):986–94.
24. Chen C, Ridzon DA, Broomer AJ, Zhou Z, Lee DH, Nguyen JT, Barbisin M, Xu NL, Mahuvakar VR, Andersen MR, Lao KQ, Livak KJ, Guegler KJ. Real-time quantification of microRNAs by stem-loop RT-PCR. Nucleic Acids Res. 2005;33(20):e179.
25. Rychlik W, Spencer WJ, Rhoads RE. Optimization of the annealing temperature for DNA amplification in vitro. Nucleic Acids Res. 1990;18(21):6409–12.
26. Schmittgen TD, Jiang J, Liu Q, Yang L. A high-throughput method to monitor the expression of microRNA precursors. Nucleic Acids Res. 2004;32(4):e43.
27. Gall JG, Pardue ML. Formation and detection of RNA-DNA hybrid molecules in cytological preparations. Proc Natl Acad Sci U S A. 1969;63(2):378–83.
28. Stumm M, Tonnies H. Fluorescence in situ hybridization techniques in medical diagnostics. Expert Opin Med Diagn. 2008;2(12):1381–90.
29. Ying SY, Chang DC, Miller JD, Lin SL. MicroRNA protocols. Perspectives. Methods Mol Biol. 2006;342:351–8. doi:1-59745-123-1:351 [pii]10.1385/1-59745-123-1:351.
30. Darnell DK, Kaur S, Stanislaw S, Davey S, Konieczka JH, Yatskievych TA, Antin PB. GEISHA: an in situ hybridization gene expression resource for the chicken embryo. Cytogenet Genome Res. 2007;117(1–4):30–5.
31. Ryan DG, Oliveira-Fernandes M, Lavker RM. MicroRNAs of the mammalian eye display distinct and overlapping tissue specificity. Mol Vis. 2006;12:1175–84.
32. Nelson PT, Baldwin DA, Kloosterman WP, Kauppinen S, Plasterk RH, Mourelatos Z. RAKE and LNA-ISH reveal microRNA expression and localization in archival human brain. RNA. 2006;12(2):187–91.
33. Habbe MN, Koorstra JBM, Mendell JT, Feldmann G, Mullendore ME, Goggins MG, Maitra A. MicroRNA miR-155 is a biomarker of early pancreatic neoplasia. Cancer Biol Ther. 2008;37(4):1–2.
34. Crew E, Tessel MA, Rahman S, Razzak-Jaffar A, Mott D, Kamundi M, Yu G, Tchah N, Lee J, Bellavia M, Zhong CJ. MicroRNA conjugated gold nanoparticles and cell transfection. Anal Chem. 2012;84(1):26–9.
35. Yin H, Zhou Y, Chen C, Zhu L, Ai S. An electrochemical signal 'off-on' sensing platform for microRNA detection. Analyst. 2012;137(6):1389–95.
36. Alhasan AH, Kim DY, Daniel WL, Watson E, Meeks JJ, Thaxton CS, Mirkin CA. Scanometric microRNA array profiling of prostate cancer markers using spherical nucleic acid-gold nanoparticle conjugates. Anal Chem. 2012;84(9):4153–60.

37. Wanunu M, Dadosh T, Ray V, Jin J, McReynolds L, Drndic M. Rapid electronic detection of probe-specific microRNAs using thin nanopore sensors. Nat Nanotechnol. 2010;5(11):807–14.
38. Peng Y, Gao Z. Amplified detection of microRNA based on ruthenium oxide nanoparticle-initiated deposition of an insulating film. Anal Chem. 2011;83(3):820–7.
39. Yang SW, Vosch T. Rapid detection of microRNA by a silver nanocluster DNA probe. Anal Chem. 2011;83(18):6935–9.
40. Butler RN, Sprott R, Warner H, Bland J, Feuers R, Forster M, Fillit H, Harman SM, Hewitt M, Hyman M, Johnson K, Kligman E, McClearn G, Nelson J, Richardson A, Sonntag W, Weindruch R, Wolf N. Biomarkers of aging: from primitive organisms to humans. J Gerontol A Biol Sci Med Sci. 2004;59(6):B560–7.
41. Simm A, Nass N, Bartling B, Hofmann B, Silber RE, Navarrete Santos A. Potential biomarkers of ageing. Biol Chem. 2008;389(3):257–65.
42. Mather KA, Jorm AF, Parslow RA, Christensen H. Is telomere length a biomarker of aging? A review. J Gerontol A Biol Sci Med Sci. 2011;66(2):202–13.
43. Blackburn EH. Telomere states and cell fates. Nature. 2000;408(6808):53–6.
44. Blasco MA, Lee HW, Hande MP, Samper E, Lansdorp PM, DePinho RA, Greider CW. Telomere shortening and tumor formation by mouse cells lacking telomerase RNA. Cell. 1997;91(1):25–34.
45. Stewart SA, Weinberg RA. Telomeres: cancer to human aging. Annu Rev Cell Dev Biol. 2006;22:531–57.
46. Aubert G, Hills M, Lansdorp PM. Telomere length measurement-caveats and a critical assessment of the available technologies and tools. Mutat Res. 2012;730(1–2):59–67.
47. Moyzis RK, Buckingham JM, Cram LS, Dani M, Deaven LL, Jones MD, Meyne J, Ratliff RL, Wu JR. A highly conserved repetitive DNA sequence, (TTAGGG)n, present at the telomeres of human chromosomes. Proc Natl Acad Sci U S A. 1988;85(18):6622–6.
48. Allshire RC, Dempster M, Hastie ND. Human telomeres contain at least three types of G-rich repeat distributed non-randomly. Nucleic Acids Res. 1989;17(12):4611–27.
49. Harley CB, Futcher AB, Greider CW. Telomeres shorten during ageing of human fibroblasts. Nature. 1990;345(6274):458–60.
50. Bryant JE, Hutchings KG, Moyzis RK, Griffith JK. Measurement of telomeric DNA content in human tissues. Biotechniques. 1997;23(3):476–8, 480, 482.
51. Norwood D, Dimitrov DS. Sensitive method for measuring telomere lengths by quantifying telomeric DNA content of whole cells. Biotechniques. 1998;25(6):1040–5.
52. Cawthon RM. Telomere measurement by quantitative PCR. Nucleic Acids Res. 2002;30(10): e47.
53. Cawthon RM. Telomere length measurement by a novel monochrome multiplex quantitative PCR method. Nucleic Acids Res. 2009;37(3):e21.
54. Baird DM, Rowson J, Wynford-Thomas D, Kipling D. Extensive allelic variation and ultra-short telomeres in senescent human cells. Nat Genet. 2003;33(2):203–7.
55. Britt-Compton B, Rowson J, Locke M, Mackenzie I, Kipling D, Baird DM. Structural stability and chromosome-specific telomere length is governed by cis-acting determinants in humans. Hum Mol Genet. 2006;15(5):725–33.
56. Martens UM, Zijlmans JM, Poon SS, Dragowska W, Yui J, Chavez EA, Ward RK, Lansdorp PM. Short telomeres on human chromosome 17p. Nat Genet. 1998;18(1):76–80.
57. Rufer N, Dragowska W, Thornbury G, Roosnek E, Lansdorp PM. Telomere length dynamics in human lymphocyte subpopulations measured by flow cytometry. Nat Biotechnol. 1998;16 (8):743–7.
58. Alter BP, Baerlocher GM, Savage SA, Chanock SJ, Weksler BB, Willner JP, Peters JA, Giri N, Lansdorp PM. Very short telomere length by flow fluorescence in situ hybridization identifies patients with dyskeratosis congenita. Blood. 2007;110(5):1439–47.
59. Baerlocher GM, Vulto I, de Jong G, Lansdorp PM. Flow cytometry and FISH to measure the average length of telomeres (flow FISH). Nat Protoc. 2006;1(5):2365–76.

60. Felsenfeld G, Davies DR, Rich A. Formation of a three-stranded polynucleotide molecule. J Am Chem Soc. 1957;79(8):2023–4.
61. Hoogsteen K. The crystal and molecular structure of a hydrogen-bonded complex between 1-methylthymine and 9-methyladenine. Acta Crystallogr. 1963;16(9):907–16.
62. Beal PA, Dervan PB. Second structural motif for recognition of DNA by oligonucleotide-directed triple-helix formation. Science. 1991;251(4999):1360–3.
63. Mirkin SM, Frank-Kamenetskii MD. H-DNA and related structures. Annu Rev Biophys Biomol Struct. 1994;23:541–76.
64. Rapozzi V, Cogoi S, Spessotto P, Risso A, Bonora GM, Quadrifoglio F, Xodo LE. Antigene effect in K562 cells of a PEG-conjugated triplex-forming oligonucleotide targeted to the bcr/abl oncogene. Biochemistry. 2002;41(2):502–10.
65. Strobel SA, Moser HE, Dervan PB. Double strand cleavage of genomic DNA at a single site by triple helix formation. J Am Chem Soc. 1988;110(23):7927–9.
66. Vasquez KM, Wilson JH. Triplex-directed modification of genes and gene activity. Trends Biochem Sci. 1998;23(1):4–9.
67. Majumdar A, Khorlin A, Dyatkina N, Lin FL, Powell J, Liu J, Fei Z, Khripine Y, Watanabe KA, George J, Glazer PM, Seidman MM. Targeted gene knockout mediated by triple helix forming oligonucleotides. Nat Genet. 1998;20(2):212–4.
68. Puri N, Majumdar A, Cuenoud B, Natt F, Martin P, Boyd A, Miller PS, Seidman MM. Targeted gene knockout by 2′-O-aminoethyl modified triplex forming oligonucleotides. J Biol Chem. 2001;276(31):28991–8.
69. Havre PA, Gunther EJ, Gasparro FP, Glazer PM. Targeted mutagenesis of DNA using triple helix-forming oligonucleotides linked to psoralen. Proc Natl Acad Sci. 1993;90(16):7879–83.
70. Vasquez KM, Wang G, Havre PA, Glazer PM. Chromosomal mutations induced by triplex-forming oligonucleotides in mammalian cells. Nucleic Acids Res. 1999;27(4):1176–81.
71. Faruqi AF, Datta HJ, Carroll D, Seidman MM, Glazer PM. Triple-helix formation induces recombination in mammalian cells via a nucleotide excision repair-dependent pathway. Mol Cell Biol. 2000;20(3):990–1000.
72. Faruqi AF, Seidman MM, Segal DJ, Carroll D, Glazer PM. Recombination induced by triple-helix-targeted DNA damage in mammalian cells. Mol Cell Biol. 1996;16(12):6820–8.
73. Vasquez KM, Marburger K, Intody Z, Wilson JH. Manipulating the mammalian genome by homologous recombination. Proc Natl Acad Sci U S A. 2001;98(15):8403–10.
74. Qiao F, Cech TR. Triple-helix structure in telomerase RNA contributes to catalysis. Nat Struct Mol Biol. 2008;15(6):634–40.
75. Ellington AD, Szostak JW. In vitro selection of RNA molecules that bind specific ligands. Nature. 1990;346(6287):818–22.
76. Tuerk C, Gold L. Systematic evolution of ligands by exponential enrichment: RNA ligands to bacteriophage T4 DNA polymerase. Science. 1990;249(4968):505–10.
77. Du Y, Li B, Wang E. "Fitting" makes "sensing" simple: label-free detection strategies based on nucleic acid aptamers. Acc Chem Res. 2013;46(2):203–13.
78. Jing Z, Jishan L, Ying J, Jianyu J, Kemin W, Ronghua Y, Weihong T. Design of aptamer-based sensing platform using triple-helix molecular switch. Anal Chem. 2011;83(17):6586–92.
79. Zheng J, Hu Y, Bai J, Ma C, Li J, Li Y, Shi M, Tan W, Yang R. Universal surface-enhanced Raman scattering amplification detector for ultrasensitive detection of multiple target analytes. Anal Chem. 2014;86(4):2205–12.
80. Ye S, Wu Y, Zhang W, Li N, Tang B. A sensitive SERS assay for detecting proteins and nucleic acids using a triple-helix molecular switch for cascade signal amplification. Chem Commun (Camb). 2014;50(66):9409–12.
81. Zhang YM, Zhang L, Liang RP, Qiu JD. DNA electronic logic gates based on metal-ion-dependent induction of oligonucleotide structural motifs. Chemistry (Easton). 2013;19 (22):6961–5.
82. Chen J, Fang Z, Lie P, Zeng L. Computational lateral flow biosensor for proteins and small molecules: a new class of strip logic gates. Anal Chem. 2012;84(15):6321–5.

83. Bi S, Yan Y, Hao S, Zhang S. Colorimetric logic gates based on supramolecular DNAzyme structures. Angew Chem Int Ed Engl. 2010;49(26):4438–42.
84. Xia F, Zuo X, Yang R, White RJ, Xiao Y, Kang D, Gong X, Lubin AA, Vallee-Belisle A, Yuen JD, Hsu BY, Plaxco KW. Label-free, dual-analyte electrochemical biosensors: a new class of molecular-electronic logic gates. J Am Chem Soc. 2010;132(25):8557–9.
85. de Silva AP, McClenaghan ND. Molecular-scale logic gates. Chemistry (Easton). 2004;10 (3):574–86.
86. Pischel U. Chemical approaches to molecular logic elements for addition and subtraction. Angew Chem Int Ed Engl. 2007;46(22):4026–40.
87. Park KS, Jung C, Park HG. "Illusionary" polymerase activity triggered by metal ions: use for molecular logic-gate operations. Angew Chem Int Ed Engl. 2010;49(50):9757–60.
88. Adleman LM. Molecular computation of solutions to combinatorial problems. Science. 1994;266(5187):1021–4.
89. Gao W, Zhang L, Zhang YM, Liang RP, Qiu JD. DNA colorimetric logic gates based on triplex–helix molecular switch. J Phys Chem C. 2014;118(26):14410–7.
90. Ling P, Lei J, Ju H. Porphyrinic metal-organic framework as electrochemical probe for DNA sensing via triple-helix molecular switch. Biosens Bioelectron. 2015;71:373–9.

Index

W. Xu, *Functional Nucleic Acids Detection in Food Safety*,
DOI 10.1007/978-981-10-1618-9

MIX

Papier | Fördert gute Waldnutzung

FSC® C083411

Kolibri 360 GmbH
Ferdinand-Jühlke-Straße 7
99095 Erfurt, Deutschland
produktsicherheit@kolibri360.de